Biosensors in Precision Medicine

From Fundamentals to Future Trends

Biosensors in Precision Medicine

From Fundamentals to Future Trends

Edited by

Laís C. Brazaca
Instituto de Química de São Carlos, Universidade de São Paulo, São Carlos, SP, Brazil

Juliane R. Sempionatto
Andrew and Peggy Cherng Department of Medical Engineering, California Institute of Technology, Pasadena, CA, United States

ELSEVIER

Elsevier
Radarweg 29, PO Box 211, 1000 AE Amsterdam, Netherlands
125 London Wall, London EC2Y 5AS, United Kingdom
50 Hampshire Street, 5th Floor, Cambridge, MA 02139, United States

Notices

Knowledge and best practice in this field are constantly changing. As new research and experience broaden our understanding, changes in research methods, professional practices, or medical treatment may become necessary.

Practitioners and researchers must always rely on their own experience and knowledge in evaluating and using any information, methods, compounds, or experiments described herein. In using such information or methods they should be mindful of their own safety and the safety of others, including parties for whom they have a professional responsibility.

To the fullest extent of the law, neither the Publisher nor the authors, contributors, or editors, assume any liability for any injury and/or damage to persons or property as a matter of products liability, negligence or otherwise, or from any use or operation of any methods, products, instructions, or ideas contained in the material herein.

ISBN: 978-0-443-15380-8

Publisher: Candice Janco
Acquisitions Editor: Maddie Wilson
Editorial Project Manager: Kyle Gravel
Production Project Manager: Sruthi Satheesh
Cover Designer: Mark Rogers

Typeset by MPS Limited, Chennai, India

Contents

Part 2
Biosensors and their applications in precision medicine

List of contributors

Amirul Fahmiin Abdullah Electrical and Electronic Engineering, Faculty of Engineering Universiti Teknologi Brunei, Brunei Darussalam

Mariana Bortholazzi Almeida São Carlos Institute of Chemistry, University of São Paulo, São Carlos, SP, Brazil; National Institute of Science and Technology in Bioanalysis - INCTBio, Campinas, SP, Brazil

Rafaela S. Andre Nanotechnology National Laboratory for Agriculture (LNNA), Embrapa Instrumentação, São Carlos, SP, Brazil

Mateus Nóbrega Aoki Laboratory for Applied Science and Technology in Health, Carlos Chagas Institute, Oswaldo Cruz Foundation (Fiocruz), Curitiba, Paraná, Brazil

Lucas Ximenes Araújo São Carlos Institute of Chemistry, University of São Paulo, São Carlos, SP, Brazil; National Institute of Science and Technology in Bioanalysis - INCTBio, Campinas, SP, Brazil

Lucas B. Ayres Department of Chemistry, Clemson University, Clemson, SC, United States

Lucas Blanes Laboratory for Applied Science and Technology in Health, Carlos Chagas Institute, Oswaldo Cruz Foundation (Fiocruz), Curitiba, Paraná, Brazil

Juliano Alves Bonacin Institute of Chemistry, University of Campinas (Unicamp), Campinas, São Paulo, Brazil

Laís C. Brazaca São Carlos Institute of Chemistry, University of São Paulo, São Carlos, SP, Brazil

Emanuel Carrilho São Carlos Institute of Chemistry, University of São Paulo, São Carlos, SP, Brazil; National Institute of Science and Technology in Bioanalysis - INCTBio, Campinas, SP, Brazil

Shirley Chen Department of Genetics, Stanford University School of Medicine, Stanford, CA, United States

Daniel S. Correa Nanotechnology National Laboratory for Agriculture (LNNA), Embrapa Instrumentação, São Carlos, SP, Brazil

Mardoqueu M. da Costa Scientific and Technological Institute of Brazil University, Universidade Brasil, São Paulo, SP, Brazil

Paulo Roberto de Oliveira Laboratory of Sensors, Nanomedicine, and Nanostructured Materials, Federal University of São Carlos (UFSCar), Araras, São Paulo, Brazil

Vinícius Guimarães Ferreira São Carlos Institute of Chemistry, University of São Paulo, São Carlos, SP, Brazil; National Institute of Science and Technology in Bioanalysis - INCTBio, Campinas, SP, Brazil

Carlos D. Garcia Department of Chemistry, Clemson University, Clemson, SC, United States

Diana Bueno Hernandez Department of Basic Sciences, Instituto Politecnico Nacional, Unidad Profesional Interdisciplinaria de Biotecnología, Mexico City, Mexico; Instituto Tecnológico de Santo Domingo, Biomedical Engineering, Santo Domingo, Dominican Republic

Bruno Campos Janegitz Laboratory of Sensors, Nanomedicine, and Nanostructured Materials, Federal University of São Carlos (UFSCar), Araras, São Paulo, Brazil

Itthipon Jeerapan Center of Excellence for Trace Analysis and Biosensor, Prince of Songkla University, Hat Yai, Songkhla, Thailand; Division of Physical Science, Faculty of Science, Prince of Songkla University, Hat Yai, Songkhla, Thailand; Center of Excellence for Innovation in Chemistry, Faculty of Science, Prince of Songkla University, Hat Yai, Songkhla, Thailand

Cristiane Kalinke Institute of Chemistry, University of Campinas (Unicamp), Campinas, São Paulo, Brazil

Suntisak Khumngern Center of Excellence for Trace Analysis and Biosensor, Prince of Songkla University, Hat Yai, Songkhla, Thailand; Division of Physical Science, Faculty of Science, Prince of Songkla University, Hat Yai, Songkhla, Thailand

Ashutosh Kumar Department of Electrical Engineering, University of Notre Dame, Notre Dame, IN, United States

Wasan Mani Kumar Stanford University School of Medicine, Stanford, CA, United States

Linda Yu-Ling Lan Department of Genetics, Stanford University School of Medicine, Stanford, CA, United States

Mylena Lemes Cunha Laboratory for Applied Science and Technology in Health, Carlos Chagas Institute, Oswaldo Cruz Foundation (Fiocruz), Curitiba, Paraná, Brazil

Syazana Abdullah Lim Center for Research on Agri-Food Science and Technology, Universiti Teknologi Brunei, Brunei Darussalam

Tiong Hoo Lim Electrical and Electronic Engineering, Faculty of Engineering Universiti Teknologi Brunei, Brunei Darussalam

Long Sha Liu Stanford University School of Medicine, Stanford, CA, United States

Kuldeep Mahato Department of Nanoengineering, University of California, San Diego, CA, United States

Luiza A. Mercante Institute of Chemistry, Federal University of Bahia (UFBA), Salvador, BA, Brazil

Rupesh Kumar Mishra Bindley Bio-Science Center, Purdue University, West Lafayette, IN, United States

Adriana Pavinatto Scientific and Technological Institute of Brazil University, Universidade Brasil, São Paulo, SP, Brazil

Aubrey Kathryn Roberts Department of Genetics, Stanford University School of Medicine, Stanford, CA, United States

Pedro V.V. Romanholo Instituto de Química, Universidade Federal de Goiás, Goiânia, GO, Brazil

Eduardo Luiz Rossini São Carlos Institute of Chemistry, University of São Paulo, São Carlos, SP, Brazil; National Institute of Science and Technology in Bioanalysis - INCTBio, Campinas, SP, Brazil

Rafaela C. Sanfelice Science and Technology Institute, Federal University of Alfenas, Poços de Caldas, MG, Brazil

Juliane R. Sempionatto Andrew and Peggy Cherng Department of Medical Engineering, California Institute of Technology, Pasadena, CA, United States

Livia F. Sgobbi Instituto de Química, Universidade Federal de Goiás, Goiânia, GO, Brazil

Atul Sharma Department of Pharmaceutical Chemistry, SGT College of Pharmacy, SGT University, Gurugram, Haryana, India

Michael Snyder Department of Genetics, Stanford University School of Medicine, Stanford, CA, United States

Swapnil Tiwari School of Studies in Chemistry, Pt. Ravishankar Shukla University, Raipur, Chhattisgarh, India

Dalila Luciola Zanette Laboratory for Applied Science and Technology in Health, Carlos Chagas Institute, Oswaldo Cruz Foundation (Fiocruz), Curitiba, Paraná, Brazil

Part 1

Biomarkers and precision medicine

Chapter 1

Biomarkers: an introduction

Vinícius Guimarães Ferreira[1,2], Eduardo Luiz Rossini[1,2], Lucas Ximenes Araújo[1,2], Mariana Bortholazzi Almeida[1,2] and Emanuel Carrilho[1,2]

[1]São Carlos Institute of Chemistry, University of São Paulo, São Carlos, SP, Brazil, [2]National Institute of Science and Technology in Bioanalysis - INCTBio, Campinas, SP, Brazil

1.1 Introduction

Gattaca, a Hollywood production from 1997, shocked the world by showing a fictional scenario where science has gone so far that it was capable of meticulously designing the DNA of each newborn, ensuring that all individuals would have the maximum capacity in a specific function (Davies, 1997; Jowett, 2009; Vogel, 1997). Besides being a fictional movie, the main idea of science attempting to link every gene to a specific function was, indeed, a major scientific goal at the beginning of the 2000s, as part of a multinational project known as the Human Genome Project (Carrasco-Ramiro et al., 2017; Emmert-Streib et al., 2017; Gibbs, 2020). From there on, scientists have always searched for links between every molecule in the human body and different conditions, such as diseases, aging, phenotypes, and behavior (Cuffe et al., 2017; Echle et al., 2021; Engel & Pitkänen, 2020; Kalia et al., 2015; Kennis et al., 2020; Kumar et al., 2006; Mitnitski, 2018; Piazza et al., 2010; Samprathi & Jayashree, 2021; Solje et al., 2021; Zhao et al., 2015). However, regardless of what is shown in Gattaca, science gives more importance to use this knowledge to predict or diagnose conditions, using the genes, proteins, and metabolites as markers or biomarkers capable of indicating their occurrence.

Several classifications can describe biomarkers. For example, the World Health Organization (WHO) defines a biomarker as: "almost any measure that reflects an interaction between a biological system and a potential hazard, which can be chemical, physical or biological" (World Health Organization, 2001). The measured response can be functional and physiological, biochemical at the cellular level, or molecular interactions (World Health Organization, 2001). On the other hand, the National Institutes of Health (NIH) Biomarkers Definitions Working Group has defined biomarkers as "a characteristic that is measured and evaluated as an indicator of

Biosensors in Precision Medicine. DOI: https://doi.org/10.1016/B978-0-443-15380-8.00001-1

normal biological processes, pathogenic processes or pharmacological responses to a therapeutic intervention" (Puntmann, 2009).

Some classifications in the scientific community divide biomarkers into three categories: exposure, effect, and susceptibility biomarkers (Maruvada et al., 2006). Terminologies and official definitions apart, we can simplify it as a biological characteristic used for the early detection, diagnosis, prognosis, and monitoring of a given condition. The most recent Food and Drug Administration (FDA) and NIH classification divide biomarkers into seven categories: susceptibility/risk, diagnostic, monitoring, prognostic, predictive, pharmacodynamic/response, and safety (Puntmann, 2009).

- Susceptibility/Risk: It indicates the potential for developing a disease or medical condition in an individual who does not currently have a disease or medical condition (Puntmann, 2009).
- Diagnostic: It is used to detect or confirm the presence of a disease, condition of interest, or to identify individuals with a subtype of the disease (Puntmann, 2009).
- Monitoring: It is characterized by repeated assessment over time and is used to assess disease progression (including new symptoms caused by disease progression) and disease response to treatment for evidence of exposure to a medical product or environmental agent (Puntmann, 2009).
- Prognostic: A prognostic biomarker indicates a likelihood of a future clinical event, recurrence, or disease progression in an identified population. Such biomarkers are often used as eligibility criteria in clinical trials to identify patients who are more likely to experience clinical events or disease progression (Puntmann, 2009).
- Predictive: Its purpose is to identify individuals with the most significant probability of responding to exposure to a particular drug or environmental agent. This response can be positive with a decrease in the symptoms of a particular disease, improvement in survival, or negative, leading to the individual developing some adverse effect or disease (Puntmann, 2009).
- Response: It is used to show that a biological response has occurred in an individual exposed to a medical product or environmental agent. Such a response can be positive or negative (Puntmann, 2009).
- Safety: A biomarker measured before or after exposure to a medical product or environmental agent to indicate the likelihood, presence, or extent of toxicity as an adverse effect. Safety biomarkers can detect or predict these adverse effects of pharmacological, drug, or environmental exposures. The detection or alteration of a biomarker is the signal that allows the modification of the dose or the interruption of the treatment (or removal from the environment) before the toxicity becomes severe (Puntmann, 2009).

A molecule or physiological measurement needs to be a measurable indicator of a specific biological state, being part of a diagnostic characteristic or from a prognostic aspect, to be considered a biomarker. The physiological measurements apart, there are a great variety of molecules present in a biological system, from DNA and RNA, proteins, and metabolites such as sugars, lipids, amino acids, nucleic acids, and other small molecules that have the potential to be biomarkers (Fig. 1.1). Therefore it is necessary to use different analytical research strategies to find the ideal biomarker for the studied condition (Maruvada et al., 2006). The following sections of this chapter will address the different approaches to biomarkers discovery and provide an overview of their use from scientific and clinical perspectives (Maruvada et al., 2006).

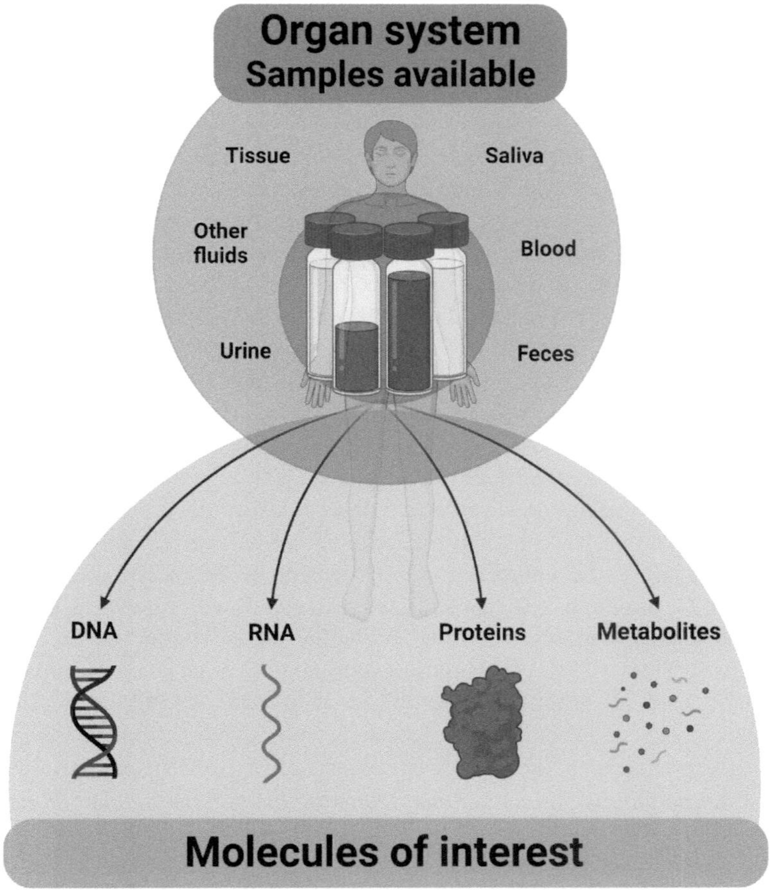

FIGURE 1.1 Types of samples and molecules that can be used for biomarker studies in various organ systems. Blood and urine are the most commonly used samples for biomarker analysis in clinical settings, but a diverse range of molecules and particles can be studied in other samples as well. These include DNA, RNA, proteins, metabolites, but also single cells, and even physiological measurements or iris patterns, all of which can be used as biomarkers for specific health hazards. *Created with Biorender.com.*

Despite the many sources and the different molecules of interest (Fig. 1.1), a specific scientific field was responsible for enormous advances in biomarker discovery in the past decades, the so-called Omics sciences. The Omics sciences were born from genomics, which focuses on monitoring patterns of expressed genes and the mechanisms by which that expression is correlated. As the cellular environment changes, gene expression patterns are also modified (Hartl, 2009; Maruvada et al., 2006). Among the leading technologies used in genomic analysis, we can mention assays based on polymerase chain reaction (PCR), fluorescence in situ hybridization (FISH), and DNA or RNA microarrays. One of the main advantages of these technologies is the robustness and high yield when amplifying specific DNAs and RNAs, even if these are in low concentrations. Another way to conduct genomic analysis is by using a genome analyzer, which typically employs next-generation sequencing (NGS), a high-throughput sequencing technique that can rapidly sequence large quantities of genetic material.

Following the central dogma of molecular biology, proteins are the next link in the information chain after the genetic material. From that perspective, the field of proteomics emerged from comparing the entire protein isoforms, their modifications, interactions, structures, and complexes between the studied groups. Protein biomarkers generally include posttranslational modifications of proteins found in body fluids or tissues (Aslam et al., 2017; Bourgoin-Voillard et al., 2016; Shen et al., 2008; Tyers & Mann, 2003; Yates et al., 2009). Proteomic analyses are commonly divided into "Bottom-Up" and "Top-Down" strategies (Veenstra & Yates III, 2019 Yates et al., 2009). The most common is the bottom-up, also known as shotgun, which consists of a process where the proteins are first hydrolyzed and then analyzed by liquid chromatography coupled to a mass spectrometer (Veenstra & Yates III, 2019; Yates et al., 2009).

Conversely, the top-down approach does not use the previous digestion of the proteins that will be analyzed, making use of specific analytical processes to analyze the entire proteins (Veenstra & Yates III, 2019; Yates et al., 2009). Although it seems less complex, top-down proteomics presents higher technical barriers, resulting in higher costs (Bader et al., 2003; Veenstra & Yates III, 2019). The most used analytical techniques applied in proteomics are electrophoresis combined with mass spectrometry, liquid chromatography coupled to mass spectrometry (LC-MS), Surface-enhanced laser desorption/ionization mass spectrometry (SELDI-MS), matrix-assisted laser desorption/ionization mass spectrometry (MALDI-MS), and Western blot (Bedia, 2018; Veenstra & Yates III, 2019; Yates et al., 2009, 2019).

With the advance in analytical platforms, scientists started to analyze not only the genetic and proteic portions of the samples but also a more challenging part: the small molecules, also called metabolites (Patti et al., 2012; Quezada et al., 2017). The newest branch of Omics sciences, metabolomics, stands on comparative studies of metabolites between two or more groups of interest, such as disease versus control (De San-Martin et al., 2020; Patti et al., 2012). Metabolites include small molecules such as glucose and lactate, amino acids, carbohydrates, hormones, steroids, vitamins, lipids, fatty acids, and an enormous variety of compounds present in an organism's cell, tissue, or body fluid (Griffiths et al., 2007; Patti et al., 2012). Metabolomics can be distinguished, in its turn, in two major approaches: (1) targeted metabolomics and (2) untargeted metabolomics. In the first approach, a few predetermined metabolites are analyzed and quantified in each sample. Typically, targeted approaches are used to validate hypothetical metabolic biomarkers (Patti et al., 2012). On the other hand, in the untargeted approach, the goal is to analyze the largest possible number of metabolites present in the sample (De San-Martin et al., 2020; Patti et al., 2012). Due to the great variety of metabolites present in biological samples, it is impossible to quantify every metabolite. Therefore untargeted metabolomics is commonly a semiquantitative analysis, focusing on discovering which metabolite is present in different amounts on the studied condition rather than specifying a quantitative threshold for the metabolite to be used as a biomarker (De San-Martin et al., 2020; Patti et al., 2012). From this approach, it is possible to identify potential biomarkers that can be linked to the hypothesis using a targeted approach (Lu et al., 2008; Patti et al., 2012). Similar to proteomics, metabolomics uses techniques such as LC-MS, CE-MS, SELDI-MS, MALDI-MS, gas chromatography-mass spectrometry (GC-MS), Fourier transform infrared (FT-IR) spectroscopy, Direct infusion mass spectrometry (DIMS) and nuclear magnetic resonance (NMR) spectroscopy (Arivaradarajan & Misra, 2019; Dunn & Ellis, 2005).

The increasing number of studies based on biomarker approach across time is notorious, followed by the growing hope of researchers for exclusive findings (Fig. 1.2). From the clinical point of view, those findings could present tremendous applications, such as the characterization, diagnosis, and monitoring of diseases, as well as individualizing and monitoring therapeutic interventions, predicting adverse drug reactions, and thus guiding the best treatment for toxic effects, diagnosing cell types, among other uses (Aronson & Ferner, 2017).

Nowadays, the race for a reliable biomarker is advantageous as they allow a quicker and more precise measurement of the patient's condition (Vernon et al., 2019). However, besides the positive impact of biomarker analysis, the true revolution began with the use of faster, more precise, sensitive, and less expensive analytical platforms for biomarker measurements, allowing the emergence of the field of personalized medicine. From that perspective, biosensors are a fundamental part of the success of biomarkers in clinics (Aronson & Ferner, 2017). In the following sections of this chapter, the reader will be introduced to the world of biomarkers discovery, the current state-of-the-art of their use in clinics, and the bright future in the era of personalized medicine.

1.2 Biomarker's discovery journey

Perhaps the biggest challenge of seeking a biomarker at this time is to find a molecule that fits the criteria of precision medicine (Ou et al., 2021), which are presumed to be as follows:

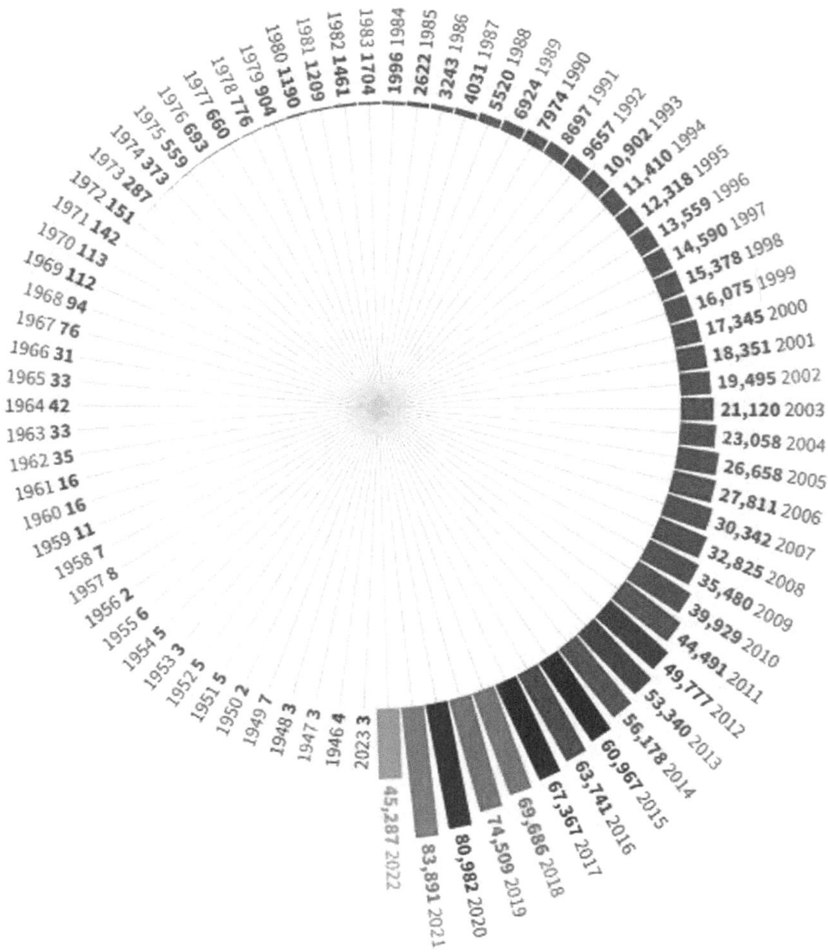

FIGURE 1.2 The number of publications that mention biomarkers on the PubMed database from 1946 to 2022. Available at PubMed.

1. Binary (e.g., present or absent)
2. Quantifiable
3. Sensitive
4. Specific
5. Capable of routine assays
6. Easily sampling

The steps involved in the biomarkers search are laborious and time consuming, being an arduous job for the scientists to translate them to clinical use. However, with the technology advance and the omic sciences, numerous tools for biomarkers discovery have been developed. As already mentioned, the most effective and unbiased detection method, so far, is MS, used as a complement of separations methods like gel and capillary electrophoresis, liquid and gas chromatography (LC and GC) (Hawkridge & Muddiman, 2009). Several studies also show a great use of NMR and vibrational spectroscopy, like infrared and Raman (Correia et al., 2022; Mitchell et al., 2014). The focus of this section is to provide a brief overview of the discovery process of biomarkers candidates. For deep information about the spectroscopy and statistical methods involved in the process, we encourage the readers to search for specific literature.

The omic discovery process of a molecule (metabolite, protein) or expression follows, in general, a pipeline (Fig. 1.3A). Researchers can employ any model systems (e.g., cell lines, animal models) or human materials (e.g., saliva, blood, plasma/serum, liquor, urine, feces, tissues) (Fig. 1.1). A binary comparison is frequently employed through a control group (healthy or normal) and the study group; however, the division of the main study group into subgroups according

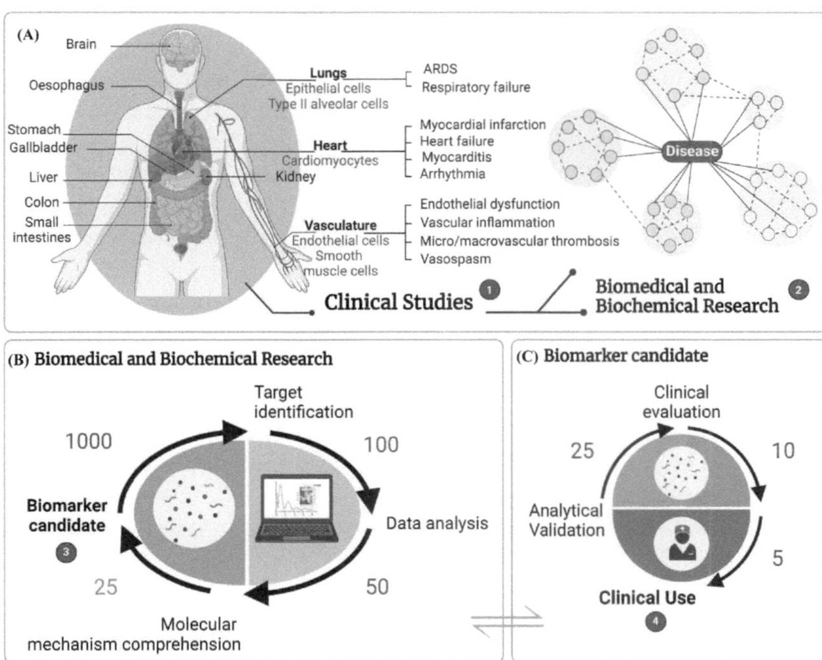

FIGURE 1.3 Biomarkers discovery overview. (A) Human body model showing several organs possible related diseases opened for further biomarker search during a clinical study (1) followed by biomedical research (2); (B) cycle of biomarker search starts with 1000 biomarker candidates (3) that decrease to 100 after target identification and to 50 after data analysis, reaching 25 candidates for molecular mechanism comprehension; (C) cycle of biomarker candidate is performed with the 25 candidates that go through an analytical validation, but during the clinical evaluation, the number decreased to 10 and only 5 are considered for clinical use (4). The number used in this figure are only representatives. *Created with Biorender.com.*

to a treatment or illness severity, for example, is also allowed. Each model system typically results in hundreds or thousands of biomarker candidates. This pipeline expects to eliminate the false-positive markers that do not contribute to the study model. Then, to point out a few biomarker candidates, the data obtained at this point, called features, should be reduced by data analysis after an extensive and solid identification considering the distinction of interest (Fig. 1.3B). As the next step, for validation purposes, the best candidates must go to a clinical evaluation, ideally applied to a higher number of samples (individuals), decreasing the number of biomarkers that can be used for clinical purposes (Fig. 1.3C) (Rifai et al., 2006).

Although the pipeline sounds virtually simple, it is worth mentioning that for biomarker qualification, a variety of trials and cohort studies are needed before the meta-analysis. Some guidances are recommended by the ICH (International Council for Harmonization of Technical Requirements for Pharmaceuticals for Human Use) and regulatory organizations, such as the FDA (FDA, 2018).

As previously mentioned, MS analysis is widely used to identify unknown substances because it allows molecular mass weight measurement with high precision and molecular formula and structure identification. However, all technologies have their own limitations. Hyphenated to GC, MS is widely used for volatile organic compound (VOC) screening, for example, in early cancer diagnosis. Nevertheless, the most affected compounds on living organisms are thermally unstable and show a wide range of polarity, compatible characteristics with an LC separation. However, in the case of macromolecules studies, like proteome analysis, the LC-MS methods can present poor reproducibility and complex operation procedures, requiring an efficient and direct soft ionization MS technology, known as matrix-assisted laser desorption ionization (MALDI)-MS. Combined application with a time-of-flight (TOF) mass analyzer provides an excellent accuracy tool for analyzing mixtures and biological macromolecules from $\sim$1000 to 3000 Da (Ge et al., 2021). All those untargeted strategies cover a large part of the biomarker composition. However, it only results in mass-to-charge ratio, time, and intensity values, usually called features, that cannot contribute quantitatively (Kohler et al., 2017).

The next stage must be identification and targeted verification. Immunoassays, in general, present high specificity; nevertheless, are laborious, pricey, and need recognition molecules that very often are not commercially available, being an impracticable disadvantage. Howsoever, multiple reaction monitoring (MRM) in MS offers a robust and reliable throughput for biomarkers candidates (Liu et al., 2017).

Despite the high performance of MS analysis, with incredible sensitivity and selectivity, this technique still requires complex sample preparation and might result in poor repeatability. In comparison, NMR is a simple manner to perform biomarker studies since it has a more straightforward or none sample preparation and presents high repeatability; nonetheless, it is less sensitive than mass spectrometry-based approaches (Cui et al., 2022).

During a biomarker discovery, associations of disease status, demographic outputs, and clinical characteristics of the subject under study (sex, age, disease stage, among others) are crucial. All this information contributes to the statistical metrics used to evaluate biomarkers candidates (Ou et al., 2021). Among the statistical approaches, generally, the use of the supervised clustering method is recommended because it provides great information and metrics for diseases and control groups, reducing the featuring and evaluating the performance. A biomarker needs to be robust. Usually, the Receiver Operating Characteristics, also known as the ROC curve, is one of the metrics used to predict the portion of false-positive and false-negative ratios. Ideally, cross-validation or bootstrapping should be performed (McDermott et al., 2013; Ou et al., 2021), and then the chosen biomarker goes to trials until it reaches a straight and fast analysis stage (Fig. 1.3) (Antoniou et al., 2019). Although the statistical analysis is the center of a biomarker finding, the six criteria mentioned earlier cannot be forgotten. A marker that is not easily sampled or unable to give an analytical response should not be considered qualified.

On the other hand, even though a protocol for biomarker discovery is presented, there are several discrepancies among the different research groups regarding the technique used for biomarker identification and following validation. Sample collection and storage present as a challenge when the full history of the patient is not collected, or the sample is limited and cannot be accessed by others to reproduce and validate the new biomarker. There are also challenges regarding the validation process. During clinical trials, for example, patients must be monitored, undergoing invasive procedures, during several visits to the hospital, or even hospitalization. To address such challenges, there are several initiatives to build a collective sample bank, where tissues, blood, and other samples are well documented and kept for biomarker discovery. Another approach to facilitate the process is point-of-care devices that allow the patient self-monitoring, decreasing the number of visits to the hospital, and the overall cost of the study.

1.3 Old and new biomarkers—an overview

Nowadays, an enormous set of biomarkers has already been discovered, and an even higher number of potential biomarkers are currently being studied. Diverse physiological measurements, several genes, a diversity of proteins, a complexity of metabolites, and even the gut microbiome are currently reported as biomarkers or, at least, potential biomarkers for different conditions (Quezada et al., 2017) and in different biofluids (Alberice et al., 2013; Di Zazzo et al., 2019; Kwon et al., 2022; Lancellotti et al., 2021; Makarov et al., 2009; Melguizo-Rodríguez et al., 2020; Nunes et al., 2021; Pham & Beauchamp, 2021; Sarshar et al., 2020) (Fig. 1.1). A few examples of genes, proteins, and metabolites are reported for various diseases (Table 1.1). Although the term "condition" may seem vague,

TABLE 1.1 Examples of reported DNA, protein and metabolite biomarkers, and biomarkers candidates for different diseases.

Disease	Sample type	Detection methods	Biomarkers	Validation status	Reference
Alzheimer disease	CSF	ELISA	Amyloid-β protein	Candidate	Shaw et al. (2009)
Bladder cancer	Urine	LC-MS	Apolipoprotein A-I, apolipoprotein A-II, heparin cofactor 2 precursor, and peroxiredoxin-2	Candidates	Chen et al. (2010)
Depression	Plasma	GC-MS	Glycerol, inositol, GABA, citrate, glycerol, oleate, stearate, hydroorotate, 9,12,octadecadienoate, α-hydroxyisobutyrate, 5-oxoproline, 3-phospho-L-serine	Candidates	Paige et al. (2007)
Primary open-angle glaucoma	Tears	LC-MS	Lipocalin-1, lysozyme C, lactotransferrin, proline-rich-protein 4, prolactin-inducibleprotein, zinc-α-2-glycoprotein, polymeric immunoglobulin receptor	Candidates	Pieragostino et al. (2012)
Intra-amniotic inflammation	Amniotic fluid	SELDI-TOF	Neutrophil defensins-2, neutrophil defensins-1, calgranulin C, and calgranulin A	Candidates	Buhimschi et al. (2005)
Lung cancer	Tumor cells	PCR	PVT1/CHD7, SCLC, and CHD5	Candidates	Pleasance et al. (2010)
Myeloid leukemia	Tumor cells	Genome analyzer	Mutations in genes NRAS, NPM1, IDH1, and AML-M1	Candidates	Mardis et al. (2009)
Obesity	Serum	LC-MS	L-Kynurenine, glycerophosphocholine (GPC), glycerol 1-phosphate, glycolic acid, tagatose, methyl palmitate, and uric acid	Candidates	Chen et al. (2015)

Pancreatic cancer	Oral fluids	2DE-MS, LC-MS	Histones, myeloperoxidase, apolipoprotein A-I precursor, transthyretin precursor, transketolase, hemopexin, lipocalin-1, and protein S100-A8	Candidates	Deutsch et al. (2020)
Parkinson disease	Plasma	LC-MS	CCDC154, TRIM3, DHH, NRP2, and CLIC1	Candidates	Dong et al. (2019)
Prostate cancer	Tumor tissue	FISH, genome analyzer	Mutations in genes PRKCI, DICER, SPTA1, SPOP, CHD1, CHD5, HDAC9, HSPA2, HSPA5, and HSP90AB1	Candidates	Berger et al. (2011)
Breast cancer	Tumor tissue and plasma	PCR	BRCA1	Validated	Denkert et al. (2012), Giskeødegård et al. (2012), and Roig et al. (2017)
Renal tubular injury	Kidney tissue	DNA and RNA arrays	Mutations in genes Kim1, Cp, Clu, Timp1, and Spp1	Candidates	Kondo et al. (2009)
Cardiovascular disease risk	Blood	Enzymatic, immunoassay	Cholesterol, C-reactive protein	Validated	Of et al. (2002)
Diabetes	Blood	Enzymatic, HPLC	Glucose, HbA_{1c}	Validated	Oliver et al. (2009) and Sultanpur et al. (2015)
Renal dysfunction	Serum, urine	HPLC	Creatinine	Validated	Herget-rosenthal et al. (2007)
Hepatitis B	Blood	Enzyme immunoassay	HBsAg	Validated	Lee and Ahn (2011)

CSF, cerebrospinal fluid; LC-MS, liquid chromatography coupled to mass spectrometry; GC-MS, gas chromatography coupled to mass spectrometry; ELISA, enzyme-linked immunosorbent assay; HPLC, high-pressure liquid chromatography; PCR, polymerase chain reaction, SELDI-TOF, surface-enhanced laser desorption/ionization coupled to time-of-flight mass spectrometer.

biomarkers are so diffused and incorporated in health sciences that it is challenging to think in a time when no biomarker has ever been proposed. For example, pregnancy tests use the levels of human chorionic gonadotrophin (hCG) hormone, secreted after a fertilized egg implants in the uterus, as a biomarker for pregnancy (Olszynko-Gryn, 2014). Another example is the prostate-specific antigen (PSA), a blood protein used as a biomarker for prostate cancer (Makarov et al., 2009; Tosoian & Loeb, 2010). If the reader is not convinced yet, different biomarkers have been used for cancer diagnosis and even cancer insurgencies, such as the BRCA1 and BRCA2 genes used in clinics as breast cancer predisposition biomarkers (Duffy et al., 2015). Therefore between the biomarkers reported in the literature, we may highlight biomarkers for cancer (Echle et al., 2021; Pleasance et al., 2010; Tosoian & Loeb, 2010), heart complications, Parkinson's disease (Chen et al., 2010), Alzheimer's disease (Bader et al., 2020), stress, aging (Piazza et al., 2010), behavior (Paige et al., 2007), mental illness (Kalia et al., 2015), and infections (Buhimschi et al., 2005; Park et al., 2020), including COVID-19. In addition to the biomarkers and biomarkers candidates presented in Table 1.1, we will further discuss the most valuable biomarkers for some specific common conditions in the following paragraphs.

Cancer is a life-threatening disease (Ritchie & Roser, 2018). Among the different cancer types, breast cancer is the most common cancer type in women, accumulating over 19 million cases in 2019 (Our world in data, 2018, Number of people with cancer, 1990 to 2017). Despite imaging-based exams being part of women's routine check-ups, about 60% of breast cancer cases are still discovered in advanced stages (stage II or higher) (Thomas et al., 2019), shortening the chances of effective treatment and decreasing the patient's survival rates (Coughlin, 2019; Thomas et al., 2019). Aiming at earlier breast cancer detection, a variety of biomarkers have been reported in the literature. Studies have shown that the measurement of certain proteins such as EMA and CK1, as well as the ratio between them, can be useful for early detection of breast cancer in serum analysis, with reported sensitivity and specificity of 72% and 76%, respectively (Attallah et al., 2014; Loke & Lee, 2018). These findings highlight the importance of monitoring these markers for improved breast cancer diagnosis. Another protein reported as a potential biomarker for early breast cancer detection is the Del-1 protein present in extracellular vesicles in blood, being able to diagnose patients with 92% of sensitivity and 86% specificity (Loke & Lee, 2018; Moon et al., 2016). Del-1 protein rearranges the p53 pathway in tumor cells, supporting the tumor cells' growth and apoptosis avoidance (Lee et al., 2019). DNA biomarkers have also been reported for breast cancer predisposition (Duffy et al., 2017). Besides the already mentioned BRCA1 and BRCA2 genes, mutations in the gene Checkpoint Kinase 2 (CHEK2) are also reported as a moderate-risk predisposition biomarker, all of them related to DNA damage repair (Britt et al., 2020; Nevanlinna & Bartek, 2006; Rainville et al., 2020). However, further research is being held to fully assess the relevance of susceptibility biomarkers for their use in clinics (Duffy et al., 2017).

Notwithstanding, effect biomarkers to predict treatment efficiency have also been reported. Trastuzumab, for example, is a monoclonal antibody used to treat breast cancer, which efficiency can be successfully evaluated by the overexpression of human epidermal growth factor receptor 2 (HER2) on tumor cells (Barzaman et al., 2020; De et al., 2013; Goossens et al., 2015; Menyhart et al., 2015). The basis behind HER2 use as a biomarker for therapy results from the trastuzumab activity itself. The drug binds directly in the extracellular portion of the HER2 receptor, inhibiting signal transduction and controlling tumor cell proliferation (Menyhart et al., 2015).

Apart from cancer, other conditions are growing in numbers, with some even more challenging to diagnose and follow up, which is the case of stress. Stress in humans can be defined as a conscious or unconscious threat to homeostasis, which is the equilibrium in physiological variables (Goldstein & Kopin, 2007; Lu et al., 2021). Therefore stressors such as exposure to low temperatures, pain, and emotional distress threaten the body's health, resulting in a series of biochemical and physiological reactions to restore equilibrium (Conway De Macario & Macario, 2000; Goldstein & Kopin, 2007). The endocrine systems are core homeostasis controllers, in particular, the sympatho-adrenomedullary (SAM) and hypothalamic–pituitary–adrenal (HPA) axes (Russell & Lightman, 2019). SAM is responsible for adrenaline and noradrenaline discharge, while HPA is linked to cortisol release (Russell & Lightman, 2019). Therefore adrenaline and cortisol are two of the most reported exposure biomarkers for stress, short-term stress, or chronic stress diagnosis (Chojnowska et al., 2021; Noushad et al., 2021; Steckl & Ray, 2018). As hair and nails are slow-growth tissues, they act as a sort of ice core, storing information from our bodies over time. Due to their inherent ability, these biological materials can be used to assess chronic stress through the dosage of cortisol in different tissue layers, resulting in cortisol timelapse (Daniel et al., 2004; Noushad et al., 2021; Phillips et al., 2021; van Ockenburg et al., 2015). Nonetheless, for active stress diagnosis, the samples must contain the cortisol and/or adrenaline release level at the time of analysis, the reason why sweat (Pearlmutter et al., 2020), blood (Reims et al., 2005; Spratt et al., 2012), urine (Babisch et al., 2001; Shimanoe et al., 2021), and saliva (Hellhammer et al., 2009) are often employed (Fig. 1.1) (Steckl & Ray, 2018; van Ockenburg et al., 2015). Physiological measurements have also been used for stress evaluation (Sharma & Gedeon, 2012; Smets et al., 2018).

Biomarkers are also used for natural conditions in humans, such as the aging process. As strange as it may sound, aging is not equal for all humans and is, in fact, highly influenced by genetic and environmental aspects (Foo et al., 2019). While some individuals may reach the late 80 completely healthy, others may present age-related diseases before 60. A possible explanation of such a scenario remains on the concept of biological age, which measures how genetics and environment affect someone's organism compared to others of the same chronological age (Horvath, 2013; Horvath & Raj, 2018; Yaneske & Angione, 2018). In that sense, the biological age can be higher or lower than the chronological age. Currently, several biomarkers have been reported in the literature for biological aging assessment, such as DNA methylation (DNAm) (Horvath & Raj, 2018). DNA methylation is associated with transcriptional silencing, which is the silencing of determining gene expression (Schübeler, 2015). Therefore the accumulation of several DNAm can lead to the loss of gene functions in cells (Benayoun et al., 2015). Previous studies showed that lower DNAm levels are not only correlated with chronological age but also with life longevity (Petkovich et al., 2017). Metabolic biomarkers in urine were also reported for aging, with glycolate, histidine, ethanolamine, and threonine concentrations reported as the main biomarkers for biological aging (Hertel et al., 2016). Nonetheless, metabolic and proteomic biomarkers are still being further studied for biological age assessment once alterations in metabolites and proteins result from several other conditions (Menni et al., 2015).

Infection biomarkers have also been largely studied and are particularly important since 20% of the deaths worldwide could be attributed to sepsis. In this condition, the body's response to an infection damages its own organism (Rudd et al., 2020). For a quick understanding, our immunological system can be divided into two types of response, (1) innate immune response and (2) adaptive immune response. When the organism undergoes an infection, the first response is from the innate immune system, which is responsible for infectious particle recognition and starts the first line of defense with molecules such as cytokines and cells such as the Natural Killer cells. Therefore cytokines such as Interleukin-6 are currently being studied for their use in infection diagnosis (Teggert et al., 2020; Faix, 2013). After the first line of defense, the adaptive immune response is triggered, bringing the B and T cells (types of white blood cells) to eliminate the infected host cells and the infectious particles. Biomarkers present in the adaptative immune system have also been reported, such as the presence of specific antibodies (created by B cells), such as the antibodies for Sars-Cov-2 diagnosis (Fang et al., 2020; Zhang & Guo, 2020). Metabolites such as lactate and creatinine are also used in clinics for infection evaluation (Correia et al., 2022; Faix, 2013; Heilmann et al., 2019). Higher concentrations of lactate and creatinine indicate organ and tissue damage; therefore, they are useful prognosis biomarkers (Correia et al., 2022; Heilmann et al., 2019).

With the advance in analytical technology, point-of-care devices, and the popularization of biomarkers, the individual does not necessarily need to be in a clinic to measure a biomarker, and often the tests display affordable prices. The accessibility of these devices can promote better patient care and self-monitoring, and it can help in early diagnosis and prevention of chronic diseases. Moreover, point-of-care devices can assist in clinical trials to ensure patient compliance and to monitor for possible side effects. The at-home COVID-19 tests are a great example of how biomarkers and biosensors can be impacting, allowing a better management of a patient's health. In parallel, even more futuristic products have been developed with the union of technology and biotech companies, such as products capable of acquiring biomarker measurements in real time, giving a starting point on the biosensor's usage for wearable devices. In the next section, the reader will be introduced to the complex and exciting biomarkers industry.

1.4 Biomarker industry

Since the 20th century, the development of bio-analysis commercial products has been driven by consumer needs, promoting advances in test methodologies. The history of one of the most used tests can illustrate this growing technology from a test tube to the digital and connected era. The analysis of glucose, a biomarker for diabetes diagnostics, is dated from 1841, when different researchers developed qualitative tests using colorimetric reagents to detect glucose in the urine, with simple instrumentation and the reagent in the test tube, in a strip of sheep's wool or a piece of paper (Clarke & Foster, 2012; Johnson, 1884). However, their commercialization and application in products are not clear (Clarke & Foster, 2012).

The market for glucose kit tests started in the 1940 decade with the introduction of Clinitest by Ames (American Chemical Society, 2010). Clinitest consists of an effervescent tablet added to a small urine volume in a test tube. The color change produced semiquantitative results. The pursuit for a more convenient and simpler device for glucose measurement brought the development of the first strip test, the Clinistix by Ames, in 1957. The biosensor used a paper impregnated with glucose oxidase and a colorimetric reagent to quantify glucose in urine as a "dip and read" method (Free et al., 1957). Following the development of new products in 1965, Ames also introduced the Dextrostix, a modified reagent strip for blood glucose, bringing a more real-time indication of glucose levels for an individual (Free & Free, 1984). In 1970 a dedicated equipment to measure the Dextrostix became available, using an analogic display. The need for at-home and self-monitoring glucose made different companies like Bayer, Abbott, and Roche develop glucometers using electrochemical detection in 1990, but still with operator-dependent steps (Clarke & Foster, 2012). In the 21st century, digital glucometers as we know them today were established, and they can be found in any pharmacy or online all over the world. More recent development in glucose analysis illustrates the new era of wearable and connected devices (Cappon et al., 2017).

Currently, a diversity of tests is available commercially to diagnose a series of diseases and medical conditions. These tests are based on the screening of different biomarkers using different technologies to identify and quantify these substances, from lateral flow assay for pregnancy (Clearblue digital pregnance test) to DNA tests for cancer predisposition (TellmeGen prostate cancer) (Table 1.2).

TABLE 1.2 Description of the biomarker detected, diagnostic for, biofluidic used, and detection technology employed for the products available in the market.

Biomarker	Diagnostic for	Biofluid	Detection technology	Reference
Protein N or S	COVID-19	Saliva and nasal swab	LFA	Abbott Panbio COVID-19 antigen self-test
IgG and IgM antibody to SARS-CoV-2		Whole blood, plasma, and serum	LFA	BioMedomics COVID-19 IgM/IgG rapid test
SARS-CoV-2's RNA		Nasal swab	LAMP	Lucira check it COVID-19 test kit
PSA	Cancer	Whole blood, serum, and plasma	LFA	Henso PSA test cassette
AFP	Cancer	Whole blood and serum	LFA	Henso Alpha Fetoprotein AFP test cassette
CEA	Cancer	Whole blood, serum, and plasma	LFA	Henso Carcinoembryonic Antigen CEA test cassette
NMP22	Cancer	Urine	LFA	Abbott Alere
BRCA1/2 and MSMB genes	Prostate cancer	Mouth swabs	–	TellmeGen prostate cancer

RNASET2, SECISBP2L, and NRG1 genes	Lung cancer	Mouth swabs	–	TellmeGen Lung cancer
HIV-1 P24 protein	HIV	Whole blood, serum, and plasma	LFA	Abbott Determine HIV early detect
HIV-1 and HIV-2 antibodies		Saliva	LFA	Oraquick HIV test at home
HBsAg	Hepatitis	Whole blood, serum, and plasma	LFA	Abbott Determine HBsAg
Antibody HBsAb		Serum or plasma	LFA	Accu-Tell HBsAb rapid test
C-reactive protein	Infection	Whole blood	LFA	Prima C-reactive protein test
Protein from the head of mature sperm	Male fertility	Semen	LFA	SpermCheck fertility test
NS1 antigen	Dengue	Serum	ELISA	Abbott Panbio dengue early ELISA
Lp-PLA_2	Vascular inflammation	Whole blood	ELISA	Imaware vascular inflammation test
Albumin	Kidney disease	Urine	Strip test	Multistix 10 SG reagent strips
Prothrombin	Coagulation	Whole blood	Electrochemical	Xprecia Stride coagulation analyzer

(*Continued*)

TABLE 1.2 (Continued)

Biomarker	Diagnostic for	Biofluid	Detection technology	Reference
IgM and IgG antibody to Treponema Pallidum	Syphilis	Whole blood, plasma, or serum	LFA	Accu-Tell Syphilis rapid test
Antibody HBsAb	Hepatitis	Serum or plasma	LFA	Accu-Tell HBsAb rapid test
IgM antibodies to *leptospira*	Leptospirosis	Serum	ELISA	Panbi Leptospira IgM ELISA
Antibodies to *Trypanosoma cruzi*	Chagas disease	Whole blood, plasma, or serum	LFA	Bioline Chagas Ab
DGP IgG, tTG IgG, DGP IgA, and tTG IgA antibodies	Celiac disease	Whole blood	ELISA	Celiac disease screening test
HLA-DQ2, HLA-DQ8 and DQA1*05 genes	Celiac disease risk	Mouth swabs	RT-PCR	EmpowerDX celiac risk gene test HLA
Nine IgE antibody reactivity to food	Food allergy	Whole blood	ELISA	Everlywell food allergy test
RF IgM, RF IgA, and CCP IgG antibodies	Rheumatoid arthritis	Whole blood	ELISA	Imaware rheumatoid arthritis test
hCG	Pregnancy	Urine	LFA	Clearblue digital pregnance test

hLH	Ovulation	Urine	LFA	Smilelab ovulation test kit
FSH	Menopause	Urine	LFA	Accu-Tell FSH rapid test strip
TSH	Hypothyroidism	Whole blood	LFA	BTNX Inc Thyroid Stimulating Hormone TSH Test Cassette
Cortisol	Stress and training effect	Saliva	LFA	Soma Biosciences Cortisol LFD
Testosterone	Bone density, libido, body fat distribution, and supporting muscle mass	Saliva	–	Everlywell testosterone test
Estradiol, progesterone, and DHEA-S	Female fertility indication, adrenal disorder, gynecomastia, ectopic pregnancy, miscarriage, ovarian tumors, breast cancer	Saliva	–	EmpowerDX women's hormones test
Prolactin	Tumors, hypothyroidism, and other diseases	Serum	LFA	CTK Biotech PRL FIA test
Glucose	Diabetes	Blood	Electrochemical	Roche Accu-Chek glucose meters and strips

(Continued)

TABLE 1.2 (Continued)

Biomarker	Diagnostic for	Biofluid	Detection technology	Reference
Triglycerides, LDL, HDL	Risk factors for liver and coronary heart disease	Blood	Electrochemical	CURO L7 professional grade testing kit for lipid blood profile
Cholesterol	Risk factors for liver and coronary heart disease	Blood	Electrochemical	CURO L5 cholesterol test kit
Uric acid	Gout	Blood	Electrochemical	UriTouch blood uric acid monitoring system
Bilirubin, ketone, specific gravity, blood, pH, protein, urobilinogen, nitrite, leukocyte, albumin, and creatinine	Urinary tract infections, kidney disorder, and diabetes	Urine	Colorimetric	Multistix 10 SG reagent strips
Vitamins, amino acids, and minerals	Health and nutrition monitoring	Blood	–	LetsGetChecked micronutrient test; Biostarks optimal nutrition test
Pathogen's DNA	Chlamydia, gonorrhea, trichomoniasis, herpes, and others STIs	Urethral, endocervical, rectal, vaginal samples, and urine	DNA hybridization on strip	Operon STD panel strip

Pathogen's DNA	Common cold, pneumonia, bronchitis, and other respiratory tract infections	Nasopharyngeal, bronchoalveolar, sputum, pleural liquid, ear smear, samples	DNA hybridization on strip	Operon respiratory bacterial panel strip
C/T 13910 and G/A 22018 genes	Lactose intolerance risk	Blood, saliva, or tissues	DNA hybridization on strip	Operon lactostrip
APOE gene	Late-onset Alzheimer's disease and dementia risk	Mouth swabs	RT-PCR	EmpowerDX APOE gene test
MTHFR gene	Mental health, depressed mood, fertility, fetal development, and migraine headaches risk	Mouth swabs	RT-PCR	EmpowerDX MTHFR gene test

hCG, human chorionic gonadotropin; LFA, lateral flow assay; ELISA, enzyme-linked immunosorbent assays; hLH, human luteinizing hormone; FSH, follicle-stimulating hormone; IgG, immunoglobulin G; IgM, immunoglobulin M; IgE, immunoglobulin E; TSH, thyroid stimulating hormone; PSA, prostate-specific antigen; AFP, alpha fetoprotein; CEA, carcinoembryonic antigen; HbsAg, hepatitis B surface antigen; HbsAb, hepatitis B surface antibody; NS1, nonstructural protein 1; LAMP, loop-mediated isothermal amplification; RT-PCR, reverse transcription polymerase chain reaction; Protein N, nucleocapsid protein; Protein S, spike protein; NMP22, nuclear matrix protein number 22; Lp-PLA_2, lipoprotein-associated phospholipase A_2; tTG, tissue transglutaminase; DGP, deamidated gliadin; RF, rheumatoid factor; CCP, cyclic citrullinated peptide; DHEA-S, dehydroepiandrosterone sulfate; LDL, low-density lipoprotein; HDL, high-density lipoprotein; HLA, human leukocyte antigen; APOE, apolipoprotein E; MTHFR, metilenotetrahidrofolato redutase; BRCA, breast cancer; MSMB, microsemino protein beta; RNASET2, ribonuclease T2; SECISBP2L, SECIS binding protein 2 like; NRG1, neuregulin 1.

The COVID-19 pandemic brought the discussion about the difference between tests, such as PCR, antibody test, and antigen test, their utilization (clinical test versus self-test) and functionality (active infection versus previous infection), and complexity. The self-test for COVID-19 is a lateral flow assay (LFA), which operates using the principle of antigen-antibody recognition, identifying part of the virus, the nucleocapsid protein (protein N) or the spike protein (protein S) present in the nasal swab or salivary sample (Jackson et al., 2022). A gold nanoparticle conjugated with a specific antibody (GNP-Ab) recognizes and bonds to the target protein. If the analyte is present in the sample, the complex formed between the antigen and the GNP-Ab is trapped in the test band, showing a positive result (Abbott Panbio COVID-19 antigen self-test) (Fig. 1.4A). Different brands and price ranges for LFA COVID-19 rapid tests are available in the market. Using the same testing principle, other products use proteins or peptides as biomarkers to detect other diseases and medical conditions: prostate-specific antigen (Henso PSA test cassette), α-fetoprotein (Henso Alpha Fetoprotein AFP test cassette), carcinoembryonic antigen (Henso Carcinoembryonic Antigen CEA test cassette), and NMP22 (Abbott Alere) as cancer markers, HIV-1 P24 protein for early HIV detection (Abbott Determine HIV early detect), HBs protein for hepatitis B assay (Abbott Determine HBsAg), C-reactive protein for infection investigation (Prima C-reactive protein test), and sperm protein for

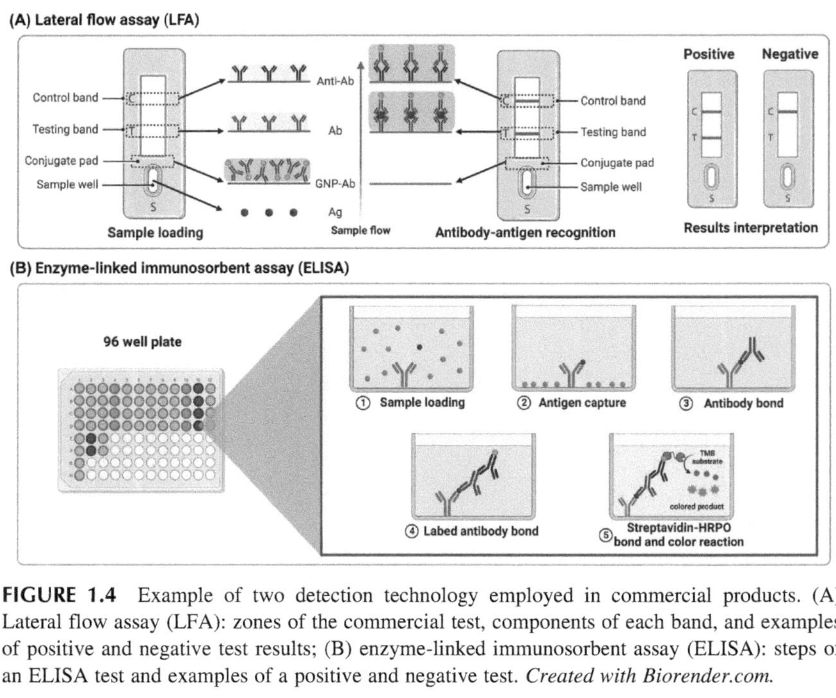

FIGURE 1.4 Example of two detection technology employed in commercial products. (A) Lateral flow assay (LFA): zones of the commercial test, components of each band, and examples of positive and negative test results; (B) enzyme-linked immunosorbent assay (ELISA): steps of an ELISA test and examples of a positive and negative test. *Created with Biorender.com.*

male fertility (SpermCheck fertility test). Other tests are based on the traditional ELISA method (Fig. 1.4B) designed to respond to NS1 antigen for dengue diagnoses (2Abbott Panbio dengue early ELISA) and Lp-PLA2 protein for vascular inflammation screening (Imaware vascular inflammation test), strip tests for albumin detection for kidney function (Multistix 10 SG reagent strips) or electrochemical sensor for prothrombin for coagulation testing (Xprecia Stride coagulation analyzer).

Another type of protein extensively used as biomarkers is antibodies. Unlike other proteins for infections described previously that examine parts of the pathogen, tests involving antibodies detection search for our immune response against the pathogen; however, there is an interval between the exposure to the initial immune response that could take days or weeks (Mortimer, 1996). After the infection, two main types of antibodies are produced: IgM, immunoglobulin M, is the first antibody to appear as a response to the pathogen exposure in the acute phase, decreasing after the exposure, whereas the IgG, immunoglobulin G, appears together or shortly after the IgM and last longer than IgM, and can be considered an indicator of the immune status to a pathogen (Mortimer, 1996). Different infectious diseases can be analyzed using antibody tests, for example, sexually transmitted infections like syphilis (Accu-Tell Syphilis rapid test), hepatitis (Accu-Tell HBsAb rapid test), and HIV (Oraquick HIV test at home) or other infectious diseases, such as COVID-19 (BioMedomics COVID-19 IgM/IgG rapid test), leptospirosis (Panbi Leptospira IgM ELISA), and Chagas disease (Bioline Chagas Ab). Besides infectious diseases, food intolerances like celiac disease (Celiac disease screening test) and other common foods allergens like egg, milk, soy, peanut, wheat, almond, shrimp, and tuna (Everlywell food allergy test), and rheumatoid arthritis (Imaware rheumatoid arthritis test) can be screened by specific antibodies found in an individual's biofluids.

Hormones are regulatory substances produced by an organism to signalize specific cells, tissues, or organs (Burtis & Bruns, 2014), so their concentration could be used to indicate someone's health. The alteration in their concentration can indicate some disease or medical condition. The most common and well-known test for the general public is the pregnancy test, an LFA screening device for hCG. Other commercially available tests analyze different hormones for other conditions: hLH (human Luteinizing Hormone) for ovulation (Smilelab ovulation test kit), FSH (follicle-stimulating hormone) for menopause (Accu-Tell FSH rapid test strip), TSH (thyroid stimulating hormone) for hypothyroidism (BTNX Inc Thyroid Stimulating Hormone TSH Test Cassette), cortisol for stress (Soma Biosciences Cortisol LFD), testosterone for men and women health (Everlywell testosterone test), estradiol, progesterone, and DHEA-S (dehydroepiandrosterone sulfate) for women's health (EmpowerDX women's hormones test), and prolactin for tumors, primary hypothyroidism, and others diseases (CTK Biotech PRL FIA test).

The glucometer is a widespread instrument found in any drugstore or online. It is a simple, fast, well-established electrochemical sensor for the quantification of glucose, an essential biomarker for diabetes diagnosis and monitoring (Roche Accu-Chek glucose meters & strips). Analysis of other organic molecules like total triglycerides, low-density lipoprotein (LDL), high-density lipoprotein (HDL) (CURO L7 professional grade testing kit for lipid blood profile), cholesterol (CURO L5 cholesterol test kit), and/or uric acid (UriTouch blood uric acid monitoring system) can be performed using an apparatus similar to the glucometer and disposable test strips. A reagent strip is a very simple, reliable, and powerful device to detect a broad range of biomarkers for urinary tract infections, kidney disorders, or diabetes. Ten important tests, including glucose, bilirubin, ketone, specific gravity, blood, pH, protein, urobilinogen, nitrite, leukocyte, albumin, and creatinine, can be performed using a single reagent strip by the "dip and read" method (Multistix 10 SG reagent strips). Other commercial products analyze vitamins, amino acids, and minerals (LetsGetChecked micronutrient test; Biostarks optimal nutrition test) to monitor health and nutrition conditions.

The detection of a pathogen can be performed using other techniques besides its antigen or the individual's antibody detection. Different pathogens associated with sexually transmitted infections like chlamydia, gonorrhea, trichomoniasis, and herpes (Operon STD panel strip) or respiratory tract infections, such as the common cold, pneumonia, and bronchitis (Operon respiratory bacterial panel strip), can be detected by their genetic material using hybridization on the strip and, more recently, COVID-19 using an at-home loop-mediated isothermal amplification (LAMP) test (Lucira check it COVID-19 test kit). A DNA test can also be used to detect genetic markers for more than 50 disease predispositions, nutrition predisposition risk, more than 95 monogenic hereditary diseases, and pharmacological compatibility (TellmeGen). Currently, genetic tests can analyze polymorphisms in certain genes associated with different medical conditions, for example, celiac disease risk (genes HLA-DQ2, HLA-DQ8, and DQA1*05) (EmpowerDX celiac risk gene test HLA), lactase persistence for lactose intolerance risk (detection of two MCM6 gene polymorphisms: C/T 13910 and G/A 22018) (Operon lactostrip), risk of developing late-onset Alzheimer's disease and dementia (apolipoprotein E, APOE, genotype) (EmpowerDX APOE gene test), risk of mental health issues, male infertility, problems in fetal development and migraine headaches associated with MTHFR gene involved in amino acid synthesis pathway (EmpowerDX MTHFR gene test), prostate cancer (BRCA1/2 and MSMB genes) (TellmeGen prostate cancer), lung cancer (RNASET2, SECISBP2L, and NRG1 genes) (TellmeGen Lung cancer), and other diseases and conditions.

Advances in technology allow upgrades in well-established devices. In LFA, for example, the usual outcome is a positive or negative result; however, for more information, some tests can produce semiquantitative results that can be tracked by a smartphone app, like some ovulation strip tests that the user can take a picture in the product's app and track the levels of the hormones in the function of the date (Smilelab ovulation test kit), or using a digital read to measure the color intensity in the test line and provide a pregnancy weeks indication (Clearblue digital pregnance test). For glucometers, different brands developed wearable sensors applied to the skin for continuous monitoring of blood glucose using a proprietary device and a smartphone (Cappon et al., 2017). Nowadays, the high demanding area is smart and wearable devices. Today smartwatches and smart bands are already capable of performing an electrocardiogram, analyzing the oxygen blood saturation, measuring heart rate, irregular heart rhythm, sleeping, stress level, and tracking hundreds of exercises, all integrated into the smartphone (Apple watch). However, the biomarker's future is beyond simple physiological measurements, future devices will be less expensive, but more accurate, sensible, and multipurpose, allowing multiple biomarkers assessment in parallel. In the next chapters, the reader will find guidance through the state-of-the-art in biomarkers analysis and will be amazed by the future perspectives of the new personalized medicine field.

References

Abbott Alere NMP22® bladderchek. <https://www.globalpointofcare.abbott/en/product-details/nmp22-bladderchek.html/> Accessed 15.07.22.

Abbott Determine HBsAg 2. <https://www.globalpointofcare.abbott/en/product-details/determine-hbsag-2.html/> Accessed 15.07.22.

Abbott Determine HIV early detect. <https://www.globalpointofcare.abbott/en/product-details/determine-hiv-early-detect.html/> Accessed 15.07.22.

Abbott Panbio COVID-19 antigen self-test. <https://www.globalpointofcare.abbott/en/product-details/panbio-covid-19-antigen-self-test.html/> Accessed 15.07.22.

Abbott Panbio dengue early ELISA. <https://www.globalpointofcare.abbott/en/product-details/panbio-dengue-early-elisa.html/> Accessed 15.07.22.

Accu-Tell FSH rapid test strip. <https://www.accubiotech.com/product-fsh-rapid-test-cassette-midstream-strip-urine.html/> Accessed 18.07.22.

Accu-Tell HBsAb rapid test. <https://www.accubiotech.com/product-hbsab-rapid-test-(serum-plasma).html/> Accessed 18.07.22.

Accu-Tell Syphilis rapid test. <https://www.accubiotech.com/product-syphilis-rapid-test-strip-cassette-(whole-blood-serum-plasma).html/> Accessed 18.07.22.

Alberice, J. V., Amaral, A. F. S., Armitage, E. G., Lorente, J. A., Algaba, F., Carrilho, E., Márquez, M., García, A., Malats, N., & Barbas, C. (2013). *Journal of Chromatography. A*, *1318*, 163–170.

American Chemical Society Historic Chemical Landmarks. (2010). Development of Diagnostic Test Strips. http://www.acs.org/content/acs/en/education/whatischemistry/landmarks/diagnosticteststrips.html. [Accessed 5 December 2022].

Antoniou, M., Kolamunnage-Dona, R., Wason, J., Bathia, R., Billingham, C., Bliss, J. M., Brown, L. C., Gillman, A., Paul, J., & Jorgensen, A. L. (2019). *Contemporary Clinical Trials Communications*, *16*, 100493.

Apple watch. <https://www.apple.com/watch/> Accessed 21.07.22.

Arivaradarajan, P., & Misra, G. (2019). *Omics approaches, technologies and applications: Integrative approaches for understanding OMICS data*. Singapore: Springer.

Aronson, J. K., & Ferner, R. E. (2017). *Current Protocols in Pharmacology, 2017*, 9.23.1–9.23.17.

Aslam, B., Basit, M., Nisar, M. A., Khurshid, M., & Rasool, M. H. (2017). *Journal of Chromatographic Science, 55*, 182–196.

Attallah, A. M., El-Far, M., Omran, M. M., Abdallah, S. O., El-desouky, M. A., El-Dosoky, I., Abdelrazek, M. A., Attallah, A. A., Elweresh, M. A., Abdel Hameed, G. E., Shawki, H. A., Salama, K. S., & El-Waseef, A. M. (2014). *Tumor Biology, 35*, 10737–10745.

Babisch, W., Fromme, H., Beyer, A., & Ising, H. (2001). *Environment International, 26*, 475–481.

Bader, G. D., Heilbut, A., Andrews, B., Tyers, M., Hughes, T., & Boone, C. (2003). *Trends in Cell Biology, 13*, 344–356.

Bader, J. M., Geyer, P. E., Müller, J. B., Strauss, M. T., Koch, M., Leypoldt, F., Koertvelyessy, P., Bittner, D., Schipke, C. G., Incesoy, E. I., Peters, O., Deigendesch, N., Simons, M., Jensen, M. K., Zetterberg, H., & Mann, M. (2020). *Molecular Systems Biology, 16*, 1–17.

Barzaman, K., Karami, J., Zarei, Z., Hosseinzadeh, A., Kazemi, M. H., Moradi-Kalbolandi, S., Safari, E., & Farahmand, L. (2020). *International Immunopharmacology, 84*, 106535.

Bedia, C. (2018). *Experimental approaches in omic sciences* (1st ed.). Elsevier B.V..

Benayoun, B. A., Pollina, E. A., & Brunet, A. (2015). *Nature Reviews. Molecular Cell Biology, 16*, 593–610.

Berger, M. F., Lawrence, M. S., Demichelis, F., Drier, Y., Cibulskis, K., Sivachenko, A. Y., Sboner, A., Esgueva, R., Pflueger, D., Sougnez, C., Onofrio, R., Carter, S. L., Park, K., Habegger, L., Ambrogio, L., Fennell, T., Parkin, M., Saksena, G., Voet, D., . . . Garraway, L. A. (2011). *Nature, 470*, 214–220.

Bioline Chagas Ab. <https://www.globalpointofcare.abbott/en/product-details/bioline-chagas-ab.html/> Accessed 18.07.22.

BioMedomics COVID-19 IgM/IgG rapid test. <https://www.biomedomics.com/products/infectious-disease/covid-19-rt/> Accessed 18.07.22.

Biostarks optimal nutrition test. <https://biostarks.com/nutrition-blood-test/#1/> Accessed 20.07.22.

Bourgoin-Voillard, S., Goron, A., Seve, M., & Moinard, C. (2016). *Proteomics, 16*, 831–846.

Britt, K. L., Cuzick, J., & Phillips, K. A. (2020). *Nature Reviews. Cancer, 20*, 417–436.

BTNX Inc Thyroid Stimulating Hormone (TSH) Test Cassette. <https://www.btnx.com/Product?id = 1716/> Accessed 18.07.22.

Buhimschi, I. A., Christner, R., & Buhimschi, C. S. (2005). *BJOG: An International Journal of Obstetrics & Gynaecology, 112*, 173–181.

Burtis, C. A., & Bruns, D. E. (2014). *Tietz fundamentals of clinical chemistry and molecular diagnostics* (7th ed.). Elsevier.

Cappon, G., Acciaroli, G., Vettoretti, M., Facchinetti, A., & Sparacino, G. (2017). *Electronics (Switzerland), 6*, 1–16.

Carrasco-Ramiro, F., Peiró-Pastor, R., & Aguado, B. (2017). *Gene Therapy, 24*, 551–561.

Celiac disease screening test. <https://www.imaware.health/at-home-blood-test/celiac-disease-screening/> Accessed 18.07.22.

Chen, H. H., Tseng, Y. J., Wang, S. Y., Tsai, Y. S., Chang, C. S., Kuo, T. C., Yao, W. J., Shieh, C. C., Wu, C. H., & Kuo, P. H. (2015). *International Journal of Obesity, 39*, 1241–1248.

Chen, Y. T., Chen, C. L., Chen, H. W., Chung, T., Wu, C. C., De Chen, C., Hsu, C. W., Chen, M. C., Tsui, K. H., Chang, P. L., Chang, Y. S., & Yu, J. S. (2010). *Journal of Proteome Research, 9*, 5803–5815.

Chojnowska, S., Ptaszyńska-Sarosiek, I., Kępka, A., Knaś, M., & Waszkiewicz, N. (2021). *Journal of Clinical Medicine, 10*, 1–11.

Clarke, S. F., & Foster, J. R. (2012). *British Journal of Biomedical Science, 69*, 83–93.

Clearblue digital pregnance test. <https://www.clearblue.com/pregnancy-tests/digital/> Accessed 13.07.22.

Conway De Macario, E., & Macario, A. J. L. (2000). *Frontiers in Bioscience, 5*, 780–786.

Correia, B. S. B., Ferreira, V. G., Piagge, P. M. F. D., Almeida, M. B., Assunção, N. A., Raimundo, J. R. S., Fonseca, F. L. A., Carrilho, E., & Cardoso, D. R. (2022). *Journal of Proteome Research.*

Coughlin, S. S. (2019). *Breast Cancer Research and Treatment, 177*, 537–548.

CTK Biotech PRL FIA test. <https://ctkbiotech.com/product/prl-fia-test/> Accessed 18.07.22.

Cuffe, J. S. M., Holland, O., Salomon, C., Rice, G. E., & Perkins, A. V. (2017). *Placenta, 54*, 104–110.

Cui, P., Li, X., Huang, C., Li, Q., & Lin, D. (2022). *Frontiers in Molecular Biosciences, 9*, 1–13.

CURO L5 cholesterol test kit. <https://curofit.com/s/blood-cholesterol/> Accessed 19.07.22.

CURO L7 professional grade testing kit for lipid blood profile. <https://curofit.com/s/lipidocare-l7-cholesterol-test-kit/> Accessed 19.07.22.

Daniel, C. R., Piraccini, B. M., & Tosti, A. (2004). *Journal of the American Academy of Dermatology, 50*, 258–261.

Davies, K. (1997). *Nature, 390*, 33.

De, P., Hasmann, M., & Leyland-Jones, B. (2013). *Cancer Treatment Reviews, 39*, 925–934.

De San-Martin, B. S., Ferreira, V. G., Bitencourt, M. R., Pereira, P. C. G. C. G., Carrilho, E., Assunção, N. A., Carvalho, L. R. S., de Assunção, N. A., & de Carvalho, L. R. S. (2020). *Archives of Endocrinology and Metabolism, 64*, 654–663.

Denkert, C., Bucher, E., Hilvo, M., Salek, R., Orešič, M., Griffin, J., Brockmöller, S., Klauschen, F., Loibl, S., Barupal, D. K., Budczies, J., Iljin, K., Nekljudova, V., & Fiehn, O. (2012). *Genome Medicine, 4*, 1–9.

Deutsch, O., Haviv, Y., Krief, G., Keshet, N., Westreich, R., Stemmer, S. M., Zaks, B., Navat, S. P., Yanko, R., Lahav, O., Aframian, D. J., & Palmon, A. (2020). *Scientific Reports, 10*, 1–13.

Di Zazzo, A., Micera, A., De Piano, M., Cortes, M., & Bonini, S. (2019). *Journal of Cellular Physiology, 234*, 9982–9993.

Dong, W., Qiu, C., Gong, D., Jiang, X., Liu, W., Liu, W., Zhang, L., & Zhang, W. (2019). *Experimental and Therapeutic Medicine, 18*, 2833–2842.

Duffy, M. J., Harbeck, N., Nap, M., Molina, R., Nicolini, A., Senkus, E., & Cardoso, F. (2017). *European Journal of Cancer, 75*, 284–298.

Duffy, M. J., Walsh, S., McDermott, E. W., & Crown, J. (2015). *In Makowski, Gregory S. (Ed.), Advances in Clinical Chemistry*, Vol. 71, pp. 1–23. Elsevier. Available from https://doi.org/10.1016/bs.acc.2015.05.001.

Dunn, W. B., & Ellis, D. I. (2005). *TrAC – Trends in Analytical Chemistry, 24*, 285–294.

Echle, A., Rindtorff, N. T., Brinker, T. J., Luedde, T., Pearson, A. T., & Kather, J. N. (2021). *British Journal of Cancer, 124*, 686–696.

Emmert-Streib, F., Dehmer, M., & Yli-Harja, O. (2017). *Frontiers in Genetics, 8*, 184.

EmpowerDX APOE gene test. <https://empowerdxlab.com/products/product/apoe-gene-test/> Accessed 21.07.22.

EmpowerDX celiac risk gene test (HLA). <https://empowerdxlab.com/products/product/celiac-risk-gene-test/> Accessed 21.07.22.

EmpowerDX MTHFR gene test. <https://empowerdxlab.com/products/product/mthfr-gene-test/ > Accessed 21.07.22.

EmpowerDX women's hormones test. <https://empowerdxlab.com/products/product/womens-hormones-test/> Accessed 18.07.22.

Engel, J., & Pitkänen, A. (2020). *Neuropharmacology*, *167*, 107735.

Everlywell food allergy test. <https://www.everlywell.com/products/food-allergy-test/> Accessed 18.07.22.

Everlywell testosterone test. <https://www.everlywell.com/products/testosterone-test/> Accessed 18.07.22.

Faix,, J. D. (2013). *Critical Reviews in Clinical Laboratory Sciences*, *50*, 23–36. Available from https://doi.org/10.3109/10408363.2013.764490.

Fang, F. C., Naccache, S. N., & Greninger, A. L. (2020). *Clinical Infectious Diseases*, *71*, 2996–3001.

FDA. (**2018**). *Guidance for industry* (pp. 1–16).

Foo, H., Mather, K. A., Thalamuthu, A., & Sachdev, P. S. (2019). *Current Opinion in Psychiatry*, *32*, 130–137.

Free, A. H., Adams, E. C., Kercher, M. L., Free, H. M., & Cook, M. H. (1957). *Clinical Chemistry*, *3*, 163–168.

Free, A. H., & Free, H. M. (1984). *Clinical Chemistry*, *30*, 829–838.

Ge, P., Luo, Y., Chen, H., Liu, J., Guo, H., Xu, C., Qu, J., Zhang, G., & Chen, H. (2021). *Frontiers in Oncology*, *11*, 1–16.

Gibbs, R. A. (2020). *Nature Reviews. Genetics*, *21*, 575–576.

Giskeødegård, G. F., Lundgren, S., Sitter, B., Fjøsne, H. E., Postma, G., Buydens, L. M. C., Gribbestad, I. S., & Bathen, T. F. (2012). *NMR in Biomedicine*, *25*, 1271–1279.

Goldstein, D. S., & Kopin, I. J. (2007). *Stress (Amsterdam, Netherlands)*, *10*, 109–120.

Goossens, N., Nakagawa, S., Sun, X., & Hoshida, Y. (2015). *Translational Cancer Research*, *4*, 256–269.

Griffiths, W. J., Karu, K., Hornshaw, M., Woffendin, G., & Wang, Y. (2007). *European Journal of Mass Spectrometry*, *13*, 45–50.

Hartl, D. L. (2009). *Essential genetics: a genomics perspective*. Jones & Bartlett Publishers.

Hawkridge, A. M., & Muddiman, D. C. (2009). *Annual Review of Analytical Chemistry*, *2*, 265–277.

Heilmann, E., Gregoriano, C., & Schuetz, P. (2019). *Seminars in Respiratory and Critical Care Medicine*, *40*, 465–475.

Hellhammer, D. H., Wüst, S., & Kudielka, B. M. (2009). *Psychoneuroendocrinology*, *34*, 163–171.

Henso Alpha Fetoprotein AFP test cassette. <https://www.hensomed.com/products/alpha-feto-protein-afp-rapid-screen-test-cassette/> Accessed 15.07.22.

Henso Carcinoembryonic Antigen CEA test cassette. <https://www.hensomed.com/products/carcinoembryonic-antigen-cea-test-cassette/> Accessed 15.07.22.

Henso PSA test cassette. <https://www.hensomed.com/products/psa-test-cassette/> Accessed 15.07.22.

Herget-rosenthal, S., Bökenkamp, A., & Hofmann, W. (2007). *Clinical Biochemistry*, *40*, 153–161.

Hertel, J., Friedrich, N., Wittfeld, K., Pietzner, M., Budde, K., Van Der Auwera, S., Lohmann, T., Teumer, A., Völzke, H., Nauck, M., & Grabe, H. J. (2016). *Journal of Proteome Research*, *15*, 400–410.

Horvath, S. (2013). *Genome Biology*, *14*, 1–20.

Horvath, S., & Raj, K. (2018). *Nature Reviews. Genetics*, *19*, 371–384.

Imaware rheumatoid arthritis test. <https://www.imaware.health/at-home-blood-test/rheumatoid-arthritis/> Accessed 18.07.22.

Imaware vascular inflammation test. <https://www.imaware.health/at-home-blood-test/vascular-inflammation-heart-health-test/> Accessed 18.07.22.

Jackson, C. B., Farzan, M., Chen, B., & Choe, H. (2022). *Nature Reviews. Molecular Cell Biology*, *23*, 3–20.

Johnson, G. (1884). *British Medical Journal*, *1*, 1 LP– 4.

Jowett, J. B. M. (2009). *Nature Reviews Endocrinology*, *5*, 187–188.

Kalia, M., Costa, J., & Silva, E. (2015). *Metabolism: Clinical and Experimental*, *64*, S11–S15.

Kennis, M., Gerritsen, L., van Dalen, M., Williams, A., Cuijpers, P., & Bockting, C. (2020). *Molecular Psychiatry*, *25*, 321–338.

Kohler, I., Hankemeier, T., van der Graaf, P. H., Knibbe, C. A. J., & van Hasselt, J. G. C. (2017). *European Journal of Pharmaceutical Sciences*, *109*, S15–S21.

Kondo, C., Minowa, Y., Uehara, T., Okuno, Y., Nakatsu, N., Ono, A., Maruyama, T., Kato, I., Yamate, J., Yamada, H., Ohno, Y., & Urushidani, T. (2009). *Toxicology*, *265*, 15–26.

Kumar, S., Mohan, A., & Guleria, R. (2006). *Biomarkers: Biochemical Indicators of Exposure, Response, and Susceptibility to Chemicals*, *11*, 385–405.

Kwon, E. H., Tennagels, S., Gold, R., Gerwert, K., Beyer, L., & Tönges, L. (2022). *Biomolecules*, *12*, 329.

Lancellotti, C., Cancian, P., Savevski, V., Kotha, S. R. R., Fraggetta, F., Graziano, P., & Di Tommaso, L. (2021). *Cells*, *10*, 787.

Lee, J. M., & Ahn, S. H. (2011). *World Journal of Gastroenterology*, *17*, 283–289.

Lee, S. J., Jeong, J. H., Kang, S. H., Kang, J., Kim, E. A., Lee, J., Jung, J. H., Park, H. Y., & Chae, Y. S. (2019). *International Journal of Molecular Sciences*, *20*, 6162.

LetsGetChecked micronutrient test. <https://www.letsgetchecked.com/home-micronutrient-test-kit/> Accessed 19.07.22.

Liu, X., Zheng, W., Wang, W., Shen, H., Liu, L., Lou, W., Wang, X., & Yang, P. (2017). *British Journal of Cancer*, *117*, 1846–1854.

Loke, S. Y., & Lee, A. S. G. (2018). *European Journal of Cancer*, *92*, 54–68.

Lu, S., Wei, F., & Li, G. (2021). *Cell Stress*, *5*, 76.

Lu, W., Bennett, B. D., & Rabinowitz, J. D. (2008). *Journal of Chromatography B: Analytical Technologies in the Biomedical and Life Sciences*, *871*, 236–242.

Lucira check it COVID-19 test kit. <https://checkit.lucirahealth.com/> Accessed 20.07.22.

Makarov, Danil V., Loeb, Stacy, Getzenberg, Robert H., & Partin, Alan W. (2009). *Annual Review of Medicine*, *60*, 139–151.

Mardis, E. R., Ding, L., Dooling, D. J., Larson, D. E., McLellan, M. D., Chen, K., Koboldt, D. C., Fulton, R. S., Delehaunty, K. D., McGrath, S. D., Fulton, L. A., Locke, D. P., Magrini, V. J., Abbott, R. M., Vickery, T. L., Reed, J. S., Robinson, J. S., Wylie, T., Smith, S. M., ... Ley, T. J. (2009). *New England Journal of Medicine*, *361*, 1058–1066.

Maruvada, P., Wang, W., Wagner, P. D., & Srivastava, S. (2006). *Biotechniques*, *15*, 4.

McDermott, J. E., Wang, J., Mitchell, H., Webb-Robertson, B. J., Hafen, R., Ramey, J., & Rodland, K. D. (2013). Expert Opinion on Medical. *Diagnostics*, *7*, 37–51.

Melguizo-Rodríguez, L., Costela-Ruiz, V. J., Manzano-Moreno, F. J., Ruiz, C., & Illescas-Montes, R. (2020). *International Journal of Molecular Sciences, 21*, 1–17.

Menni, C., Kiddle, S. J., Mangino, M., Viñuela, A., Psatha, M., Steves, C., Sattlecker, M., Buil, A., Newhouse, S., Nelson, S., Williams, S., Voyle, N., Soininen, H., Kloszewska, I., Mecocci, P., Tsolaki, M., Vellas, B., Lovestone, S., Spector, T. D., ... Valdes, A. M. (2015). *The Journals of Gerontology: Series A, 70*, 809–816.

Menyhart, O., Santarpia, L., & Gyorffy, B. (2015). *Current Cancer Drug Targets, 15*, 665–683.

Mitchell, A. L., Gajjar, K. B., Theophilou, G., Martin, F. L., & Martin-Hirsch, P. L. (2014). *Journal of Biophotonics, 7*, 153–165.

Mitnitski, A. B. (2018). *Epigenetics of aging and longevity* (pp. 153–170). Academic Press.

Moon, P. G., Lee, J. E., Cho, Y. E., Lee, S. J., Jung, J. H., Chae, Y. S., Bae, H. I., Kim, Y. B., Kim, I. S., Park, H. Y., & Baek, M. C. (2016). *Clinical Cancer Research, 22*, 1757–1766.

Mortimer, P. P. (1996). *Clinical and Diagnostic Virology, 5*, 131–136.

Multistix 10 SG reagent strips. <https://www.siemens-healthineers.com/nl/urinalysis-products/urinalysis-reagents/multistix-10-sg-reagent-strips/> Accessed 20.07.22.

Nevanlinna, H., & Bartek, J. (2006). *Oncogene, 25*, 5912–5919.

Noushad, S., Ahmed, S., Ansari, B., Mustafa, U.-H., Saleem, Y., & Hazrat, H. (2021). *International Journal of Health Sciences, 15*, 46.

Nunes, M. J., Cordas, C. M., Moura, J. J. G., Noronha, J. P., & Branco, L. C. (2021). *Sports Medicine – Open, 7*, 1–9.

Of, C., Cholesterol, C. P. L., Prediction, L., First, O. F., & Events, C. (2002). *New England Journal of Medicine, 347*, 1557–1565.

Oliver, N. S., Toumazou, C., Cass, A. E. G., & Johnston, D. G. (2009). *Diabetic Medicine, 26*, 197–210.

Olszynko-Gryn, J. (2014). *Studies in History and Philosophy of Science Part C:Studies in History and Philosophy of Biological and Biomedical Sciences, 47*, 233–247.

Operon lactostrip. <https://operondx.com/lactostrip/> Accessed 21.07.22.

Operon respiratory bacterial panel strip. <https://operondx.com/operon-respiratory-bacterial-panel-strip/> Accessed 20.07.22.

Operon STD panel strip. <https://operondx.com/std-panelstrip/> Accessed 20.07.22.

Oraquick HIV test at home. <https://oraquick.com/> Accessed 18.07.22.

Ou, F. S., Michiels, S., Shyr, Y., Adjei, A. A., & Oberg, A. L. (2021). *Journal of Thoracic Oncology, 16*, 537–545.

Our world in data, 2018, Number of people with cancer, 1990 to 2017. <https://ourworldindata.org/grapher/number-of-people-with-cancer-by-type/>.

Paige, L. A., Mitchell, M. W., Krishnan, K. R. R., Kaddurah-Daouk, R., & Steffens, D. C. (2007). *International Journal of Geriatric Psychiatry, 22*, 418–423.

Panbi Leptospira IgM ELISA. <https://www.globalpointofcare.abbott/en/product-details/panbio-leptospira-igm-elisa.html/> Accessed 18.07.22.

Park, H., Kim, J., Kim, S. Y., Kim, Y., Lee, J. S., Dan, K., Seong, M. W., & Han, D. (2020). *Scientific Reports, 10*, 1–10.

Patti, G. J., Yanes, O., & Siuzdak, G. (2012). *Nature Reviews. Molecular Cell Biology, 13*, 263–269.

Pearlmutter, P., DeRose, G., Samson, C., Linehan, N., Cen, Y., Begdache, L., Won, D., & Koh, A. (2020). *Scientific Reports, 10*, 1–11.

Petkovich, D. A., Podolskiy, D. I., Lobanov, A. V., Lee, S. G., Miller, R. A., & Gladyshev, V. N. (2017). *Cell Metabolism, 25*(954-960), e6.

Pham, Y. L., & Beauchamp, J. (2021). *Molecules (Basel, Switzerland), 26*, 5514.

Phillips, R., Kraeuter, A. K., McDermott, B., Lupien, S., & Sarnyai, Z. (2021). *Psychoneuroendocrinology*, *123*, 104903.

Piazza, J. R., Almeida, D. M., Dmitrieva, N. O., & Klein, L. C. (2010). *The Journals of Gerontology: Series B*, *65B*, 513–525.

Pieragostino, D., Bucci, S., Agnifili, L., Fasanella, V., D'Aguanno, S., Mastropasqua, A., Ciancaglini, M., Mastropasqua, L., Di Ilio, C., Sacchetta, P., Urbani, A., & Del Boccio, P. (2012). *Molecular Biosystems*, *8*, 1017–1028.

Pleasance, E. D., Stephens, P. J., O'Meara, S., McBride, D. J., Meynert, A., Jones, D., Lin, M. L., Beare, D., Lau, K. W., Greenman, C., Varela, I., Nik-Zainal, S., Davies, H. R., Ordõez, G. R., Mudie, L. J., Latimer, C., Edkins, S., Stebbings, L., Chen, L., ... Campbell, P. J. (2010). *Nature*, *463*, 184–190.

Prima C-reactive protein test. <https://primalabsa.ch/en/home/infectivology/c-reactive-protein-test/> Accessed 19.07.22.

PubMed N.I.H., <https://pubmed.ncbi.nlm.nih.gov/?term = biomarker/>.

Puntmann, V. O. (2009). *Postgraduate Medical Journal*, *85*, 538–545.

Quezada, H., Guzmán-Ortiz, A. L., Díaz-Sánchez, H., Valle-Rios, R., & Aguirre-Hernández, J. (2017). *Boletín Médico Del Hospital Infantil de México (English Edition)*, *74*, 219–226.

Rainville, I., Hatcher, S., Rosenthal, E., Larson, K., Bernhisel, R., Meek, S., Gorringe, H., Mundt, E., & Manley, S. (2020). *Breast Cancer Research and Treatment*, *180*, 503–509.

Reims, H. M., Sevre, K., Fossum, E., Mellem, H., Eide, I. K., & Kjeldsen, S. E. (2005). *Blood Pressure*, *14*, 217–226.

Rifai, N., Gillette, M. A., & Carr, S. A. (2006). *Nature Biotechnology*, *24*, 971–983.

Ritchie, H., & Roser, M. (2018). *Our World in Data.*

Roche Accu-Chek glucose meters and strips. <https://www.accu-chek.com/product-category/> Accessed 19.07.22.

Roig, B., Rodríguez-Balada, M., Samino, S., Lam, E. W. F., Guaita-Esteruelas, S., Gomes, A. R., Correig, X., Borràs, J., Yanes, O., & Gumà, J. (2017). *Scientific Reports*, *2017*(7), 1, *7*, 1–9.

Rudd, K. E., Johnson, S. C., Agesa, K. M., Shackelford, K. A., Tsoi, D., Kievlan, D. R., Colombara, D. V., Ikuta, K. S., Kissoon, N., Finfer, S., Fleischmann-Struzek, C., Machado, F. R., Reinhart, K. K., Rowan, K., Seymour, C. W., Watson, R. S., West, T. E., Marinho, F., Hay, S. I., ... Naghavi, M. (2020). *The Lancet*, *395*, 200–211.

Russell, G., & Lightman, S. (2019). *Nature Reviews Endocrinology 2019 15:9*, *15*, 525–534.

Samprathi, M., & Jayashree, M. (2021). *Frontiers in Pediatrics*, *8*, 972.

Sarshar, M., Scribano, D., Ambrosi, C., Palamara, A. T., & Masotti, A. (2020). *Cancers*, *12*, 2174, *2020, Vol. 12, Page 2174.*

Schübeler, D. (2015). *Nature*, *517*, 321–326.

Sharma, N., & Gedeon, T. (2012). *Computer Methods and Programs in Biomedicine*, *108*, 1287–1301.

Shaw,, L. M., Vanderstichele,, H., Knapik-Czajka,, M., Clark,, C. M., Aisen, P. S., Petersen, R. C., Blennow, K., Soares, H., Simon, A., Lewczuk, P., Dean, R., Siemers, E., Potter, W., Lee, V. M. Y., & Trojanowski, J. Q. (2009). *Annals of Neurology*, *65*, 403–413.

Shen, L., Farid, H., & Mcpeek, M. A. (2008). *OMICS applications in biomedical, agricultural, and environmental sciences*. Boca Raton: *CRC Press.*

Shimanoe, C., Matsumoto, A., Hara, M., Akao, C., Nishida, Y., Horita, M., Nanri, H., Higaki, Y., & Tanaka, K. (2021). *Psychoneuroendocrinology*, *125*, 105125.

Smets, E., Rios Velazquez, E., Schiavone, G., Chakroun, I., D'Hondt, E., De Raedt, W., Cornelis, J., Janssens, O., Van Hoecke, S., Claes, S., Van Diest, I., & Van Hoof, C. (2018). *Npj Digital Medicine*, *1*, 1–10.

Smilelab ovulation test kit. <https://www.smilelab.co/smilereader/> Accessed 18.07.22.

Solje, E., Benussi, A., Buratti, E., Remes, A. M., Haapasalo, A., & Borroni, B. (2021). *Diagnostics*, *11*, 788.

Soma Biosciences Cortisol LFD. <http://somabioscience.com/?page_id = 191/> Accessed 18.07.22.

SpermCheck fertility test. <https://spermcheck.com/fertility/> Accessed 19.07.22.

Spratt, E. G., Nicholas, J. S., Brady, K. T., Carpenter, L. A., Hatcher, C. R., Meekins, K. A., Furlanetto, R. W., & Charles, J. M. (2012). *Journal of Autism and Developmental Disorders*, *42*, 75–81.

Steckl, A. J., & Ray, P. (2018). *ACS Sensors*, *3*, 2025–2044.

Sultanpur, C. M., Deepa, K., & Kumar, S. V. (2015). *International Journal of Pharmaceutical Sciences Review and Research*, *3*, 119–122.

Teggert, A., Datta, H., & Ali, Z. (2020). *Micromachines*, *11*, 286.

TellmeGen. <https://www.tellmegen.com/> Accessed 20.07.22.

TellmeGen Lung cancer. <https://www.tellmegen.com/pt/resultados/test-adn-salud/cancer-pulmao/> Accessed 21.07.22.

TellmeGen prostate cancer. <https://www.tellmegen.com/pt/resultados/test-adn-salud/cancer-prostata/> Accessed 21.07.22.

Thomas, A., Rhoads, A., Pinkerton, E., Schroeder, M. C., Conway, K. M., Hundley, W. G., McNally, L. R., Oleson, J., Lynch, C. F., & Romitti, P. A. (2019). *JNCI Cancer Spectrum*, 3.

Tosoian, J., & Loeb, S. (2010). *The Scientific World Journal*, *10*, 1919–1931.

Tyers, M., & Mann, M. (2003). *Nature*, *422*, 193–197.

UriTouch blood uric acid monitoring system. <https://shop.goutandyou.com/collections/uric-acid/products/uritouch-blood-uric-acid-monitoring-system/> Accessed 19.07.22.

van Ockenburg, S. L., Booij, S. H., Riese, H., Rosmalen, J. G. M., & Janssens, K. A. M. (2015). *Psychoneuroendocrinology*, *62*, 189–199.

Veenstra, T. D., & Yates III, J. R. (Eds.), (2019). *Proteomics for biological discovery* (2nd ed). John Wiley & Sons.

Vernon, S. T., Hansen, T., Kott, K. A., Yang, J. Y., O'Sullivan, J. F., & Figtree, G. A. (2019). *Microcirculation (New York, N.Y.: 1994)*, *26*, e12488.

Vogel, G. (1997). *Science (New York, N.Y.)*, *277*, 1753–1754.

World Health Organization. (2001). *Environmental Health*, 144.

Xprecia Stride coagulation analyzer. <https://www.siemens-healthineers.com/coagulation/coagulation/xprecia-stride-analyzer/> Accessed 18.07.22.

Yaneske, E., & Angione, C. (2018). *BMC Bioinformatics*, *19*, 83–96.

Yates, J. R. (2019). *F1000Research*, 8.

Yates, J. R., Ruse, C. I., & Nakorchevsky, A. (2009). *Annual Review of Biomedical Engineering*, *11*, 49–79.

Zhang, L., & Guo, H. (2020). *Advances in Biomarker Sciences and Technology*, *2*, 1–23.

Zhao, X., Modur, V., Carayannopoulos, L. N., & Laterza, O. F. (2015). *Clinical Chemistry*, *61*, 1343–1353.

Chapter 2

Biomarkers in precision medicine

Linda Yu-Ling Lan[1], Wasan Mani Kumar[2], Long Sha Liu[2], Aubrey Kathryn Roberts[1], Shirley Chen[1] and Michael Snyder[1]
[1]*Department of Genetics, Stanford University School of Medicine, Stanford, CA, United States,* [2]*Stanford University School of Medicine, Stanford, CA, United States*

2.1 Applications of biomarkers in medicine

Precision medicine is an approach to medical care considering individual differences in people's genes, environments, and lifestyles. Precision medicine aims to tailor medical treatments and interventions to the specific characteristics of each patient, to improve health outcomes and reduce the risk of side effects. As of result, biomarkers play a crucial role in precision medicine, as they are used to identify specific biological features of a disease that can inform the most appropriate treatment strategy for an individual patient. Biomarkers can be used in several ways in precision medicine, including (1) diagnostic biomarkers—These markers are used to identify an individual's disease or condition. For example, biomarkers can diagnose cancers, heart disease, or neurological disorders. (2) Monitoring biomarkers—These markers are used for monitoring a disease's progression or the response to treatment. (3) Prognostic biomarkers—These markers are used to predict the likely outcome of a disease or condition in an individual. (4) Therapeutic biomarkers—These markers select the most appropriate treatment for an individual based on their specific biological features. (5) Predictive biomarkers—These biomarkers can provide critical information about the probability of a patient benefiting from a particular therapeutic intervention or the likelihood of experiencing adverse reactions. We will provide examples of each category and the identification of appropriate biomarkers in a variety of biological samples, including blood, urine, tissue, and saliva.

2.1.1 Biomarkers for disease prevention—for diagnostic

A diagnostic biomarker is an instrument used to identify or verify the existence of a particular illness or condition, or to distinguish different versions

Biosensors in Precision Medicine. DOI: https://doi.org/10.1016/B978-0-443-15380-8.00003-5

of a disease. With the rise of precision medicine, these biomarkers are becoming increasingly sophisticated and used to identify people with diseases such as cancer, heart disease, and neurological disorders in their early stages or reclassify it.

For example, cancer is increasingly classified based on molecular and imaging markers rather than just the organ where it occurs. Imaging biomarkers (IBs) in oncology are used for tumour, node, metastasis (TNM) metastasis staging, bone mineral density (DEXA scan T-score), breast mammography (BI-RADS), and left ventricular ejection fraction (LVEF), among others (Califf, 2018; O'Connor et al., 2017).

When a diagnostic biomarker is used in specific clinical situations, it is important to consider the context of use. A biomarker in one situation may not be useful in another. For instance, a biomarker must have an extremely low false-positive rate in detecting rare diseases such as pancreatic or ovarian cancer. The diagnosis can be emotionally traumatic and may necessitate further invasive tests. However, in the case of prevalent conditions like hypertension or high cholesterol, where retesting poses minimal risk, a higher false-positive rate might be more tolerable. This has led to some controversy, such as using the serum prostate-specific antigen (PSA) screening test for prostate cancer, the predominant biomarker to screen for prostate cancer. The development of the PSA test in 1986 led to a doubling in the number of men in the United States diagnosed with prostate cancer from 1985 to 1995 and a sixfold increase in invasive procedures for prostate cancer treatment over this period (Prensner et al., 2012). This has led to questioning if patients with benign conditions, such as inflammation or hyperplasia, were being overdiagnosed and overtreated.

Validation of diagnostic biomarkers is a complex process, with one goal being to identify a method that ensures the biomarker can be measured accurately, consistently, and inexpensively.

The process of validating a diagnostic biomarker typically involves several steps, including discovery and preliminary testing, preclinical validation, clinical validation, Regulatory approval such as the US Food and Drug Administration (FDA) or the European Medicines Agency (EMA), and postmarket monitoring. Unfortunately, many assays are not properly validated, leading to false assumptions about their value. This is especially true for biomarkers like troponin, which is important for diagnosing cardiac injury (Babuin & Jaffe, 2005). Different assays for measuring troponin have varying accuracy and reliability levels, which can lead to confusion and misdiagnosis. As detection methods continue to improve, our understanding of the value of diagnostic biomarkers will likely evolve as well. Receiver-operating characteristic (ROC) curves are a commonly used statistical method in validating diagnostic biomarkers. A ROC curve is a graphical representation of the performance of a diagnostic test, and it provides a visual way of comparing the test's ability to distinguish between patients with and without a specific disease or condition. Even though ROC curves are valuable in assessing the effectiveness of diagnostic biomarkers and in contrasting the performance of various biomarkers or diagnostic exams, it can be challenging. This is because there usually is not a historical benchmark to ascertain the presence or absence of a disease or condition, and deciding the cutoff point where a biomarker level is deemed positive or negative can be complicated (Ou et al., 2021). Therefore this involves obtaining regulatory approval from agencies to ensure that the biomarker meets the necessary standards for use in clinical practice.

2.1.2 Biomarkers for monitoring

A monitoring biomarker is reassessed to evaluate disease progression, additional disease complications, or changes in disease severity. In the case of HIV treatment, one of the most important biomarkers for monitoring HIV treatment is the viral load, with a decrease in viral load, indicating that the treatment is working and that the virus is being suppressed (Teigler et al., 2018). Another important biomarker for monitoring HIV treatment is the CD4 cell count, as HIV infection causes a significant decline in the number of these immune cells. The decrease in the CD4 cell count can indicate that HIV is damaging the patient's immune system. An increase in the CD4 cell count can indicate that the patient's immune system is recovering with the help of antiretroviral therapy (Kagan et al., 2015). While the concept of monitoring biomarkers is clear, determining how changes in these markers should impact treatment decisions, such as when to perform additional testing or intervene, can be challenging and may not always be precise. Moreover, it is frequently necessary to empirically establish the best time gap between measurements or the length of the clinical progression during which these measurements should be recorded.

Several biomarkers employed in clinical practice possess ambiguous operational characteristics, necessitating clinical judgment for interpretation (Califf, 2018). In developing medical treatments, diagnostic procedures, and devices, variations in biomarker levels are used to identify if significant thresholds have been achieved. These changes indicate if the treatment is influencing the biological target sufficiently to justify continued development. "On-treatment" monitoring biomarkers have been used frequently in cancer therapeutics, such as ctDNA levels, lymphocyte count, and CD8 + T cell count in melanoma (Tarhini & Kudchadkar, 2018). However, the ability to measure off-target effects on biological systems will bring more biomarkers and systems measurements into play to evaluate intermediate findings in medical product development.

Monitoring biomarkers plays a vital role in safeguarding the well-being of human research participants. This can involve tracking liver functionality in drug trials with potential liver toxicity or examining troponins for heart-related events.

Additionally, these biomarkers are beneficial for observing pharmacodynamic effects, identifying early responses to treatment, and detecting complications from a disease or therapy. An intriguing aspect of monitoring biomarkers is the widely held view among researchers and clinicians that alterations in biomarker measurements provide the most accurate prediction of a patient or population's likely outcome. Nevertheless, in many scenarios, the actual measurement, rather than the change, serves as the best outcome forecast, even though the change might be the best method for tracking the therapy's impact. For instance, an angiotensin converting enzyme (ACE) inhibitor could increase serum creatinine and/or potassium levels. Yet, the primary risk to the patient or research participant is determined mainly by the creatinine or potassium level itself rather than the change in the level (Cohen & Townsend, 2008). Incorporating biomarkers into treatment monitoring offers crucial insights to healthcare providers and plays a pivotal role in ensuring patients receive optimal and personalized care.

2.1.3 Prognostic biomarkers

Prognostic biomarkers serve to estimate the chance of a clinical incident, disease recurrence, or progression of a health condition in patients. These biomarkers are distinct from susceptibility/risk biomarkers such as genomic profiling, which forecast a patient's likelihood of developing certain illnesses or conditions.

For instance, in the context of cancer, prognostic biomarkers like tumor markers can be utilized to anticipate the probability of disease progression, recurrence, or death. Similarly, in cardiovascular disease scenarios, biomarkers like high-sensitivity C-reactive protein (hsCRP) and lipoprotein(a) (Lp(a)) can be used to predict the chance of a patient having a heart attack or stroke.

Clinical trials frequently employ prognostic biomarkers to determine participant eligibility and identify high-risk populations. The statistical strength of a trial depends more on the occurrence of events than the sample size. Through strategic recruitment of participants based on prognostic biomarkers, the event rate increases, which magnifies the quantitative effect of the treatment but not the qualitative aspect.

Moreover, prognostic biomarkers are crucial in predicting an individual's risk of an event or adverse outcome, an essential factor for hospitalization and intensive care decisions. They also have significant implications in allocating resources to population health, pinpointing patients who require a more rigorous evaluation and those who can bypass unnecessary additional diagnostic tests or medical interventions (Simon, 2010).

2.1.4 Therapeutic biomarkers

Therapeutic biomarkers are biological markers used to evaluate a specific treatment's effectiveness. They can provide information about the pharmacodynamics (how a drug works) and pharmacokinetics (how a drug is absorbed, distributed, metabolized, and eliminated) of treatment. They can be used to optimize dosing, monitor drug exposure, and assess treatment response (Califf, 2018). For instance, in the context of cancer, biomarkers like tumor markers are employed to assess the efficacy of chemotherapy. Fluctuations in these marker levels can reveal whether the treatment successfully impacts the cancer cells. This information aids healthcare professionals in making knowledgeable decisions about modifying the treatment strategy. Similarly, when dealing with cardiovascular disease, biomarkers such as high-density lipoprotein (HDL) cholesterol and low-density lipoprotein (LDL) cholesterol are utilized to evaluate the effectiveness of cholesterol-reducing medications (Weverling-Rijnsburger et al., 2003). Monitoring changes in these markers can indicate the treatment's effectiveness in reducing cholesterol levels as intended. These changes provide valuable information to healthcare providers, empowering them to make informed decisions regarding adjustments to the treatment plan.

In addition to traditional biomarkers, newer technologies, such as imaging techniques, can be therapeutic biomarkers. For instance, when it comes to cancer, imaging methods like magnetic resonance imaging (MRI) or computed tomography (CT) scans can be utilized to assess the efficacy of cancer therapies. This is done by monitoring the alterations in a tumor's size over a certain period. In conclusion, therapeutic biomarkers have a crucial role in assessing the effectiveness of treatments, contributing to the administration of appropriate and impactful interventions for patients. These biomarkers offer insights into the pharmacodynamics and pharmacokinetics of treatment, enabling optimization of dosing, monitoring of drug exposure, and evaluation of treatment response. Ultimately, leveraging therapeutic biomarkers improves health outcomes and better patient care.

2.1.5 Biomarkers in disease prediction

A predictive biomarker is a marker that anticipates how a person or group will react to a medical product or environmental factor. To demonstrate that a biomarker is beneficial, a thorough approach to clinical studies is required. Patients with varying biomarker levels are randomly distributed to different treatments, and the results are compared.

Predictive biomarkers are critical for designing and conducting clinical trials, as they enable a more tailored approach to treatment by enlisting individuals who are likely to respond effectively to the therapy (Antman & Loscalzo, 2016; Califf, 2018). This concept is also employed in clinical practice, where treatments are prescribed based on biomarkers suggesting a patient is likely to respond to the therapy.

Recently, genetic and genomic markers have positioned themselves at the forefront of predictive biomarker development, offering novel and powerful methods for individual risk stratification and therapeutic decision-making.

These markers encompass various molecular signatures, ranging from single nucleotide polymorphisms (SNPs), copy number variations (CNVs), to more complex gene expression profiles and epigenetic modifications. Genetic markers include BRCA1/BRCA2 mutations for predicting breast and ovarian cancer risk (Kuchenbaecker et al., 2017), and *CFTR* gene mutations for cystic fibrosis (Veit et al., 2016). On a genomic scale, next-generation sequencing (NGS) technologies provide a comprehensive landscape of an individual's genomic constitution, capturing nuances in gene–gene and gene–environment interactions, which might confer disease susceptibility. These genetic and genomic markers have found substantive applications in precision medicine, enabling physicians to prescribe the right treatment at the right time for the right patient. They aid in tailoring treatment strategies based on an individual's genetic predisposition, reducing unnecessary treatments and associated side effects. Moreover, they serve as valuable tools for developing novel, targeted therapies, fueling the growth of the burgeoning field of pharmacogenomics.

Despite these advancements, numerous challenges persist. These include the need for larger, diverse patient cohorts for validation studies, ethical issues surrounding genetic testing and data privacy, and logistical hurdles in integrating these technologies into clinical workflows. Despite the challenges, the potential of predictive biomarkers, especially genetic and genomic markers, is indisputable. Their capability to transform healthcare is truly remarkable.

2.1.6 Other types of biomarkers

A safety biomarker is a measurement that aids in assessing the probability, presence, or degree of toxicity as a harmful event, before or after exposure to a medical procedure or an environmental agent. Monitoring toxicity in organs like the liver, kidneys, or cardiovascular system is crucial for many treatments to ensure their safe application. Safety biomarkers are beneficial in recognizing patients experiencing side effects from treatment, and they are also utilized to keep track of a population's exposure to environmental risks or after such exposure. A key component in developing safety biomarkers is balancing safety and the potential therapeutic benefits. For instance, the use of QT intervalprolongation as a safety biomarker for antiarrhythmic drugs has been controversial and discussed. This is because it may have resulted in the discontinuation of drugs where the benefits outweigh the risks (Sager et al., 2014).

Surrogate biomarkers are test values that can be used instead of clinically meaningful endpoints that predict therapeutic effectiveness. In instances where a clinical trial may not effectively demonstrate the detection of clinical benefit from the treatment, a surrogate biomarker may be used to avoid delays in approval for indications that are considered clinically important.

In Alzheimer's disease, imaging markers such as total brain volume or hippocampal volume have been proposed as surrogate markers (Katz, 2004). Blood pressure can be a surrogate biomarker for cardiovascular disease (Lassere et al., 2012). HbA1c, which can be a surrogate biomarker for diabetes control (Baechle et al., 2022). Harnessing the power of surrogate biomarkers has the potential to fast-track the creation of novel therapies and interventions. These handy tools offer insight into a treatment's possible efficacy even before attaining the ultimate clinical goal. However, it is crucial to exercise caution when using biomarkers as stand-ins for significant clinical outcomes. Even when a biomarker and clinical outcome are in sync, their consistent alignment should not be taken for granted. A thorough and critical evaluation is vital to ensure we are not investing too much faith in these biomarkers and overlooking other key clinical data. Fleming and DeMets described that several pathophysiological processes intersect in disease progression, and surrogate biomarkers may be present in some—still, none of these pathways (Fleming & DeMets, 1996).

2.2 Pharmacogenomics and biomarkers

Imagine a world where a single pill could work differently for different people, like characters in a story each reacting in their own unique way to the same plot twist. Welcome to the fascinating realm of pharmacogenomics. It's a field of study that explores the interplay between our genetic blueprint and how we respond to medication. In this tale of genes and drugs, multiple characters have unique traits. Similar to different personalities, genetic variations could influence how effective a medication is, what dosage is needed, and what side effects may emerge, much like a plot twist being perceived differently by different characters. Moreover, this variation in response can be seen across different drug indications (Fig. 2.1).

Pharmacogenomics directly influences two significant acts in the pharmaceutical story—pharmacokinetics and pharmacodynamics. The first, pharmacokinetics, is the grand journey of the drug within the body, its absorption, its spread, and its ultimate metabolism—like a character traveling through the landscapes of the story, leaving an impact everywhere they go. Alternatively, pharmacodynamics pertains to the process of how the drug, now a major protagonist, interacts with the body. This interaction generates biochemical and physiological changes, akin to dramatic plot twists that propel the story forward.

Adding further intrigue to this story are two key drug–drug and drug–disease interactions. In the case of drug–drug interactions, it is akin to two characters influencing each other's story arcs, where one drug alters the effect of another, leading to changes in effectiveness or even unexpected plot twists in the form of new side effects. On the other hand, drug–disease interactions are where a drug character's arrival worsens or triggers a new disease

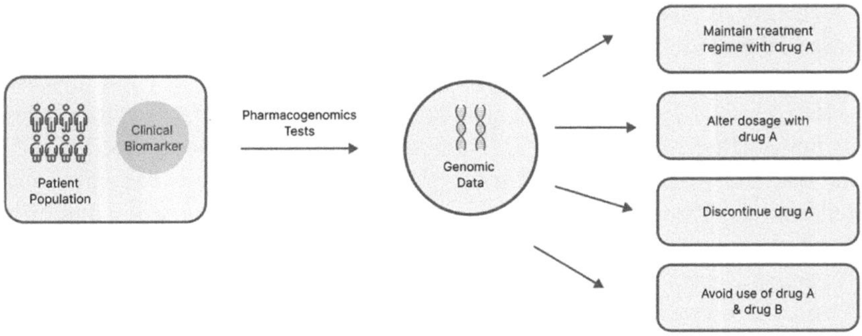

FIGURE 2.1 Pharmacogenomics is the field that explores how a person's genetic makeup affects their reaction to medications. This can cause changes in the drug's effectiveness, required dosage, and the occurrence of side effects, and these effects can vary depending on the specific drug used.

subplot, like introducing a twist that stirs the pot in an ongoing storyline. In this continuously evolving narrative of pharmacogenomics, it is a captivating reminder of how every individual is unique and reacts differently to medication, much like characters in a story each responding uniquely to plot developments.

Currently, pharmacogenomic testing is used to understand the impact of allelic variations on drug metabolism, adverse events, dosing, mechanistic effects, polymorphic drug targets, and clinical trial design. The first FDA-approved pharmacogenomic test was the AmliChip Test in 2005 for genotyping CYP2D6 alleles and CYP2C19 alleles, which are involved in the oxidation of antidepressants and antipsychotics. Patients with polymorphisms in this allele have additional side effects to the antipsychotic risperidone, and patients with a duplicated allele do not respond to antidepressant treatment. CYP2C19 alleles alter the response to clopidogrel, a blood thinner, and alter the metabolism of voriconazole, an antifungal agent (de Leon, 2006).

Pharmacogenomic testing has been applied to many fields of medicine, including cardiology, oncology, neurology, psychiatry, pain medicine, rheumatology, and infectious diseases. For instance, in oncology, thiopurine drugs are a class of chemotherapeutic drugs, including agents such as 6-mercaptopurine. Thiopurine methyltransferase (TPMT) is an enzyme involved in the metabolism of thiopurine drugs. Polymorphisms in this gene can lead to enhanced bone marrow toxicity due to the accumulation of unmetabolized thiopurine drugs. TPMT variants in children on cisplatin (chemotherapy) have been associated with ototoxicity. For patients with colorectal cancer, Kirsten rat sarcoma virus (KRAS) mutation status is a predictor for cetuximab therapy success, while for patients with lung cancer, epidermal growth factor receptor (EGFR) mutations predict gefitinib therapy outcomes (Lièvre et al., 2006; Sordella et al., 2004).

For cardiovascular conditions, the response to clopidogrel and CYP2C19 for antiplatelet therapy after invasive coronary artery procedures such as percutaneous coronary intervention is well documented (Malki & Pearson, 2020). Drugs like warfarin (blood thinner) have poor metabolism from CYP2C9 alleles, leading to drug toxicity from heightened drug accumulation. A categorization of ultrarapid metabolizer, normal metabolizer, intermediate metabolizer, and poor metabolizer has been developed to describe the metabolic response to medications by an individual's genome (Squassina et al., 2010). Additionally, these interactions can themselves interact, resulting in the emergence of drug–drug–gene interactions, where drug–drug interactions are further modified by individual genetic variation. These broadly fall into three categories, drug and genome inhibit metabolism leading to prolonged effects, increase metabolism leading to reduced efficacy, and new onset of off-target effects (Relling & Evans, 2015) (Fig. 2.2).

2.3 Postmarket drug surveillance and monitoring of adverse drug reactions

Biomarkers are integral to postmarket drug surveillance and monitoring of adverse drug reactions, aiding in detecting and understanding unexpected drug effects within a diverse patient population. The use of monitoring biomarkers, such as regular tracking of blood glucose levels in response to new diabetes medication, enables the observation of the drug's effectiveness. Pharmacodynamic biomarkers offer insights into a drug's impact on its intended target, and any unexpected alterations could suggest an adverse reaction. Predictive biomarkers, including certain genetic markers, help identify patients more likely to experience specific medication adverse effects.

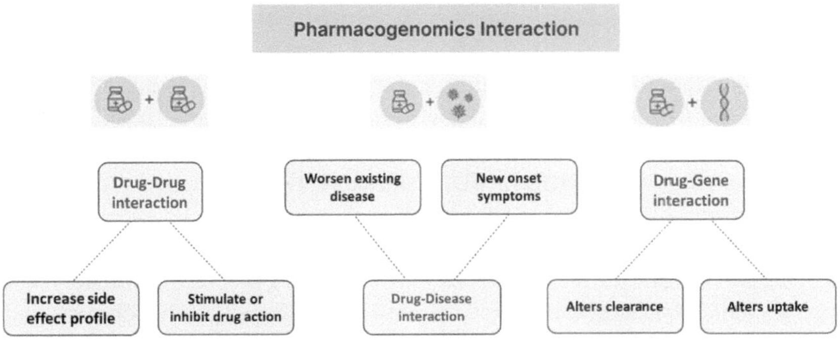

FIGURE 2.2 Pharmacogenomic interactions can be divided into three main types: drug–drug, drug–disease, and drug–gene interactions. Drug–drug interactions occur when one medication impacts the effectiveness or side effects of another. Drug–disease interactions happen when a drug aggravates an existing condition or triggers new symptoms. Drug–gene interactions refer to a person's genetic makeup's influence on how they respond to medication.

This understanding enhances safety as drugs reach broader populations post-approval, where rare or long-term side effects not identified in clinical trials may surface.

Even with the extensive and meticulous premarket assessment carried out by the Center for Drug Evaluation and Research (CDER), an agency within the FDA in charge of guaranteeing the safety and effectiveness of drugs, postmarketing surveillance (PMS) remains vital. This surveillance keeps track of marketed medications for adverse drug reactions (ADRs) after they have been approved and are in use (Reyes et al., 2022). Often considered a Phase IV clinical trial, PMS occurs in a real-world setting outside the confines and limitations of other clinical trials that preceded it, enabling (1) the identification of adverse events (e.g., adverse reactions and poisonings) that did not otherwise appear during the drug approval process, (2) clinical effectiveness comparison of new products with existing options and standard of care (SOC), and (3) utilization analysis and potential updating of marketing labels as certain uses or populations find greater benefit than others (Spelsberg et al., 2017).

The real-world insights gleaned through the PMS process. Identification of adverse events bears particular mention as even with a series of clinical trials drawn out over several years, the scope of safety and efficacy of new medicines would only provide limited information on discovering "rare" (1 in 1000) and "very rare" (1 in 10,000) ADRs due to relatively small enrollment size, narrow patient selection criteria, and often the short duration of studies (Watanabe & Narukawa, 2016). Total patient enrollment throughout the phases of the premarket approval process, for instance, ranges from several hundred to only about 3000 individuals. The selection of patients for a trial or another is often contradictory in requirements as study designs are intended to establish a homogenous group of individuals to enable a better statistical comparison, reduce response variability, and decrease the risk of bias (Bourin, 2017). Moreover, rapid results are often needed to minimize the already lengthy cost involved in drug development and to reduce time to market.

All these factors, understandably, contribute to the substantial challenge of attaining all necessary safety data from within the limited scope of premarket clinical trials and thus make PMS all the more important in determining drug safety and efficacy. FDA subsequently then uses information provided through PMS and other risk assessment programs to update drug labeling periodically and, if needed, to reevaluate the approval or use of the medication (Alomar et al., 2020). Notably, due to PMS, approximately 4% of new medications had their approval withdrawn due to safety concerns and ~20% were issued a black box warning—the strictest and most serious warning given by the FDA to medications with serious or life-threatening side effects or risks (Zhang et al., 2016). A table with select medications with serious ADRs identified postmarketing and FDA regulatory action is populated in Table 2.1.

TABLE 2.1 Select medications with serious ADRs identified during the postmarketing period and FDA regulatory action (in https://www.fda.gov/drugs/surveillance/postmarket-drug-and-biologic-safety-evaluations).

Drug (active ingredient)	Primary use	ADRs reported	FDA regulatory action
Avandia (rosiglitazone)	Diabetes Management	Increased risk of heart failure or heart attack	DA warning issued November 2007 Black Box Warning 2023 List of Medications
Cipro/Levaquin/ Avelox (Fluoroquinolone)	Antibiotic for bacterial infections	Increased risk of tendinitis and tendon rupture	FDA warning issued July 2008
Aptiom (eslicarbazepine acetate)	Treatment for partial-onset seizures	Increased risk of hematologic events, syndrome of inappropriate antidiuretic hormone secretion (SIADH), and atrioventricular block	FDA warning issued September 2017
Opdivo (nivolumab)	Treatment for metastatic melanoma, advanced renal cell carcinoma, and others	Increased risk of retinal detachment, severe vision loss, and tumor lysis syndrome	FDA warning issued September 2017
Velphoro (sucroferric oxyhydroxide)	Serum phosphorus levels control in patients with chronic kidney disease on dialysis	Increased risk of tooth discoloration and rash	FDA labeling supplement added in August 2017 in the "Adverse Reactions; Postmarketing Experience" section

PMS actions for drugs involve the principles of pharmacovigilance and pharmacoepidemiology (Faillie et al., 2016). While the emergence of the technological era has ushered in new ways to conduct PMS via computer systems and electronic medical records actively, the majority of data collection occurs through spontaneous reporting systems (SRSs), a passive method that is largely regulated by federal agencies to spontaneously collect reports of ADRs from clinicians, consumers, and pharmaceutical organizations (Ibrahim et al., 2016). A key example of a SRS is the US MedWatch program that allows healthcare professionals and the public to voluntarily report any "serious adverse event, product quality problem, product use/medication error, or therapeutic inequivalence/failure" with FDA-regulated medical products. The US FDA has also created a computerized information database for PMS of all approved drugs and therapeutic biological agents, officially referred to as the US FDA Adverse Event Reporting System (FAERS), to provide the best and most up-to-date resources for storing, referencing, and analyzing ADR reports. Annually, FAERS receives anywhere from 300,000 to 500,000 new safety reports (Ibrahim et al., 2016). More actively, the FDA also conducts periodic postmarked evaluations of drugs either 18 months post-FDA approval or after their use by 10,000 individuals, whichever is later. Overall, through the combination of FAERS, MedWatch, and periodic drug production and facilities inspections, the FDA, or more specifically the CDER office, can take appropriate regulatory actions to improve product safety and thus better protect the health of the broader public.

To conclude, PMS provides real-world insights that are useful, if not arguably critical, to the homeostasis of the organized entropy that defines the healthcare system. Varying stakeholders ultimately derive different benefits through PMS to optimize care or service delivery. For the FDA, PMS provides a way to continuously monitor and validate the safety and efficacy of postmarket-approved drugs. For pharmaceutical companies, PMS provides evidence-driven data to compare the relative performance of their product to the disease and enable the development of new strategies as appropriate for their clinical, marketing, and pricing angles. For healthcare providers (HCPs) and hospitals, PMS informs on and enables better decision-making regarding treatment options as appropriate for the particular patient, including using off-label or generic products. And lastly, for payers and regulatory agencies, PMS can inform on the cost-efficacy of drugs to better educate on decisions related to insurance coverage and product reimbursement. However, as healthcare develops and becomes more intricate, there is a parallel requirement for PMS and SRSs to progress and increase their effectiveness. Here, we urge a call to action for greater awareness of these programs among HCPs and a push for the integration of pharmacogenetic data (or, put another way, patients' genetic data) as part of the pharmacovigilance process (Fig. 2.3).

After exploring the intricate world of pharmacogenomics and biomarkers, we have gained insight into how individual genetic makeup can influence drug response and guide personalized medical treatment. Similarly, we have delved into postmarket drug surveillance and its crucial role in monitoring ADRs, ensuring the safety and efficacy of medications even after they have been introduced to the market.

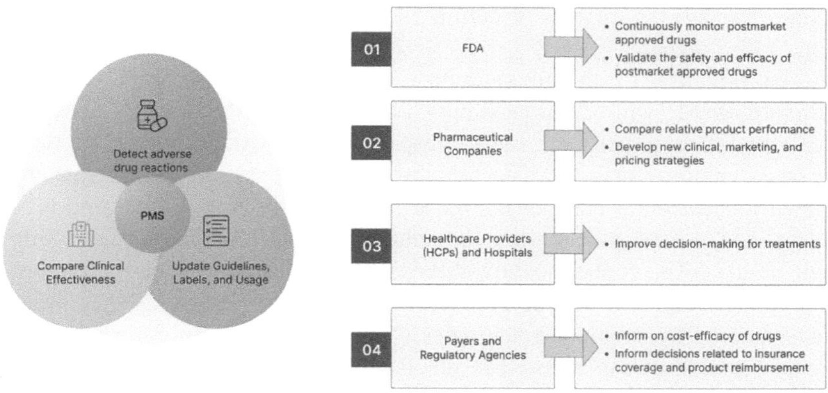

FIGURE 2.3 Postmarketing surveillance (PMS) is part of pharmacovigilance and involves monitoring a pharmaceutical drug's or medical device's safety and efficacy after its release on the market. PMS is crucial in detecting adverse drug reactions (ADRs) not found during clinical trials and can be used to compare the effectiveness of a drug in real-world practice versus its effectiveness in clinical trials. In addition, the data gathered from PMS can lead to updates in the usage instructions of a drug, its labeling, or even the clinical guidelines related to its use. So, PMS is essential in ensuring drug safety, efficacy, and appropriate usage once a medication or device is available to the public.

As we transition into the next phase of our discussion, let's embark on an exciting journey into digital biomarkers. This cutting-edge technology promises to revolutionize our understanding of health and disease and how we monitor and respond to them. As we move into the digital age, digital biomarkers offer an innovative approach to health tracking and potential synergy with traditional biomarkers and pharmacogenomics to enhance personalized healthcare.

2.4 Exploring the frontier of digital biomarkers

As digitization sweeps across the healthcare landscape, rapid advancements in digital health technologies have revolutionized medical therapeutics, diagnostics, and research. Now, more than ever before, the shifts have enabled an unprecedented ability to measure, collect, and store large amounts of "digital biomarkers"—health-related objective, quantifiable, and continuous data outside of the traditional clinical setting (Babrak et al., 2019). From smartphones to wearables, implantable, and other sensor-based devices, the collected health-related phenomena we now refer to as digital biomarkers provide various aspects of an individual's health, including physiological and behavioral characteristics to explain, influence, and/or predict health outcomes. With a global market valuation expected to reach \$27.2 billion USD by 2030 and a compound annual growth rate of 35.3%, the emergence of digital biomarkers enables ushering in an era of personalized medicine (Fig. 2.4).

The principle of the measurement of digital biomarkers is relatively straightforward, with communication occurring across different hardware technologies (e.g., sensors) and corresponding software that then captures, analyzes, and/or stores the data to subsequently be used to extract patterns related to behavioral and physiological health. While vital monitoring technology has existed for decades, the value of digital biomarkers stems from their ability to be continuously and remotely monitored even in the comfort of a patient's home, work, or even travel, thereby supplementing the clinical data gathered by order of magnitudes. Through this wealth of information, digital biomarkers can be used in precision health to create a "personalized baseline" of health for each patient—rather than comparing to an arbitrary threshold of biomarkers—and thus inform and improve the accuracy of disease diagnosis, prognosis, and clinical decision-making. Continuous vital

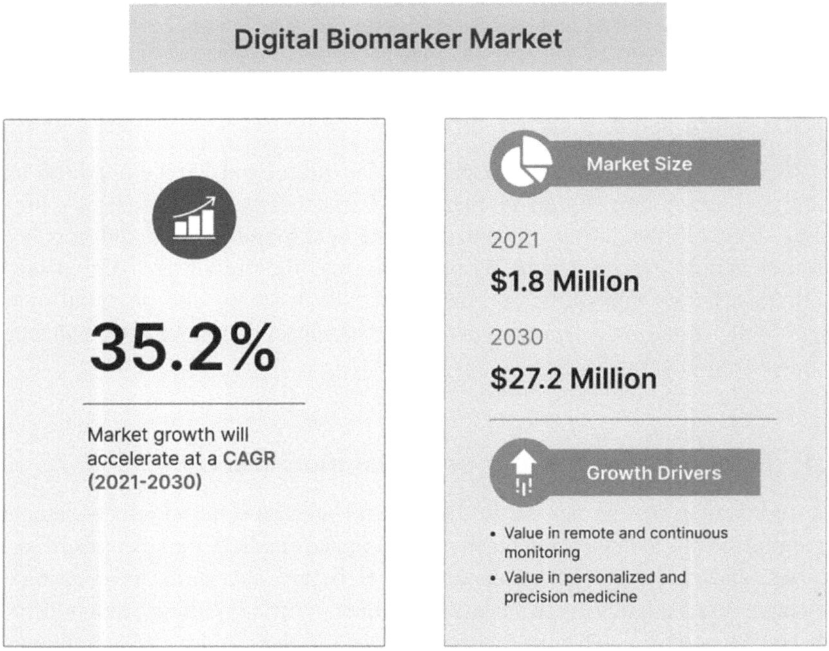

FIGURE 2.4 The digitization of healthcare has led to significant advancements in digital health technologies, transforming medical therapeutics, diagnostics, and research. This shift has enabled collecting and storing large amounts of "digital biomarkers," which are objective, quantifiable, and continuous health-related data gathered outside of traditional clinical settings. The global market for digital biomarkers is projected to reach a valuation of $27.2 billion USD by 2030, with a compound annual growth rate of 35.3%. The emergence of digital biomarkers is paving the way for personalized medicine and transforming the healthcare industry. CAGR - compound annual growth rate.

tracking of a person's heart rate, oxygen saturation, or blood pressure through a wearable, for instance, can provide valuable information about their cardiovascular health that then could be used by a healthcare provider to diagnose a potential heart condition and to tailor a treatment plan to the individual's needs. Continuous blood glucose monitoring through an implantable sensor, for another, can inform on a patient's physical activity and food intake to then allow clinicians to adjust and tailor a more appropriate insulin dosage as well as establish better predictive alarms (via personalized baseline) for alerting on critical low blood glucose levels (Vettoretti et al., 2020).

Digital biomarkers have also significantly benefited the field of neurology, where there still exists many unmet clinical needs and challenges in identifying noninvasive, objective, and predictive biomarkers for different neurological diseases. One example of a digital biomarker being studied in neurological disease is the electroencephalogram (EEG), which measures brain activity. EEG data can be collected using wearable devices or portable sensors, providing valuable information about a person's neurological health. This data can be used to diagnose conditions such as epilepsy or to monitor the effectiveness of treatments for conditions such as Alzheimer's disease or Parkinson's disease (Kourtis et al., 2019; Lipsmeier et al., 2018). Another example is the use of mobile phones and other wearable devices to track physical activity and sleep patterns, which can provide valuable insights into a person's overall health and fitness and be used to diagnose conditions such as sleep disorders or to monitor the effectiveness of treatments for conditions such as obesity (Depner et al., 2020).

Last but not least, there is even potential in the use of digital biomarkers, as measured by consumer wearable devices, to supplement and even more reliably measure and depict vital signs data than current clinical standards, with some studies observing lower prediction error for clinical laboratory measurements made with data from consumer wearables than those made with clinically obtained vital signs data (Dunn et al., 2021). Moreover, the wireless nature of new wearable sensors enables a versatile potential to considerably improve patient monitoring, comfort, and mobility both within and outside of clinical/hospital settings (Kroll et al., 2017) (Fig. 2.5).

Many challenges still remain, however, from electronic health record integration of wearable data to FDA approval and on-off-label usage of wearable data, security of data storage and privacy, high device cost and cross-platform device incompatibility, and lack of data standards that limit broader use, interpretation of validity, and ubiquity (Dunn et al., 2018). Perhaps more importantly, the accuracy and evidence of health outcome improvements must become more established before the adoption and acceptance of digital biomarkers by healthcare professionals and the broader public can occur. In particular, false negatives, which can lead to missing potentially fatal health outcomes, and false positives, which can lead to

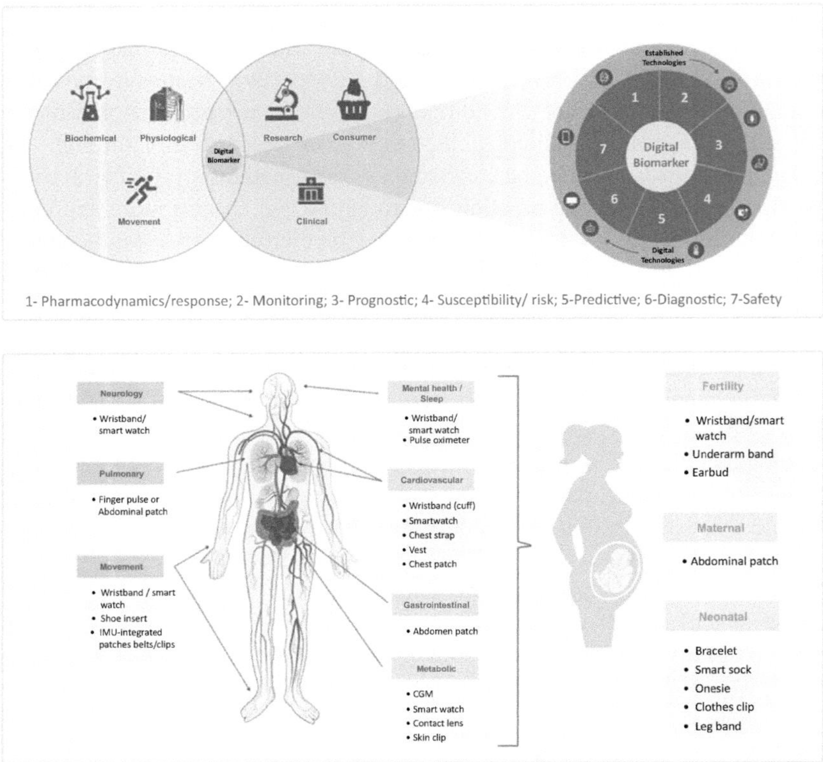

FIGURE 2.5 Digital biomarkers are measurable, objective physiological and behavioral data collected via digital devices. These devices can include wearables, implantable devices, or other mobile health (mHealth) platforms. The application of digital biomarkers is becoming increasingly important in various aspects of healthcare, from research and clinical trials to everyday patient care.

psychological paranoia, overtreatment, and anxiety, must be minimized (Farrington, 2016). However, with each new large-scale initiative studying longitudinal wearable data with other clinical correlates and information, the hope expands that our medical paradigm can shift from predominantly a reactive health system to one that is proactive and preventative.

In summary, digital biomarkers hold the power to radically transform the realm of precision medicine. They can provide objective, continuous, and customized data regarding an individual's health and overall well-being. These data can be used to improve the accuracy of disease diagnosis, reduce clinical errors, support personalized clinical decisions, and tailor treatments to the individual's needs, ultimately leading to better health outcomes for patients.

Digital biomarkers are measures of health-related phenomena that can be collected using digital technologies, such as smartphones, wearable devices, and other sensors. In contrast to clinical biomarkers infrequently collected at routine patient visits, digital biomarker data collection is continuous and remote. Wearable devices can provide objective and continuous data on various aspects of a person's health, including physiological and behavioral characteristics—such as heart rate, skin temperature, or activity levels (Table 2.2). Wearables can also provide these continuous measurements at a high resolution, often collecting millions of measurements for a single individual each day; for example, the Polar H10 Chest Strap can measure the heart rate of its user once every second. We will subsequently discuss examples of digital biomarkers and how they can supplement traditional clinical care, provide novel information about patient lifestyles, and how they are helping to advance personalized medicine.

Digital biomarkers can be used in precision health and augment traditional clinical care by improving the accuracy of disease diagnosis and monitoring the effectiveness of treatments. For example, a digital biomarker could track a person's heart rate or blood pressure over time, providing

TABLE 2.2 Digital biomarkers and relevant devices.

Digital biomarker	Device
Heart rate, heart rate variability	Fitbit, Apple Watch, Garmin or Polar smartwatches
Skin temperature	Fitbit Sens, Oura Ring, Tempdrop, iThermonitor, Empatica Embrace2, Ava Bracelet, BioButton
Galvanic skin response	Empatica E4 wristband, Fitbit Sense, Feel wristband, Garmin wearables, Biofeedback rings such as Moodmetric, Microsoft Band, Apple Watch
Sleep quality, steps, physical activity	Fitbit, Garmin Devices, Apple Watch, Samsung Galaxy Watch Series, Ouraring, Withings Devices, Xiaomi Mi Band, Polar Devices, Huawei Watch Series
Glucose	Dexcom G6, FreeStyle Libre CGM systems, Eversense CGM System, Medtronic Guardian Connect CGM
Calories, nutrition	MyFitnessPal, Cronometer
Electrocardiogram (ECG or EKG)	Apple Watch Series 4, Withings Move ECG, Fitbit Sense, Samsung Galaxy Watch3 and Watch Active2, AliveCor KardiaMobile
Mood or stress levels	Empatica E4 wristband, Feel wristband, Spire Health Tags, Muse headband
Blood pressure	Omron HeartGuide, FITVII GT5 SmartWatch, MorePro Fitness Tracker, Samsung Galaxy Watch3 and Watch Active2

valuable information about their cardiovascular health. This data could then be used by a healthcare provider to diagnose a potential heart condition and to tailor a treatment plan to the individual's needs.

Digital health biomarkers have many promising advantages for the evolution of personalized medicine. Digital health biomarkers can be advantageous because they allow clinicians and patients to measure biomarkers related to lifestyle components that are often overlooked. Clinicians often focus on measuring biomarkers related to illness in the clinic, and key lifestyle components such as nutrition, sleep, and physical activity are usually not measured nor discussed in depth. In fact, clinicians receive little training about how lifestyle can be optimized to promote health (Crowley et al., 2019) and spend most of their training in medical school focusing on pharmacological treatments and curing disease after it is already present.

Mobile phones and other wearables can collect physical activity, nutrition, and sleep data, which traditionally would not be measured in the clinic. For instance, mobile phones and smartwatches can record steps taken and heart rate, as well as sleep patterns, disturbances, and timing of sleep and wake. The mobile applications Cronometer and MyFitnessPal enable the measurement of longitudinal calorie and nutrient intake that could readily be shared with clinicians and/or dieticians. Continuous glucose monitors could also enable patients to understand what types of foods cause undesirable spikes in blood sugar; foods that trigger spikes have been seen to be variable among individuals, and this would enable patients to optimize their diet.

Digital biomarkers are also important because they can help shift the focus of medicine toward disease prevention by incorporating lifestyle into the equation of health—enabling physicians and patients to quantify disease risk and create action plans early on. A landmark study previously found that lifestyle intervention was significantly more effective than metformin, a pharmacological SOC, in reducing the incidence of diabetes in high-risk patients; (Reduction in the Incidence of Type 2, 2002) yet physical activity and lifestyle measurements are not the focus of clinical care. This data can provide valuable insights into a person's overall health and fitness, and it can be used to diagnose conditions such as sleep disorders or to monitor the effectiveness of treatments for conditions such as obesity as well.

Lastly, the continuous and frequent measurements provided by digital biomarkers, in comparison to traditional clinical biomarkers, can help to advance personalized medicine by shifting the medical paradigm toward disease prevention rather than retroactive treatment. Wearables can continuously monitor resting heart rate, and researchers have employed this technology to promote early diagnosis of COVID-19 (Alavi et al., 2021). Resting heart rate often elevates when a person acquires an infection and wearables can identify when resting heart rate deviates from a person's normal, healthy range (Alavi et al., 2021). Wearables have been used to see if deviations in a person's resting heart rate could predict the future onset of

COVID-19. This has important public health implications; if COVID-19 could be detected early, before the onset of symptoms, communities could greatly mitigate the spread of viral infections. In addition, individuals could prepare for the onset of COVID-19 and seek care early on. Similarly, other devices such as Apple watches have created algorithms to help identify heart arrhythmias in users, which enable and empower patients to seek care earlier.

Digital biomarkers can enable patients to focus on optimizing their health and physiological performance rather than responding to disease after its onset. Elite athletes are using wearable devices to monitor sleep status, physical exertion, and their recovery (Seshadri et al., 2019). Wearables also allow individuals to optimize their health by enabling people to establish their own personal baselines for health metrics. Clinical standards for blood pressure or resting heart rate are only "standards" based on preexisting data and may not apply to people of all genders, race/ethnicities, or age. A deviation in blood pressure or resting heart rate may be in the "normal range" at the clinic but may be abnormal for an individual given their healthy baseline range. Wearable devices providing continuous measures can help patients establish their normal physiological range and compare measures to their own historical, longitudinal data—enabling more precise identification of what is abnormal or not optimal for their own health (Fig. 2.6).

Overall, digital biomarkers have the potential to revolutionize the field of precision health by providing objective, continuous, and personalized data on a person's health and well-being. This data can allow for earlier disease diagnosis and enable clinicians to focus on disease prevention rather than retroactive responses to illness. Continuous and remote physiological measurements can enable clinicians to gain valuable data and tailor treatments to the individual's needs, ultimately leading to better health outcomes and empowering patients to participate in their own medical care.

2.5 Discovering biomarkers with microsampling technology

Beyond wearable health/fitness tracking devices, the Snyder Laboratory at Stanford University is also looking to collect extensive information about an individual's health through a single drop of blood. His lab is developing a microsampling device that can be used by an individual at home to collect a drop of blood and then ship it off to the lab for analysis. Researchers can collect information about 22 analytes—metabolites, lipids, cytokines, and proteins—from this single drop of blood and do so much more often than most people have routine blood tests (Shen et al., 2023). The data collected can help paint a complete picture of an individual's health and how it changes over time. It can also potentially help identify biomarkers for specific diseases. This approach is reliable and scalable and may be combined

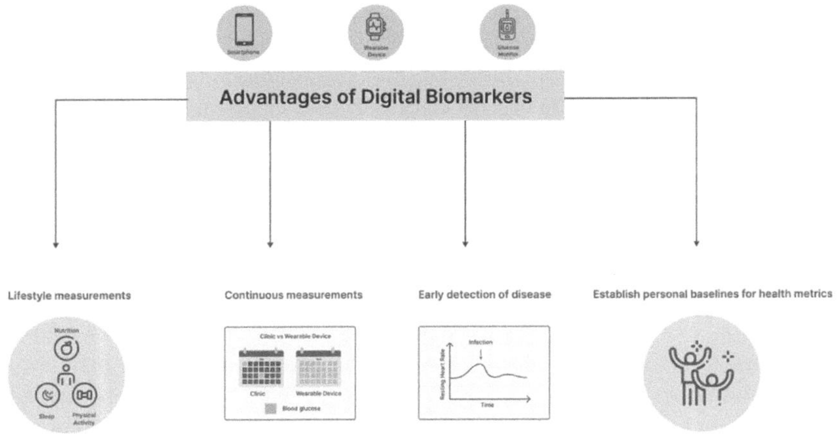

FIGURE 2.6 Wearable devices and mobile applications play a crucial role in collecting real-time health data, supplying an abundance of valuable information. This constant tracking and analysis empower users with a deeper comprehension of their health trends, paving the way for informed lifestyle modifications. These tools can help in the early detection of health issues, including illnesses like COVID-19, providing a significant advantage in healthcare management. Individuals can establish personalized health baselines rather than solely depending on generic clinical benchmarks, thus incorporating a more tailored approach to their healthcare. The unique advantage of these technologies lies in their ability to compare current health data with historical personal records, fostering a highly precise and individualized approach to healthcare and wellness.

with data from wearable devices to provide an unprecedented amount of health profiling that should improve the prevention and treatment of disease.

2.6 Future directions

We have seen that biomarkers have many applications in clinical medicine. From assisting with disease diagnosis to monitoring and predicting disease progression to helping to assess pharmacodynamic responses, biomarkers can be obtained for various diseases and technologies. Digital biomarkers can also help shape a new, more personalized world of medicine and allow patients to participate in their medical care.

We can obtain a wealth of data about the human body, and in the future, we can only expect to continue collecting more data more frequently. With new multiomics technologies that enable genome, proteome, and metabolome sequencing, for instance, as well as digital biomarkers, we can expect to begin to collect large volumes of data for individuals. Collecting large volumes of data from individuals and integrating into large databases will provide a gold mine for searching for new candidate biomarkers—whether these are molecules, features in medical images, or behavioral. We will be able to use datasets to look at genetic information, disease outcomes, drug responses, and molecules; algorithms will need to be engineered to help us to discover new types of biomarkers in real time, which may be a faster, cheaper, and more rigorous alternative to conducting clinical trials.

This emerging era of "big data" will not be without challenges. Scientists and computational researchers will need to make sense of all of this data with bioinformatics approaches to identify new biomarker candidates. Researchers will also have to find ways to share data while maintaining privacy. It will also be important to ensure equity so all populations can be represented in genome sequencing, digital datasets, and any other type of data collection that would facilitate biomarker discovery. The field of digital health with wearables and new multiomic analyses are promising for identifying new candidate biomarkers; however, we can expect that the integration of these new types of data into clinical practice will take time, yet may help to eventually shift clinical practice toward a new era of personalized medicine.

References

Alavi, N., et al. (2021). Delivering an online cognitive behavioral therapy program to address mental health challenges faced by correctional workers and other public safety personnel: Protocol for a mixed methods study. *JMIR Research Protocols, 10*, e30845.

Alomar, M., Tawfiq, A. M., Hassan, N., & Palaian, S. (2020). Post marketing surveillance of suspected adverse drug reactions through spontaneous reporting: Current status, challenges and the future. *Therapeutic Advances in Drug Safety, 11*, 2042098620938595.

Antman, E. M., & Loscalzo, J. (2016). Precision medicine in cardiology. *Nature Reviews Cardiology, 13*, 591–602.

Babrak, L. M., et al. (2019). Traditional and digital biomarkers: Two worlds apart? in. *Digital Biomarkers, 3*, 92–102.

Babuin, L., & Jaffe, A. S. (2005). Troponin: The biomarker of choice for the detection of cardiac injury. *CMAJ: Canadian Medical Association Journal = Journal de l'Association Medicale Canadienne, 173*, 1191–1202.

Baechle, C., Scherler, W., Lang, A., Filla, T., & Kuss, O. (2022). Is HbA1c a valid surrogate for mortality in type 2 diabetes? Evidence from a meta-analysis of randomized trials. *Acta Diabetologica, 59*, 1257–1263.

Bourin, M. (2017). Criteria of patient selection in a clinical trial. *SOJ Pharmacy & Pharmaceutical Sciences, 4*, 1–4.

Califf, R. M. (2018). Biomarker definitions and their applications. *Experimental Biology and Medicine (Maywood, N.J.), 243*, 213–221.

Cohen, D. L., & Townsend, R. R. (2008). What should the physician do when creatinine increases after starting an angiotensin-converting enzyme inhibitor or an angiotensin receptor blocker? *Journal of Clinical Hypertension (Greenwich, Conn.), 10*, 803–804.

Crowley, J., Ball, L., & Hiddink, G. J. (2019). Nutrition in medical education: A systematic review. *The Lancet Planetary Health, 3*, e379–e389.

Depner, C. M., et al. (2020). Wearable technologies for developing sleep and circadian biomarkers: A summary of workshop discussions. *Sleep, 43*.

de Leon, J. (2006). AmpliChip CYP450 test: Personalized medicine has arrived in psychiatry. *Expert Review of Molecular Diagnostics*, *6*, 277–286.

Dunn, J., et al. (2021). Wearable sensors enable personalized predictions of clinical laboratory measurements. *Nature Medicine*, *27*, 1105–1112.

Dunn, J., Runge, R., & Snyder, M. (2018). Wearables and the medical revolution. *Personalized Medicine*, *15*, 429–448.

Faillie, J. L., Montastruc, F., Montastruc, J. L., & Pariente, A. (2016). Pharmacoepidemiology and its input to pharmacovigilance. *Therapie*, *71*, 211–216.

Farrington, C. (2016). Wearable technologies and stigma in diabetes: The role of medical aesthetics. *Lancet Diabetes & Endocrinology*, *4*, 566.

Fleming, T. R., & DeMets, D. L. (1996). Surrogate end points in clinical trials: Are we being misled? *Annals of Internal Medicine*, *125*, 605–613.

Ibrahim, H., Saad, A., Abdo, A., & Sharaf Eldin, A. (2016). Mining association patterns of drug-interactions using post marketing FDA's spontaneous reporting data. *Journal of Biomedical Informatics*, *60*, 294–308.

Kagan, J. M., Sanchez, A. M., Landay, A., & Denny, T. N. (2015). A brief chronicle of CD4 as a biomarker for HIV/AIDS: A tribute to the memory of John L. Fahey. *For Immunopathol Dis Therap*, *6*, 55–64.

Katz, R. (2004). Biomarkers and surrogate markers: An FDA perspective. *NeuroRx: The Journal of the American Society for Experimental NeuroTherapeutics*, *1*, 189–195.

Kourtis, L. C., Regele, O. B., Wright, J. M., & Jones, G. B. (2019). Digital biomarkers for Alzheimer's disease: The mobile/wearable devices opportunity. *npj Digital Medicine*, *2*, 9.

Kroll, R. R., et al. (2017). Use of wearable devices for post-discharge monitoring of ICU patients: A feasibility study. *Journal of Intensive Care*, *5*, 64.

Kuchenbaecker, K. B., et al. (2017). Evaluation of polygenic risk scores for breast and ovarian cancer risk prediction in BRCA1 and BRCA2 mutation carriers. *JNCI: Journal of the National Cancer Institute*, *109*.

Lassere, M. N., Johnson, K. R., Schiff, M., & Rees, D. (2012). Is blood pressure reduction a valid surrogate endpoint for stroke prevention? An analysis incorporating a systematic review of randomised controlled trials, a by-trial weighted errors-in-variables regression, the surrogate threshold effect (STE) and the biomarker-surrogacy (BioSurrogate) evaluation schema (BSES). *BMC Medical Research Methodology*, *12*, 27.

Lipsmeier, F., et al. (2018). Evaluation of smartphone-based testing to generate exploratory outcome measures in a phase 1 Parkinson's disease clinical trial. *Movement Disorders: Official Journal of the Movement Disorder Society*, *33*, 1287–1297.

Lièvre, A., et al. (2006). KRAS mutation status is predictive of response to cetuximab therapy in colorectal cancer. *Cancer Research*, *66*, 3992–3995.

Malki, M. A., & Pearson, E. R. (2020). Drug–drug–gene interactions and adverse drug reactions. *The Pharmacogenomics Journal*, *20*, 355–366.

O'Connor, J. P. B., et al. (2017). Imaging biomarker roadmap for cancer studies. *Nature Reviews Clinical Oncology*, *14*, 169–186.

Ou, F. S., Michiels, S., Shyr, Y., Adjei, A. A., & Oberg, A. L. (2021). Biomarker discovery and validation: Statistical considerations. *Journal of Thoracic Oncology: Official Publication of the International Association for the Study of Lung Cancer*, *16*, 537–545.

Prensner, J. R., Rubin, M. A., Wei, J. T., & Chinnaiyan, A. M. (2012). Beyond PSA: The next generation of prostate cancer biomarkers. *Science Translational Medicine*, *4*, 127rv123.

Reduction in the Incidence of Type 2. (2002). Diabetes with lifestyle intervention or metformin. *New England Journal of Medicine*, *346*, 393–403.

Relling, M. V., & Evans, W. E. (2015). Pharmacogenomics in the clinic. *Nature*, *526*, 343–350.

Reyes, M., Kortepeter, C., & Muñoz, M. (2022). Postmarket assessment for drugs and biologics used in dermatology and cutaneous adverse drug reactions. *Dermatologic Clinics*, *40*, 265–277.

Sager, P. T., Gintant, G., Turner, J. R., Pettit, S., & Stockbridge, N. (2014). Rechanneling the cardiac proarrhythmia safety paradigm: A meeting report from the cardiac safety research consortium. *American Heart Journal*, *167*, 292–300.

Seshadri, D. R., et al. (2019). Wearable sensors for monitoring the internal and external workload of the athlete. *NPJ Digital Medicine*, *2*, 71.

Shen, X., et al. (2023). Multi-omics microsampling for the profiling of lifestyle-associated changes in health. *Nature Biomedical Engineering*.

Simon, R. (2010). Clinical trial designs for evaluating the medical utility of prognostic and predictive biomarkers in oncology. *Personalized Medicine*, *7*, 33–47.

Sordella, R., Bell, D. W., Haber, D. A., & Settleman, J. (2004). Gefitinib-sensitizing EGFR mutations in lung cancer activate anti-apoptotic pathways. *Science (New York, N.Y.)*, *305*, 1163–1167.

Spelsberg, A., et al. (2017). Contribution of industry funded post-marketing studies to drug safety: Survey of notifications submitted to regulatory agencies. *BMJ (Clinical Research ed.)*, *356*, j337.

Squassina, A., et al. (2010). Realities and expectations of pharmacogenomics and personalized medicine: Impact of translating genetic knowledge into clinical practice. *Pharmacogenomics*, *11*, 1149–1167.

Tarhini, A., & Kudchadkar, R. R. (2018). Predictive and on-treatment monitoring biomarkers in advanced melanoma: Moving toward personalized medicine. *Cancer Treatment Reviews*, *71*, 8–18.

Teigler, J. E., et al. (2018). Distinct biomarker signatures in HIV acute infection associate with viral dynamics and reservoir size. *JCI Insight*, *3*.

Veit, G., et al. (2016). From CFTR biology toward combinatorial pharmacotherapy: Expanded classification of cystic fibrosis mutations. *Molecular Biology of the Cell*, *27*, 424–433.

Vettoretti, M., Cappon, G., Facchinetti, A., & Sparacino, G. (2020). Advanced diabetes management using artificial intelligence and continuous glucose monitoring sensors. *Sensors (Basel)*, *20*.

Watanabe, T., & Narukawa, M. (2016). Characteristics of safety information obtained from postmarketing observational studies for re-examination in Japan. *SpringerPlus*, *5*, 905.

Weverling-Rijnsburger, A. W. E., Jonkers, I. J. A. M., van Exel, E., Gussekloo, J., & Westendorp, R. G. J. (2003). High-density vs low-density lipoprotein cholesterol as the risk factor for coronary artery disease and stroke in old age. *Archives of Internal Medicine*, *163*, 1549–1554.

Zhang, X., et al. (2016). Overview of phase IV clinical trials for postmarket drug safety surveillance: A status report from the ClinicalTrials.gov registry. *BMJ Open*, *6*, e010643.

Part 2

Biosensors and their applications in precision medicine

Chapter 3

Biosensors: an introduction

Cristiane Kalinke[1], Paulo Roberto de Oliveira[2], Juliano Alves Bonacin[1] and Bruno Campos Janegitz[2]

[1]*Institute of Chemistry, University of Campinas (Unicamp), Campinas, São Paulo, Brazil,* [2]*Laboratory of Sensors, Nanomedicine, and Nanostructured Materials, Federal University of São Carlos (UFSCar), Araras, São Paulo, Brazil*

3.1 Overview

The term "biosensor" has emerged in a work developed by Clark and Lyons in the 1960s (Turner et al., 1987), when they used an enzyme-based electrode to measure the amount of oxygen consumed during the reaction of glucose with the glucose oxidase enzyme (Clark & Lyons, 1962). Since then, significant advances were made in biosensor methodologies, including the development of enzyme-based biosensors for several uses, such as drugs (Wang, 1999) and pollutants monitoring (Ma et al., 2022.), and in the food industry (Villalonga et al., 2022). In addition, they are widely used in precision medicine, mainly in medical diagnostics (Min et al., 2018), focusing on exploring new transducer materials, detection methodology strategies, and potential targets to improve their range of applications. Usually, biosensors gather selective quantitative or semiquantitative analytical information using a specific biological recognition element (Karunakaran et al., 2015; Thévenot et al., 2001). We can associate the functioning of a biosensor with that of our brain, for example, when we feel external *stimuli* through our senses, such as taste. Automatically the taste receptors make the recognition and the transduction of the taste. Synaptic signals occur between the neurons, and the brain processes the received information. Thus the principle of operating a biosensor includes three main elements: (1) biological recognition by the bioreceptor, (2) transduction, and (3) signal amplification and/or detection (Fig. 3.1).

Firstly, biological recognition occurs between the analyte and the biological element (i.e., enzymes, antibodies, proteins, and nucleic acids, among others) immobilized on a surface (of an electrode, for example) (Lahcen & Amine, 2022.). This interaction is selective, for example, by enzyme–substrate or antigen–antibody interactions, which ensures the selectivity of the

Biosensors in Precision Medicine. DOI: https://doi.org/10.1016/B978-0-443-15380-8.00004-7

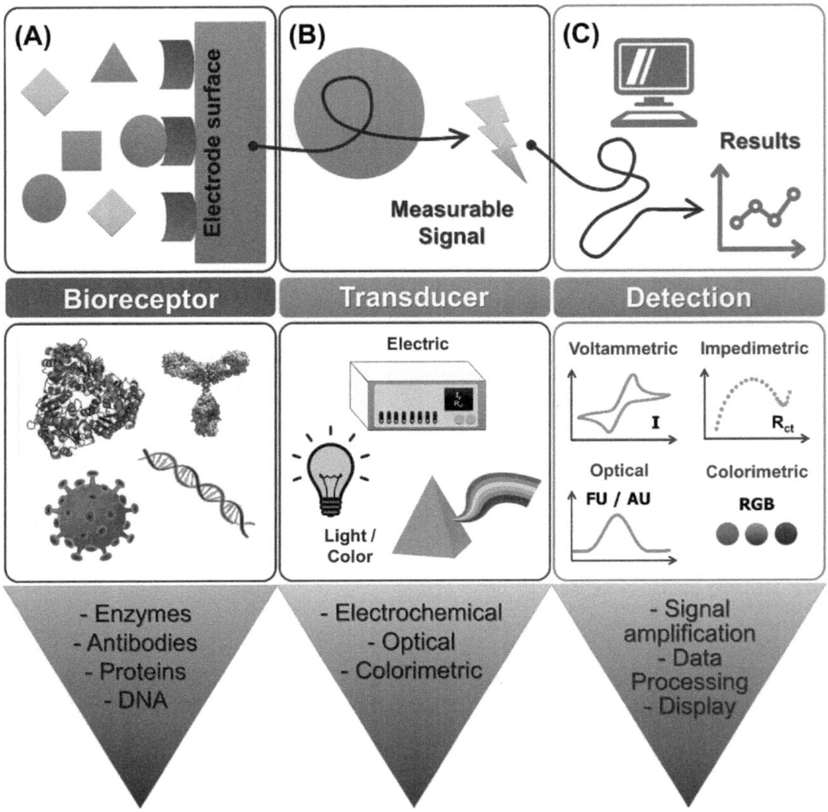

FIGURE 30.1 Representation of the principle of operation of a biosensor, including (A) the biorecognition element (bioreceptor), immobilized on a surface (e.g., electrode surface), (B) the transducer element, that converts the biological interaction into a measurable signal (e.g., voltage, current, etc.), and (C) the amplification and/or detection of the signal related to the specific interaction of a target with the bioreceptor.

biosensor. The biological interaction is then converted to a measurable signal by the transducer, including changes in electrical properties, emission or absorption of light, or color change, for example (Singh et al., 2020). The electronic component detects and transmits the registered signal. Biosensors can be classified by their transducer types, in which several analytical techniques have been used for this purpose, mainly electrochemical (i.e., voltammetric and impedimetric techniques), optical (i.e., fluorescence and colorimetric techniques), and spectroelectrochemical (Costa-Rama & Fernández-Abedul, 2022.). Thus depending on the immobilized element, biosensors can be applied for the detection of different biological targets, including viruses (Abubakar Sadique et al., 2021; Ahangar & Mehrgardi, 2017; Bhardwaj et al., 2019; Brazaca et al., 2021; Diouani et al., 2008; Figueiredo et al., 2015; Martins et al., 2021), specific genes and biomarkers related to relevant diseases (i.e., cancer, Alzheimer's, Parkinson's, and diabetes, among others) (Arkan et al., 2015; Carneiro et al., 2020; De Carvalho Oliveira et al., 2022; Doucey & Carrara, 2019; Kalinke et al., 2023; Karaman et al., 2022; Kim et al., 2022; Lee et al., 2022; Mobed et al., 2021; Pacheco et al., 2018; Soni et al., 2022; Thenrajan & Wilson, 2022), allergens and other molecules related to food safety (Akgönüllü & Denizli, 2022; Eissa et al., 2013; Peng et al., 2023; Ribeiro et al., 2022; Son et al., 2023; Wan et al., 2022), and environmental contaminants (Chandel et al., 2022; Ojeda et al., 2012).

3.2 Bioreceptors: immobilization and recognition events

As previously mentioned, bioreceptors compose the biological recognition layer of the biosensor, allowing specific recognition of the target. Selectivity and sensitivity at low-concentration levels are characteristics desired of the biosensors, which are associated with this specific biorecognition (Lahcen & Amine, 2022.). In this sense, the immobilization of bioreceptors and biorecognition events on the electrode surface are crucial to ensure the proper functioning and long-term stability of a biosensor.

Biosensors can be classified based on the bioreceptor, with the terminology being based on specific prefixes (i.e., immunosensor, genosensor, and aptasensor). In this case, biosensors based on immunoglobulins (antibodies) as bioreceptors are called immunosensors, while those based on genetic materials and aptamers are called genosensors and aptasensors. Next, the main types of bioreceptors used for the construction of biosensors will be presented, as well as the main methods and linkers employed for their immobilization on the electrode surface.

3.2.1 Enzymatic biosensors

The term "enzyme electrode" was originally mentioned by Clark and Lyons in 1962, which used a Clark oxygen electrode containing a membrane with glucose oxidase (GOx) enzyme for the indirect determination of glucose by the decrease in oxygen levels (Clark & Lyons, 1962; Newman & Setford, 2006). This is recognized as the first reported biosensor. After, Updike and Hicks also described the development of an enzyme electrode, in 1967 (Updike & Hicks, 1967). Since the pioneering work, enzyme-based biosensors have probably been the most known and reported in the literature until now. This combination of electrochemical techniques with the biological activity of enzymes was a milestone for science, leading to great advancement in precision medicine.

Enzymes are often composed of amino acids with side chains that bind to primary amine groups ($-NH_2$), allowing their immobilization on the electrode surface through hydroxyl ($-OH$), carbonyl ($C=O$), and carboxyl groups (−COOH) (Wohlfahrt et al., 2004). The main principle of the enzyme biosensors is the specific interaction between the enzyme (i.e., dehydrogenases, oxidases, monooxygenases, and peroxygenases) immobilized on the electrode and the respective targets, for example, glucose oxidase-glucose, urease-urea, cholesterol oxidase-cholesterol, tyrosinase-tyrosine, lactate dehydrogenase-lactate, creatininase-creatinine, polyphenol oxidase-phenol (Sadani et al., 2022).

Among different enzymes, oxidases have been widely used due to their high specificity, sensitivity, and low cost of production (Jugović et al., 2016; Wilson & Turner, 1992). However, the enzymatic determination shows some disadvantages, mainly due to the stability of the enzyme, which can affect the usability of the biosensor (Yang et al., 2009). Therefore it is often necessary to monitor the activity of the enzymes to predict the lifespan of the biosensor. In addition, these enzymes can catalyze the oxidation or reduction of the target in a reaction involving molecular oxygen as an electron acceptor. GOx is an oxidoreductase enzyme that has been widely used for the development of biosensors due to its fast mechanism of action, high stability, and specificity (Bauer et al., 2022). It catalyzes the oxidation of glucose to gluconate and hydrogen peroxide, in which the product of this reaction (hydrogen peroxide) can be quantified for the indirect determination of glucose levels. A practical example of this biosensor is the glucometer, a commercial device commonly used for the rapid self-diagnosis of glucose in blood aiming at the monitoring of diabetes disease (Newman & Setford, 2006).

3.2.2 Immunosensors: antibodies or antigens bioreceptors

Immunosensors are devices based on specific antibodies−antigens recognition (Mollarasouli et al., 2019). Antibodies, or glycoproteins, belong to the category of immunoglobulins and are produced by B lymphocytes, essential components of the immune system (Avrameas et al., 1978). When the human organism is subject to an infection, immunoglobulins are produced to act against the antigen of the infection or disease. Initially, immunoglobulins M (IgM) are produced, which bind to some antigenic sites. Meanwhile, the immune system recognizes the foreign species and begins to produce specific memory tolerant, immunoglobulin G (IgG), to combat the antigen (Alberts et al., 2002). The antigens specifically bind to the amine ($-NH_2$) terminations located in the heavy chain of the Y-shaped antibody (epitope), in the fragment antigen-binding site (Lahcen & Amine, 2022).

One of the main applications of immunosensors concerns the detection of antigens for the diagnosis of diseases. Commonly, antibodies are immobilized on the electrode surface for the detection of antigens; however, the opposite approach is also reported. In this way, antibodies can be detected in biological samples, allowing their determination by the immobilization of the specific antigen on the immunosensor surface (Alireza Hashemi et al., 2021; Gogola et al., 2019). Recently, immunosensors have been widely reported for both applications (antibodies or antigens detections) related to the diagnosis of SARS-CoV-2 infection, aiming even the detection of immunological response or viral antigens, depending on the stage of the disease. This same argument is valid for the diagnosis of other diseases (i.e., cardiovascular, neurodegenerative, bacterial, viruses, cancer, and others) (Leva-Bueno et al., 2020; Mollarasouli et al., 2019). In addition to the clinical analysis, immunosensors have been applied for other applications such as environmental monitoring and food analysis (Aydin et al., 2021; Felix & Angnes, 2017; Gavrilaş et al., 2022; Krisnawati et al., 2022); hence, they are very useful for several approaches.

3.2.3 Genosensors: genetic materials bioreceptors (DNA and RNA)

Genosensors are based on genetic materials (i.e., DNA or RNA nucleic acids) as recognition elements, in which the specific hybridization reaction takes place by DNA − DNA or DNA − RNA molecular recognition (Liu et al., 2009). It is important to mention that when the desired target is a complementary DNA (cDNA) an additional step is required, which is obtaining a copy of DNA from the RNA, a process named reverse transcription. In reverse transcriptase polymerase chain reaction (RT-PCR), for example, great amounts of cDNA are synthesized by the use of the reverse transcriptase enzyme on the sample RNA, and the target (cDNA) is then replicated by PCR reaction under automated thermal cyclers (Pray, 2008).

For DNA − DNA recognition, single-stranded DNA (ssDNA) sequences are immobilized on the electrode. The immobilization of the probe on the electrode can be performed by van der Waals interactions; however, covalent bonding or crosslinking approaches provide a more stable interaction (Brazaca et al., 2021). In some cases, the ssDNA sequence can be modified with functional groups terminations, such as amines or thiols, which can facilitate the binding with specific groups or species found on the electrode surface, for example, by amine-carboxylic groups or thiol−gold interactions. The use of nanomaterials, such as gold nanoparticles, is also widely reported for the modification of electrodes and/or as labels, allowing the increase in the electroactive area, facilitating the immobilization of the DNA sequence probe, improving the hybridization rate and the selectivity of the biosensor (Abu-Salah et al., 2010; Brazaca et al., 2017). The choice of terminations or modifiers is directly related to the strategy adopted for the immobilization of the probe. More details on immobilization approaches are going to be shown in the next.

The monitoring is typically performed by a hybridization reaction between DNA sequences. The DNA probe recognizes its respective complementary target sequence, forming a double-stranded DNA (dsDNA). DNA probes (single-stranded DNA, ssDNA) can be labeled to provide detectable signals for DNA hybridization (Pividori et al., 2000). Usually, fragments of genetic material are used as probes (between 20 and 50 pairs of bases) to avoid possible DNA hairpins. The determination of DNA or RNA nucleic acids is attractive from the clinical analysis point of view because it allows the monitoring of genetic mutations associated with diseases (Brazaca et al., 2017). In addition, genetic materials from viruses, genetically modified foods, and bacteria can also be determined by genosensors (Alzate et al., 2020; Kalogianni et al., 2006; Silva et al., 2022; Wang et al., 2011).

3.2.4 Aptasensors: aptamers bioreceptors

Aptamers are fragments of single-stranded DNA or RNA selected to have high affinity and specificity to the desired targets (Zou et al., 2019). They are artificially obtained by Systematic Evolution of Ligands by Exponential Enrichment (SELEX), a combinatorial method based on consecutive rounds of selection and amplification, which uses a library of nucleic acid sequences for the selection of the ideal probe (Bayat et al., 2018; Ellington & Szostak, 1990; Sampson, 2003). Due to their high capacity for interaction, aptamers have been versatile probes for the binding and determination of several different targets, such as biomolecules, viruses, proteins, bacteria, and whole cells (Wu & Kwon, 2016). This also makes aptamers candidates for applications in multiplex systems, allowing highly specific distinctions between different targets of diseases. Aptamers also have advantages over other biosensor configurations in terms of temperature stability; comparatively, antibodies undergo denaturation at elevated temperatures (Tombelli et al., 2007), being also easy to prepare and able to provide great signal amplification (Cai et al., 2023; Wu & Kwon, 2016).

The immobilization of DNA or RNA aptamers can be easier, compared to antibodies, for example. Due to their similarity with genosensors, the immobilization steps and characteristics of the aptamers-based biosensors are also comparable to the first ones. It is normally carried out by electrostatic interactions, covalent bonds, and $\pi-\pi$ stacking (Zou et al., 2019). Aptamers have been successfully applied for clinical diagnostics, due to their ease of adapting to different platform systems, such as sandwich, and lateral-flow assays (Reid et al., 2020). In addition, aptamers are usually labeled with gold nanoparticles; however, label-free approaches are also reported in the literature. Aptamers have already been successfully applied for the detection of hormones, proteins, toxins, viruses, bacteria, and for the diagnosis of several diseases (i.e., cancer, cholera, cardiovascular, Alzheimer's, and other neurodegenerative diseases) (Li et al., 2019; Reid et al., 2020).

It is important to mention that new and improved bioreceptors are also found in the literature, with a detailed discussion about their types, nature and uses being found in Chapter 6 of this book.

3.3 Bioreceptors immobilization approaches

Biorecognition is one of the keys in the development of biosensors since specific and high-affinity recognition can predict the accuracy and success of a biosensor. Also, the complexity of the immobilization process needs to be considered. The immobilization and/or recognition of biomolecules on the electrode surface can be performed by different reaction approaches, such as covalent, crosslinking, coordinate, and hydrogen bonding, among others. Usually, electrodes can show no active sites for the binding of bioreceptors, which makes necessary the modification of the surface electrodes. The choice of the linkers and methods depends on the electrode material and the bioreceptor functional groups; therefore, knowledge regarding the biosensor system is required.

3.3.1 Covalent bond

The covalent binding of receptors to the biosensor surface is commonly carried out by its surface activation using multifunctional reagents, followed by bioreceptor coupling to the activated support. Carbodiimide, glutaraldehyde, and mercapto-based acids can be highlighted as the most used reagents for this approach. The covalent bond strategy presents the advantages of good stability, high binding strength, and long-term use (Kim & Kang, 2008). Carbodiimides allow the linking between carboxyl and amino groups (Sassolas et al., 2012). A widely reported carbodiimide for this concern is 1-ethyl-3-(3-dimethylaminopropyl)carbodiimide (EDC), which is normally used in combination with NHS (*N*-hydroxysuccinimide). In this approach, EDC commonly promotes the activation of carboxylic groups on the sensor surface. Then, EDC is replaced by NHS, forming an intermediary that allows the immobilization of biomolecules covalently to the electrode through the functional groups ($-NH_2$) present in these molecules (Fig. 3.2A) (He et al., 2009; Kalinke et al., 2023; Shen & Liu, 2007).

Another commonly used covalent bond strategy for probe immobilization is based on thiol–gold interaction (Fig. 3.2B) (Brazaca et al., 2021; Manzano et al., 2018). Gold-based electrodes are chemically inert and biocompatible, desirable characteristics for the development of biosensors. Besides that, gold nanoparticles can be immobilized on other materials, improving the catalytic, surface area, conductivity, and sensitivity properties of the sensor (Jiang et al., 2018). Gold substrates can be used for the immobilization of biomolecules by thiol or amine bonds (i.e., Au–S or Au–N), which can involve multiple bonding scenarios depending on the location of the hydrogen atom, possibiliting the formation of Au–SH groups coordinate bonds (Xue et al., 2014). Parallel to this, self assembly monolayer (SAM) structures can be employed working as a bridge to attach biomolecules or DNA to the gold substrate (Wang et al., 2012). This strategy allows achieving monolayer films on the gold electrode, which can be performed by using COOH-terminated reagents, for example, 3-mercaptopropionic acid (MPA), 11-mercaptoundecanoic acid (MUA), and 4-mercaptobenzoic acid (MBA).

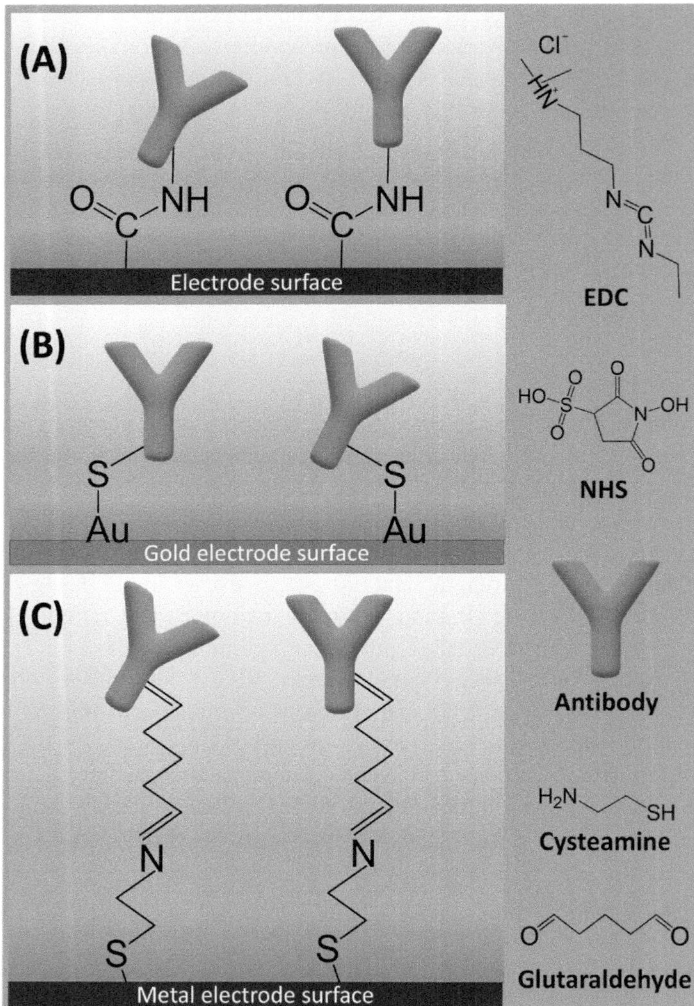

FIGURE 30.2 Schematic examples of bioreceptor immobilization approaches based on (A) EDC/NHS; (B) thiol–gold; and (C) cysteamine and glutaraldehyde interaction.

Thus the bioreceptor can be attached to these terminations of the SAM (Wang et al., 2012).

Cysteamine is another example of a mercaptamine, that is, it presents thiol and amine groups in its structure. Due to its bifunctional characteristic, cysteamine can be explored by different immobilization strategies, for example, to bind with gold electrodes or nanoparticles through thiol groups, or by binding between its amine groups with carboxylic groups from bioreceptors. Cysteamine is commonly used in combination with glutaraldehyde. In that instance, cysteamine amine groups are available to interact with carbonyl groups of glutaraldehyde, which presents aldehyde groups to bind to amine groups from bioreceptors (Fig. 3.2C) (Cai et al., 2019; Oliveira et al., 2020). This allows several possibilities for the stable immobilization of bioreceptors, leading to the development of more sensitive biosensors.

3.3.2 Crosslinking

The immobilization of biomaterials by covalent bonds can also be accomplished using crosslinking agents, such as streptavidin/biotin, glyoxal, and hexamethylenediamine, in addition to glutaraldehyde and EDC/NHS. Crosslinking agents, or crosslinkers, are molecules that contain two or more reactive groups (e.g., carboxyl, amine, hydroxyl, aldehyde, among others) that allow crosslinkages between biomolecules by covalent bond (Nguyen & Kim, 2017). This is considered a simple and effective method of the strong chemical binding achieved (Sassolas et al., 2012).

The crosslinking between streptavidin (or avidin) and biotin is very strong and creates close interactions, providing a highly stable immobilization (Wong et al., 1999). Avidin and streptavidin proteins show a considerable affinity for the cofactor biotin. In this case, biotin is enclosed in streptavidin active sites by eight hydrogen bonds, as well as van der Waals interaction among nonpolar groups (Hyre et al., 2006). Avidin can bind up to four biotin molecules, forming an avidin–biotin complex, which can immobilize enzymes, antibodies, or proteins (McMahon, 2008). An example of this application is observed by the immobilization of biotinylated bioreceptors on electrodes modified with streptavidin (Carinelli et al., 2017).

Glutaraldehyde is also used as a crosslinker due to its homobifunctional characteristic, containing two aldehyde groups that can be coupled to primary amino groups forming a Schiff base (–N = C–). Two molecules or biomolecules can be attached to these ends (Li et al., 2020). EDC also plays this role allowing the direct interaction between carboxylic groups and primary amines, forming amide bonds with the bioreceptors. NHS or sulfo-NHS are frequently used as the pair of EDC, improving stable immobilization, as shown previously.

3.3.3 Other immobilization approaches

Physical strategies have also been used for the immobilization of biomolecules, which include electrostatic interaction (Shukla et al., 2017). This strategy can be explored when the electrode and the bioreceptor show different surface charges (i.e., positively or negatively charged). In this case, it is important to know the isoelectric point of the biomolecule, to predict when it is positively or negatively charged, allowing the interaction with opposite charges on the electrode surface. An example of this approach is observed for the development of biosensors by layer-by-layer deposition or electrochemical doping method based on electrostatic adsorption (Sassolas et al., 2012). Another strategy for immobilization can be performed by entrapment, for example, by the incorporation of an enzyme within a gel or a polymer (e.g., polypyrrole, polyaniline, or polyphenol) (Cosnier, 1999; Gupta & Chaudhury, 2007; Sassolas et al., 2012).

3.4 Transduction and detection techniques

The biorecognition and detection on a biosensor can be performed indirectly by using a labeled bioreceptor. Gold, silver, and magnetic nanoparticles (Cajigas et al., 2020; Eguílaz et al., 2011; Wen et al., 2014; Yen et al., 2015), and carbon quantum dots (Freitas et al., 2020; Zhao et al., 2021; Zhou et al., 2020) have mostly been used for this purpose. Usually, these species have high electrical conductivity, and due to their small size, they also have a high electroactive area, which contributes to signal amplification, improving the sensitivity of the biosensor. Regarding quantum dots, an additional advantage can be achieved due to their luminescent properties, allowing optical detections (Chatterjee et al., 2019; Kalkal et al., 2021).

On the other side, when labels are not used, biosensors are classified as label-free. In this case, direct recognition can be carried out by physical phenomena, for example, when the antibody recognizes the antigen (Samanman et al., 2015). Some advantages of label-free biosensors can be acquired mainly with the elimination of the enzyme or marker incorporation step, simplifying the method, and promoting the reduction of costs, handling, and response time (Dong et al., 2013; Felix & Angnes, 2017). Countless studies report the use of label-free immunosensors based on different materials for the incorporation of biological materials (Eissa et al., 2013; Vestergaard et al., 2007).

In both cases, the signal transduction and measurement are performed as a function of changes that occur on the electrode surface after the specific interaction between the bioreceptor and the target (Costa-Rama & Fernández-Abedul, 2022.). After biorecognition, these changes can be converted or transduced into measurable signals (i.e., electrical, optical, and thermal). Target detection can be done quantitatively, evaluating the relationship between different target concentrations and the obtained signal. This permits the monitoring of biomarker levels related to a disease, for example. On the other hand, qualitative analyzes are based on obtaining positive/negative or on/off responses and have been increasingly used in clinical analyzes for the determination of biomarkers or pathogens (i.e., virus DNA/RNA, bacteria, proteins, biomolecules, and others). The transduction of biorecognition is typically performed by electrochemical, optical, or spectroelectrochemical techniques, which are further described below.

Recent advancements in detection techniques are described in detail in Chapter 7 of this book.

3.4.1 Optical detection

Optical detection is based on the measurement of electromagnetic radiation that emanates from matter or interacts with it by absorbance/transmittance, reflectance, or fluorescence (Costa-Rama & Fernández-Abedul, 2022.). In this case, the signal can be recorded by a photodetector and transformed into an electrical signal (Borisov & Wolfbeis, 2008). Optical biosensors present some advantages, such as high sensitivity, robustness, reliability, and stability. In addition, these biosensors can be integrated into miniaturized and microfluidic systems for the development of point-of-care devices (Chen & Wang, 2020).

Fluorescence spectrometry is one of the most applied optical methods, in which the emitted light or fluorescence is proportional to the concentration of the target. Fluorescence detection has become one of the most sensitive methods for bioanalytical and medical tests (Le Gac et al., 2006). Species with fluorescing properties (i.e., fluorophores) have been used allowing the detection in different ways, under white light or radiation sources (e.g., UV) (Costa-Rama & Fernández-Abedul, 2022.). Glucose oxidase, in addition to other enzymes, is an interesting example of fluorescent species due to the presence of a fluorophore (FAD, flavine adenine dinucleotide). In this case, after glucose oxidase reaction with glucose, a changing in the optical property is observed, which allows its detection (Odaci et al., 2009). Carbon quantum dots have also been applied as a label for the development of optical biosensors based on fluorescence/luminescence due to their excellent photoluminescence, brightness, and electrical conductivity properties (Mintz et al., 2019). In this case, the detection is performed by the fluorescence quenching of quantum dots. Regarding clinical analysis, fluorescence techniques have been applied for the detection of biomarkers related to viral, bacterial, neurodegenerative, and autoimmune diseases (Borisov & Wolfbeis, 2008; Chen & Wang, 2020).

Colorimetric methods are probably some of the most used for the detection of targets in medical analysis due to the simple construction and manipulation of devices, low operational costs, and fast results (Brazaca et al., 2021). Colorimetric detection is characterized as the change or appearance of color, that is, the wavelengths of the visible region that are not absorbed (Costa-Rama & Fernández-Abedul, 2022.). The use of labels is often reported for this purpose, mainly gold nanoparticles, allowing the interpretation of the results by the naked eye (Yu et al., 2020). RGB (red, green, and blue) is probably the most commonly used system of colors, allowing the correlation between the intensity of the colors and the concentration of the target. In this case, the sensor can be coupled to smartphones and software for data analysis, increasing the reliability of the results (Capitán-Vallvey et al., 2015). Both qualitative and quantitative detection can be achieved (De Carvalho Oliveira et al., 2022; Qian et al., 2022; Son et al., 2023).

Lateral-flow assay is one of the most known examples of point-of-care diagnostic systems based on optical detection. This became more expressive after the recent COVID-19 outbreak, where people could use tests for self-diagnosis. In addition, colorimetric tests have also been developed for the monitoring and fast diagnosis of several viral diseases (Brazaca et al., 2021). The use of lateral-flow assays has also been described for multiplex systems allowing the multiple diagnosis and distinction of diseases (Wu et al., 2020). An example of this application is reported for the point-of-care diagnostic of dengue, yellow fever, and ebola virus by using a lateral-flow assay based on silver nanoparticles as a label (Yen et al., 2015). This is an important advance for precision medicine, as many viral diseases have similar symptoms, making accurate diagnosis difficult. In addition, prior diagnosis allows adequate treatment to be started as soon as possible, increasing the chances of curing the disease.

3.4.2 Electrochemical detection (voltammetric and impedimetric)

The development of electrochemical biosensors has been widely reported in the literature, mainly due to the low cost, simple operation, and real-time assessment attributed to electrochemical techniques (Labib et al., 2016). Electrochemical biosensors are versatile devices and can be constructed by using different materials (i.e., carbon, metals, nanomaterials), shapes, and approaches. Furthermore, there is the possibility of miniaturization and portability aiming at the development of point-of-care devices, and coupling with microfluidic systems and other electronic devices (Mostafa et al., 2022). In addition, wearable electrochemical biosensors have been applied for the noninvasive or minimally invasive monitoring of biomarkers in biological samples, such as sweat and tears (Drexelius et al., 2022). Wearable biosensors are discussed in detail in Chapter 9 of this book.

The electrochemical detection is based on a mensurable electrical signal (i.e., electrical current, potential, and resistance, among others) from redox reactions that occur on the electrode surface. Various electrochemical techniques can be used, such as voltammetric, potentiometric, amperometric, and impedimetric (Labib et al., 2016). For biosensors, voltammetric- and impedimetric-based techniques are currently the most used, and the difference in the electrical signal is recorded before and after the specific recognition between the bioreceptor and the target. This can be correlated, for example, to the formation of the antigen–antibody immunocomplex or the RNA/DNA strands hybridization reaction (genosensors) that behave like a barrier causing an isolating effect and blocking the electron transfer on the electrode surface (Martins et al., 2021; Mojsoska et al., 2021). For voltammetric techniques (e.g., cyclic voltammetry, differential pulse voltammetry, square wave voltammetry, and others), the monitoring is performed by the change of anodic and/or cathodic current peaks; while for impedimetric detection by electrochemical impedance spectroscopy, the variation of the charge transfer resistance is evaluated (Kalinke et al., 2023). Both labeled and label-free electrochemical biosensors are reported, in which a difference in the functioning mechanism of the biosensor is observed, similar to other previously described techniques. Regarding precision medicine, electrochemical biosensors have been applied for the detection of biomarkers related to several cancer types (Damiati et al., 2017; Lahcen et al., 2021; Mostafa et al., 2022), neurodegenerative diseases (Brazaca et al., 2020; Kalinke et al., 2023), viral (Brazaca et al., 2021; Stefano et al., 2022; Tancharoen et al., 2019), and bacterial (Singh et al., 2018) diseases, and among others.

3.4.3 Other techniques

The combination of techniques can be advantageous for biosensors for medical applications, since often the reaction mechanism may not be well defined or known. Spectroelectrochemistry allows a more complete description of the processes that occur on the electrode surface, aiming at the combination of advantages of both techniques (spectroscopy and electrochemistry). The monitoring of redox processes, catalysis, kinetics, and quantitative analysis can be accompanied by spectroscopic techniques (i.e., Raman, Infrared, UV–vis, and others), allowing the monitoring of details and changes during the reactions and contributing to the comprehension of the electrochemical interfacial processes (León & Mozo, 2018; Pereira et al., 2022). This detection is simultaneously performed by using the same electrochemical cell or device (i.e., in situ method), which avoids misinterpretation of the results, and improves the selectivity of the biosensor (Eguílaz et al., 2011).

Spectroelectrochemical biosensors have been described for precision medicine applications. Examples include the monitoring of specific hepatitis C RNA hybridization reactions by using Raman spectroscopy and square wave voltammetry techniques (El-Said & Choi, 2019), and for the SARS-CoV-2 S protein detection by surface-enhanced Raman scattering (SERS), cyclic and square wave voltammetry techniques (El-Said et al., 2022). In addition, spectroelectrochemical measurements can also be used to obtain additional information about the materials and biomolecules, such as to evaluate the mobility of enzymes and the electron transfer in carbon-based materials, used for the development of biosensors (He et al., 2009).

3.5 Transduction and detection approaches

The assembly of a biosensor can be carried out in different ways, aiming at the immobilization of the biological recognition element or labeling species with specific properties, as previously discussed. The modification way, as well as the design of the approach employed, aims not only to ensure the stability and chemical activity of the recognition element but also to allow better selectivity and sensitivity. From the point of view of sensitivity and the decrease of the limit of detection (Baeumner et al., 2004), some assembling strategies have been widely used for different types of biosensors with different transducers. Probably, one of the most widespread approaches found in the literature is the use of sandwich assembling type, in addition to the use of signaling molecules, among others.

3.5.1 Sandwich

A biosensor sandwich-based is a device that uses two biological elements, one for capture, and one for detection, in which the analyte is sandwiched between them (Seo & Gu, 2017; Xia et al., 2018). The type of biosensor using sandwich assay will depend on the biological recognition element used, for example: for sandwich-based immunosensors, antibodies are used as capture and detection elements; for nanosensors, genetic materials are used; for aptasensors, aptamers (Seo & Gu, 2017), and so on. Fig. 3.2A shows a generic immunosensor scheme of the simplest layout of the sandwich approach to help the understanding. First, the capture antibody is attached to the surface of the deposition platform, generally by covalent bonds (Fig. 3.3A, a); then, the addition and selective capture of the antigen (target) is performed (Fig. 3.3A, b); and finally, a new incorporation of the detection antibody on the antigen is performed (Fig. 3.3A, c). The detection antibody takes this denomination because it indicates the presence of the target through the generation or induction of a signal. Most of the time, this biological element is linked to a label material, which an organic or inorganic material could be used for signal generation, such as electrochemical, optical, or others (Fig. 3.3B).

The biological recognition element is linked to an enzyme, for example, peroxidases, which allows the identification of the formation of the antigen/antibody complex in the presence of hydrogen peroxide by redox reaction and, consequently, an electrochemical signal generation. Similarly, this detection antibody can be linked to a particle or a nanoparticle with a high-contrast color, such as a gold nanoparticle; or a fluorescent specie, which will generate a signal, in these cases, colorimetric and fluorescent, respectively. The possibility of incorporating different types of probes on the biological recognition element makes this biosensor assembling strategy versatile from the point of view of the type of signal generation and transduction.

Besides the versatility of biosensors sandwich-based, maybe the most attractive quality of these devices is the amplification of the generated signal, which in turn promotes a higher sensitivity, accuracy, and specificity (Jang et al., 2009). This enhancement can be achieved by the type of label specie used to generate the signal. For example, in the development of a colorimetric biosensor, it is essential to use a marker with high contrast compared to

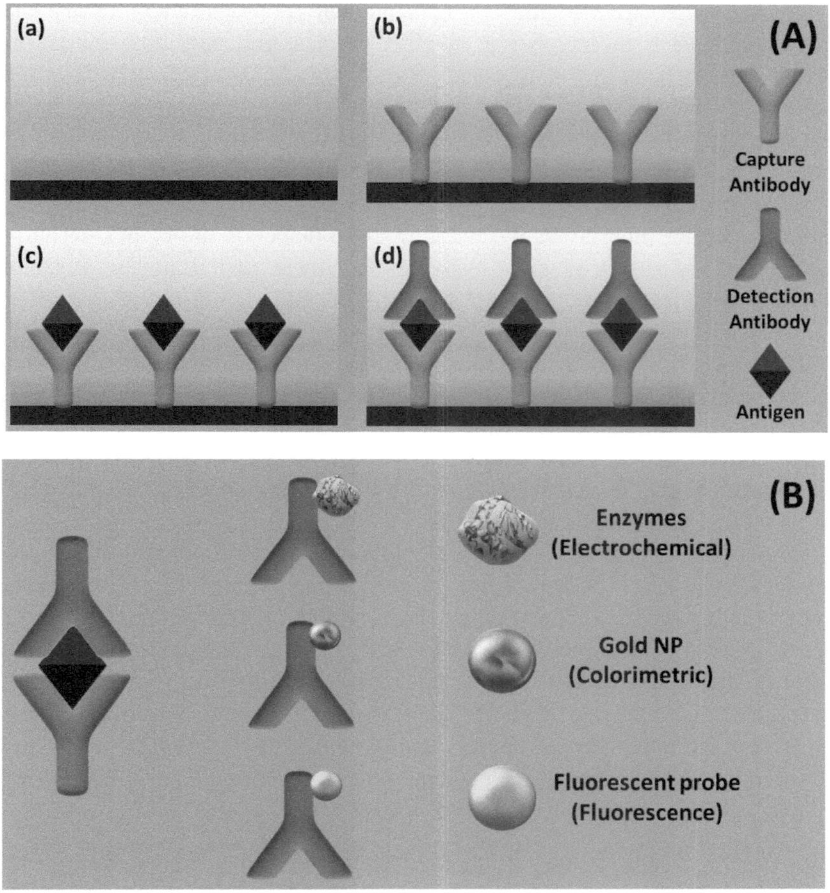

FIGURE 30.3 (A) Generic scheme for sandwich biosensor, in this case, an immunosensor assembly; (B) examples of label species linked to a biological element: enzymes, gold nanoparticles (gold NP), and fluorescent probe.

the analysis platform. The greater this contrast, the greater the possibility of distinguishing between a positive and negative signal, even at low concentrations. For example, rapid lateral-flow diagnostics, such as pregnancy, COVID-19, and other disease tests. In these devices, recognition elements marked with gold nanoparticles (red color) are commonly used (Chang et al., 2019; Oh et al., 2022). In the absence of AuNPs or by using another nanoparticle with less color contrast, the differentiation between the control and test signal would be unfeasible. Similarly, for electrochemical or fluorescence applications, the characteristics of the label species are very important to provide good sensitivity to the biosensor, more precisely by using label species with high electron transfer and with a high photon emission capacity, respectively (Qi et al., 2009).

Another alternative to the use of sandwich approaches to improve the detection signal is the number of labeling species, generally used for multiplexed sandwich assays (Cao et al., 2006). This consists of increasing the number of labeled species in the same "sandwich-type" arrangement containing the target. In other words, for each unit of the complex target, two or more species will generate the signal. This enhancement can be achieved by introducing more label species in the same detection antibody (secondary antibody); using multilayers by inserting more detection antibodies, tertiary, quaternary, and so on; or even by combining these two strategies. Fig. 3.4 demonstrates these alternatives for assembling a generic immunosensor using a fluorescent probe.

Regarding this strategy promoting the introduction of more label species per complex target, its use is unfeasible for systems that employ label-free determination. Another disadvantage, particularly for multilayer construction is the number of steps that take place in the assembly process. Each additional layer or modification increases the time of manufacturing, which can also lead to a decrease in repeatability.

In general, obtaining a biosensor employing sandwich approaches helps in identifying and monitoring the target specie, regardless of its nature, whether it is an enzyme, antigen, aptamer, or DNA and RNA (Mandli et al., 2017). Of course, in some cases, the presence only of the target can generate a signal, indicating its presence in the sample. However, in most cases, the target does not generate a sensor signal change, or the measuring device does not have adequate selectivity, and, in this case, any signal generated can be obtained by any species present in the sample. Another possibility of

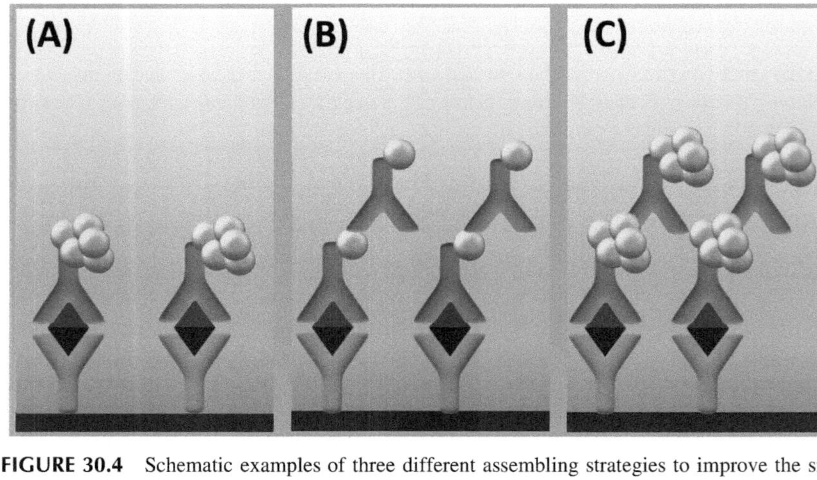

FIGURE 30.4 Schematic examples of three different assembling strategies to improve the signal in a generic immunosensor using a fluorescent probe (A) more label species in the same detection antibody; (B) multilayers of labeled antibodies; and (C) combination of the used strategies in A and B.

identifying the target specie is through label-free measurements, for example by increasing or decreasing the mass in quartz crystal microbalance measurements (Kim et al., 2009; Kurosawa et al., 2003), or by electrochemical signal inhibition using electrochemical impedance spectroscopy (Brazaca et al., 2017; Roberto de Oliveira et al., 2023) or voltammetry (Blasques et al., 2023; Brazaca et al., 2021). However, these measures, when compared to the use of label species, have less sensitivity.

3.5.2 Signaling molecules

In general, major of living organisms work as complex machines, in which all the machine components are integrated for better performance. The communication between different life systems in an organism, such as complex or simple cells typically is made using chemical signals over to neighboring cells. In this case, a sending cell release a substance (signaling molecule or ligand) to the cell target, in general by organic molecules, that it is transduced in a specific signal (Jeon et al., 2018; Wang et al., 2013). From the medical point of view, this signal can be correlated to the presence of a disease by the monitoring of different levels of the signaling molecules or can indicate the presence of determinate organisms, such as bacteria or viruses. Cell communication can occur in different ways, depending on the mechanism and distance between the sending and the target, such as paracrine, autocrine, endocrine, and direct contact signaling (Perbal, 2003). Some examples of signaling molecules consist of neurotransmitters such as norepinephrine and histamine secreted and received between neurons (Hansson & Rönnbäck, 2003); communication between cells in bacteria, for example, gram negative by the release of *N*-acyl-homoserine lactones (AHLs) autoinducers in the process of "quorum sensing" (QS) (Cooley et al., 2008); release of molecules known as a mating factor in yeasts secretion (Michaelis & Barrowman, 2012); and hormonal protein release in plant responses to stresses (Larrieu et al., 2015).

The determination of most of these molecules can be performed by chromatography in blood and urine (Lwin & Patel, 2010), as it presents high efficiency and low limit of detection; however, it demands rigorous sample preparation, high expertise labor, and high purity of the reagents and gases, increasing the cost and time of analysis. Yet, with recent advances in technologies, such as more advanced instrumentation and new materials, the determination of signaling molecules in living cells has been possible to be carried out (Wells et al., 2013). This fact increases the range of possibilities in these analyses, such as the possibility of using miniaturized and portable devices, which allow the reduction of sample volume, operational simplicity, and in some cases, in situ determination, which allows real-time monitoring (Struss et al., 2010).

One example of this application consists in the development of biosensors for monitoring signaling based on QS including AHLs in the pathogenesis of bacteria-related disorders, in gram-negative bacteria (Steindler & Venturi, 2007). The high number of works for this purpose is understandable because most cases of microbial infections are associated with the presence of these organisms. Numerous works use conventional physical-chemical monitoring techniques, such as liquid or gas chromatography, and mass spectroscopy, usually using a cell-free system (Claassens et al., 2019). Fig. 3.5A shows an example of a biosensor based on a signaling molecule (QS) for monitoring the concentration of AHL, more precisely a genetically engineered Escherichia *coli* system. In this system, the presence of AHL (signaling

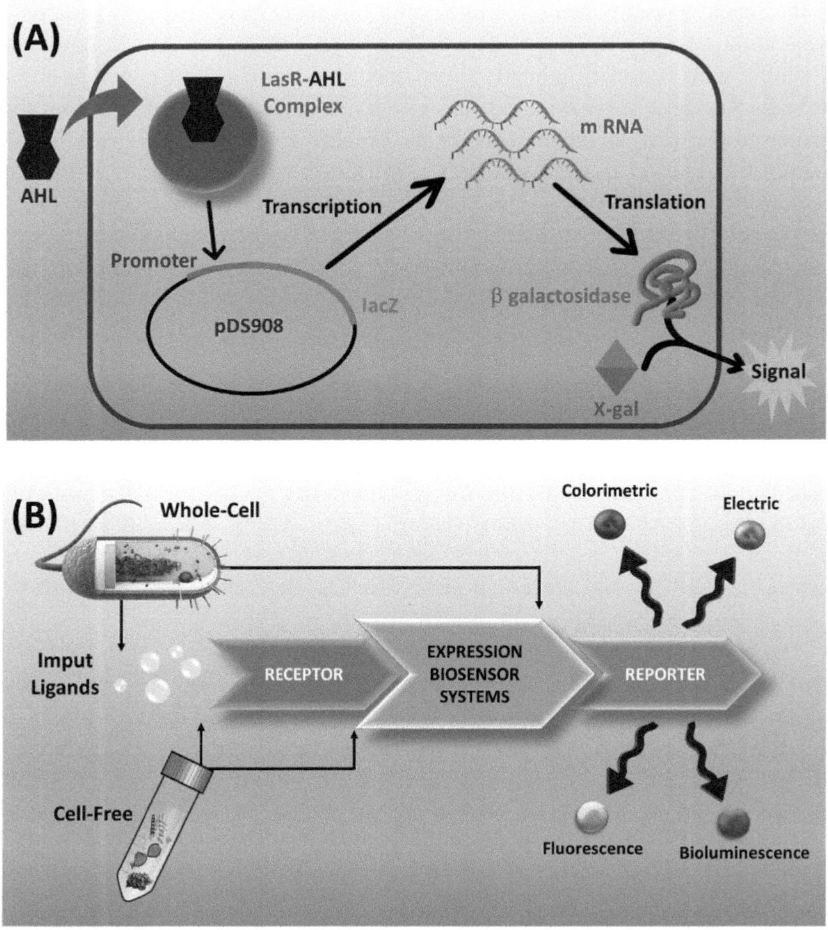

FIGURE 30.5 (A) Schematic representation for an engineered *E. coli* colorimetric system based on quorum sensing (QS). (B) Comparison between the expression biosensor systems for a whole-cell and cell-free. Abbreviations: AHL—N-acyl-homoserine lactone; X-gal—5-bromo-4-chloro-3-indolyl-β-D-galactopyranoside.

molecule or ligand) promotes the formation of a complex together with the regulatory protein LasR (cell target), present in the cell-free incorporated into the modification platform or in solution. The formation of this regulatory complex activates the lacZ reporter gene transcription, which generates the expression of the reporter protein, β-galactosidase, which in turn is cleaved to X-gal (5-bromo-4-chloro-3-indolyl-β-D-galactopyranoside), generating a colorimetric signal by producing an insoluble blue product.

There is a growing trend in the development of biosensors or biosystems based on whole-cell bacterial sensing (Claassens et al., 2019; Dudley et al., 2016). This array consists of devices containing a receptor cell incorporated into the solid platform, which when in contact with the signaling molecule is stimulated and generates a measurable signal, directly or indirectly by subsequent reactions. Fig. 3.5B shows the difference between cell-free and whole-cell systems.

The main advantage of the use of biosensor assays, based on signaling molecules, can be attributed to the fact of it does not require sophisticated instrumentation. When these devices make use of whole-cell biosensing systems, the advantage reaches out to other factors, for example, the ability to withstand different conditions: pH, ionic strengths, and temperature, among others. In addition, in whole-cell systems, a great part of a biological production system is attached to endogenous metabolism, in other words, a more complete biological system (Claassens et al., 2019). On the other hand, for biosystems based on cell-free, the complete conversion of a substrate with low losses of biomass can be achieved, in contrast to whole-cell systems (Claassens et al., 2019). These characteristics make the development of biosensors based on signaling molecules highly feasible in different types of samples with simple construction and application, low cost, portability, and miniaturization, both for the use of whole cells and cell-free systems.

3.5.3 Other detection approaches

Dendrimers are monodispersed macromolecules with a high number of branches and high surface groups density, that present three distinct components in their structure: the core, the interior dendron, and the exterior, which is highly functionalized (Evtyugin & Stoikova, 2015; Naresh & Lee, 2021). This structural configuration, star-shape based, and the presence of functionalized groups in the edges make the dendrimers highly used in biomedicine applications (Naresh & Lee, 2021). The three main components end their common structure can be shown in Fig. 3.6A.

Despite other vast applications, such as in diagnostics, gene and drug delivery, biomedical imaging, and biomimetic systems, among others; dendrimers are highly used in biomolecule conjunction (Naresh & Lee, 2021; Satija et al., 2011). For any applications, especially for biological ones, dendrimers are used to link biological species by covalent or electrostatic

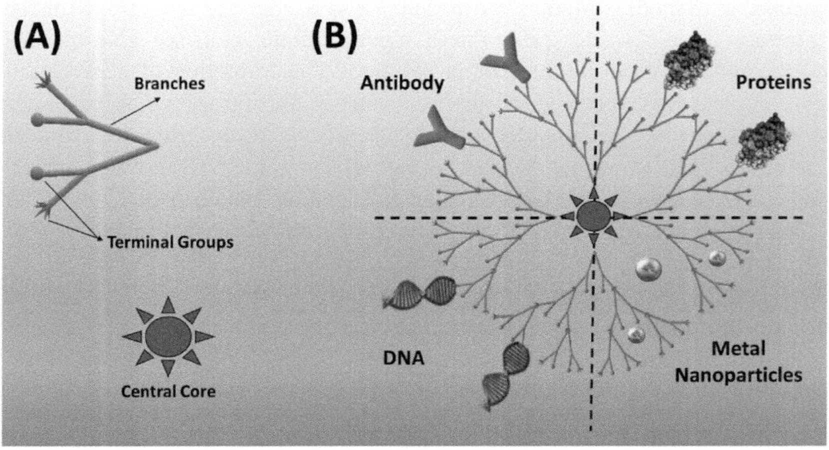

FIGURE 30.6 (A) Classification of different parts of a dendrimer components. (B) Strategies for biosensor application using dendrimers-based approaches.

interactions, forming complexes with excellent stability, by the functional groups present on the dendrimer structure, such as carboxyl, hydroxyl, and/or amine groups (Triulzi et al., 2008).

Functional groups can be present not only in the peripherical surface but also in different layers of the dendrimer molecule (Fig. 3.6B). This characteristic allows the organic or inorganic materials to attach to the whole dendrimer in several kinds of applications, such as the encapsulation of biomaterials (Evtyugin & Stoikova, 2015; Sato et al., 2012). Hence, there is a vast application of dendrimer-based biosensors, which can be introduced the biological asset, as well as the analytical probes, for example, fluorescent (Rios & Garcia, 2008), electrochemical (Zhu et al., 2007), magnetic (Doaga et al., 2020), and colorimetric (Fu et al., 2023).

The insertion of different kinds of biological materials can be carried out in the same assembly system, such as the attachment of enzymes together to genetic materials, or the combination of antibodies and proteins in the same dendrimer structure, and others (Singh, 2007). Finally, this assembly strategy also allows the construction of multiarrays of dendrimers to increase the sensitivity of the biosensor, mainly by increasing the number of recognition sites or analytical probe units, in association with other modification strategies, such as by the formation of multilayered dendrimers by layer by layer (LbL).

Another interesting example of a detection approach consists of molecular imprinted polymers (MIPs). As the name suggests, they consist of the construction of 3D polymeric materials, in general, forming rigid materials that have the ability of specific molecular recognition. This fact is due to the construction or printing of this material being carried out in the presence of the target, be it a simple organic molecule, enzyme, large protein, antibody, or another more complex molecule (Ayankojo et al., 2018). This impression is performed by the copolymerization of monomers and cross-linked polymers, which in the presence of the target, there is the formation of selective recognition sites between them (Saxena et al., 2022). After polymer assembly, the target is removed from the structure matrix, generating imprinted cavities.

In certain situations, this biosensor assembly strategy can be confused with the classification of the type of material used as a biological recognition site. However, obtaining MIPs can be performed in many approaches, and in this way, the discussion about MIPs assembly is inserted in this topic: deposition by coating or spin coating on electrodes: self-assembled monolayers, electropolymerization or light-induced polymerization, microcontact imprinting (Saxena et al., 2022).

The cavities present specific interaction sites, in terms of size, chemical affinity, shape, and orientation, like the original biological receptor structure (Rebelo et al., 2021). Therefore this detection approach strategy allows the construction of robust artificial receivers, which presents great mechanical characteristics, tolerance to high heat and pressure, chemical stability, and more simplicity regarding the obtaining and storage steps in comparison to the biological ones (Cui et al., 2020; Karaman et al., 2022).

3.6 Targets in precision medicine

As already discussed before, the use of biosensors can provide real-time and noninvasive monitoring of different types of diseases with high specificity and accuracy by monitoring a specific molecule or biomolecule in real samples. These specific monitored species are called targets, which can be small or big molecules, such as proteins, genetic materials, cells, extracellular vesicles, or microorganisms. The use of these species as targets in precision medicine can benefit the early diagnosis of any disease, as well as determining personalized treatments, reducing healthcare costs, and improving the quality of care and cure chances for patients.

3.6.1 Proteins

Proteins are large and complex molecules, which can be made up of long chains of amino acids and linked together by peptide bonds (Light & Smith, 1963). The different types and configurations of amino acids are used to build different kinds of proteins, and the sequence of these amino acids determines the structure and function of the protein, depending on the "body necessities." These macromolecules have a wide range of functions, such as catalyzing biochemical reactions, transporting molecules, providing structural support to cells and tissues in general, and playing a role in the immune system (Buxbaum, 2007; Light & Smith, 1963). Besides their presence in the human body constitution, they also can be found in other living organisms, from bacteria to plants to animals, and are essential for their proper functioning of them (Buchner & Rudolph, 1991; Cassab, 1998).

Proteins are widely explored in precision medicine, for example, as targeted cancer therapy that is based on specific proteins that are overexpressed or mutated in cancer cells (Persad et al., 2000; Strohmaier et al., 2001). Their use is vast also in pharmacogenomics, providing information about how a genetic material of an individual makeup affects their response to drugs, for example (Hardie, 2007). This approach involves analyzing specific proteins that are involved in different functions, such as drug metabolism, transport, and target pathways. Another important application of proteins in precision medicine is their use in personalized vaccines, for example by analyzing the tumor cells from patients, which it is possible to identify unique proteins that are expressed by this specific tumor and, so design a vaccine that targets these proteins (Wilson & Nie, 2006). And finally, we can highlight the use of proteins as biomarkers in the development of biosensors. Proteins can serve as biomarkers for various diseases, such as cancer, Alzheimer, cardiovascular, and, more recently, for SARS-CoV-2 disease.

The use of protein as a biomarker is vast, and it has been increasing and maybe, one of the most classical examples is the use of enzymes in the biosensor's development, which are responsible for catalyzing chemical reactions, present in almost all living creatures (Karunakaran et al., 2015; Newman & Setford, 2006; Sassolas et al., 2012). Therefore their properties are very known and several types of enzymes can be used (Bull et al., 2015). One of these examples is the G-protein–coupled receptors (GPCRs), which represent the superfamily of enzymatic receptors (Gomes et al., 2016). The great importance of GPCRs in precision medicine applications is regarding their ability to control the initiation and progression process of tumors (Dorsam & Gutkind, 2007). Furthermore, some GPCRs may be used as prognostic biomarkers in certain diseases, by monitoring different levels to be considered normal for a healthy individual (Suteau et al., 2021).

Another type of protein that is usually investigated as a target in precision medicine is those that have ion channels, more precisely, proteins that have membranes capable of mediating passive ion fluxes (Arcangeli et al., 2009). This characteristic allows the control of their activity between nonconducting conformational states and conductive. Monitoring of protein activity usually is performed by electrical potential difference, generated from the ionic flux, and provides information about the tumor progression (Ouadid-Ahidouch & Ahidouch, 2008). On the other hand, qualitative analysis of ions allows the differentiation between a sample contaminated by influenza A, which has highly selective channel components for K^+ ions, and HIV-1, which has a structure permeable to both Na^+ and K^+ ions (Wang, Yang, et al., 2011).

Kinases are another important type of protein used in precision medicine. This kind of protein plays a crucial role in cellular signaling, catalyzing the transfer of phosphate groups from ATP molecules to other proteins (phosphorylation) (Nordin et al., 2005; Toledo et al., 1999). Many different types of kinases can be found, and each one has unique substrate specificity and cellular behavior. They are involved in the metabolism of several diseases, as well as important targets for drug development and precision medicine. Kinases are widely used in the development of biosensors, generally attached to the sensor surface, and the addition of the inhibitor, in this case, the target, promotes changes in the signal, due to the decrease in kinase activity (Greenwald et al., 2018; Joshi et al., 2022). However, there are works in the literature that use this kind of protein as a target, such as their use as a biomarker in cancer and Alzheimer's disease monitoring (Guo et al., 2019; Lauretti et al., 2020).

Another great and important family of proteins that are used in precision medicine is the protease proteins. They play an important role in human body processes, such as in physiological processes, including digestion, blood clotting, and immune response, and also, they are crucial for the maintenance of normal cellular functions (Canbay & auf dem Keller, 2021; López-Otín & Bond, 2008). They are attributed to regulating the protease activity process; therefore, they are widely used in medicine, mainly in the treatment of diseases caused by excessive protease activity, such as cancer, HIV, and SARS-CoV-2, monitoring the proteases in biological fluids such as blood and saliva (Borberg et al., 2022; Emeje et al., 2019; Feng et al., 2019). Finally, a not so much known protein used in biosensors protein-based is the green fluorescent protein (Claassens et al., 2019; Mita et al., 2019). It consists of a protein with natural fluorescence when exposed to light. The advantage of using this protein, besides the other inherent ones from protein in general, is the nonlabel material needs, which can decrease the cost and the number of assembly steps. Monitoring the levels of calcium ions in a cell is one example of this application (Dana et al., 2019).

There are numerous advantages to biosensors protein-based use, such as low limits of detection; in some cases, these limits can achieve a parts-per-trillion range (Tan et al., 2010). Based on the foregoing, biosensors protein-based are useful for detecting very small quantities of these biomarkers in biological fluids. Devices based on protein biosensors also present the ability to monitor the sample in real time and in loco, depending on the used transductor. Also, they can be used as noninvasive devices in comparison to traditional blood tests, since some biosensors can detect target molecules in breath, sweat, or saliva. Hence, we can consider the protein-based biosensors useful for a wide range of applications, from environmental monitoring to medical diagnostics, using proteins as a target is a versatile methodology in disease monitoring. Also, depending on the analytical technique, protein-based biosensors can offer high sensitivity, specificity, and versatility, making them valuable tools for a wide range of applications.

3.6.2 Genetic materials

Genetic materials are important molecules that, according to almost all biologists, are present in all living creatures, which contain the genetic information required for the development and reproduction of these living organisms (Hiyoshi et al., 2011). There are three forms of genetic material based on long chains of nucleotide: DNA (deoxyribonucleic acid), consisting of four nucleotides: adenine (A), guanine (G), cytosine (C), and thymine (T); RNA (ribonucleic acid), that presents similar in structure to DNA but using uracil (U) instead of thymine (T); and aptamers, that are single-stranded DNA or RNA ligands (Mintz et al., 2019).

The use of genetic materials to identify specific genetic mutations and biomarkers can provide powerful information that helps in early diagnosis, as well as in targeted therapies and personalized treatment for patients (Yamin Wang & Chen, 2020). For example, the use of these materials as a target to identify possible mutations in genes, in genomic profiling, that are associated with cancer, consequently, helping to find a good candidate for treatment with an inhibitor (Xie et al., 2020). Another powerful tool in precision medicine is pharmacogenomics by analysis of genetic materials from a patient to identify possible changes that could give information about how a specific drug is metabolized or interacts with the body (Drozda et al., 2014; Ji et al., 2016). This information can be used to guide the dosing and selection of medications that help in super doses prevention, for example. The circulating tumor also can be identified by analysis of the genetic material, in liquid biopsies for example analyzing samples of body fluids (Alix-Panabières & Pantel, 2013). Considering cancer monitoring, these tests can be classified as a noninvasive methodology to monitor disease progression and treatment response in cancer patients.

In terms of biosensing using genetic materials as targets we can consider many determination strategies, such as PCR, maybe one of the most known (Petralia & Conoci, 2017); clustered regularly interspaced short palindromic repeats (CRISPR) (Uddin et al., 2020); whole genome sequencing (Park & Kim, 2016); and nucleic acid-based biosensors (Bora et al., 2013). Regarding the last application, using genetic material as a target, the detection can take place by either direct (label-free) or indirect (labeling) strategies.

One of the simplest applications is to consider the direct monitoring of the nucleotide signal, which presents electroactivity. For example, the electrochemical monitoring of the oxidation of guanine or adenine in DNA strands (Pedrero et al., 2011; Pividori et al., 2000). The directly genetic materials monitoring, although a simple and interesting monitoring strategy, can present less selectivity since a vast kind of gene fragments could present a similar number of the same nucleotide. However, it is simpler than other strategies and is very useful to elucidate the electrochemical mechanism of oxidation and reduction for nucleotides (Oliveira & Oliveira-Brett, 2010).

Another simple possibility of direct genetic material monitoring can be performed by evaluating the hybridization event without any label material. In this case, when the target genetic material is brought into contact with the sensor surface, already containing the capture species, the target molecules bind to them by extremely affinity interaction (Šípová & Homola, 2013). Depending on the used transductor, this bonding is detected by increasing the current signal or capacitance changes, by the electrochemical application using a redox indicator (Wang, 2002); increasing the refractive index, by surface plasmon resonance (SPR) sensing (Šípová & Homola, 2013); and by increasing of mass using microgravimetric strategy, for example by quartz crystal microbalance (QCM) analysis (Liu, 2019).

Genetic materials hybridization reactions, present high specificity, which can be leveraged to design highly selective and sensitive devices (Bachelin et al., 1998). In general, the hybridization process occurs much more quickly than traditional methods, in just a few minutes or hours, whereas traditional ones, for example, culturing and plating techniques, may take days or even weeks to take place (Flauzino et al., 2022; Hashem et al., 2022). Another advantage in comparison to traditional methods is the use of biosensors based on genetic material as a target presents lower cost than PCR methods (Shariati et al., 2022), for example, making this strategy more suitable for large-scale production, consequently, more accessibility to all of us. Therefore the high specificity to a particular organism or disease, rapid detection, and low cost make genetic material an excellent target for biosensors.

3.6.3 Other molecules

Besides the use of proteins and genetic materials as targets in precision medicine, probably more commonly applied, it is possible to highlight two different biomarkers as targets: extracellular vesicles (EVs) and circulating tumor cells (CTCs). EVs are small tiny particles that are released by cells into the extracellular space, which can contain several molecules and ions, such as proteins, polysaccharides, lipids, and nucleic acids (György et al., 2011; Van der Pol et al., 2012). The contained biological materials inside the EVs can reflect their characteristics, as well as the molecular profile of the cell that released them. Analyzing EVs, or their internal compounds, allows identifying of specific biomarkers for diseases in body fluids, for example for monitoring of cancer (Becker et al., 2016; Kuslich et al., 2010), metabolic syndrome (Martínez & Andriantsitohaina, 2017; Sabe et al., 2023), and sepsis (Hwang et al., 2022; Schiavello et al., 2023). Also, these targets can be used for earlier and more accurate diagnosis, as well as more effective monitoring of disease progression and treatment response, as well as for regulation of immune responses (Robbins & Morelli, 2014). Hence, there are considered promising biomarkers for several diseases (Shao et al., 2018).

Another interesting possibility for EVs use in precision medicine is the fact that they can be engineered to carry specific molecules, such as drugs, and nucleic acids, and targeted to specific cells or tissues (Tian et al., 2022), reducing the risk of side effects, and improving treatment efficacy for drug delivery application. In regenerative medicine, stem cells can differentiate a variety of types of cells; however, in traditional methods, this analysis can be limited by ethical concerns or difficulty in obtaining enough cells (Kirankumar et al., 2022). In this concern, the use of EVs can be an interesting alternative, as EVs are considered potential sources of stem cells for regenerative medicine.

In precision medicine, the monitoring of potential tumor cells is crucial, besides evaluating whether they are spreading and forming new tumors in other parts of the body. These target cells are known as CTCs, in other words, they are cancer cells that have detached from a tumor and entered the bloodstream (Plaks et al., 2013). The importance of these cells as targets consists of the detection of metastasis, helping in an effective treatment and patient outcomes, determining the effectiveness of cancer treatment, for example, if the patient needs chemotherapy or radiation therapy, determining genomic aberrations that is an index of tumors susceptible to treatments, or monitoring the disease progression in real time, helping the adjustment of the treatment (Foster et al., 2017; Rupp et al., 2022).

Traditional methodologies make use of blood analyses to identify these cells, so it could be considered a simpler sampling than other liquid biopsy markers (Rupp et al., 2022). However, it is necessary the use specialized and expensive devices to promote the capture and isolation of CTCs from the blood sample (Paterlini-Brechot, Benali, 2007). At this point, other strategies emerge as alternatives to perform the monitoring of CTCs with lower cost and more simplicity in comparison to the traditional ones, for example, the use of biosensors (Hwang et al., 2022), microfilters (Zheng et al., 2011), or magnetic extraction (Paterlini-Brechot & Benali, 2007).

3.7 Conclusion

As discussed in this chapter, biosensors are widely used in precision medicine, leading to exciting new developments in several fields, especially for medical diagnostics. One of these new trends consists of exploring the use of CRISPR in biosensors development, using it to engineer biomarkers that can detect specific DNA sequences (Kumaran et al., 2023; Tian & Zhou, 2023). The capacity of miniaturization of these devices can permit using biosensors as wearable devices, allowing for continuous monitoring of biomarkers by biosensors worn on the skin or even implanted in the body (Zichen, 2023). In the same way, point-of-care biosensors have been explored, providing rapid, accurate diagnostic data (Khor et al., 2022). Another exciting strategy is the use of multiplexed biosensors for multiple target species monitoring at the same time, which in association with machine learning algorithms can permit the application of biosensors to identify patterns or trends in reliable measures (Gandhi et al., 2022; Zhou et al., 2022).

References

Abu-Salah, K. M., Alrokyan, S. A., Khan, M. N., & Ansari, A. A. (2010). Nanomaterials as analytical tools for genosensors. *Sensors, 10*(1), 963–993. Available from https://doi.org/10.3390/s100100963, http://www.mdpi.com/1424-8220/10/1/963/pdf.

Abubakar Sadique, M., Yadav, S., Ranjan, P., Akram Khan, M., Kumar, A., & Khan, R. (2021). Rapid detection of SARS-CoV-2 using graphene-based IoT integrated advanced electrochemical biosensor. *Materials Letters, 305*. Available from https://doi.org/10.1016/j.matlet.2021.130824.

Ahangar, L. E., & Mehrgardi, M. A. (2017). Amplified detection of hepatitis B virus using an electrochemical DNA biosensor on a nanoporous gold platform. *Bioelectrochemistry (Amsterdam, Netherlands), 117*, 83–88. Available from https://doi.org/10.1016/j.bioelechem.2017.06.006, http://www.elsevier.com/locate/bioelechem.

Akgönüllü, S., & Denizli, A. (2022). (1st ed., pp. 357–371). *Recent advances in plasmonic biosensors for the detection of food allergens encyclopedia of sensors and biosensors*, (Vols. 1–4, pp. 357–371). Turkey: Elsevier. https://www.sciencedirect.com/referencework/9780128225493/encyclopedia-of-sensors-and-biosensors; Available from https://doi.org/10.1016/B978-0-12-822548-6.00143-6.

Alberts,B., Johnson, A., Lewis, J., Walter, P., Raff, M., & Roberts, K. (2002). *Molecular biology of the cell* (4th ed.). International Student Edition. Routledge.

Alireza Hashemi, S., Bahrani, S., Mojtaba Mousavi, S., Omidifar, N., Ghaleh Golab Behbahan, N., Arjmand, M., Ramakrishna, S., Bagheri Lankarani, K., Moghadami, M., Shokripour, M., Firoozsani, M., & Chiang, W.-H. (2021). Ultra-precise label-free nanosensor based on integrated graphene with Au nanostars toward direct detection of IgG antibodies of SARS-CoV-2 in blood. *Journal of Electroanalytical Chemistry, 894*. Available from https://doi.org/10.1016/j.jelechem.2021.115341.

Alix-Panabières, C., & Pantel, K. (2013). Circulating tumor cells: Liquid biopsy of cancer. *Clinical Chemistry, 59*(1), 110–118. Available from https://doi.org/10.1373/clinchem.2012.194258.

Alzate, D., Cajigas, S., Robledo, S., Muskus, C., & Orozco, J. (2020). Genosensors for differential detection of Zika virus. *Talanta, 210*. Available from https://doi.org/10.1016/j.talanta.2019.120648.

Arcangeli, A., Crociani, O., Lastraioli, E., Masi, A., Pillozzi, S., & Becchetti, A. (2009). Targeting ion channels in cancer: A novel frontier in antineoplastic therapy. *Current Medicinal Chemistry, 16*(1), 66–93. Available from https://doi.org/10.2174/092986709787002835.

Arkan, E., Saber, R., Karimi, Z., & Shamsipur, M. (2015). A novel antibody–antigen based impedimetric immunosensor for low level detection of HER2 in serum samples of breast cancer patients via modification of a gold nanoparticles decorated multiwall carbon nanotube-ionic liquid electrode. *Analytica Chimica Acta, 874*, 66–74. Available from https://doi.org/10.1016/j.aca.2015.03.022.

Avrameas, S., Ternynck, T., & Guesdon, J.-L. (1978). Coupling of enzymes to antibodies and antigens. *Scandinavian Journal of Immunology, 8*, 7–23. Available from https://doi.org/10.1111/j.1365-3083.1978.tb03880.x.

Ayankojo, A. G., Reut, J., Öpik, A., Furchner, A., & Syritski, V. (2018). Hybrid molecularly imprinted polymer for amoxicillin detection. *Biosensors and Bioelectronics, 118*, 102–107. Available from https://doi.org/10.1016/j.bios.2018.07.042. http://www.elsevier.com/locate/bios.

Aydin, M., Aydin, E. B., & Sezgintürk, M. K. (2021). Advances in immunosensor technology. *Advances in Clinical Chemistry, 102*, 1–62. Available from https://doi.org/10.1016/bs.acc.2020.08.001. http://www.elsevier.com/wps/find/bookdescription.cws_home/705213/description#description.

Bachelin, M., Hessler, G., Kurz, G., Hacia, J. G., Dervan, P. B., & Kessler, H. (1998). Structure of a stereoregular phosphorothioate DNA/RNA duplex. *Nature Structural Biology, 5*(4), 271–276. Available from https://doi.org/10.1038/nsb0498-271.

Baeumner, A. J., Jones, C., Wong, C. Y., & Price, A. (2004). A generic sandwich-type biosensor with nanomolar detection limits. *Analytical and Bioanalytical Chemistry, 378*(6), 1587–1593. Available from https://doi.org/10.1007/s00216-003-2466-0.

Bauer, J. A., Zámocká, M., Majtán, J., & Bauerová-Hlinková, V. (2022). Glucose oxidase, an enzyme "Ferrari": Its structure, function, production and properties in the light of various industrial and biotechnological applications. *Biomolecules, 12*(3). Available from https://doi.org/10.3390/biom12030472. https://www.mdpi.com/2218-273X/12/3/472/pdf.

Bayat, P., Nosrati, R., Alibolandi, M., Rafatpanah, H., Abnous, K., Khedri, M., & Ramezani, M. (2018). SELEX methods on the road to protein targeting with nucleic acid aptamers. *Biochimie, 154*, 132–155. Available from https://doi.org/10.1016/j.biochi.2018.09.001.

Becker, A., Thakur, B. K., Weiss, J. M., Kim, H. S., Peinado, H., & Lyden, D. (2016). Extracellular vesicles in cancer: Cell-to-cell mediators of metastasis. *Cancer Cell, 30*(6), 836–848. Available from https://doi.org/10.1016/j.ccell.2016.10.009, https://www.journals.elsevier.com/cancer-cell.

Bhardwaj, J., Sharma, A., & Jang, J. (2019). Vertical flow-based paper immunosensor for rapid electrochemical and colorimetric detection of influenza virus using a different pore size sample pad. *Biosensors and Bioelectronics, 126*, 36–43. Available from https://doi.org/10.1016/j.bios.2018.10.008.

Blasques, R. V., de Oliveira, P. R., Kalinke, C., Brazaca, L. C., Crapnell, R. D., Bonacin, J. A., Banks, C. E., & Janegitz, B. C. (2023). Flexible label-free platinum and bio-PET-based immunosensor for the detection of SARS-CoV-2. *Biosensors, 13*(2). Available from https://doi.org/10.3390/bios13020190. http://www.mdpi.com/journal/biosensors/.

Bora, U., Sett, A., & Singh, D. J. B. J. (2013). Nucleic acid based biosensors for clinical applications. *Biosensors Journal, 2*(1), 1–8.

Borberg, E., Granot, E., & Patolsky, F. (2022). Ultrafast one-minute electronic detection of SARS-CoV-2 infection by 3CLpro enzymatic activity in untreated saliva samples. *Nature Communications, 13*(1). Available from https://doi.org/10.1038/s41467-022-34074-2. https://www.nature.com/ncomms/.

Borisov, S. M., & Wolfbeis, O. S. (2008). Optical biosensors. *Chemical Reviews, 108*(2), 423–461. Available from https://doi.org/10.1021/cr068105t.

Brazaca, L. C., Ribovski, L., Janegitz, B. C., & Zucolotto, V. (2017). *Nanostructured materials and nanoparticles for point of care (POC) medical biosensors medical biosensors for point of care (POC) applications* (pp. 229–254). Elsevier Inc., Brazil. http://www.sciencedirect.com/science/book/9780081000724. https://doi.org/10.1016/B978-0-08-100072-4.00010-1.

Brazaca, L. C., Sampaio, I., Zucolotto, V., & Janegitz, B. C. (2020). Applications of biosensors in Alzheimer's disease diagnosis. *Talanta, 210*. Available from https://doi.org/10.1016/j.talanta.2019.120644.

Brazaca, L. C., dos Santos, P. L., de Oliveira, P. R., Rocha, D. P., Stefano, J. S., Kalinke, C., Abarza Muñoz, R. A., Bonacin, J. A., Janegitz, B. C., & Carrilho, E. (2021). Biosensing strategies for the electrochemical detection of viruses and viral diseases – A review.

Analytica Chimica Acta, *1159*. Available from https://doi.org/10.1016/j.aca.2021.338384. http://www.journals.elsevier.com/analytica-chimica-acta/.

Buchner, J., & Rudolph, R. (1991). Routes to active proteins from transformed microorganisms. *Current Opinion in Biotechnology*, *2*(4), 532–538. Available from https://doi.org/10.1016/0958-1669(91)90077-I.

Bull, S. C., Doig, A. J., & Yamanishi, Y. (2015). Properties of protein drug target classes. *PLoS One*, *10*(3). Available from https://doi.org/10.1371/journal.pone.0117955.

Buxbaum, E. (2007). *Fundamentals of protein structure and function fundamentals of protein structure and function* (pp. 1–367). Trinidad and Tobago: Springer US. http://www.springerlink.com/openurl.asp?genre = book&isbn = 978-0-387-26352-6; https://doi.org/10.1007/978-0-387-68480-2.

Cai, J., Gou, X., Sun, B., Li, W., Li, D., Liu, J., Hu, F., & Li, Y. (2019). Porous graphene-black phosphorus nanocomposite modified electrode for detection of leptin. *Biosensors and Bioelectronics*, *137*, 88–95. Available from https://doi.org/10.1016/j.bios.2019.04.045.

Cai, R., Chen, X., Zhang, Y., Wang, X., & Zhou, N. (2023). Systematic bio-fabrication of aptamers and their applications in engineering biology. *Systems Microbiology and Biomanufacturing*, *3*(2), 223–245. Available from https://doi.org/10.1007/s43393-022-00140-5.

Cajigas, S., Alzate, D., & Orozco, J. (2020). Gold nanoparticle/DNA-based nanobioconjugate for electrochemical detection of Zika virus. *Microchimica Acta*, *187*(11). Available from https://doi.org/10.1007/s00604-020-04568-1.

Canbay, V., & auf dem Keller, U. (2021). New strategies to identify protease substrates. *Current Opinion in Chemical Biology*, *60*, 89–96. Available from https://doi.org/10.1016/j.cbpa.2020.09.009. http://www.elsevier.com/locate/cbi.

Cao, C., Kim, J. P., Kim, B. W., Chae, H., Yoon, H. C., Yang, S. S., & Sim, S. J. (2006). A strategy for sensitivity and specificity enhancements in prostate specific antigen-α1-antichymotrypsin detection based on surface plasmon resonance. *Biosensors and Bioelectronics*, *21*(11), 2106–2113. Available from https://doi.org/10.1016/j.bios.2005.10.014.

Capitán-Vallvey, L. F., López-Ruiz, N., Martínez-Olmos, A., Erenas, M. M., & Palma, A. J. (2015). Recent developments in computer vision-based analytical chemistry: A tutorial review. *Analytica Chimica Acta*, *899*, 23–56. Available from https://doi.org/10.1016/j.aca.2015.10.009. http://www.journals.elsevier.com/analytica-chimica-acta/.

Carinelli, S., Kühnemund, M., Nilsson, M., & Pividori, M. I. (2017). Yoctomole electrochemical genosensing of Ebola virus cDNA by rolling circle and circle to circle amplification. *Biosensors and Bioelectronics*, *93*, 65–71. Available from https://doi.org/10.1016/j.bios.2016.09.099.

Carneiro, P., Morais, S., & do Carmo Pereira, M. (2020). Biosensors on the road to early diagnostic and surveillance of Alzheimer's disease. *Talanta*, *211*. Available from https://doi.org/10.1016/j.talanta.2019.120700.

Cassab, G. I. (1998). Plant cell wall proteins. *Annual Review of Plant Biology*, *49*, 281–309. Available from https://doi.org/10.1146/annurev.arplant.49.1.281. http://www.annualreviews.org/journal/arplant.

Chandel, H., Wang, B., & Verma, M. L. (2022). (1st ed., pp. 742–754). *Nanomaterial-based fluorescent biosensors for monitoring environmental pollutants encyclopedia of sensors and biosensors*, (Vols. 1–4, pp. 742–754). India: Elsevier. https://www.sciencedirect.com/referencework/9780128225493/encyclopedia-of-sensors-and-biosensors; https://doi.org/10.1016/B978-0-12-822548-6.00146-1.

Chang, C. C., Chen, C. P., Wu, T. H., Yang, C. H., Lin, C. W., & Chen, C. Y. (2019). Gold nanoparticle-based colorimetric strategies for chemical and biological sensing applications. *Nanomaterials*, *9*(6). Available from https://doi.org/10.3390/nano9060861. https://www.mdpi.com/2079-4991/9/6/861/pdf.

Chatterjee, S., Fujimoto, M. S., Cheng, Y. H., Kargupta, R., Soltis, J. A., Motkuri, R. K., & Basuray, S. (2019). Improving the sensitivity of electrochemical sensors through a complementary luminescent mode: A new spectroelectrochemical approach. *Sensors and Actuators, B: Chemical*, *284*, 663–674. Available from https://doi.org/10.1016/j.snb.2018.10.093. https://www.journals.elsevier.com/sensors-and-actuators-b-chemical.

Chen, C., & Wang, J. (2020). Optical biosensors: An exhaustive and comprehensive review. *Analyst*, *145*(5), 1605–1628. Available from https://doi.org/10.1039/c9an01998g. http://pubs.rsc.org/en/journals/journal/an.

Claassens, N. J., Burgener, S., Vögeli, B., Erb, T. J., & Bar-Even, A. (2019). A critical comparison of cellular and cell-free bioproduction systems. *Current Opinion in Biotechnology*, *60*, 221–229. Available from https://doi.org/10.1016/j.copbio.2019.05.003. http://www.elsevier.com/locate/copbio.

Clark, L. C., & Lyons, C. (1962). Electrode systems for continuous monitoring in cardiovascular surgery. *Annals of the New York Academy of Sciences*, *102*(1), 29–45. Available from https://doi.org/10.1111/j.1749-6632.1962.tb13623.x.

Cooley, M., Chhabra, S. R., & Williams, P. (2008). N-acylhomoserine lactone-mediated quorum sensing: A twist in the tail and a blow for host immunity. *Chemistry & Biology*, *15*(11), 1141–1147. Available from https://doi.org/10.1016/j.chembiol.2008.10.010.

Cosnier, S. (1999). Biomolecule immobilization on electrode surfaces by entrapment or attachment to electrochemically polymerized films. A review. *Biosensors and Bioelectronics*, *14*(5), 443–456. Available from https://doi.org/10.1016/s0956-5663(99)00024-x.

Costa-Rama, E., & Fernández-Abedul, M. T. (2022). *Wearable Physical, Chemical and Biological Sensors* (pp. 71–122). Elsevier Inc.

Cui, F., Zhou, Z., & Zhou, H. S. (2020). Molecularly imprinted polymers and surface imprinted polymers based electrochemical biosensor for infectious diseases. *Sensors*, *20*(4). Available from https://doi.org/10.3390/s20040996.

Damiati, S., Küpcü, S., Peacock, M., Eilenberger, C., Zamzami, M., Qadri, I., Choudhry, H., Sleytr, U. B., & Schuster, B. (2017). Acoustic and hybrid 3D-printed electrochemical biosensors for the real-time immunodetection of liver cancer cells (HepG2). *Biosensors and Bioelectronics*, *94*, 500–506. Available from https://doi.org/10.1016/j.bios.2017.03.045.

Dana, H., Sun, Y., Mohar, B., Hulse, B. K., Kerlin, A. M., Hasseman, J. P., Tsegaye, G., Tsang, A., Wong, A., Patel, R., Macklin, J. J., Chen, Y., Konnerth, A., Jayaraman, V., Looger, L. L., Schreiter, E. R., Svoboda, K., & Kim, D. S. (2019). High-performance calcium sensors for imaging activity in neuronal populations and microcompartments. *Nature Methods*, *16*(7), 649–657. Available from https://doi.org/10.1038/s41592-019-0435-6. http://www.nature.com/nmeth/.

De Carvalho Oliveira, G., Machado, C. C. S., Inácio, D. K., Silveira Petruci, J. Fd, & Silva, S. G. (2022). RGB color sensor for colorimetric determinations: Evaluation and quantitative analysis of colored liquid samples. *Talanta*, *241*. Available from https://doi.org/10.1016/j.talanta.2022.123244. https://www.journals.elsevier.com/talanta.

Diouani, M. F., Helali, S., Hafaid, I., Hassen, W. M., Snoussi, M. A., Ghram, A., Jaffrezic-Renault, N., & Abdelghani, A. (2008). Miniaturized biosensor for avian influenza virus detection. *Materials Science and Engineering: C*, *28*(5-6), 580–583. Available from https://doi.org/10.1016/j.msec.2007.10.043.

Doaga, R., McCormac, T., & Dempsey, E. (2020). Functionalized magnetic nanomaterials for electrochemical biosensing of cholesterol and cholesteryl palmitate. *Microchimica Acta, 187* (4). Available from https://doi.org/10.1007/s00604-020-4203-1. http://www.springer/at/mca.

Dong, J., Zhao, H., Xu, M., Ma, Q., & Ai, S. (2013). A label-free electrochemical impedance immunosensor based on AuNPs/PAMAM-MWCNT-Chi nanocomposite modified glassy carbon electrode for detection of Salmonella typhimurium in milk. *Food Chemistry, 141*(3), 1980–1986. Available from https://doi.org/10.1016/j.foodchem.2013.04.098.

Dorsam, R. T., & Gutkind, J. S. (2007). G-protein-coupled receptors and cancer. *Nature Reviews. Cancer, 7*(2), 79–94. Available from https://doi.org/10.1038/nrc2069.

Doucey, M. A., & Carrara, S. (2019). Nanowire sensors in cancer. *Trends in Biotechnology, 37* (1), 86–99. Available from https://doi.org/10.1016/j.tibtech.2018.07.014. http://www.elsevier.com/locate/tibtech.

Drexelius, A., Yuan, Y., Friedel, M., DeBrosse, M., & Heikenfeld, J. (2022). *Wearable biosensors wearable physical, chemical and biological sensors: Fundamentals, materials and applications* (pp. 235–254). United States: Elsevier. https://www.sciencedirect.com/book/9780128216613; https://doi.org/10.1016/B978-0-12-821661-3.00001-X.

Drozda, K., Müller, D. J., & Bishop, J. R. (2014). Pharmacogenomic testing for neuropsychiatric drugs: Current status of drug labeling, guidelines for using genetic information, and test options. *Pharmacotherapy: The Journal of Human Pharmacology and Drug Therapy, 34*(2), 166–184. Available from https://doi.org/10.1002/phar.1398.

Dudley, Q. M., Anderson, K. C., & Jewett, M. C. (2016). Cell-free mixing of *Escherichia coli* crude extracts to prototype and rationally engineer high-titer mevalonate synthesis. *ACS Synthetic Biology, 5*(12), 1578–1588. Available from https://doi.org/10.1021/acssynbio.6b00154. http://pubs.acs.org/journal/asbcd6.

Eguílaz, M., Villalonga, R., Yáñez-Sedeño, P., & Pingarrón, J. M. (2011). Designing electrochemical interfaces with functionalized magnetic nanoparticles and wrapped carbon nanotubes as platforms for the construction of high-performance bienzyme biosensors. *Analytical Chemistry, 83*(20), 7807–7814. Available from https://doi.org/10.1021/ac201466m.

Eissa, S., L'Hocine, L., Siaj, M., & Zourob, M. (2013). A graphene-based label-free voltammetric immunosensor for sensitive detection of the egg allergen ovalbumin. *Analyst, 138*(15), 4378–4384. Available from https://doi.org/10.1039/c3an36883a. http://pubs.rsc.org/en/journals/journal/an.

El-Said, W. A., & Choi, J. W. (2019). High selective spectroelectrochemical biosensor for HCV-RNA detection based on a specific peptide nucleic acid. *Spectrochimica Acta – Part A: Molecular and Biomolecular Spectroscopy, 217*, 288–293. Available from https://doi.org/10.1016/j.saa.2019.03.115. https://www.journals.elsevier.com/spectrochimica-acta-part-a-molecular-and-biomolecular-spectroscopy.

El-Said, W. A., Al-Bogami, A. S., & Alshitari, W. (2022). Synthesis of gold nanoparticles@reduced porous graphene-modified ITO electrode for spectroelectrochemical detection of SARS-CoV-2 spike protein. *Spectrochimica Acta Part A: Molecular and Biomolecular Spectroscopy, 264*. Available from https://doi.org/10.1016/j.saa.2021.120237.

Ellington, A. D., & Szostak, J. W. (1990). In vitro selection of RNA molecules that bind specific ligands. *Nature, 346*(6287), 818–822. Available from https://doi.org/10.1038/346818a0.

Emeje., Onyenekwe, C. C., Ukibe, N. R., & Ahaneku, J. E. J. A. P. P. (2019). Prospective evaluation of p24 antigen and HIV-1 protease assays at 6 months and 12 months initiation of antiretroviral therapy in HIV infected participants at. *Federal Medical Center, 10*(2), 72–80.

Evtyugin, G. A., & Stoikova, E. E. (2015). Electrochemical biosensors based on dendrimers. *Journal of Analytical Chemistry, 70*(5), 517–534. Available from https://doi.org/10.1134/s1061934815050044.

Felix, F. S., & Angnes, L. (2017). Electrochemical immunosensors—a powerful tool for analytical applications. *Biosensors and Bioelectronics*.

Feng, Y., Li, Q., Chen, J., Yi, P., Xu, X., Fan, Y., Cui, B., Yu, Y., Li, X., Du, Y., Chen, Q., Zhang, L., Jiang, J., Zhou, X., & Zhang, P. (2019). Salivary protease spectrum biomarkers of oral cancer. *International Journal of Oral Science, 11*(1). Available from https://doi.org/10.1038/s41368-018-0032-z.

Figueiredo, A., Vieira, N. C. S., Dos Santos, J. F., Janegitz, B. C., Aoki, S. M., Junior, P. P., Lovato, R. L., Nogueira, M. L., Zucolotto, V., & Guimarães, F. E. G. (2015). Electrical detection of dengue biomarker using egg yolk immunoglobulin as the biological recognition element. *Scientific Reports, 5*. Available from https://doi.org/10.1038/srep07865. http://www.nature.com/srep/index.html.

Flauzino, J. M. R., Nguyen, E. P., Yang, Q., Rosati, G., Panáček, D., Brito-Madurro, A. G., Madurro, J. M., Bakandritsos, A., Otyepka, M., & Merkoçi, A. (2022). Label-free and reagentless electrochemical genosensor based on graphene acid for meat adulteration detection. *Biosensors and Bioelectronics, 195*. Available from https://doi.org/10.1016/j.bios.2021.113628.

Foster, C. W., Down, M. P., Zhang, Y., Ji, X., Rowley-Neale, S. J., Smith, G. C., Kelly, P. J., & Banks, C. E. (2017). 3D printed graphene based energy storage devices. *Scientific Reports, 7*. Available from https://doi.org/10.1038/srep42233. http://www.nature.com/srep/index.html.

Freitas, M., Neves, M. M. P. S., Nouws, H. P. A., & Delerue-Matos, C. (2020). Quantum dots as nanolabels for breast cancer biomarker HER2-ECD analysis in human serum. *Talanta, 208*. Available from https://doi.org/10.1016/j.talanta.2019.120430.

Fu, X., Chen, Z., Ma, W., Zhang, H., Mo, W., Li, J., et al. (2023). Entropy-driven dynamic self-assembled DNA dendrimers for colorimetric detection of African swine fever virus. *Analytical and Bioanalytical Chemistry*, 1–11.

Gandhi, C., Ahmad, S. S., Mehbodniya, A., Webber, J. L., Hemalatha, S., Elwahsh, H., & Tiwari, B. (2022). Biosensor-assisted method for abdominal syndrome classification using machine learning algorithm. *Computational Intelligence and Neuroscience, 2022*. Available from https://doi.org/10.1155/2022/4454226. http://www.hindawi.com/journals/cin.

Gavrilaş, S., Ursachi, C. Ş., Perţa-Crişan, S., & Munteanu, F. D. (2022). Recent trends in biosensors for environmental quality monitoring. *Sensors, 22*(4). Available from https://doi.org/10.3390/s22041513. https://www.mdpi.com/1424-8220/22/4/1513/pdf.

Gogola, J. L., Martins, G., Caetano, F. R., Ricciardi-Jorge, T., Duarte dos Santos, C. N., Marcolino-Junior, L. H., & Bergamini, M. F. (2019). Label-free electrochemical immunosensor for quick detection of anti-hantavirus antibody. *Journal of Electroanalytical Chemistry, 842*, 140–145. Available from https://doi.org/10.1016/j.jelechem.2019.04.066.

Gomes, I., Ayoub, M. A., Fujita, W., Jaeger, W. C., Pfleger, K. D. G., & Devi, L. A. (2016). G protein-coupled receptor heteromers. *Annual Review of Pharmacology and Toxicology, 56*, 403–425. Available from https://doi.org/10.1146/annurev-pharmtox-011613-135952. http://arjournals.annualreviews.org/loi/pharmtox.

Greenwald, E. C., Mehta, S., & Zhang, J. (2018). Genetically encoded fluorescent biosensors illuminate the spatiotemporal regulation of signaling networks. *Chemical Reviews, 118*(24), 11707–11794. Available from https://doi.org/10.1021/acs.chemrev.8b00333. http://pubs.acs.org/journal/chreay.

Guo, J., Dai, X., Laurent, B., Zheng, N., Gan, W., Zhang, J., Guo, A., Yuan, M., Liu, P., Asara, J. M., Toker, A., Shi, Y., Pandolfi, P. P., & Wei, W. (2019). AKT methylation by SETDB1 promotes AKT kinase activity and oncogenic functions. *Nature Cell Biology, 21*(2), 226–237. Available from https://doi.org/10.1038/s41556-018-0261-6.

Gupta, R., & Chaudhury, N. K. (2007). Entrapment of biomolecules in sol–gel matrix for applications in biosensors: Problems and future prospects. *Biosensors and Bioelectronics*, *22*(11), 2387–2399. Available from https://doi.org/10.1016/j.bios.2006.12.025.

György, B., Szabó, T. G., Pásztói, M., Pál, Z., Misják, P., Aradi, B., László, V., Pállinger, E., Pap, E., Kittel, A., Nagy, G., Falus, A., & Buzás, E. I. (2011). Membrane vesicles, current state-of-the-art: Emerging role of extracellular vesicles. *Cellular and Molecular Life Sciences*, *68*(16), 2667–2688. Available from https://doi.org/10.1007/s00018-011-0689-3.

Hansson, E., & Rönnbäck, L. (2003). Glial neuronal signaling in the central nervous system. *The FASEB Journal*, *17*(3), 341–348. Available from https://doi.org/10.1096/fj.02-0429rev.

Hardie, D. G. (2007). AMP-activated protein kinase as a drug target. *Annual Review of Pharmacology and Toxicology*, *47*, 185–210. Available from https://doi.org/10.1146/annurev.pharmtox.47.120505.105304.

Hashem, A., Hossain, M. A. M., Marlinda, A. R., Mamun, M. A., Sagadevan, S., Shahnavaz, Z., Simarani, K., & Johan, M. R. (2022). Nucleic acid-based electrochemical biosensors for rapid clinical diagnosis: Advances, challenges, and opportunities. *Critical Reviews in Clinical Laboratory Sciences*, *59*(3), 156–177. Available from https://doi.org/10.1080/10408363.2021.1997898. http://www.tandfonline.com/loi/ilab20.

He, P., Wang, Z., Zhang, L., & Yang, W. (2009). Development of a label-free electrochemical immunosensor based on carbon nanotube for rapid determination of clenbuterol. *Food Chemistry*, *112*(3), 707–714. Available from https://doi.org/10.1016/j.foodchem.2008.05.116.

Hiyoshi, A., Miyahara, K., Kato, C., & Ohshima, Y. (2011). Does a DNA-less cellular organism exist on Earth? *Genes to Cells*, *16*(12), 1146–1158. Available from https://doi.org/10.1111/j.1365-2443.2011.01558.x. http://onlinelibrary.wiley.com/journal/10.1111/(ISSN)1365-2443.

Hwang, W., Shimizu, M., & Lee, J. W. (2022). Role of extracellular vesicles in severe pneumonia and sepsis. *Expert Opinion on Biological Therapy*, *22*(6), 747–762. Available from https://doi.org/10.1080/14712598.2022.2066470. https://www.tandfonline.com/loi/iebt20.

Hyre, D. E., Le Trong, I., Merritt, E. A., Eccleston, J. F., Green, N. M., Stenkamp, R. E., & Stayton, P. S. (2006). Cooperative hydrogen bond interactions in the streptavidin-biotin system. *Protein Science*, *15*(3), 459–467. Available from https://doi.org/10.1110/ps.051970306. http://www.proteinscience.org/cgi/reprint/15/3/459.

Jang, H. S., Park, K. N., Kang, C. D., Kim, J. P., Sim, S. J., & Lee, K. S. (2009). Optical fiber SPR biosensor with sandwich assay for the detection of prostate specific antigen. *Optics Communications*, *282*(14), 2827–2830. Available from https://doi.org/10.1016/j.optcom.2009.03.078.

Jeon, H., Lee, E., Kim, D., Lee, M., Ryu, J., Kang, C., Kim, S., & Kwon, Y. (2018). Cell-based biosensors based on intein-mediated protein engineering for detection of biologically active signaling molecules. *Analytical Chemistry*, *90*(16), 9779–9786. Available from https://doi.org/10.1021/acs.analchem.8b01481.

Ji, Y., Skierka, J. M., Blommel, J. H., Moore, B. E., Vancuyk, D. L., Bruflat, J. K., Peterson, L. M., Veldhuizen, T. L., Fadra, N., Peterson, S. E., Lagerstedt, S. A., Train, L. J., Baudhuin, L. M., Klee, E. W., Ferber, M. J., Bielinski, S. J., Caraballo, P. J., Weinshilboum, R. M., & Black, J. L. (2016). Preemptive pharmacogenomic testing for precision medicine: A comprehensive analysis of five actionable pharmacogenomic genes using next-generation DNA sequencing and a customized CYP2D6 genotyping cascade. *Journal of Molecular Diagnostics*, *18*(3), 438–445. Available from https://doi.org/10.1016/j.jmoldx.2016.01.003, http://www.sciencedirect.com/science/journal/15251578.

Jiang, P., Wang, Y., Zhao, L., Ji, C., Chen, D., & Nie, L. (2018). Applications of gold nanoparticles in non-optical biosensors. *Nanomaterials*, *8*(12). Available from https://doi.org/10.3390/nano8120977.

Joshi, P. N., Mervinetsky, E., Solomon, O., Chen, Y. J., Yitzchaik, S., & Friedler, A. (2022). Electrochemical biosensors based on peptide-kinase interactions at the kinase docking site. *Biosensors and Bioelectronics, 207.* Available from https://doi.org/10.1016/j.bios.2022.114177. http://www.elsevier.com/locate/bios.

Jugović, B., Grgur, B., Antov, M., Knežević-Jugović, Z., Stevanović, J., & Gvozdenović, M. (2016). Polypyrrole-based enzyme electrode with immobilized glucose oxidase for electrochemical determination of glucose. *International Journal of Electrochemical Science, 11*(2), 1152–1161. http://www.electrochemsci.org/papers/vol11/110201152.pdf.

Kalinke, C., De Oliveira, P. R., Banks, C. E., Janegitz, B. C., & Bonacin, J. A. (2023). 3D-printed immunosensor for the diagnosis of Parkinson's disease. *Sensors and Actuators B: Chemical, 381.* Available from https://doi.org/10.1016/j.snb.2023.133353. https://www.journals.elsevier.com/sensors-and-actuators-b-chemical.

Kalkal, A., Kadian, S., Pradhan, R., Manik, G., & Packirisamy, G. (2021). Recent advances in graphene quantum dot-based optical and electrochemical (bio)analytical sensors. *Materials Advances, 2*(17), 5513–5541. Available from https://doi.org/10.1039/d1ma00251a.

Kalogianni, D. P., Koraki, T., Christopoulos, T. K., & Ioannou, P. C. (2006). Nanoparticle-based DNA biosensor for visual detection of genetically modified organisms. *Biosensors and Bioelectronics, 21*(7), 1069–1076. Available from https://doi.org/10.1016/j.bios.2005.04.016.

Karaman, C., Karaman, O., Atar, N., & Yola, M. L. (2022). A molecularly imprinted electrochemical biosensor based on hierarchical Ti2Nb10O29 (TNO) for glucose detection. *Microchimica Acta, 189*(1). Available from https://doi.org/10.1007/s00604-021-05128-x, http://www.springer/at/mca.

C. Karunakaran, R. Rajkumar, K. Bhargava, Introduction to biosensors biosensors and bioelectronics. Elsevier Inc., India 2015 1–68, http://www.sciencedirect.com/science/book/9780128031001; Available from https://doi.org/10.1016/B978-0-12-803100-1.00001-3.

Karunakaran, C., Madasamy, T., & Sethy, N. K. (2015). *Enzymatic biosensors biosensors and bioelectronics.* Elsevier Inc., India. http://www.sciencedirect.com/science/book/9780128031001; Available from https://doi.org/10.1016/B978-0-12-803100-1.00003-7.

Khor, S. M., Choi, J., Won, P., & Ko, S. H. (2022). Challenges and strategies in developing an enzymatic wearable sweat glucose biosensor as a practical point-of-care monitoring tool for type II diabetes. *Nanomaterials, 12*(2). Available from https://doi.org/10.3390/nano12020221. https://www.mdpi.com/2079-4991/12/2/221/pdf.

Kim, D., & Kang, D. (2008). Molecular recognition and specific interactions for biosensing applications. *Sensors, 8*(10), 6605–6641. Available from https://doi.org/10.3390/s8106605.

Kim, H. J., Ahn, H., Kim, H., Park, D., Lee, J. S., Lee, B. C., Kim, J., Yoon, D. S., & Hwang, K. S. (2022). Nanoparticle-based multiplex biosensor utilising dual dielectrophoretic forces for clinical diagnosis of Alzheimer's disease. *Sensors and Actuators B: Chemical, 355.* Available from https://doi.org/10.1016/j.snb.2021.131288.

Kim, N., Kim, D. K., & Cho, Y. J. (2009). Development of indirect-competitive quartz crystal microbalance immunosensor for C-reactive protein. *Sensors and Actuators, B: Chemical, 143*(1), 444–448. Available from https://doi.org/10.1016/j.snb.2009.09.046.

Kirankumar, S., Gurusamy, N., Rajasingh, S., Sigamani, V., Vasanthan, J., Perales, S. G., & Rajasingh, J. (2022). Modern approaches on stem cells and scaffolding technology for osteogenic differentiation and regeneration. *Experimental Biology and Medicine, 247*(5), 433–445. Available from https://doi.org/10.1177/15353702211052927. https://journals.sagepub.com/home/EBM.

Krisnawati, D. I., Alimansur, M., Atmojo, D. S., Rahmawati, E. Q., Rahayu, D., Susilowati, E., & Kuo, T. R. (2022). Food allergies: Immunosensors and management. *Applied Sciences*

(Switzerland), *12*(5). Available from https://doi.org/10.3390/app12052393. https://www.mdpi.com/2076-3417/12/5/2393/pdf.

Kumaran, A., Jude Serpes, N., Gupta, T., James, A., Sharma, A., Kumar, D., Nagraik, R., Kumar, V., & Pandey, S. (2023). Advancements in CRISPR-based biosensing for next-gen point of care diagnostic application. *Biosensors*, *13*(2). Available from https://doi.org/10.3390/bios13020202.

Kurosawa, S., Aizawa, H., Tozuka, M., Nakamura, M., & Park, J. W. (2003). Immunosensors using a quartz crystal microbalance. *Measurement Science and Technology*, *14*(11), 1882–1887. Available from https://doi.org/10.1088/0957-0233/14/11/005. http://www.iop.org/EJ/journal/0957-0233.

Kuslich, C. D., Pawlowski, T., Kimbrough, J., Deng, T., Tinder, T., Kim, J., et al. (2010). Plasma exosomes are a robust biosignature for prostate cancer. *Cancer Research*, *70* (8_Supplement).

Labib, M., Sargent, E. H., & Kelley, S. O. (2016). Electrochemical methods for the analysis of clinically relevant biomolecules. *Chemical Reviews*, *116*(16), 9001–9090. Available from https://doi.org/10.1021/acs.chemrev.6b00220. http://pubs.acs.org/journal/chreay.

Lahcen, A. A., Rauf, S., Aljedaibi, A., de Oliveira Filho, J. I., Beduk, T., Mani, V., Alshareef, H. N., & Salama, K. N. (2021). Laser-scribed graphene sensor based on gold nanostructures and molecularly imprinted polymers: Application for Her-2 cancer biomarker detection. *Sensors and Actuators B: Chemical*, *347*. Available from https://doi.org/10.1016/j.snb.2021.130556.

Lahcen, A. A., & Amine, A. (2022). Chapter 3: Biorecognition elements. In E. Morales-Narvaez, & C. Dincer (Eds.), *Wearable physical, chemical and biological sensors* (1st ed., pp. 41–70). Elsevier Inc.

Larrieu, A., Champion, A., Legrand, J., Lavenus, J., Mast, D., Brunoud, G., Oh, J., Guyomarc'h, S., Pizot, M., Farmer, E. E., Turnbull, C., Vernoux, T., Bennett, M. J., & Laplaze, L. (2015). A fluorescent hormone biosensor reveals the dynamics of jasmonate signalling in plants. *Nature Communications*, *6*(1). Available from https://doi.org/10.1038/ncomms7043.

Lauretti, E., Dincer, O., & Praticò, D. (2020). Glycogen synthase kinase-3 signaling in Alzheimer's disease. *Biochimica et Biophysica Acta (BBA) - Molecular Cell Research*, *1867* (5). Available from https://doi.org/10.1016/j.bbamcr.2020.118664.

Le Gac, S., Vermes, I., & van den Berg, A. (2006). Quantum dots based probes conjugated to annexin V for photostable apoptosis detection and imaging. *Nano Letters*, *6*(9), 1863–1869. Available from https://doi.org/10.1021/nl060694v.

Lee, J., Ji, J., Hyun, K., Lee, H., & Kwon, Y. (2022). Flexible, disposable, and portable self-powered glucose biosensors visible to the naked eye. *Sensors and Actuators B: Chemical*, *372*. Available from https://doi.org/10.1016/j.snb.2022.132647.

Leva-Bueno, J., Peyman, S. A., & Millner, P. A. (2020). A review on impedimetric immunosensors for pathogen and biomarker detection. *Medical Microbiology and Immunology*, *209*(3), 343–362. Available from https://doi.org/10.1007/s00430-020-00668-0. link.springer.de/link/service/journals/00430/index.htm.

León, L., & Mozo, J. D. (2018). Designing spectroelectrochemical cells: A review. *TrAC Trends in Analytical Chemistry*, *102*, 147–169. Available from https://doi.org/10.1016/j.trac.2018.02.002.

Li, L., Wang, S., Xiao, Y., & Wang, Y. (2020). Recent advances in immobilization strategies for biomolecules in sensors using organic field-effect transistors. *Transactions of Tianjin University*, *26*(6), 424–440. Available from https://doi.org/10.1007/s12209-020-00234-y.

Li, Z., Mohamed, M. A., Vinu Mohan, A. M., Zhu, Z., Sharma, V., Mishra, G. K., & Mishra, R. K. (2019). Application of electrochemical aptasensors toward clinical diagnostics, food, and environmental monitoring: Review. *Sensors (Switzerland), 19*(24). Available from https://doi.org/10.3390/s19245435. https://www.mdpi.com/1424-8220/19/24/5435/pdf.

Light, A., & Smith, El (1963). *Amino acid analysis of peptides and proteins* (pp. 1–44). Elsevier BV. Available from https://doi.org/10.1016/b978-0-12-395723-8.50007-6.

Liu, L. (2019). *Nucleic acid amplification strategies based on QCM nucleic acid amplification strategies for biosensing, bioimaging and biomedicine* (pp. 197–209). Singapore, China: Springer. http://www.springer.com/in/book/9789811370434; Available from https://doi.org/10.1007/978-981-13-7044-1_10.

Liu, S. J., Nie, H. G., Jiang, J. H., Shen, G. L., & Yu, R. Q. (2009). Electrochemical sensor for mercury(II) based on conformational switch mediated by interstrand cooperative coordination. *Analytical Chemistry, 81*(14), 5724–5730. Available from https://doi.org/10.1021/ac900527fChina. http://pubs.acs.org/doi/pdfplus/10.1021/ac900527f.

Lwin, A., & Patel, B. A. (2010). High performance liquid chromatography method for the detection of released purinergic and biogenic amine signaling molecules from in vitro ileum tissue. *Journal of Separation Science, 33*(11), 1538–1545. Available from https://doi.org/10.1002/jssc.200900853. http://www3.interscience.wiley.com/cgi-bin/fulltext/123420386/PDFSTART.

López-Otín, C., & Bond, J. S. (2008). Proteases: Multifunctional enzymes in life and disease. *Journal of Biological Chemistry, 283*(45), 30433–30437. Available from https://doi.org/10.1074/jbc.R800035200. http://www.jbc.org/cgi/reprint/283/45/30433.

Ma, Z., Meliana, C., Munawaroh, H. S. H., Karaman, C., Karimi-Maleh, H., Low, S. S., & Show, P. L. (2022). Recent advances in the analytical strategies of microbial biosensor for detection of pollutants. *Chemosphere, 306*, 135515. Available from https://doi.org/10.1016/j.chemosphere.2022.135515.

Mandli, J., Mohammadi, H., & Amine, A. (2017). Electrochemical DNA sandwich biosensor based on enzyme amplified microRNA-21 detection and gold nanoparticles. *Bioelectrochemistry (Amsterdam, Netherlands), 116*, 17–23. Available from https://doi.org/10.1016/j.bioelechem.2017.03.002. http://www.elsevier.com/locate/bioelechem.

Manzano, M., Viezzi, S., Mazerat, S., Marks, R. S., & Vidic, J. (2018). Rapid and label-free electrochemical DNA biosensor for detecting hepatitis A virus. *Biosensors and Bioelectronics, 100*, 89–95. Available from https://doi.org/10.1016/j.bios.2017.08.043.

Martins, G., Gogola, J. L., Budni, L. H., Janegitz, B. C., Marcolino-Junior, L. H., & Bergamini, M. F. (2021). 3D-printed electrode as a new platform for electrochemical immunosensors for virus detection. *Analytica Chimica Acta, 1147*, 30–37. Available from https://doi.org/10.1016/j.aca.2020.12.014. http://www.journals.elsevier.com/analytica-chimica-acta/.

Martínez, M. C., & Andriantsitohaina, R. (2017). Extracellular vesicles in metabolic syndrome. *Circulation Research, 120*(10), 1674–1686. Available from https://doi.org/10.1161/CIRCRESAHA.117.309419. http://circres.ahajournals.org.

McMahon, R. J. (2008). *Avidin–biotin interactions: Methods and applications*. Springer Science & Business Media.

Michaelis, S., & Barrowman, J. (2012). Biogenesis of the sccharomyces cerevisiae pheromone a-factor, from yeast mating to human disease. *Microbiology and Molecular Biology Reviews, 76*(3), 626–651. Available from https://doi.org/10.1128/MMBR.00010-12. http://mmbr.asm.org/content/76/3/626.full.pdf + html.

Min, J., Nothing, M., Coble, B., Zheng, H., Park, J., Im, H., Weber, G. F., Castro, C. M., Swirski, F. K., Weissleder, R., & Lee, H. (2018). Integrated biosensor for rapid and point-of-care sepsis diagnosis. *ACS Nano, 12*(4), 3378–3384. Available from https://doi.org/10.1021/acsnano.7b08965.

Mintz, K. J., Zhou, Y., & Leblanc, R. M. (2019). Recent development of carbon quantum dots regarding their optical properties, photoluminescence mechanism, and core structure. *Nanoscale*, *11*(11), 4634–4652. Available from https://doi.org/10.1039/c8nr10059d. http://pubs.rsc.org/en/journals/journal/nr.

Mita, M., Ito, M., Harada, K., Sugawara, I., Ueda, H., Tsuboi, T., & Kitaguchi, T. (2019). Green fluorescent protein-based glucose indicators report glucose dynamics in living cells. *Analytical Chemistry*, *91*(7), 4821–4830. Available from https://doi.org/10.1021/acs.analchem.9b00447.

Mobed, A., Razavi, S., Ahmadalipour, A., Shakouri, S. K., & Koohkan, G. (2021). Biosensors in Parkinson's disease. *Clinica Chimica Acta*, *518*, 51–58. Available from https://doi.org/10.1016/j.cca.2021.03.009. http://www.elsevier.com/locate/clinchim.

Mojsoska, B., Larsen, S., Olsen, D. A., Madsen, J. S., Brandslund, I., & Alatraktchi, F. A. 'a (2021). Rapid SARS-CoV-2 detection using electrochemical immunosensor. *Sensors*, *21*(2). Available from https://doi.org/10.3390/s21020390.

Mollarasouli, F., Kurbanoglu, S., & Ozkan, S. A. (2019). The role of electrochemical immunosensors in clinical analysis. *Biosensors*, *9*(3), 86. Available from https://doi.org/10.3390/bios9030086.

Mostafa, I. M., Tian, Y., Anjum, S., Hanif, S., Hosseini, M., Lou, B., & Xu, G. (2022). Comprehensive review on the electrochemical biosensors of different breast cancer biomarkers. *Sensors and Actuators B: Chemical*, *365*. Available from https://doi.org/10.1016/j.snb.2022.131944.

Naresh, V., & Lee, N. (2021). A review on biosensors and recent development of nanostructured materials-enabled biosensors. *Sensors*, *21*(4). Available from https://doi.org/10.3390/s21041109.

Newman, J. D., & Setford, S. J. (2006). Enzymatic biosensors. *Molecular Biotechnology*, *32*(3), 249–268. Available from https://doi.org/10.1385/MB:32:3:249.

Nguyen, H. H., & Kim, M. (2017). An overview of techniques in enzyme immobilization. *Applied Science and Convergence Technology*, *26*(6), 157–163. Available from https://doi.org/10.5757/asct.2017.26.6.157.

Nordin, H., Jungnelius, M., Karlsson, R., & Karlsson, O. P. (2005). Kinetic studies of small molecule interactions with protein kinases using biosensor technology. *Analytical Biochemistry*, *340*(2), 359–368. Available from https://doi.org/10.1016/j.ab.2005.02.027, http://www.elsevier.com/inca/publications/store/6/2/2/7/8/1/index.htt.

Odaci, D., Gacal, B. N., Gacal, B., Timur, S., & Yagci, Y. (2009). Fluorescence sensing of glucose using glucose oxidase modified by PVA-pyrene prepared via "Click" chemistry. *Biomacromolecules*, *10*(10), 2928–2934. Available from https://doi.org/10.1021/bm900755y.

Oh, H. K., Kim, K., Park, J., Im, H., Maher, S., & Kim, M. G. (2022). Plasmon color-preserved gold nanoparticle clusters for high sensitivity detection of SARS-CoV-2 based on lateral flow immunoassay. *Biosensors and Bioelectronics*, *205*. Available from https://doi.org/10.1016/j.bios.2022.114094. http://www.elsevier.com/locate/bios.

Ojeda, I., López-Montero, J., Moreno-Guzmán, M., Janegitz, B. C., González-Cortés, A., Yáñez-Sedeño, P., & Pingarrón, J. M. (2012). Electrochemical immunosensor for rapid and sensitive determination of estradiol. *Analytica Chimica Acta*, *743*, 117–124. Available from https://doi.org/10.1016/j.aca.2012.07.002.

Oliveira, G. C. Md, Carvalho, J. Hd. S., Brazaca, L. C., Vieira, N. C. S., & Janegitz, B. C. (2020). Flexible platinum electrodes as electrochemical sensor and immunosensor for Parkinson's disease biomarkers. *Biosensors and Bioelectronics*, *152*. Available from https://doi.org/10.1016/j.bios.2020.112016. http://www.elsevier.com/locate/bios.

Oliveira, S. C. B., & Oliveira-Brett, A. M. (2010). Boron doped diamond electrode pretreatments effect on the electrochemical oxidation of dsDNA, DNA bases, nucleotides, homopolynucleotides and biomarker 8-oxoguanine. *Journal of Electroanalytical Chemistry, 648*(1), 60–66. Available from https://doi.org/10.1016/j.jelechem.2010.06.020.

Ouadid-Ahidouch, H., & Ahidouch, A. (2008). K + channel expression in human breast cancer cells: Involvement in cell cycle regulation and carcinogenesis. *Journal of Membrane Biology, 221*(1), 1–6. Available from https://doi.org/10.1007/s00232-007-9080-6.

Pacheco, J. G., Rebelo, P., Freitas, M., Nouws, H. P. A., & Delerue-Matos, C. (2018). Breast cancer biomarker (HER2-ECD) detection using a molecularly imprinted electrochemical sensor. *Sensors and Actuators, B: Chemical, 273*, 1008–1014. Available from https://doi.org/10.1016/j.snb.2018.06.113.

Park, S. T., & Kim, J. (2016). Trends in next-generation sequencing and a new era for whole genome sequencing. *International Neurourology Journal, 20*(Suppl 2), S76–S83. Available from https://doi.org/10.5213/inj.1632742.371.

Paterlini-Brechot, P., & Benali, N. L. (2007). Circulating tumor cells (CTC) detection: Clinical impact and future directions. *Cancer Letters, 253*(2), 180–204. Available from https://doi.org/10.1016/j.canlet.2006.12.014. http://www.elsevier.com/locate/canlet.

Pedrero, M., Campuzano, S., & Pingarrón, J. M. (2011). Electrochemical genosensors based on PCR strategies for microorganisms detection and quantification. *Analytical Methods, 3*(4), 780–789,. Available from https://doi.org/10.1039/c0ay00755b.

Peng, L., Li, P., Chen, J., Deng, A., & Li, J. (2023). Recent progress in assembly strategies of nanomaterials-based ultrasensitive electrochemiluminescence biosensors for food safety and disease diagnosis. *Talanta, 253*. Available from https://doi.org/10.1016/j.talanta.2022.123906.

Perbal, B. (2003). Communication is the key. *Cell Communication and Signaling, 1*. Available from https://doi.org/10.1186/1478-811X-1-3France, http://www.biosignaling.com/content/1/1/3.

Pereira, M. V., Datti, E., Alvarenga, G. R., Janegitz, B. C., & Bonacin, J. A. (2022). 3D printing and its applications in spectroelectrochemistry. *Microchemical Journal, 183*. Available from https://doi.org/10.1016/j.microc.2022.108083.

Persad, S., Attwell, S., Gray, V., Delcommenne, M., Troussard, A., Sanghera, J., & Dedhar, S. (2000). Inhibition of integrin-linked kinase (ILK) suppresses activation of protein kinase B/Akt and induces cell cycle arrest and apoptosis of PTEN-mutant prostate cancer cells. *Proceedings of the National Academy of Sciences, 97*(7), 3207–3212. Available from https://doi.org/10.1073/pnas.97.7.3207.

Petralia, S., & Conoci, S. (2017). PCR technologies for point of care testing: Progress and perspectives. *ACS Sensors, 2*(7), 876–891. Available from https://doi.org/10.1021/acssensors.7b00299, http://pubs.acs.org/journal/ascefj.

Pividori, M. I., Merkoçi, A., & Alegret, S. (2000). Electrochemical genosensor design: Immobilisation of oligonucleotides onto transducer surfaces and detection methods. *Biosensors and Bioelectronics, 15*(5–6), 291–303. Available from https://doi.org/10.1016/S0956-5663(00)00071-3.

Plaks, V., Koopman, C. D., & Werb, Z. (2013). Circulating tumor cells. *Science (New York, N. Y.), 341*(6151), 1186–1188. Available from https://doi.org/10.1126/science.1235226, http://www.sciencemag.org/content/341/6151/1186.full.pdf.

Pray, L. (2008). The biotechnology revolution: PCR and the use of reverse transcriptase to clone expressed genes. *Nature Education, 1*.

Qi, H., Peng, Y., Gao, Q., & Zhang, C. (2009). Applications of nanomaterials in electrogenerated chemiluminescence biosensors. *Sensors, 9*(1), 674–695. Available from https://doi.org/10.3390/s90100674China, http://www.mdpi.com/1424-8220/9/1/674/pdf.

Qian, S., Cui, Y., Cai, Z., & Li, L. (2022). Applications of smartphone-based colorimetric biosensors. *Biosensors and Bioelectronics: X, 11*. Available from https://doi.org/10.1016/j.biosx.2022.100173.

Rebelo, T. S. C. R., Miranda, I. M., Brandão, A. T. S. C., Sousa, L. I. G., Ribeiro, J. A., Silva, A. F., & Pereira, C. M. (2021). A disposable saliva electrochemical MIP-based biosensor for detection of the stress biomarker α-amylase in point-of-care applications. *Electrochem, 2*(3), 427–438. Available from https://doi.org/10.3390/electrochem2030028.

Reid, R., Chatterjee, B., Das, S. J., Ghosh, S., & Sharma, T. K. (2020). Application of aptamers as molecular recognition elements in lateral flow assays. *Analytical Biochemistry, 593*. Available from https://doi.org/10.1016/j.ab.2020.113574.

Ribeiro, B. V., Ferreira, L. F., & Franco, D. L. (2022). Advances in biosensor development for the determination of antibiotics in cow's milk – A review. *Talanta Open, 6*. Available from https://doi.org/10.1016/j.talo.2022.100145.

Rios, L., & Garcia, A. A. (2008). Dendrimer based non-competitive fluoroimmunoassay for analysis of cortisol. *Reactive and Functional Polymers, 68*(1), 307–314. Available from https://doi.org/10.1016/j.reactfunctpolym.2007.08.006.

Robbins, P. D., & Morelli, A. E. (2014). Regulation of immune responses by extracellular vesicles. *Nature Reviews. Immunology, 14*(3), 195–208. Available from https://doi.org/10.1038/nri3622.

Roberto de Oliveira, P., Crapnell, R. D., Garcia-Miranda Ferrari, A., Wuamprakhon, P., Hurst, N. J., Dempsey-Hibbert, N. C., Sawangphruk, M., Janegitz, B. C., & Banks, C. E. (2023). Low-cost, facile droplet modification of screen-printed arrays for internally validated electrochemical detection of serum procalcitonin. *Biosensors and Bioelectronics, 228*. Available from https://doi.org/10.1016/j.bios.2023.115220. http://www.elsevier.com/locate/bios.

Rupp, B., Ball, H., Wuchu, F., Nagrath, D., & Nagrath, S. (2022). Circulating tumor cells in precision medicine: Challenges and opportunities. *Trends in Pharmacological Sciences, 43*(5), 378–391. Available from https://doi.org/10.1016/j.tips.2022.02.005. http://www.elsevier.com/locate/tips.

Sabe, S. A., Scrimgeour, L. A., Karbasiafshar, C., Sabra, M., Xu, C. M., Aboulgheit, A., Abid, M. R., & Sellke, F. W. (2023). 5 1 2023/05/01 Journal of Thoracic and Cardiovascular Surgery 10.1016/j.jtcvs.2022.07.016 1097685X 5 e225-e236 Elsevier Inc. United States Extracellular vesicles modulate inflammatory signaling in chronically ischemic myocardium of swine with metabolic syndrome http://www.elsevier.com/inca/publications/store/6/2/3/1/5/1/index.htt165.

Sadani, K., Nag, P., Thian, X. Y., & Mukherji, S. (2022). Enzymatic optical biosensors for healthcare applications. *Biosensors and Bioelectronics: X, 12*. Available from https://doi.org/10.1016/j.biosx.2022.100278.

Samanman, S., Numnuam, A., Limbut, W., Kanatharana, P., & Thavarungkul, P. (2015). Highly-sensitive label-free electrochemical carcinoembryonic antigen immunosensor based on a novel Au nanoparticles–graphene–chitosan nanocomposite cryogel electrode. *Analytica Chimica Acta, 853*(1), 521–532. Available from https://doi.org/10.1016/j.aca.2014.10.006.

Sampson, T. (2003). Aptamers and SELEX: The technology. *World Patent Information, 25*(2), 123–129. Available from https://doi.org/10.1016/s0172-2190(03)00035-8.

Sassolas, A., Blum, L. J., & Leca-Bouvier, B. D. (2012). Immobilization strategies to develop enzymatic biosensors. *Biotechnology Advances, 30*(3), 489–511. Available from https://doi.org/10.1016/j.biotechadv.2011.09.003.

Satija, J., Sai, V. V. R., & Mukherji, S. (2011). Dendrimers in biosensors: Concept and applications. *Journal of Materials Chemistry, 21*(38), 14367–14386,. Available from https://doi.org/10.1039/c1jm10527b, http://www.rsc.org/Publishing/Journals/jm/index.asp.

Sato, K., Takahashi, S., & Anzai, Ji (2012). Layer-by-layer thin films and microcapsules for biosensors and controlled release. *Analytical Sciences, 28*(10), 929–938. Available from https://doi.org/10.2116/analsci.28.929. https://www.jstage.jst.go.jp/article/analsci/28/10/28_929/_pdf.

Saxena, K., Murti, B. T., Yang, P. K., Malhotra, B. D., Chauhan, N., & Jain, U. (2022). Fabrication of a molecularly imprinted nano-interface-based electrochemical biosensor for the detection of CagA virulence factors of *H. pylori*. *Biosensors*, *12*(12). Available from https://doi.org/10.3390/bios12121066. http://www.mdpi.com/journal/biosensors/.

Schiavello, M., Vizio, B., Bosco, O., Pivetta, E., Mariano, F., Montrucchio, G., & Lupia, E. (2023). Extracellular vesicles: New players in the mechanisms of sepsis- and COVID-19-related thromboinflammation. *International Journal of Molecular Sciences*, *24*(3). Available from https://doi.org/10.3390/ijms24031920.

Seo, H. B., & Gu, M. B. (2017). Aptamer-based sandwich-type biosensors. *Journal of Biological Engineering*, *11*(1). Available from https://doi.org/10.1186/s13036-017-0054-7. http://www.jbioleng.org/.

Shao, H., Im, H., Castro, C. M., Breakefield, X., Weissleder, R., & Lee, H. (2018). New technologies for analysis of extracellular vesicles. *Chemical Reviews*, *118*(4), 1917–1950. Available from https://doi.org/10.1021/acs.chemrev.7b00534. http://pubs.acs.org/journal/chreay.

Shariati, M., Sadeghi, M., & Shojaei, S. H. R. (2022). Sensory analysis of hepatitis B virus DNA for medicinal clinical diagnostics based on molybdenum doped ZnO nanowires field effect transistor biosensor; a comparative study to PCR test results. *Analytica Chimica Acta*, *1195*. Available from https://doi.org/10.1016/j.aca.2022.339442.

Shen, J., & Liu, C. C. (2007). Development of a screen-printed cholesterol biosensor: Comparing the performance of gold and platinum as the working electrode material and fabrication using a self-assembly approach. *Sensors and Actuators, B: Chemical*, *120*(2), 417–425. Available from https://doi.org/10.1016/j.snb.2006.02.035.

Shukla, M., Pramila., Palani, I. A., & Singh, V. (2017). Effect of immobilization technique on performance ZnO nanorods based enzymatic electrochemical glucose biosensor. *Journal of Physics: Conference Series*, *924*(1). Available from https://doi.org/10.1088/1742-6596/924/1/012013.

Silva, L. R. G., Stefano, J. S., Orzari, L. O., Brazaca, L. C., Carrilho, E., Marcolino-Junior, L. H., Bergamini, M. F., Munoz, R. A. A., & Janegitz, B. C. (2022). Electrochemical biosensor for SARS-CoV-2 cDNA detection using AuPs-modified 3D-printed graphene electrodes. *Biosensors*, *12*(8). Available from https://doi.org/10.3390/bios12080622. http://www.mdpi.com/journal/biosensors/.

Singh, C., Ali, M. A., Reddy, V., Singh, D., Kim, C. G., Sumana, G., & Malhotra, B. D. (2018). Biofunctionalized graphene oxide wrapped carbon nanotubes enabled microfluidic immunochip for bacterial cells detection. *Sensors and Actuators, B: Chemical*, *255*, 2495–2503. Available from https://doi.org/10.1016/j.snb.2017.09.054.

Singh, P. (2007). Dendrimers and their applications in immunoassays and clinical diagnostics. *Biotechnology and Applied Biochemistry*, *48*(1), 1–9,. Available from https://doi.org/10.1042/BA20070019.

Singh, S., Kumar, V., Dhanjal, D. S., Datta, S., Prasad, R., & Singh, J. (2020). *Biological biosensors for monitoring and diagnosis* (pp. 317–335). Springer Science and Business Media LLC, 10.1007/978-981-15-2817-0_14.

Son, M. H., Park, S. W., Sagong, H. Y., & Jung, Y. K. (2023). Recent advances in electrochemical and optical biosensors for cancer biomarker detection. *Biochip Journal*, *17*(1), 44–67. Available from https://doi.org/10.1007/s13206-022-00089-6. https://www.springer.com/journal/13206.

Soni, G. K., Saima., Manhas, P., & Sharma, R. K. (2022). Peptide-based optical biosensors: A promising approach for early-stage cancer detection. *Biosensors and Bioelectronics: X*, *12*. Available from https://doi.org/10.1016/j.biosx.2022.100259.

Stefano, J. S., Guterres e Silva, L. R., Rocha, R. G., Brazaca, L. C., Richter, E. M., Abarza Muñoz, R. A., & Janegitz, B. C. (2022). New conductive filament ready-to-use for 3D-printing electrochemical (bio)sensors: Towards the detection of SARS-CoV-2. *Analytica Chimica Acta, 1191*. Available from https://doi.org/10.1016/j.aca.2021.339372. http://www.journals.elsevier.com/analytica-chimica-acta/.

Steindler, L., & Venturi, V. (2007). Detection of quorum-sensing *N*-acyl homoserine lactone signal molecules by bacterial biosensors. *FEMS Microbiology Letters, 266*(1), 1–9. Available from https://doi.org/10.1111/j.1574-6968.2006.00501.x.

Strohmaier, H., Spruck, C. H., Kaiser, P., Won, K. A., Sangfelt, O., & Reed, S. I. (2001). Human F-box protein hCdc4 targets cyclin E for proteolysis and is mutated in a breast cancer cell line. *Nature, 413*(6853), 316–322. Available from https://doi.org/10.1038/35095076.

Struss, A., Pasini, P., Ensor, C. M., Raut, N., & Daunert, S. (2010). Paper strip whole cell biosensors: A portable test for the semiquantitative detection of bacterial quorum signaling molecules. *Analytical Chemistry, 82*(11), 4457–4463. Available from https://doi.org/10.1021/ac100231a.

Suteau, V., Seegers, V., Munier, M., Ben Boubaker, R., Reyes, C., Gentien, D., Wery, M., Croué, A., Illouz, F., Hamy, A., Rodien, P., & Briet, C. (2021). G protein–coupled receptors in radioiodine-refractory thyroid cancer in the era of precision medicine. *The Journal of Clinical Endocrinology & Metabolism, 106*(8), 2221–2232. Available from https://doi.org/10.1210/clinem/dgab343.

Šípová, H., & Homola, J. (2013). Surface plasmon resonance sensing of nucleic acids: A review. *Analytica Chimica Acta, 773*, 9–23. Available from https://doi.org/10.1016/j.aca.2012.12.040.

Tan, W., Huang, Y., Nan, T., Xue, C., Li, Z., Zhang, Q., & Wang, B. (2010). Development of protein A functionalized microcantilever immunosensors for the analyses of small molecules at parts per trillion levels. *Analytical Chemistry, 82*(2), 615–620. Available from https://doi.org/10.1021/ac901937g.

Tancharoen, C., Sukjee, W., Thepparit, C., Jaimipuk, T., Auewarakul, P., Thitithanyanont, A., & Sangma, C. (2019). Electrochemical biosensor based on surface imprinting for zika virus detection in serum. *ACS Sensors, 4*(1), 69–75. Available from https://doi.org/10.1021/acssensors.8b00885.

Thenrajan, T., & Wilson, J. (2022). Biosensors for cancer theranostics. *Biosensors and Bioelectronics: X, 12*. Available from https://doi.org/10.1016/j.biosx.2022.100232.

Thévenot, D. R., Toth, K., Durst, R. A., & Wilson, G. S. (2001). Electrochemical biosensors: Recommended definitions and classification. *Biosensors and Bioelectronics, 16*(1–2), 121–131. Available from https://doi.org/10.1016/S0956-5663(01)00115-4.

Tian, F., Liu, C., Deng, J., & Sun, J. (2022). Microfluidic separation, detection, and engineering of extracellular vesicles for cancer diagnostics and drug delivery. *Accounts of Materials Research, 3*(5), 498–510. Available from https://doi.org/10.1021/accountsmr.1c00276, pubs.acs.org/journal/amrcda.

Tian, T., & Zhou, X. (2023). CRISPR-based biosensing strategies: Technical development and application prospects. *Annual Review of Analytical Chemistry, 16*(1). Available from https://doi.org/10.1146/annurev-anchem-090822-014725.

Toledo, L. M., Lydon, N. B., & Elbaum, D. (1999). The structure-based design of ATP-site directed protein kinase inhibitors. *Current Medicinal Chemistry, 6*(9), 775–805.

Tombelli, S., Minunni, M., & Mascini, M. (2007). Aptamers-based assays for diagnostics, environmental and food analysis. *Biomolecular Engineering, 24*(2), 191–200. Available from https://doi.org/10.1016/j.bioeng.2007.03.003.

Triulzi, R. C., Micic, M., Orbulescu, J., Giordani, S., Mueller, B., & Leblanc, R. M. (2008). 1 2008/01 Analyst 10.1039/b718730k 00032654 5 667-672 Royal Society of Chemistry United States Antibody-gold quantum dot-PAMAM dendrimer complex as an immunoglobulin immunoassay http://pubs.rsc.org/en/journals/journal/an133.

Turner, A. P. F., Karube, I., Wilson, G. S., & Worsfold, P. J. (1987). Biosensors: Fundamentals and applications. *Analytica Chimica Acta, 201*, 363–364. Available from https://doi.org/10.1016/s0003-2670(00)85361-1.

Uddin, F., Rudin, C. M., & Sen, T. (2020). CRISPR gene therapy: Applications, limitations, and implications for the future. *Frontiers in Oncology, 10*. Available from https://doi.org/10.3389/fonc.2020.01387. http://www.frontiersin.org/Oncology/about.

Updike, S. J., & Hicks, G. P. (1967). The enzyme electrode. *Nature, 214*(5092), 986–988. Available from https://doi.org/10.1038/214986a0.

Van der Pol, E., Böing, A. N., Harrison, P., Sturk, A., & Nieuwland, R. (2012). Classification, functions, and clinical relevance of extracellular vesicles. *Pharmacological Reviews, 64*(3), 676–705. Available from https://doi.org/10.1124/pr.112.005983. http://pharmrev.aspetjournals.org/content/64/3/676.full.pdf + html.

Vestergaard, M. \'d, Kerman, K., & Tamiya, E. (2007). An overview of label-free electrochemical protein sensors. *Sensors, 7*(12), 3442–3458. Available from https://doi.org/10.3390/s7123442.

Villalonga, A., Sánchez, A., Mayol, B., Reviejo, J., & Villalonga, R. (2022). Electrochemical biosensors for food bioprocess monitoring. *Current Opinion in Food Science, 43*, 18–26. Available from https://doi.org/10.1016/j.cofs.2021.09.006. http://www.journals.elsevier.com/current-opinion-in-food-science/.

Wan, C., Qu, A., Deng, L., Liu, X., & Wu, C. (2022). Preparation of an electrochemical biosensor based on indium tin oxide and its performance in detecting antibiotic resistance genes. *Microchemical Journal, 182*. Available from https://doi.org/10.1016/j.microc.2022.107953.

Wang, B., Barahona, M., & Buck, M. (2013). A modular cell-based biosensor using engineered genetic logic circuits to detect and integrate multiple environmental signals. *Biosensors and Bioelectronics, 40*(1), 368–376. Available from https://doi.org/10.1016/j.bios.2012.08.011.

Wang, H., Ohnuki, H., Endo, H., & Izumi, M. (2012). Effects of self-assembled monolayers on amperometric glucose biosensors based on an organic–inorganic hybrid system. *Sensors and Actuators B: Chemical, 168*, 249–255. Available from https://doi.org/10.1016/j.snb.2012.04.018.

Wang, J. (1999). Amperometric biosensors for clinical and therapeutic drug monitoring: A review. *Journal of Pharmaceutical and Biomedical Analysis, 19*(1–2), 47–53. Available from https://doi.org/10.1016/S0731-7085(98)00056-9. http://www.elsevier.com/locate/jpba.

Wang, J. (2002). Electrochemical nucleic acid biosensors. *Analytica Chimica Acta, 469*(1), 63–71. Available from https://doi.org/10.1016/S0003-2670(01)01399-X.

Wang, K., Xie, S., & Sun, B. (2011). Viral proteins function as ion channels. *Biochimica et Biophysica Acta - Biomembranes, 1808*(2), 510–515. Available from https://doi.org/10.1016/j.bbamem.2010.05.006.

Wang, Q., Yang, L., Yang, X., Wang, K., He, L., & Zhu, J. (2011). Electrochemical biosensors for detection of point mutation based on surface ligation reaction and oligonucleotides modified gold nanoparticles. *Analytica Chimica Acta, 688*(2), 163–167. Available from https://doi.org/10.1016/j.aca.2011.01.004.

Wang, Y., & Chen, Z. (2020). Mutation detection and molecular targeted tumor therapies. *STEMedicine, 1*(1). Available from https://doi.org/10.37175/stemedicine.v1i1.11.

Wells, D. M., Laplaze, L., Bennett, M. J., & Vernoux, T. (2013). Biosensors for phytohormone quantification: Challenges, solutions, and opportunities. *Trends in Plant Science, 18*(5), 244–249. Available from https://doi.org/10.1016/j.tplants.2012.12.005.

Wen, T., Zhu, W., Xue, C., Wu, J., Han, Q., Wang, X., Zhou, X., & Jiang, H. (2014). Novel electrochemical sensing platform based on magnetic field-induced self-assembly of Fe3O4@Polyaniline nanoparticles for clinical detection of creatinine. *Biosensors and Bioelectronics, 56*, 180–185. Available from https://doi.org/10.1016/j.bios.2014.01.013.

Wilson, M. S., & Nie, W. (2006). Multiplex measurement of seven tumor markers using an electrochemical protein chip. *Analytical Chemistry, 78*(18), 6476–6483. Available from https://doi.org/10.1021/ac060843u.

Wilson, R., & Turner, A. P. F. (1992). Glucose oxidase: An ideal enzyme. *Biosensors and Bioelectronics, 7*(3), 165–185. Available from https://doi.org/10.1016/0956-5663(92)87013-f.

Wohlfahrt, G., Trivić, S., Zeremski, J., Peričin, D., & Leskovac, V. (2004). The chemical mechanism of action of glucose oxidase from *Aspergillus niger. Molecular and Cellular Biochemistry, 260*(1), 69–83. Available from https://doi.org/10.1023/B:MCBI.0000026056.75937.98.

Wong, J., Chilkoti, A., & Moy, V. T. (1999). Direct force measurements of the streptavidin-biotin interaction. *Biomolecular Engineering, 16*(1–4), 45–55. Available from https://doi.org/10.1016/S1050-3862(99)00035-2. http://www.elsevier.com/inca/publications/store/5/0/5/7/6/2.

Wu, Y., Zhou, Y., Leng, Y., Lai, W., Huang, X., & Xiong, Y. (2020). Emerging design strategies for constructing multiplex lateral flow test strip sensors. *Biosensors and Bioelectronics, 157*. Available from https://doi.org/10.1016/j.bios.2020.112168.

Wu, Y. X., & Kwon, Y. J. (2016). Aptamers: The "evolution" of SELEX. *Methods (San Diego, Calif.), 106*, 21–28. Available from https://doi.org/10.1016/j.ymeth.2016.04.020. http://www.elsevier.com/inca/publications/store/6/2/2/9/1/4/index.htt.

Xia, F., Zhang, X., Lou, X., & Yuan, Q. (2018). *Biosensors based on sandwich assays* (pp. 1–216). Singapore, China: Springer. http://www.springer.com/in/book/9789811078347; https://doi.org/10.1007/978-981-10-7835-4.

Xie, Y. H., Chen, Y. X., & Fang, J. Y. (2020). Comprehensive review of targeted therapy for colorectal cancer. *Signal Transduction and Targeted Therapy, 5*(1). Available from https://doi.org/10.1038/s41392-020-0116-z. http://www.nature.com/sigtrans.

Xue, Y., Li, X., Li, H., & Zhang, W. (2014). Quantifying thiol–gold interactions towards the efficient strength control. *Nature Communications, 5*(1). Available from https://doi.org/10.1038/ncomms5348.

Yang, G., Liu, E., Khun, N. W., & Jiang, S. P. (2009). Direct electrochemical response of glucose at nickel-doped diamond like carbon thin film electrodes. *Journal of Electroanalytical Chemistry, 627*(1-2), 51–57. Available from https://doi.org/10.1016/j.jelechem.2008.12.019.

Yen, C. W., De Puig, H., Tam, J. O., Gómez-Márquez, J., Bosch, I., Hamad-Schifferli, K., & Gehrke, L. (2015). Multicolored silver nanoparticles for multiplexed disease diagnostics: Distinguishing dengue, yellow fever, and Ebola viruses. *Lab on a Chip, 15*(7), 1638–1641. Available from https://doi.org/10.1039/c5lc00055f. http://pubs.rsc.org/en/journals/journal/lc.

Yu, L., Song, Z., Peng, J., Yang, M., Zhi, H., & He, H. (2020). Progress of gold nanomaterials for colorimetric sensing based on different strategies. *TrAC Trends in Analytical Chemistry, 127*. Available from https://doi.org/10.1016/j.trac.2020.115880.

Zhao, Y., Li, Y., Zhang, P., Yan, Z., Zhou, Y., Du, Y., Qu, C., Song, Y., Zhou, D., Qu, S., & Yang, R. (2021). Cell-based fluorescent microsphere incorporated with carbon dots as a sensitive immunosensor for the rapid detection of Escherichia coli O157 in milk. *Biosensors and Bioelectronics, 179*. Available from https://doi.org/10.1016/j.bios.2021.113057.

Zheng, S., Lin, H. K., Lu, B., Williams, A., Datar, R., Cote, R. J., & Tai, Y. C. (2011). 3D microfilter device for viable circulating tumor cell (CTC) enrichment from blood. *Biomedical Microdevices*, *13*(1), 203–213. Available from https://doi.org/10.1007/s10544-010-9485-3.

Zhou, L., Wang, T., Bai, Y., Li, Y., Qiu, J., Yu, W., & Zhang, S. (2020). Dual-amplified strategy for ultrasensitive electrochemical biosensor based on click chemistry-mediated enzyme-assisted target recycling and functionalized fullerene nanoparticles in the detection of microRNA-141. *Biosensors and Bioelectronics*, *150*. Available from https://doi.org/10.1016/j.bios.2019.111964.

Zhou, Z., Wang, L., Wang, J., Liu, C., Xu, T., & Zhang, X. (2022). Machine learning with neural networks to enhance selectivity of nonenzymatic electrochemical biosensors in multianalyte mixtures. *ACS Applied Materials & Interfaces*, *14*(47), 52684–52690. Available from https://doi.org/10.1021/acsami.2c17593.

Zhu, Y., Zhu, H., Yang, X., Xu, L., & Li, C. (2007). Sensitive biosensors based on (dendrimer encapsulated Pt nanoparticles)/enzyme multilayers. *Electroanalysis*, *19*(6), 698–703. Available from https://doi.org/10.1002/elan.200603802. http://onlinelibrary.wiley.com/journal/10.1002/(ISSN)1521-4109.

Zichen, G. (2023). The implantable and wearable biosensors for blood sugar level detection. *ProcSPIE*.

Zou, X., Wu, J., Gu, J., Shen, L., & Mao, L. (2019). Application of aptamers in virus detection and antiviral therapy. *Frontiers in Microbiology*, *10*(July). Available from https://doi.org/10.3389/fmicb.2019.01462.

Chapter 4

Validation of biosensors

Pedro V.V. Romanholo and Livia F. Sgobbi
Instituto de Química, Universidade Federal de Goiás, Goiânia, GO, Brazil

4.1 Introduction

Biosensors have emerged as a powerful tool for continuous healthcare monitoring to allow disease identification, prevention, rehabilitation, patient health surveillance, and management. Advances in microengineering and the Internet of Things have enabled users to access physiological information anywhere; then, collect this information on mobile phone apps and share it with a doctor or caregivers (Haleem et al., 2021). Biosensors represent a true revolution in personalized healthcare, with an ever-increasing demand for automatized devices able to perform a myriad of clinical analyses. The design of a new sensor needs to combine clinical requirements with analytical performance (Pendley & Lindner, 2017). The presence of or the variation in the analyte concentration must be connected to the disease evolution, and the sensor must be able to determine its concentration precisely in the matrix of interest (e.g., blood, urine, saliva, etc.). The concentration range of a sensor and its accuracy are pivotal parameters for its medical applications. When a new sensor is developed, it is important to assess if it is performing as expected. In this scenario, analytical validation plays a paramount role in sensor applicability. The number of commercial sensors for quantitative analyses applied to the medical field is still scarce and concentrated on glucose detection (Pereira da Silva Neves et al., 2018). This fact contradicts the huge volume of publications involving new sensing platforms. One of the explanations is related to the superficial or inefficient analytical validation of such methods (Bakker, 2018). Typically, the figures of merits, such as limit of detection, selectivity, and measuring concentration range, are provided by most sensor publications. However, the real test consists of applying the sensor to the intended problem, meaning that the new sensor needs to be tested in the concerned complex matrix. Another limitation of biosensor validation encompasses the comparison between the results obtained with the biosensor and those obtained with an established method, a gold standard (Bakker, 2018), which is seldom reported in most publications. During the validation

Biosensors in Precision Medicine. DOI: https://doi.org/10.1016/B978-0-443-15380-8.00005-9

process, researchers commonly collect data from only a few realistic samples to demonstrate the sensor potential for the proposed application (Masson, 2022). The lack of standard validation protocols for the development of biosensors makes reproducibility in different laboratories a hard task, since even through the use of raw matrixes, there is still a current challenge of correlating the results with a gold standard.

So, what is the best way to validate your new sensor? This chapter discusses the main figures of merit related to the analytical validation of sensors for medical analyses, the strategies to achieve a more robust device, issues involving complex matrices, and the main challenges and limitations found in sensor applications for real analyses. Only a critical mindset about sensors' analytical validation will move the field forward. So, bear in mind that the most appropriate way to validate sensors includes a wide range of experiments of unchanged, unspiked samples in addition to the spiked ones, and its correlation with a reference method (Bakker, 2018).

4.2 Analytical validation

When we see the word validation it immediately evokes a perception of confirmation, of a set of actions and procedures that aim at guaranteeing that something is robust and reliable. Beyond this general idea, validation has a much more technical and specialized meaning, translated as a well-established and governed compendium of assessments that must be performed to prove that a method is suitable for its intended use and fulfills the necessary quality requirements for the purpose of acquiring results. In the chemical realm, validation can be employed in many contexts. For instance, the development of a new molecule, or of a new drug, must be scrutinized by a validation process that will study the overall behavior of the substance from cradle to grave in a slew of different conditions (Green, 1996). Also, the development of a new method that brings advantages over a traditional one for pharmaceutical manufacturing must, too, be validated to ensure it will bring out the same results expected from the reference method. The list of possible types of validation approaches goes on and on, but in as much as the topic this chapter is concerned with, this text is going to mainly address the validation course for the development and application of analytical devices called biosensors.

Analytical validation, formally defined as the systematic, and documented tool for the assessment of the capability of an analytical method to return reliable, accurate, and reproducible results, is majorly conducted aligned with the worldwide guidelines by the International Council for Harmonization (ICH), as well as guided by the protocols proposed by other technical bodies such as the Food and Drug Administration (FDA) and the World Health Organization (WHO) (Marson et al., 2020; Swartz & Krull, 2012). Although most of the proposed guidelines apply to analytical methods aimed at drug design and pharmaceutical manufacturing, many analytical procedures are analogous to those employed in the development of biosensors, which may also find technical foundations in the International Union of Pure and Applied Chemistry (IUPAC) chemical protocols (Thompson & Wood, 1993). Briefly recollecting, a biosensor is defined as a sensor that employs a biological recognition element that will carry out a selective reaction with the analyte of interest, being the output signal intensity related to the concentration of the studied substance (Bhalla et al., 2016; Kim et al., 2019; Ronkainen et al., 2010). For more information, please refer to Chapter 3—Biosensors: an introduction. Consequently, the validation of a biosensor must see to all assembly stages, testing protocols, and the quality of the biological elements used in the analytical platform.

When it comes to the validation of biosensors, most of the published literature in the form of articles shows a similar trend for the parameters that are assessed, including linear range, the limit of detection, selectivity, and so on. Investigating such features of a biosensor is of fundamental importance, but it should not be limited to the application of the devices (Justino et al., 2010). For the fully assembled platform to be obtained there are many steps before the finished sensor, raising the question of whether authors should provide the work with some robust analytical insights on the validation of the whole manufacturing process of the biosensor, and not only for the final application of the device. Such a matter should be further addressed by future works that may propose the analytical validation of biosensors aimed at precise medicine, especially because by opting for this course there could be a better comparison among different works that employ similar fabrication methods.

Returning to the validation of the finished biosensor, it is important to pay attention to several aspects. A great number of works devoted to the development of biosensors considerably lack in providing the readers with sufficiently useful figures of merit, assays of target analytes in real samples of unknown composition, and cross-correlation studies against reference techniques employing unmodified matrices (avoiding the use of dilution, centrifugation, and filtering) (Bakker, 2018). In other words, the analytical validation employed for the application of biosensors must rely on figures of merit that are derived from the actual range of the investigated concentrations, as well as obtained originally from the intended matrix. The matrix itself should be analyzed with the least alteration as possible, which will raise the practical impact of the device. Ultimately, the proposed biosensor should be validated against a consolidated analytical method for the target analyte, aiming at certifying the biodevice is valid to be employed as an alternative analytical method. As a guiding summary, we propose, in Fig. 4.1, a flowchart that can be quickly used for the validation of a biosensor that is ready to be applied for the analysis of the desired target analyte.

For biosensors to be successfully incorporated into the thriving objectives of precision medicine, it is of utmost importance that the device is designed

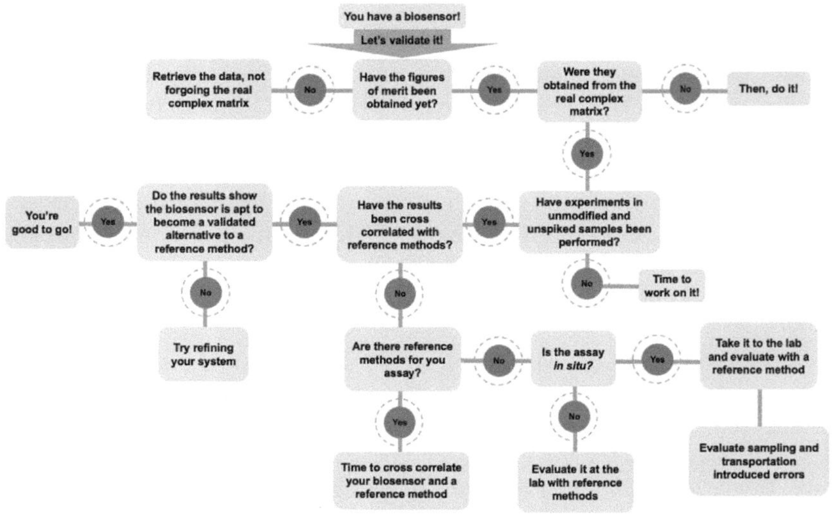

FIGURE 4.1 Flowchart of the proposed analytical validation steps.

and tested aiming to tackle not only markers for simple diseases, but also aid with the diagnosis and the prescription of treatments for complicated and refractory diseases (Shukla et al., 2015). Targeting such complex scenarios requires analytical devices to be able to assimilate broad-spectrum applicability employing single samples, which in turn calls for a whole shift in the paradigms that guide diagnostic assay regulation. The analytical validation of biosensors employing technologies capable of screening a huge number of analytes, including rare, difficult to culture, or hazardous organisms, in a single sample must not simply be extended from the current guidelines (Metzgar et al., 2013), but instead, it needs to accommodate modern concepts of green chemistry, wearability, bioinformatics, and machine learning (Chan et al., 2016; Cui et al., 2020; del Valle, 2021; Kim et al., 2019).

4.3 Validating a biosensor—assessment of its figure of merits

When analytical validation must be put into practice, it is necessary to define the range of parameters that shall be employed for such a goal. There is a wide variety of parameters that can be used, but it will mainly depend on the study design and on the intention of the validation of the biosensor. If the primary goal is to introduce the biosensor to the market, then a wider range of parameters is probably going to be required for the device to be approved by the regulatory agencies through a clinical trial. The whole process involved in a clinical trial can be rather challenging, calling for a thorough study design including defined arms and interventions, measured main and secondary outcomes, and eligibility criteria (Bandodkar et al., 2015; Studying the Effectiveness of Non-Invasive Glucose Sensors in Patients With Diabetes: The SENSOR Study SENSOR, 2020). If, on the other hand, the biosensor is going to be validated only for academic purposes, a more compact set of parameters may suffice. Generally, there is an intrinsic mandatory choice for selectivity, calibration model (or linearity), stability, accuracy, and limits of detection and quantification to be present in the analytical validation of a sensor (Peters & Maurer, 2002; Swartz & Krull, 2012). There are many other parameters and analyses that can be evaluated when a biosensor is under validation scrutiny, such as precision (wherein repeatability, intermediate precision, reproducibility, and ruggedness may be analyzed), experimental statistical design assessment, sensor specification study, and out-of-specification investigation. A detailed description of the most important validation parameters is presented in the following lines.

4.3.1 Calibration model

The most fundamental step when developing a new biosensor is to assess how a given concentration of the target analyte correlates to the output signal employed as the readout of the device. For this purpose, a series of different concentration values must be measured with the biosensor, which in turn must give out the corresponding output signal intensity. There is no definite recommendation on the number of concentration levels that need to be employed while validating a biosensor, but most researchers, professionals, and statutory bodies agree on a minimum of five concentration levels with at least three replicates per level (Peters & Maurer, 2002). In terms of the data acquisition, it is crucial that the calibration curve is retrieved under conditions as closest as possible to the real sample, preferably without using separation, filtration, and dilution steps.

The first step when choosing a mathematical calibration model for the developed biosensor is to always inspect by eye the plotted calibration curve, since the straight application of statistics without prior knowledge of the relationship between signal intensity and concentration values may lead to erroneous interpretation. Such a mistake might happen when using the product-moment correlation coefficient (r), which only informs about linear correlations. If the plotted graph clearly indicates a nonlinear correlation between points, a near zero r value only indicates the lack of linearity but does not indicate a complete lack of correlation. Nonetheless, the great majority of analytes exhibit a linear correlation between studied concentrations and the signal readouts generated by commonly employed transductors. A last word of caution when plotting and inspecting the calibration curve is that the experimentalist must always strive to include the blank signal in the data set, since the absence of the analyte does not remove the other matrix components, such as the solvent and additional substances.

Once a linear correlation has been identified, the experimentalist must choose a calibration method to fit the data at hand. The simple least squares method is frequently the calibration model of choice, which yields a calibration line with the algebraic form of $y = ax + b$. The b term is called intercept and is computed through Eq. (4.1). The a term is known as the slope (or angular coefficient), also used as a figure of merit called sensitivity (Currie, 1995). Sensitivity is defined as the ability of the method to discriminate small differences in concentrations of the analyte (de Mello, 2022). It is important to highlight that this quantity has units of "signal units per unit of concentration" or "signal unit per unit mass," and it must be accompanied by a statement of the linear concentration range associated with it. It is important not to misinterpret the term sensitivity used in diagnostic tests, which refers to the probability of a test yielding a positive result if the tested individual does have the disease (de Mello, 2022). The slope can be mathematically defined as shown in Eq. (4.2). In both equations, i stands for the i-indexed concentration and its respective output signal intensity, whereas $\overline{x}$ and $\overline{y}$ mean is the average of all concentration values and the average of all signal output intensities, respectively.

$$b = \overline{y} - a\overline{x} \tag{4.1}$$

$$a = \frac{\sum_i [(x_i - \overline{x})(y_i - \overline{y})]}{\sum_i (x_i - \overline{x})^2} \tag{4.2}$$

Eqs. (4.1) and (4.2) are valid for a set of data that exhibit constant variance over the experimental range, that is, homoscedastic data. When the calibration curve shows a statistically significant difference between the variance of lower and higher concentrations, it is said that the data are heteroscedastic (Fig. 4.2). The mathematical method used to check for heteroscedasticity is given by creating the residuals plot of the analytical data, where a plot randomly scattered around a residual value of zero is interpreted as a homoscedastic model, and a systematical increase or decrease in the residuals along with an increase in the variable x means the analytical data behave as a heteroscedastic model (Miller & Miller, 2010a, 2010b). This

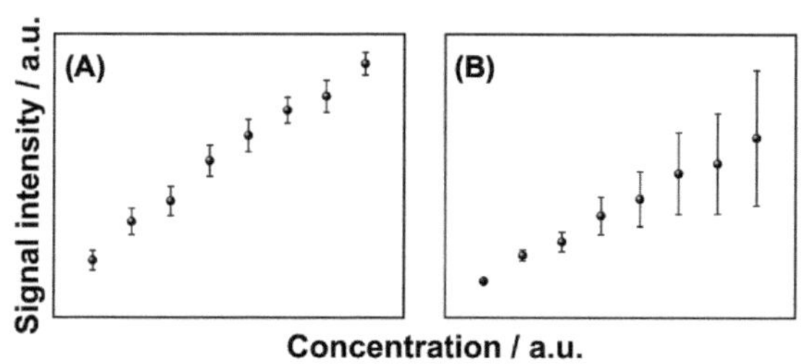

FIGURE 4.2 Calibration plots exemplifying the different types of errors that may arise in the case of (A) homoscedasticity and (B) heteroscedasticity.

latter situation entails an adaptation in the calibration model, which is commonly called a weighted model. Although most works employ an unweighted calibration model, it has become more frequent for researchers to use weighted models and for regulatory agencies to require it to validate a biosensor. The weights employed for this model can be computed as shown in Eq. (4.3), whereas the weighted slope and intercept are calculated through Eqs. (4.4) and (4.5), respectively. The coordinates of the weighted centroid in the weighted calibration model, $\bar{x}_w$ and $\bar{y}_w$, are given by Eqs. (4.6) and (4.7), respectively. Eq. (4.3), s_i stands for the standard deviation of the i-indexed concentration (Long & Ervin, 2000; Miller & Miller, 2010a; 2010b; Onifade & Olanrewaju, 2020).

When testing for heteroscedasticity in a data set, it might be that at times the comparison between the weighted and unweighted models provides very similar results, which may be one of the reasons why it has only more recently been more frequently adopted. However, the weighted model for heteroscedastic data is of fundamental importance when evaluating the errors and the confidence limits of the samples analyzed with a biosensor that is to be validated (Miller & Miller, 2010a; 2010b; Peters & Maurer, 2002). Thus although a bit less practical to use than the unweighted model, the value of the weighted calibration method should not be underestimated. Also, predicting which system is going to follow which model is extremely difficult and unfeasible since multiple methods that on first look may seem similar commonly display very different error behavior. Ultimately, this difficulty leads to a long-term experience that the analyst will have to rely on to interpret if a weighted or unweighted model will be best employed (Miller & Miller, 2010a, 2010b).

$$w_i = \frac{{s_i}^{-2}}{\sum_i {s_i}^{-2}/n} \tag{4.3}$$

$$a_w = \frac{\sum_i w_i x_i y_i - n\bar{x}_w \bar{y}_w}{\sum_i w_i {x_i}^2 - n{\bar{x}_w}^2} \tag{4.4}$$

$$b_w = \bar{y}_w - a\bar{x}_w \tag{4.5}$$

$$\bar{x}_w = \frac{\sum_i w_i x_i}{n} \tag{4.6}$$

$$\bar{y}_w = \frac{\sum_i w_i y_i}{n} \tag{4.7}$$

Frequently, the terms "sensitivity" and "limit of detection" are used in published articles about sensor development interchangeably, not considering their real meanings. It is important to bear in mind that they represent different parameters and cannot be reported in an inconsistent manner (de Mello, 2022).

4.3.2 Limits of detection and quantification

Other validation parameters that must be determined when a new biosensor is proposed are the analytical limits that can be reached with the device. Such limits, measured in concentration units, translate what statistically significant minimum (and maximum) values can be accurately read. The lowest, or highest, values that can be experimentally calculated for the limits of detection of a biosensor are intimately connected to the sensor sensitivity, that is, how quickly the signal output changes due to concentration variations. If the signal output measured at a certain analyte concentration is too low it will blend in with the inherent noise of the platform, rendering the signal nondifferentiating from the background signal. The background signal, or the platform inherent noise, is particular of the sensor build-up process, caused by the materials employed during the fabrication, the circuitry used for the signal capturing and processing, the solvent/environment characteristics, the chemical interaction between the analyte and the transducing element, and other minor considerations (Lavín et al., 2018). In terms of an increased limit of detection, common parameters that directly influence a higher limit include the sensor sampling/reacting area, especially for electrochemical sensors in which a higher geometrical area results in a higher output signal, and the turnover number for an enzyme employed in the sensor, in which the higher the frequency the higher the number of available molecules to react/be detected by the sensor. Thus the first limit that needs to be defined is the limit of detection (LOD). The LOD can be formally defined as the smallest concentration that can be measured with the sensor that is, with the proper confidence level, different from the blank signal (de Mello, 2022; Swartz & Krull, 2012). Establishing a LOD for a sensor implies a qualitative minimum value rather than a quantitative one. For instance, the LOD can be determined, although less accurately, by visual inspection, such as when there is a signal peak when the target analyte is present in the sample, or when there is a color change when a colorimetric approach is used by the device. Although such a strategy can be used for the determination of the LOD, it serves better as a didactic method than a robust analytical validation parameter determination.

Among other more adequate methods to establish the LOD of a sensor, there is the S/N approach (signal/noise), which is well suited for assays that exhibit baseline noise (such as chromatographic methods). However, a more comprehensive way to deal with the LOD is to derive it straight from the calibration model employed for the sensor. There is little meaning to a calculated LOD that does not fall into the analytical range evaluated by the calibration model (Bakker, 2018). This method requires the LOD to be determined using Eq. (4.8), where $y_{@LOD}$ stands for the output signal intensity at the limit of detection, y_B for the blank signal, and S_B the standard deviation of the blank.

$$y_{@LOD} = y_B + 3S_B \tag{4.8}$$

The value of y_B is calculated as the average value of 20–30 blank samples (de Mello, 2022). However, such an approach can be rather morose and time-consuming. Thus an alternative to estimate the value y_B is to use the intercept of the calibration plot, b, as shown in Eq. (4.1). Such a method provides a more accurate estimate for the blank signal than carrying out just a few blank measurements (as opposed to the extensive 20–30 signal acquisitions) (Miller & Miller, 2010a, 2010b). Likewise, the standard deviation of the intercept, S_B, can be used to estimate the standard deviation of the blank. Its value can be computed from Eq. (4.9), where $\hat{y}_i$ is the i-th point calculated from the regression line corresponding to the i-th x value. The first square root term in Eq. (4.9) can be called $S_{y/x}$, which estimates the random errors in the y-direction.

$$S_B = \sqrt{\frac{\sum_i (y_i - \hat{y}_i)^2}{n-2}} \sqrt{\frac{\sum_i x_i^2}{n \sum_i (x_i - \overline{x})^2}} = S_{y/x} \sqrt{\frac{\sum_i x_i^2}{n \sum_i (x_i - \overline{x})^2}} \tag{4.9}$$

Once both y_B and S_B have been estimated, it is possible to use Eq. (4.8) to compute the output signal intensity at the limit of detection, $y_{@LOD}$. Once this value has been calculated, the calibration curve equation can be used to retrieve the concentration at the limit of detection, which is simply known as LOD.

Another analytical limit that is quite handy is the limit of quantification, which translates to the lowest concentration limit that can be quantitatively measured with suitable precision and accuracy (Peters & Maurer, 2002). The limit of quantification (LOQ), also known as the limit of determination, is expressed by Eq. (4.10). As can be noted, the mathematical procedures employed for the determination of the LOD lead straight to that of the LOQ.

$$y_{@LOQ} = y_B + 10S_B \tag{4.10}$$

The approach presented here for the determination of the limits of detection and quantification mainly applies to homoscedastic systems. As previously discussed in the calibration model section, heteroscedastic systems need to be evaluated through adjusted mathematical models, reflecting on how the LOD can be calculated when heteroscedasticity is confirmed. Desimoni and collaborators (Desimoni & Brunetti, 2009) have proposed an evaluation of alternative methods for the estimation of the LOD in heteroscedastic analytical systems, concluding that, for such systems, the LOD may not be enough to establish the performance of an analytical method. It seems that LOD is too much data-dependent and affected by an uncertainty that is never estimated. Therefore it could be problematic to use LOD as a feasible criterion in demonstrating the performance of a method. Instead, the LOQ, including the uncertainty of measurement, could be a better criterion for end-users.

4.3.3 Selectivity

The selectivity parameter comes to aid in a validation protocol by evaluating the matrix interference on the output signal of the sensor. In other words, the selectivity assesses if, and by how much, other substances naturally present in the analytical matrix may interfere with the analyte sensor readout. Formally speaking, a selective sensor is a device that can discriminate the analyte signal among other substances present in the matrix, but that does not necessarily detect only a single analyte. To differentiate such a scenario, there comes into play a second parameter that may be featured by the designed sensor: specificity. A specific sensor is also deemed selective, but differently from the latter, as the former entails the sensor being only capable of detecting one single signal, shunning away its analytical readout to any other substance present in the sample.

When it comes to biosensors, there are many available tools that can be effectively employed to achieve a highly selective, even specific, device. Such elements encompass different types of enzymes (Rocchitta et al., 2016), highly specific antibodies (Sharma et al., 2016), engineerable aptamers (Song et al., 2008), and even the audacious new approaches using modified microorganisms, such as viruses and bacteria, to detect proteins and peptides (Mao et al., 2009). We recommend the excellent short overview published by Peveler and colleagues (Peveler et al., 2016) on the challenges and advantages of employing synthetic arrays in the development of biosensors aiming at high selectivity and specificity.

The evaluation of the selectivity needs to be performed for as many important interferents as possible. An adequate selectivity study must consider what substances shall be present in the matrix the biosensor is designed for, taking into account which of those rules over the others in terms of both analytical and chemical importance. Chemically speaking, the interferents found in the sample can interact with the biorecognition elements mounted on the biosensor, or even with other building pieces present in the sensing platform, leading to a compromised signal readout. Further, the interferents may interact directly with the analyte, leading to poorer detection and quantification. Thus it is of utmost importance that the selectivity study is carried out considering such issues, especially because several works found in the literature employ lab-controlled matrixes with added interferents instead of real matrix samples (Justino et al., 2010). The selection of the interfering molecules must obey an adequate approach in terms of the final application of the sensor, that is, it must include substances that are expected to be present in the real samples, at reasonable concentrations. If there are similar molecules to the studied analyte that are capable of interfering with its adequate detection but are not expected to be present in the real sample, there is no need to carry out a whole validation for that group of interferents once their presence is mostly unlikely. If, on the other hand, there is a known possibility of such a similar group of molecules to be present in the real sample, then the sensor must be validated against such interferents at their real occurring concentration range. Ultimately, if there is enough interference, the sensor must be tuned to avoid the detection of undesired signals. In terms of the real matrix composition, it is virtually impossible to know beforehand its exact compositing when it comes to biological fluids. Nonetheless, this dynamic change of the matrix exact composition does not hinder a validation process to take place. A good practice must include realistic limits of concentration for the real matrix components if the validation goes through artificial fluids, as well as a prior knowledge of the possible chemical interactions that may play a role with the desired analyte. Completely unknown biological matrixes are virtually inexistent nowadays, but in case the validation study comes across such a case there would be a need to thoroughly investigate its composition before the specificity assays are deployed and analyzed. Furthermore, some species, such as inhibitors, activators, or substrates of the enzyme component could affect the biosensor signal by influencing the enzyme's activity, and therefore calibration needs to be performed by taking in account these conditions. Additionally, nonspecific adsorption is a significant issue that negatively compromises biosensors, decreasing their sensitivity, reproducibility, and specificity (Bucur et al., 2021; Lichtenberg et al., 2019).

An adequate selectivity assay should test both the real matrix, without any added substance, and the modified real matrix, where the concentration of the target analytes and that of the most important interferents are carefully studied over a significant range that translates into a real-life scenario in which the biosensor is designed to be used. Graphically wise, the results of a selectivity assay may be presented in the forms shown in Fig. 4.3. Many

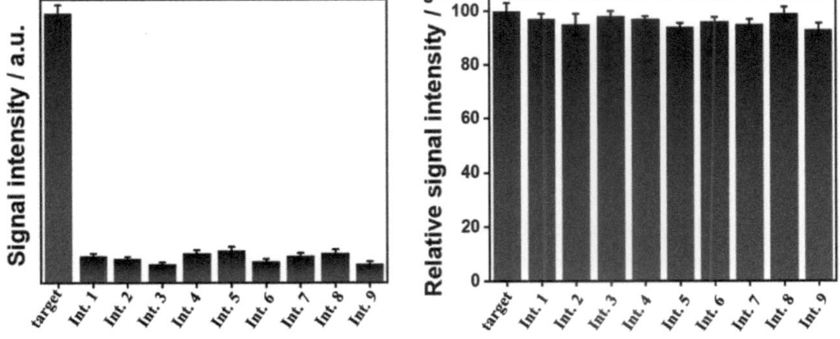

FIGURE 4.3 Possible selectivity study graphical results. (A) The target analyte is tested in the absence of any interferent, and the individual interferents are tested in the absence of the target analyte. (B) The target analyte is tested in the real matrix, where its output signal intensity is compared to that resulting from the individual increase in concentration for each individual interferent.

works carry out the study such as shown in Fig. 4.3A, where the biosensor is tested with multiple samples, each containing either the target analyte or a single interferent at a time. However, such an approach may lead to dubious, or even erroneous, conclusions, since the real matrix will not meet such artificial conditions. Rather, the assay should consider the original matrix, increasing the concentration of the interferents to the highest possible, and reality-matched, value. Such a method is illustrated in Fig. 4.3B, where the output signal intensity of the target analyte (in the original matrix), in the absence of any extra concentration of the studied interferents, is compared to the signal readout when the concentration of each interferent is raised to a determined value.

4.3.4 Robustness

Robustness is the validation parameter that is going to be responsible for the evaluation of the finished biosensor in terms of how well the device holds its operability when small disturbances are applied to it. An important feature of this parameter is that the external perturbations are deliberate, that is, the parameters considered for the robustness assay are intentionally varied in a given range that slightly exceeds the variations expected in a real-life application (Swartz & Krull, 2012). Which parameters should be tested may vary considerably depending on the validation experimental design and on the desired application scenarios, including pH, temperature, the concentration of the matrix components, transportation and storage conditions, manual handling, sampling, and many others.

Robustness is seldom reported in peer-reviewed journals, probably due to the experimental aspect shown in many of the designed biosensors. While plenty of these works end up in projects that validly operate strictly in laboratory-controlled environments, one must be extra careful when proposing a biosensor that aims at hitting the market, or at substituting a current analytical method, especially when designing a device for precision medicine. Analytical platforms that are developed for the monitoring of health status must be meticulously evaluated in terms of robustness since such technologies are majorly intended for use by untrained workers and for point-of-care applications. We invite the reader to visit the excellent reference work published by Heyden and colleagues (Vander Heyden et al., 2001) that proposes a comprehensive guide in setting up and interpreting robustness tests.

4.3.5 Precision

Precision is defined as the closeness of agreement among individual test results from repeated analyses of a homogeneous sample under the prescribed conditions (Peters & Maurer, 2002; Swartz & Krull, 2012). A precision study is usually carried out considering three different levels, namely repeatability, intermediate precision, and reproducibility.

The repeatability level is concerned with whether the sensor is able to produce several signal readouts as close as possible to each other over a short time interval under the same conditions (this is commonly referred to as intraassay precision). Usually, the repeatability is tested with a minimum of nine measurements, preferably spanning the analytical range of concentrations at which the desired analyte is expected to be determined. This can be done, for instance, through testing three, or more, analyte concentrations in triplicates. The test results can be evaluated by means of the relative standard deviation, where values below 2% are desirable (Miller & Miller, 2010a; 2010b; Swartz & Krull, 2012).

At the intermediate precision level, a biosensor is going to be validated in terms of how external and random events may affect its signal readout. Among several different external parameters that may be checked, for a biosensor aimed at precision medicine, different days, analysts, and equipment are some of the most important ones that should be carefully evaluated. Typically, an intermediate precision test involves the sensor assessment when it is operated on different days (usually two or three days), by different workers (two or three analysts). The repeatability should also be checked before an intermediate precision test is carried out, so it should be included in the experimental design. The statistical difference among a given set of days, or workers, is then evaluated through the percent difference between results, commonly followed by a statistical test to check for significance (typically Student's *t*-test) (Swartz & Krull, 2012).

Finally, reproducibility comes as a critical parameter to be evaluated when a biosensor is intended for the market since it needs to be validated through different laboratories. The main reason for this is that the device needs to produce results that do not display significant differences in the output signal when mounted and operated in different laboratory sets (considering that all stages, from cradle to operation are strictly the same in both environments). Similar to the tests conducted for intermediate precision, the reproducibility stage needs at least two analysts to carry out the experiment, but differently from the former, the latter determines that the workers realize the tests in their respective laboratories and that they are not the same individuals picked for the intermediate precision tests. The percent difference should be evaluated for the mean value of the signal readout, as well as statistical hypothesis tests should be employed to test for significant differences (Peters et al., 2007; Swartz & Krull, 2012). Regarding biosensors literature, reproducibility is frequently evaluated by testing the signal readout variation through several distinct devices, especially devices originating from the same batch and those originating from different ones.

Lastly, a new biosensor should be validated against a reference method that is consolidated in the literature—usually a gold-standard method for the determination of the desired analyte. Sometimes, especially in the face of byproducts or new substance detection, there might not be a reference method that can be referred to. However, whenever there is such a possibility, all newly developed biosensors should provide a validation protocol with the statistical correlation between the proposed method and the reference one.

To obtain these data, a range of analyte concentrations should be selected so that it falls within the analytical limits of both the reference method and the proposed biosensor. The number of replicates should be at least 3 but may vary according to the study design. Both methods must then provide the read concentration and be plotted against each other, as shown in Fig. 4.4. Linear regression is then calculated for the resulting graph, indicating how well-matched the two methods are. Also, an *F*-test is usually performed to check if there are any statistically significant differences between the reference method and the new biosensor. A value of $p > 0.05$ (for 95% confidence level) combined with a coefficient of determination (R^2) close to 1 validates the biosensor in terms of the comparison of its performance over that of a gold-standard method.

4.3.6 Stability

The last validation parameter that this chapter is devoted to exploring is stability. This parameter is concerned with the time-dependent stability of a given analyte/substance present in the sensing platform or in the sample medium. Since precision medicine-oriented devices are primarily intended for the monitoring of biomolecules and biosignals, stability is imperative to attest to the validity of a biosensor.

Problems with stability in biosensors are an old challenge since most devices rely on the incorporation of enzymes, antibodies, DNA, and other biomolecules in the biosensor platform. Such biomolecules can be extremely

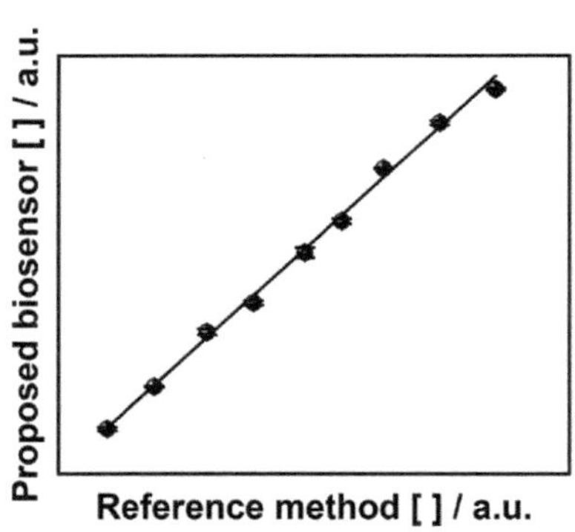

FIGURE 4.4 Example of a statistical correlation between the proposed biosensor (*y*-axis) and a reference method for the same analyte (*x*-axis). The black line represents the linear regression calculated for the data.

sensitive to temperature, pH, and ionic force changes, easily compromising their function in biological recognition. This can come out as a considerable problem, especially when the biosensor is designed for point-of-care and wearable applications, since such environments may bring about considerable fluctuations in these parameters (Gao et al., 2020; Ray et al., 2019). Some recent advances have brought potential solutions for such difficulties, especially with the advances seen in the synthetic chemistry field, wherein a new class of materials has emerged in the last decades, namely biomimetic materials (Hou et al., 2011; Naik & Singamaneni, 2017; Zhang et al., 2017). Such materials have been shown to address the fragile stability found in enzymes and antibodies, allowing biomimetic materials to work over a wider range of pH values and temperatures without compromising their functions. Biomimetic biosensors are a thriving area in analytical chemistry, especially in the development of colorimetric (Wu et al., 2019) and electrochemical biosensors (Romanholo et al., 2021).

Overall, a stability assay must be conducted considering the different stages through which a biosensor is going to be tested. For instance, for a device that intends to reach the market, it needs to assess whether the whole platform is going to produce the same signal readout when transported from the laboratory/industry to the commercialization hub, as well as for how long the output signal remains in acceptable values when stored in different conditions (also known as shelf stability). The stability test, then, should provide the analytical validation evaluator with a precise picture of how long the device keeps its analytical response within acceptable limits and under what circumstances it preserves its functions and characteristics (see Fig. 4.5 for a frequently used template to report stability assay results). Also, the sampling and analysis stages should also be subject to stability assays, since some systems may produce a short-lived stable signal readout that must be optimized to guarantee an adequate validation process (known as operational stability). Finally, when proposing a wearable device tailored at precise medicine applications, one must pay close attention to the operational stability of the device

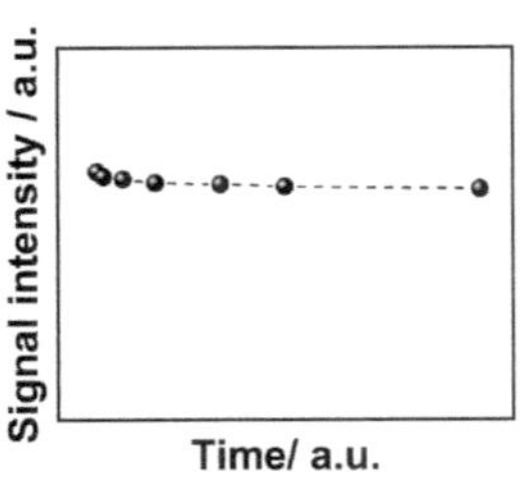

FIGURE 4.5 A common template frequently employed to report the results of a stability assay. The graph depicts how stable the signal readout of the biosensor is throughout a specified timeframe, which can be given in hours, days, weeks, or whichever other range the study finds relevant for the proposed device and application.

in terms of the transducing surface integrity since biofouling can easily affect the signal output during the operation of the sensor. This becomes critical when continuous analyte monitoring is sought after and there must be an adequate surface renewal for extended signal acquisition. In those cases, extra engineering might be needed to guarantee a long-lived surface that will allow continuous measurements without compromising the signal stability (Daboss et al., 2022; Sempionatto et al., 2021).

4.3.7 Recovery assays

Recovery assays are designed to provide the analyst with a ratio between the analyte concentration found in the matrix of interest (C_F) and the theoretical concentration (or the stated concentration, C_T). Mathematically, the recovery ratio (R) is given by Eq. (4.11).

$$R = C_F/C_T \times 100\% \quad (4.11)$$

There are different methods widely employed in the literature for several different types of analytes and matrixes. However, the general terminology should be followed throughout these different studies (without it, confusion with the lack of standardization may arise). First, a recovery assay should be understood as the process of extracting the analyte of interest from the real matrix and further transferring it to a simpler solution employed for instrumental determination. The recovery, then, will be obtained by Eq. (4.11), considering the stated amount of the transferred analyte as C_T and the actual concentration read by instrumental analysis as C_F (Thompson et al., 1999). A common error related to such a strategy is that there is frequent analyte loss during transference procedures, leading to discrepant results. A solution to such an obstacle comes in the form of correcting the quantitative results for recovery (Holst et al., 2002; Thompson et al., 1999). Nonetheless, recovery studies as stated here, as well as any correction that may be needed, can be bypassed when an empirical method is being evaluated. In other words, biosensors in the making can be evaluated with an alternative method before they can be consolidated and then, be subject to a thorough recovery assay (Thompson et al., 1999).

This alternative is known as surrogate recovery assay, in which the pure compound (pure analyte of interest) is added (or spiked) to the matrix under study. The pure substance must match the chemical and physical behaviors of the analyte of interest in the matrix. Thus the theoretical spiked concentration can be evaluated against the measured one employing Eq. (4.11), resulting in the surrogate recovery (Thompson et al., 1999).

Commonly, the results of a recovery assay are presented in the form of a table, such as Table 4.1, indicating the added (or spiked) concentrations of the analyte to the matrix under study (such as a blank matrix, if available), the concentration found by instrumental analysis, the recovered concentration

TABLE 40.1 Table template of recovery assay results as commonly presented in the biosensors literature.

Spiked concentration	Matrix 1			Matrix 2		
	Found	Recovered (%)	RSD (%)	Found	Recovered (%)	RSD (%)
C_{S1}	C_{F1}	R_1	RSD_1	C_{F4}	R_4	RSD_4
C_{S2}	C_{F2}	R_2	RSD_2	C_{F5}	R_5	RSD_5
C_{S3}	C_{F3}	R_3	RSD_3	C_{F6}	R_6	RSD_6

ratio (R), and the relative standard deviation (RSD). The concentrations to be used in the assay must be previously analyzed, since too low, or high, values may lead to biased data (Thompson et al., 1999). Also, sometimes the experimentalist may find it important to study different types of matrixes for the same analyte.

$C_{S:}$ Added analyte concentration
C_F: Concentration found through instrumental analysis
R: Recovery rate
RSD: Relative standard deviation

4.4 Validating with real-world samples

Some crucial points in the development of precise medicine biosensors are the portability, ease of use, low overall costs associated with manufacturing and operation, and desirably, the wearability of the device by the individual to be monitored/screened. To meet such goals entails a sensing platform that can be operated with as few steps as possible, allowing the analysts (or the patients on their own) to carry out a measurement without needing to go through troublesome sample preparation stages, or difficult sample acquisition protocols. A useful biosensor needs to be capable of functioning in the proper way when working with complex samples, such as biological fluids (Bakker, 2018). A biosensor that requires a too strict protocol to be employed for health monitoring may achieve ultra-trace detections or even display extraordinary selectivity, but as long as it needs cumbersome stages for sampling and data processing it will not be adequate for the incorporation in point-of-care applications. Ultimately, the biosensor must achieve a high degree of robustness, once the device will be expected to be handled by unskilled/untrained hands during its use in the field of wearable devices aimed at precise and personalized medicine.

Guaranteeing a biosensor is able to work with minimally processed samples is the job of analytical validation, where the device is going to be tested for several sampling and data processing conditions. But before the biosensor validation is performed, it is necessary that the sensing platform is built upon strategies that make such features possible. In that sense, the major challenge to be tackled when developing an analytical device that allows for unprocessed sampling is surface fouling. This troublesome phenomenon takes place when the transduction surface becomes contaminated with nonspecifically adsorbed substances present in the sample matrix, leading to misleading signal output, increased background noise, and impaired selectivity/biological recognition (Barfidokht & Gooding, 2014; Gooding, 2018). When dealing with biological complex matrixes, such as serum, urine, sweat, and saliva, the biofouling of a sensor is commonly caused by substances present in the medium that is naturally found in much higher concentrations than that of the analyte of interest. Traditional, and commonly employed gold-standard analytical techniques, (such as mass spectrometry, chromatography, and electrophoresis) often employ a series of sample treatments to either remove or reduce the interference caused by these competing substances. But, then again, such strategies hamper untrained and unskilled workers/patients from correctly operating the biosensor. Thus a more tactful approach would be to modify the surface of the device (Jiang et al., 2020), allowing it to bypass any molecule that may pose a threat to an adequate signal readout, or to design sample treatment stages that are inbuilt in the biosensor, avoiding the extra sample processing outside the sensing platform (Gong & Sinton, 2017). Although biofouling plays a major role in the operation of devices under continuous operation, it has a lower impact on single-use biosensors since the latter is not prone to cumulative effect of interferents adsorption.

There are several strategies that can be employed to achieve a biosensor that is resistant to biofouling. Jiang and collaborators (Jiang et al., 2020) have proposed three major categories in which most antifouling strategies can be pooled, namely chemical, physical, and biological. A chemical antifouling strategy relies on the modification of the sensing interface with highly polar/hydrated materials that may prevent the adsorption of unwanted substances onto the sensing surface. Some materials that show strong hydration features are polyethylene-glycol (Castiello & Tabrizian, 2018), zwitterionic phosphorylcholine (He et al., 2008), carboxybetaine (Leng et al., 2015), some peptides based on glutamic or aspartic acid that possess randomly mixed charge (Sakala & Reches, 2018), and some polysaccharides (Liu et al., 2014). The sensor modification with these chemical groups/materials promotes a tightly bound water layer at the sensing interface, which in turn forms a very efficient physical and energetic barrier to the adsorption of proteins (commonly found in abundant concentrations in biofluids) (Chen et al., 2010). The biggest challenge related to these chemical antifouling strategies is the delicate compromise between the optimal performances of the antifouling agents and that of the biorecognition elements (Jiang et al., 2020). Physical strategies especially benefit electrochemical biosensors, where the conductive electrode material can be tuned in terms of porosity, allowing for a diffusional barrier to be formed. Such a barrier acts like a sieve, restricting the passage of larger molecules to the sensor surface while allowing access to smaller analytes to be electrochemically detected (Patel et al., 2013). The surface modification with membrane filters also stands as a very prominent strategy to filter out the electrochemical signal of interferents while imparting a high output signal intensity for the analytes of interest (Wu et al., 2012). Some disadvantages of this approach are related to the analyte size, which works best for very small molecules (Jiang et al., 2020). Finally, the biological approach proposes some interesting alternatives for off-sensor sample treatments. Some approaches employ antibody-functionalized materials that can be used to remove potential proteins that may promote sensor fouling (Kongsuphol et al., 2014), while others employ enzymes that are capable of degrading biomolecules that may cause biofouling (Shi et al., 2011). However, most of the biological strategies rely on off-sensor sample treatment steps, which inserts extra difficulties in the biosensor operation.

Thus designing and manufacturing a device that is capable of surmounting the challenges of biofouling, typically present in real-world samples, is still an ambitious task. Many strategies can be resorted to, allowing the analyst to evaluate the optimal tradeoffs between selectivity and sensitivity. Once the sensing platform is assembled, the analytical validation must be performed in the intended matrix, avoiding the use of artificial laboratory-controlled samples that are enriched with selected interferents.

4.5 Sampling and validating biofluids assays

The bloodstream is one of the richest biofluids that can be sampled to assess and monitor the health of an individual. Containing several molecules that can be used in the active monitoring of the health status of an individual, as well as for the investigation of numerous diseases, this biofluid has been the main choice for bioanalytical assays for decades. Traditionally, blood samples are acquired and processed at large medical facilities, where robust equipment needed for gold-standard analytical methods is readily available for sample treatment and assay deployment. However, the last decades have seen a rise in the development and application of biosensors that are designed for point-of-care purposes, allowing for lab-on-a-chip platforms capable of providing sampling, processing, and measuring the analyte of interest from minute amounts of blood (Labib et al., 2016; Sena-Torralba et al., 2022). Some famous examples of sensors that have made all their way to the market are the platforms aimed at monitoring the levels of glucose, ketone, and lactate in blood, which have opened new venues for new strategies to monitor biomarkers in alternative biofluids at lower costs while building new integrated platforms capable of simultaneously measuring several analytes at once (Moon et al., 2022; Moonla et al., 2023; Teymourian et al., 2020). Although blood provides extremely important data for precise medicine-oriented biosensors, it is not free of challenges. To sample it, invasive techniques must be employed, such as the use of syringes and microneedles. Invasive approaches may lead to several issues, such as bruising, pain, and infections. Also, most devices demand the use of sample treatment stages, which may increase the cost, the time of operation, and the skills level needed to manipulate the device, adding obstacles to the sensor commercialization. To avoid such complications, different types of biofluids may be investigated for biomarkers and other substances that may show a link to the overall health of an individual. Among different types of biofluids, sweat and saliva are some of the least invasive and most readily accessible matrixes that can be sampled for clinical purposes. Nonetheless, they all have their own limitations and obstacles to be surmounted.

When a biosensor is developed and aimed at a noninvasive operation, one must assess for which matrix the device is going to be employed. There are multiple works reported in the literature that investigate the use of such platforms for the sampling of sweat (Chung et al., 2019) and saliva (Malon et al., 2014). Nonetheless, there is still a big lack of validation studies regarding the specificities of each biofluid. Thus the following lines will bring some advice on how to comply with the basics of the analytical validation of a biosensor.

Sweat, for instance, is a physiological mechanism mainly devoted to regulating body temperature. The excretion of sweat happens by means of three different types of glands: eccrine, apoeccrine, and apocrine (Kobielak et al., 2015). Eccrine glands are vastly distributed over the skin, eliminating mainly water and electrolytes helping with body temperature maintenance. Apocrine glands, on the other hand, are only found in regions where hair follicles are present, and they excrete oily sweat containing lipids, proteins, and steroids. Differently from the eccrine glands, the apocrine ones mainly respond to emotional stimuli. Lastly, the apoeccrine glands also eliminate watery sweat, like their eccrine counterpart, but they only appear in the same area of apocrine glands. Thus it becomes apparent that sweat cannot be regarded as a single biological matrix to be sampled. Instead, its sampling must account for the different compositions that may arise due to different types of stimuli. (and location)

When validating a biosensor that uses sweat as a biological matrix one should be aware of the type of mechanism that is going to be employed for the excretion of the sweat to be sampled by the device. If the sweat is thermally produced, it should be noted that its composition is going to be mainly based on water and dissolved electrolytes, whereas if produced by emotional stress it would come out with a totally different composition (Kobielak et al., 2015). Thus carefully choosing the external stimulus is imperative for adequate sampling. Adding to that, the experimentalist must draw the correlation between the concentration of the target analyte in the sweat and its concentration in the bloodstream, which usually is the concentration that can be traced back to a diagnostic concentration range. Most works fail to show such correlation, missing a more thorough approach to how stable the concentrations of the desired analyte through time in the sweat matrix are. Chemically induced sweat (by employing iontophoresis) (Wang et al., 2022; Yang et al., 2022), commonly used in many biosensor approaches, also needs to be carefully validated, once it is locally produced and its composition may vary significantly when compared to naturally occurring sweat. Moreover, the volume of sweat produced by the chosen stimuli, as well as the body part it is going to be sampled from, must be part of the validation protocol, since the resulting volume varies widely regarding both parameters (Nyein et al., 2021; Yuan et al., 2019). Lastly, sweat may not carry the desired analyte, or may carry it at a very different concentration from the one found in the blood, adding an extra obstacle when validating a sweat-based biosensor against a blood-based gold-standard method.

Saliva is another biofluid that can be easily accessed through oral biosensors. Nonetheless, its composition is prone to alteration depending on several factors that range from emotional stimuli to food characteristics. For instance, certain types of food, such as dairy and some fruits, may cause considerable changes in the saliva production rate and in the activity of the alpha-amylase enzyme (Bozorgi et al., 2020; Simões et al., 2021). Some substances that may be present in saliva can also be detected at much higher titters than those usually detected in the blood, which is especially true for amphetamines, while some others may be detected at much lower levels than those found in blood, such as benzodiazepines (Drummer, 2005). Thus biosensors that are designed for sampling the oral cavity must be tested and validated against numerous scenarios, including the right moment of sampling (before, after, or even during food ingestion), the eating and drinking habits of the user, the presence or absence of intraoral objects (such as braces), and the strategy employed for the assay data gathering and processing. Also, the sensor must show stable signal acquisition during operation, which might come out to be challenging due to the frequent movement of the oral cavity structures during most of the daily activities (which is a considerable challenge for wearable sensors). Low susceptibility to biofouling is highly desirable since saliva is rich in different biochemical compounds and the mouth gets in contact with several external substances throughout the day.

Overall, saliva-based biosensors show promising results for precision medicine applications, especially in the prevention of mouth diseases (Wang et al., 2013) and in the monitoring of general biomolecules (Arakawa et al., 2016). However, most works still lack robust validation studies, vastly relying on artificial saliva for the greater part of the analytical studies. Hence, it is imperative that future works provide the biosensors literature with more a realistic and thorough analytical validation treaty.

4.6 Current challenges and outlook

Analytical validation of a biosensor for clinical analyses is a paramount matter regarding its purpose. Improvements in analytical figures of merit have motivated the field to achieve more sensitive and selective devices to perform more accurate and robust tests. However, several publications related to new biosensors present superficial or inefficient analytical validation. The successful performance of a biosensor relies on applying it in biological matrices. Many biosensors are only evaluated in buffer media simulating biological conditions (Pendley & Lindner, 2017). The outcomes obtained in artificial conditions do not always match with real ones. After the development of biosensors on the laboratory scale, it is important to provide further enhancements at the clinical level for real applications (Justino et al., 2010). When a new biosensor is developed, it is expected that its results to be compared with gold-standard methods, which are poorly reported in the literature. If you develop a new biosensor, have in mind that you need to deeply assess it for its intended application.

References

Arakawa, T., Kuroki, Y., Nitta, H., Chouhan, P., Toma, K., Sawada, S.-I., Takeuchi, S., Sekita, T., Akiyoshi, K., Minakuchi, S., & Mitsubayashi, K. (2016). Mouthguard biosensor with telemetry system for monitoring of saliva glucose: A novel cavitas sensor. *Biosensors and Bioelectronics*, 106–111. Available from https://doi.org/10.1016/j.bios.2015.12.014, https://www.sciencedirect.com/science/article/pii/S0956566315306709.

Bakker, E. (2018). So, you have a great new sensor. How will you validate it? *ACS Sensors*, *3* (8), 1431. Available from https://doi.org/10.1021/acssensors.8b00798, https://doi.org/10.1021/acssensors.8b00798.

Bandodkar, A. J., Jia, W., Yardımcı, C., Wang, X., Ramirez, J., & Wang, J. (2015). Tattoo-based noninvasive glucose monitoring: A proof-of-concept study. *Analytical Chemistry*, *87* (1), 394–398. Available from https://doi.org/10.1021/ac504300n, https://doi.org/10.1021/ac504300n.

Barfidokht, A., & Gooding, J. J. (2014). Approaches toward allowing electroanalytical devices to be used in biological fluids. *Electroanalysis*, *26*(6), 1182–1196. Available from https://doi.org/10.1002/elan.201400097.

Bhalla, N., Jolly, P., Formisano, N., & Estrela, P. (2016). Introduction to biosensors. *Essays in Biochemistry*, *60*(1), 1–8. Available from https://doi.org/10.1042/ebc20150001.

Bozorgi, C., Holleufer, C., & Wendin, K. (2020). Saliva secretion and swallowing—The impact of different types of food and drink on subsequent intake. *Nutrients*, *12*(1), 2072–6643. Available from https://doi.org/10.3390/nu12010256.

Bucur, B., Purcarea, C., Andreescu, S., & Vasilescu, A. (2021). Addressing the selectivity of enzyme biosensors: Solutions and perspectives. *Sensors*, *21*(9), 1424–8220. Available from https://doi.org/10.3390/s21093038.

Castiello, F. R., & Tabrizian, M. (2018). Multiplex surface plasmon resonance imaging-based biosensor for human pancreatic islets hormones quantification. *Analytical Chemistry*, *90*(5), 3132–3139. Available from https://doi.org/10.1021/acs.analchem.7b04288.

Chan, S. H., Lee, S. Y., Fang, Q., & Ma, H. (2016). Integration of bioelectronics and bioinformatics: Future direction of bioengineering research. *Journal of Medical and Biological Engineering*, *36*(6), 751–754. Available from https://doi.org/10.1007/s40846-016-0185-1.

Chen, S., Li, L., Zhao, C., & Zheng, J. (2010). Surface hydration: Principles and applications toward low-fouling/nonfouling biomaterials. *Polymer*, *51*(23), 5283–5293. Available from https://doi.org/10.1016/j.polymer.2010.08.022.

Chung, M., Fortunato, G., & Radacsi, N. (2019). Wearable flexible sweat sensors for healthcare monitoring: a review. *Journal of The Royal Society Interface*, *16*(159), 20190217. Available from https://doi.org/10.1098/rsif.2019.0217, https://doi.org/10.1098/rsif.2019.0217.

Cui, F., Yue, Y., Zhang, Y., Zhang, Z., & Zhou, H. S. (2020). Advancing Biosensors with Machine Learning. *ACS Sensors*, *5*(11), 3346–3364. Available from https://doi.org/10.1021/acssensors.0c01424.

Currie, L. A. (1995). Nomenclature in evaluation of analytical methods including detection and quantification capabilities. *Pure Applied Chemistry*, *67*, 1699–1723.

Daboss, E. V., Tikhonov, D. V., Shcherbacheva, E. V., & Karyakin, A. A. (2022). Ultrastable lactate biosensor linearly responding in whole sweat for noninvasive monitoring of hypoxia. *Analytical Chemistry*, *94*(25), 9201–9207. Available from https://doi.org/10.1021/acs.analchem.2c02208, https://doi.org/10.1021/acs.analchem.2c02208.

del Valle, M. (2021). Sensors as green tools in analytical chemistry. *Current Opinion in Green and Sustainable Chemistry*, *31*. Available from https://doi.org/10.1016/j.cogsc.2021.100501.

de Mello, A. J. (2022). I'm sensitive about sensitivity. *ACS Sensors*, *7*(5), 1235–1236. Available from https://doi.org/10.1021/acssensors.2c00982.

Desimoni, E., & Brunetti, B. (2009). About estimating the limit of detection of heteroscedastic analytical systems. *Analytica Chimica Acta*, *655*(1–2), 30–37. Available from https://doi.org/10.1016/j.aca.2009.09.036.

Drummer, O. H. (2005). *Encyclopedia of forensic and legal medicinetoxicology* (pp. 315–321). Oxford: Elsevier.

Gao, Y., Yu, L., Yeo, J. C., & Lim, C. T. (2020). Flexible hybrid sensors for health monitoring: Materials and mechanisms to render wearability. *Advanced Materials*, *32*(15), e1902133. Available from https://doi.org/10.1002/adma.201902133.

Gong, M. M., & Sinton, D. (2017). Turning the page: Advancing paper-based microfluidics for broad diagnostic application. *Chemical Reviews*, *117*(12), 8447–8480. Available from https://doi.org/10.1021/acs.chemrev.7b00024.

Gooding, J. J. (2018). What is a "Real Sample"? *ACS Sensors*, *3*(9), 1609. Available from https://doi.org/10.1021/acssensors.8b00956.

Green, J. M. (1996). Peer reviewed: A practical guide to analytical method validation. *Analytical Chemistry*, *68*(9), 305A–309A. Available from https://doi.org/10.1021/ac961912f.

Haleem, A., Javaid, M., Singh, R. P., Suman, R., & Rab, S. (2021). Biosensors applications in medical field: A brief review. *Sensors International*, *2*, 100100. Available from https://doi.org/10.1016/j.sintl.2021.100100.

He, Y., Hower, J., Chen, S., Bernards, M. T., Chang, Y., & Jiang, S. (2008). Molecular simulation studies of protein interactions with zwitterionic phosphorylcholine self-assembled monolayers in the presence of water. *Langmuir: the ACS Journal of Surfaces and Colloids*, *24*(18), 10358–10364. Available from https://doi.org/10.1021/la8013046.

Holst, C. V., Stroka, J., & Anklam, E. (2002). Correction of analytical results for recovery: a comparison of the method performance characteristics from recent collaborative trials studies for aflatoxin quantification using conventional and robust statistics. *Food Additives & Contaminants*, *19*(8), 701–708. Available from https://doi.org/10.1080/02652030210136982, https://doi.org/10.1080/02652030210136982.

Hou, X., Guo, W., & Jiang, L. (2011). Biomimetic smart nanopores and nanochannels. *Chemical Society Reviews*, *40*(5), 2385–2401. Available from https://doi.org/10.1039/c0cs00053a.

Jiang, C., Wang, G., Hein, R., Liu, N., Luo, X., & Davis, J. J. (2020). Antifouling strategies for selective in vitro and in vivo sensing. *Chemical Reviews*, *120*(8), 3852–3889. Available from https://doi.org/10.1021/acs.chemrev.9b00739.

Justino, C. I. L., Rocha-Santos, T. A., Duarte, A. C., & Rocha-Santos, T. A. (2010). Review of analytical figures of merit of sensors and biosensors in clinical applications. *TrAC Trends in Analytical Chemistry*, *29*(10), 1172–1183. Available from https://doi.org/10.1016/j.trac.2010.07.008.

Kim, J., Campbell, A. S., de Avila, B. E., & Wang, J. (2019). Wearable biosensors for healthcare monitoring. *Nature Biotechnology*, *37*(4), 389–406. Available from https://doi.org/10.1038/s41587-019-0045-y.

Kobielak, J., Kandyba, E., & Leung, Y. (2015). *Translational regenerative medicine chapter 22 – skin and skin appendage regeneration* (pp. 269–292). Boston: Academic Press.

Kongsuphol, P., Ng, H. H., Pursey, J. P., Arya, S. K., Wong, C. C., Stulz, E., & Park, M. K. (2014). EIS-based biosensor for ultra-sensitive detection of TNF-alpha from non-diluted human serum. *Biosensors & Bioelectronics*, *61*, 274–279. Available from https://doi.org/10.1016/j.bios.2014.05.017.

Labib, M., Sargent, E. H., & Kelley, S. O. (2016). Electrochemical methods for the analysis of clinically relevant biomolecules. *Chemical Reviews*, *116*(16), 9001–9090. Available from https://doi.org/10.1021/acs.chemrev.6b00220, https://doi.org/10.1021/acs.chemrev.6b00220.

Lavín, Á., Vicente, J. D., Holgado, M., Laguna, M. F., Casquel, R., Santamaría, B., Maigler, M. V., Hernández, A. L., & Ramírez, Y. (2018). On the determination of uncertainty and limit of detection in label-free biosensors. *Sensors*, *18*(7), 1424–8220. Available from https://doi.org/10.3390/s18072038.

Leng, C., Hung, H.-C., Sieggreen, O. A., Li, Y., Jiang, S., & Chen, Z. (2015). Probing the surface hydration of nonfouling zwitterionic and poly(ethylene glycol) materials with isotopic dilution spectroscopy. *The Journal of Physical Chemistry C*, *119*(16), 8775–8780. Available from https://doi.org/10.1021/acs.jpcc.5b01649.

Lichtenberg, J. Y., Ling, Y., & Kim, S. (2019). Non-specific adsorption reduction methods in biosensing. *Sensors*, *19*(11), 1424–8220. Available from https://doi.org/10.3390/s19112488.

Liu, X., Huang, R., Su, R., Qi, W., Wang, L., & He, Z. (2014). Grafting hyaluronic acid onto gold surface to achieve low protein fouling in surface plasmon resonance biosensors. *ACS Applied Materials Interfaces*, *6*(15), 13034–13042. Available from https://doi.org/10.1021/am502921z.

Long, J. S., & Ervin, L. H. (2000). Using heteroscedasticity consistent standard errors in the linear regression model. *The American Statistician*, *54*(3), 217–224. Available from https://doi.org/10.2307/2685594, http://www.jstor.org/stable/2685594.

Malon, R. S. P., Sadir, S., Balakrishnan, M., & Córcoles, E. P. (2014). Saliva-based biosensors: Noninvasive monitoring tool for clinical diagnostics. *BioMed Research International*, *2014*, 20.

Mao, C., Liu, A., & Cao, B. (2009). Virus-based chemical and biological sensing. *Angewandte Chemie (International Ed. in English)*, *48*(37), 6790–6810. Available from https://doi.org/10.1002/anie.200900231.

Marson, B. M., Concentino, V., Junkert, A. M., Fachi, M. M., Vilhena, R. O., & Pontarolo, R. (2020). Validation of analytical methods in a pharmaceutical quality system: an overview focused on hplc methods. *Química Nova*, *43*(8), 1190–1203. Available from https://doi.org/10.21577/0100-4042.20170589.

Masson, J.-F. (2022). Lessons learned from the validation of sensors with large cohorts of samples. *ACS Sensors*, *7*(11), 3225–3227. Available from https://doi.org/10.1021/acssensors.2c02301, https://doi.org/10.1021/acssensors.2c02301.

Metzgar, D., Sampath, R., Rounds, M. A., & Ecker, D. J. (2013). The value and validation of broad spectrum biosensors for diagnosis and biodefense. *Virulence*, *4*(8), 752–758. Available from https://doi.org/10.4161/viru.26652.

Miller, J. N., & Miller, J. C. (2010a). *Statistics and chemometrics for analytical chemistry* (6th edn). Harlow: Pearson.

Miller, J. N., & Miller, J. C. (2010b). *Calibration methods in instrumental analysis: Regression and correlation. Statistics and Chemometrics for Analytical Chemistry* (pp. 110–153). England: Pearson.

Moon, J.-M., Del Caño, R., Moonla, C., Sakdaphetsiri, K., Saha, T., Francine Mendes, L., Yin, L., Chang, A.-Y., Seker, S., & Wang, J. (2022). Self-testing of ketone bodies, along with glucose, using touch-based sweat analysis. *ACS Sensors*, *7*(12), 3973–3981. Available from https://doi.org/10.1021/acssensors.2c02369, https://doi.org/10.1021/acssensors.2c02369.

Moonla, C., Del Caño, R., Sakdaphetsiri, K., Saha, T., De la Paz, E., Düsterloh, A., & Wang, J. (2023). Disposable screen-printed electrochemical sensing strips for rapid decentralized measurements of salivary ketone bodies: Towards therapeutic and wellness applications. *Biosensors and Bioelectronics*, *220*, 114891. Available from https://doi.org/10.1016/j.bios.2022.114891, https://www.sciencedirect.com/science/article/pii/S0956566322009319.

Naik, R. R., & Singamaneni, S. (2017). Introduction: Bioinspired and biomimetic materials. *Chemical Reviews*, *117*(20), 12581–12583. Available from https://doi.org/10.1021/acs.chemrev.7b00552.

Nyein, H. Y. Y., Bariya, M., Tran, B., Ahn, C. H., Brown, B. J., Ji, W., Davis, N., & Javey, A. (2021). A wearable patch for continuous analysis of thermoregulatory sweat at rest. *Nature Communications*, *12*(1), 1823. Available from https://doi.org/10.1038/s41467-021-22109-z, https://doi.org/10.1038/s41467-021-22109-z.

Onifade, O. C., & Olanrewaju, S. O. (2020). Investigating performances of some statistical tests for heteroscedasticity assumption in generalized linear model: A monte carlo simulations study. *Open Journal of Statistics*, *10*(3), 453–493.

Patel, J., Radhakrishnan, L., Zhao, B., Uppalapati, B., Daniels, R. C., Ward, K. R., & Collinson, M. M. (2013). Electrochemical properties of nanostructured porous gold electrodes in biofouling solutions. *Analytical Chemistry*, *85*(23), 11610–11618. Available from https://doi.org/10.1021/ac403013r.

Pendley, B. D., & Lindner, E. (2017). Medical sensors for the diagnosis and management of disease: The physician perspective. *ACS Sensors*, *2*(11), 1549–1552. Available from https://doi.org/10.1021/acssensors.7b00642.

Pereira da Silva Neves, M. M., González-García, M. B., Hernández-Santos, D., & Fanjul-Bolado, P. (2018). Future trends in the market for electrochemical biosensing. *Current Opinion in Electrochemistry*, *10*, 107–111. Available from https://doi.org/10.1016/j.coelec.2018.05.002.

Peters, F. T., Drummer, O. H., & Musshoff, F. (2007). Validation of new methods. *Forensic Science International*, *165*(2), 216–224. Available from https://doi.org/10.1016/j.forsciint.2006.05.021.

Peters, F. T., & Maurer, H. H. (2002). Bioanalytical method validation and its implications for forensic and clinical toxicology – A review. *Accreditation and Quality Assurance*, *7*, 441–449. Available from https://doi.org/10.1007/s00769-002-0516-5.

Peveler, W. J., Yazdani, M., & Rotello, V. M. (2016). Selectivity and specificity: Pros and cons in sensing. *ACS Sens*, *1*(11), 1282–1285. Available from https://doi.org/10.1021/acssensors.6b00564.

Ray, T. R., Choi, J., Bandodkar, A. J., Krishnan, S., Gutruf, P., Tian, L., & Rogers, J. A. (2019). Bio-integrated wearable systems: A comprehensive review. *Chemical Reviews*, *119*(8), 5461–5533. Available from https://doi.org/10.1021/acs.chemrev.8b00573.

Rocchitta, G., Spanu, A., Babudieri, S., Latte, G., Madeddu, G., Galleri, G., & Serra, P. A. (2016). Enzyme biosensors for biomedical applications: Strategies for safeguarding analytical performances in biological fluids. *Sensors (Basel)*, *16*(6). Available from https://doi.org/10.3390/s16060780.

Romanholo, P. V. V., Razzino, C. A., Raymundo-Pereira, P. A., Prado, T. M., Machado, S. A. S., & Sgobbi, L. F. (2021). Biomimetic electrochemical sensors: New horizons and challenges in biosensing applications. *Biosensors & Bioelectronics*, *185*, 113242. Available from https://doi.org/10.1016/j.bios.2021.113242.

Ronkainen, N. J., Halsall, H. B., & Heineman, W. R. (2010). Electrochemical biosensors. *Chemical Society Reviews*, *39*(5), 1747–1763. Available from https://doi.org/10.1039/B714449K.

Sakala, G. P., & Reches, M. (2018). Peptide-based approaches to fight biofouling. *Advanced Materials Interfaces*, *5*(18). Available from https://doi.org/10.1002/admi.201800073.

Sempionatto, J. R., Lin, M., Yin, L., De la paz, E., Pei, K., Sonsa-ard, T., de Loyola Silva, A. N., Khorshed, A. A., Zhang, F., Tostado, N., Xu, S., & Wang, J. (2021). An epidermal patch for the simultaneous monitoring of haemodynamic and metabolic biomarkers. *Nature Biomedical Engineering*, *5*(7), 737–748. Available from https://doi.org/10.1038/s41551-021-00685-1, https://doi.org/10.1038/s41551-021-00685-1.

Sena-Torralba, A., Álvarez-Diduk, R., Parolo, C., Piper, A., & Merkoçi, A. (2022). Toward next generation lateral flow assays: Integration of nanomaterials. *Chemical Reviews*, *122*(18), 14881–14910. Available from https://doi.org/10.1021/acs.chemrev.1c01012, https://doi.org/10.1021/acs.chemrev.1c01012.

Sharma, S., Byrne, H., & O'Kennedy, R. J. (2016). Antibodies and antibody-derived analytical biosensors. *Essays in Biochemistry*, *60*(1), 9–18. Available from https://doi.org/10.1042/EBC20150002.

Shi, Q., Su, Y., Ning, X., Chen, W., Peng, J., & Jiang, Z. (2011). Trypsin-enabled construction of anti-fouling and self-cleaning polyethersulfone membrane. *Bioresource Technology*, *102*(2), 647–651. Available from https://doi.org/10.1016/j.biortech.2010.08.030.

Shukla, S. K., Murali, N. S., & Brilliant, M. H. (2015). Personalized medicine going precise: from genomics to microbiomics. *Trends in Molecular Medicine*, *21*(8), 461–462. Available from https://doi.org/10.1016/j.molmed.2015.06.002.

Simões, C., Caeiro, I., Carreira, L., Silva, F. C., & Lamy, E. (2021). How different snacks produce a distinct effect in salivary protein composition. *Molecules (Basel, Switzerland)*, *26*(9), 1420–3049. Available from https://doi.org/10.3390/molecules26092403.

Song, S., Wang, L., Li, J., Fan, C., & Zhao, J. (2008). Aptamer-based biosensors. *TrAC Trends in Analytical Chemistry*, *27*(2), 108–117. Available from https://doi.org/10.1016/j.trac.2007.12.004.

Studying the effectiveness of non-invasive glucose sensors in patients with diabetes: the SENSOR Study (SENSOR). (2020). 2023 3 30. https://clinicaltrials.gov/ct2/show/NCT02800044.

Swartz, M. E., & Krull, I. S. (2012). *Introduction to analytical method validation. Handbook of analytical validation* (pp. 1–22). CRC Press.

Teymourian, H., Moonla, C., Tehrani, F., Vargas, E., Aghavali, R., Barfidokht, A., Tangkuaram, T., Mercier, P. P., Dassau, E., & Wang, J. (2020). Microneedle-based detection of ketone bodies along with glucose and lactate: Toward real-time continuous interstitial fluid monitoring of diabetic ketosis and ketoacidosis. *Analytical Chemistry*, *92*(2), 2291–2300. Available from https://doi.org/10.1021/acs.analchem.9b05109, https://doi.org/10.1021/acs.analchem.9b05109.

Thompson, M., Ellison, S. L. R., Fajgelj, A., Willetts, P., & Wood, R. (1999). Harmonized guidelines for the use of recovery information in analytical measurement. Available from https://doi.org/10.1351/pac199971020337.

Thompson, M., & Wood, R. (1993). The international harmonized protocol for the proficiency testing of (chemical) analytical laboratories (technical Report). *Pure and Applied Chemistry*, *65*(9), 2123–2144. Available from https://doi.org/10.1351/pac199365092123.

Vander Heyden, Y., Nijhuis, A., Smeyers-Verbeke, J., Vandeginste, B. G. M., & Massarta, D. L. (2001). Guidance for robustness/ruggedness tests in method validation. *Journal of Pharmaceutical and Biomedical Analysis*, *24*(5), 723–753. Available from https://doi.org/10.1016/S0731-7085(00)00529-X.

Wang, M., Yang, Y., Min, J., Song, Y., Tu, J., Mukasa, D., Ye, C., Xu, C., Heflin, N., McCune, J. S., Hsiai, T. K., Li, Z., & Gao, W. (2022). A wearable electrochemical biosensor for the monitoring of metabolites and nutrients. *Nature Biomedical Engineering.*, *6*(11), 1225–1235. Available from https://doi.org/10.1038/s41551-022-00916-z, https://doi.org/10.1038/s41551-022-00916-z.

Wang, Z. W., Zhang, J., Guo, Y., Wu, X. Y., Yang, W. J., Xu, L. J., Chen, J. H., & Fu, F. F. (2013). A novel electrically magnetic-controllable electrochemical biosensor for the ultra sensitive and specific detection of attomolar level oral cancer-related microRNA. *Biosensors and Bioelectronics*, 108–113. Available from https://doi.org/10.1016/j.bios.2013.02.007, https://www.sciencedirect.com/science/article/pii/S0956566313000857.

Wu, J., Wang, X., Wang, Q., Lou, Z., Li, S., Zhu, Y., & Wei, H. (2019). Nanomaterials with enzyme-like characteristics (nanozymes): next-generation artificial enzymes (II). *Chemical Society Reviews*, *48*(4), 1004–1076. Available from https://doi.org/10.1039/c8cs00457a.

Wu, S., Lan, X., Huang, F., Luo, Z., Ju, H., Meng, C., & Duan, C. (2012). Selective electrochemical detection of cysteine in complex serum by graphene nanoribbon. *Biosensors & Bioelectronics*, *32*(1), 293–296. Available from https://doi.org/10.1016/j.bios.2011.12.006.

Yang, Q., Rosati, G., Abarintos, V., Aroca, M. A., Osma, J. F., & Merkoçi, A. (2022). Wearable and fully printed microfluidic nanosensor for sweat rate, conductivity, and copper detection with healthcare applications. *Biosensors and Bioelectronics*, *202*, 114005. Available from https://doi.org/10.1016/j.bios.2022.114005, https://www.sciencedirect.com/science/article/pii/S0956566322000458.

Yuan, Z., Hou, L., Bariya, M., Nyein, H. Y. Y., Tai, L.-C., Ji, W., Li, L., & Javey, A. (2019). A multi-modal sweat sensing patch for cross-verification of sweat rate, total ionic charge, and Na^+ concentration. *Lab on a Chip*, *19*, 3179–3189. Available from https://doi.org/10.1039/C9LC00598F, http://doi.org/10.1039/C9LC00598F.

Zhang, S., Geryak, R., Geldmeier, J., Kim, S., & Tsukruk, V. V. (2017). Synthesis, assembly, and applications of hybrid nanostructures for biosensing. *Chemical Reviews*, *117*(20), 12942–13038. Available from https://doi.org/10.1021/acs.chemrev.7b00088.

Chapter 5

The application of biosensors in precision medicine

Laís C. Brazaca[1] **and Juliane R. Sempionatto**[2]
[1]São Carlos Institute of Chemistry, University of São Paulo, São Carlos, SP, Brazil, [2]Andrew and Peggy Cherng Department of Medical Engineering, California Institute of Technology, Pasadena, CA, United States

5.1 Introduction

In previous chapters of this book, biomarkers and biosensors were introduced. Specifically, Chapters 1 and 2 were dedicated to biomarkers discovery and its applications in precision medicine, while Chapters 3 and 4 discussed the construction, functioning, and validation of biosensors for medical purposes. In this chapter, we will consider the application of biosensors to precision medicine, encompassing its advantages and main challenges for regular implementation in medical practices. We will also discuss the traditional analytical techniques applied to biomarkers determination, along with the main reasons for the continuous search for improved methods.

5.2 Precision medicine and its applications

As previously discussed, precision medicine is a new approach to tailored disease treatment and prevention. In this case, the patient's genetic profile, biomarker levels, lifestyle, and environment are taken into account to promote more assertive medical decisions. As described by the FDA, the goal of precision medicine is "to target the right treatment to the right patients at the right time" (U.S. Food & Drug Administration FDA, 2023). Precision medicine can be beneficial to a variety of conditions, being especially highlighted in oncology, chronic diseases, and when narrow therapeutic ranges are required, among others.

Cancer is, currently, one of the main focuses for new therapeutic strategies. The combination of high lethality rates, the toxicity of existing treatments, and the expectation of the rise in case numbers due to population aging make it an urgent challenge to address. The variety of cancer types

Biosensors in Precision Medicine. DOI: https://doi.org/10.1016/B978-0-443-15380-8.00006-0

and subtypes combined with the heterogeneity of patients and their characteristics requires personalized treatment strategies to provide improved prognostics. In this case, precision medicine can aid in determining medical management, mitigating adverse effects caused by conventional treatments, and improving treatment efficacy, as well as in the identification of individual susceptibility to different types of cancer, allowing preventive actions (Naithani, Atal, et al., 2021).

Recently, the use of specific anticancer drugs (e.g., molecularly targeted agents and immunotherapies) has shown to be extremely effective for tumors with particular molecular alterations (Le Tourneau et al., 2019). For example, some cancer cells display abnormal expression of receptors and can, therefore, be targeted by receptor-directed monoclonal antibodies as a more assertive therapy. One of the most relevant examples is the use of trastuzumab to treat HER2-positive breast cancer (Garattini & Padula, 2019; Garattini et al., 2015). The use of new approaches besides DNA sequencing alone (e.g., circulating tumor DNA [ctDNA] and RNA sequencing, etc.) has been providing even more information about the targeted cancer, allowing to tailor therapies in specific cases, including metastatic and recurrent cancers (Le Tourneau et al., 2019).

Regarding the identification of cancer predispositions, an important example relies on the use of genetic profiling for identifying mutations in BRCA genes (Economopoulou et al., 2015; Paul & Paul, 2014). The combination of this data with family history is important for identifying individuals with a higher chance of developing breast and/or ovarian cancer. For some patients, prophylactic actions (e.g., mastectomy or the removal of the ovaries and fallopian tubes) might be indicated, ensuring the disease will not develop and threaten the patient's life in the future (Economopoulou et al., 2015; Paul & Paul, 2014). One more important oncology application relies on pharmacogenomics. In this case, genomics is used to understand the interaction between the patients' body and the drugs administered, aiding in the identification of its efficacy and toxicity. For example, genetic profiling and biomarker levels can be used to characterize drug or metabolite clearance, and levels of enzyme activity (Ahmed et al., 2016; Wheeler et al., 2013).

Chronic diseases can also be especially benefited through the use of precision medicine due to the typical high complexity and heterogeneity of the disorders (Boccia et al., 2020; Ilan, 2022; Subramanian et al., 2020). Diseases such as asthma, chronic obstructive pulmonary disease, heart failure, chronic kidney disease, and diabetes are some of the current targets. AstraZeneca, for example, is investing in precision medicine for improving chronic disease treatment by better understanding the possible disease mechanisms to be targeted. Regarding asthma, there are different leukotrienes that can drive inflammation and other disease-related processes—making it important for the disease to be precisely classified and allowing the treatment to be matched to it (AstraZeneca, 2022).

As the last main example to be cited, precision medicine has also important uses when narrow therapeutic index drugs (NTIs) are applied (Manzari et al., 2021; Zhou et al., 2008). These are defined as drugs that can lead to serious therapeutic failures or adverse effects when administered in even slightly different doses or when blood concentrations are not in the ideal range (U.S. Food & Drug Administration, 2017). One of the main challenges in this case is addressing the subject variability of the clinical response upon the administration of NTIs. Precision medicine might aid in defining more assertive drug dosages and administration schedules through the understanding of drug metabolization, and the specifics of the disease, for example.

Although we have addressed some of the most important targets of precision medicine, its applications are numerable and also include tests to detect enzyme polymorphisms related to the metabolism of antidepressant drugs, anticoagulants, and proton pump inhibitors, as well as the metabolism of drugs that treat autoimmune diseases, neurodegenerative, infectious diseases, and more (Center for Research on Science, 2023).

To replace the one-size-fits-all strategy, precision medicine relies on new areas of knowledge, including big data, artificial intelligence, and machine learning (as discussed in Chapter 10). In this scenario, the generation of reliable patients' data is a core aspect for achieving more reliable diagnosis: precision medicine can only be as good as the data it relies on. Clinical information, biomarkers levels, laboratory and radiological investigations, and environmental and social data need to be gathered and analyzed with validated methods, assuring truly assertive outcomes (Kraus et al., 2018; Naithani, Sinha, et al., 2021).

In this book, we focus on the molecular factors that contribute to precision medicine. Currently, most biomarkers are determined using laboratory tests that require specialized workers and bulky equipment and provide delayed results, hampering their accessibility. Biosensors are an alternative for promoting rapid, low cost, and decentralized analysis of proteins, genes, small molecules, and more. Recent advancements allowed the miniaturization of devices and the integration of multiple tasks in a single platform (addressed in Chapter 8), its integration into the human body (addressed in Chapter 9), and improved analytical performance to be reached (Chapters 6 and 7), expanding its uses in healthcare applications.

5.3 Traditional analytical techniques for biomarker detection

Currently, most biomarkers are routinely monitored using a limited range of techniques, including mass spectrometry, polymerase chain reaction (PCR), enzyme-linked immunosorbent assay (ELISA), and western blot (WB), among others. Methods are selected depending on the nature of the biomarker (protein, nucleic acid, small molecule, cell, etc.), equipment availability, available funds, precision or limit of detection required, and the time required for obtaining results. Some of the main methods used are described in the following sections, including their principles, advantages, and disadvantages.

5.3.1 Mass spectrometry

As described in previous chapters, mass spectrometry (MS) is a technique fundamentally important for different medical purposes, including not only diagnosis itself but also biomarkers discovery, the understanding of disease development, and the study of new and improved drugs, among others (Domon & Aebersold, 2006; Glish & Vachet, 2003). Therefore it is a key technique for precision medicine (Ang et al., 2019; Hristova & Svinarov, 2022). Clinical MS is commonly used to diagnose metabolism deficiencies (including in newborn screening), to determine whether biomarkers are present and for toxicology testing, for example (Fernández-Lainez et al., 2012). It can be used for the analysis of proteins, metabolites, lipids, and more, being an extremely versatile technique.

The technique enables the direct determination of molecules and atoms based on their mass-to-charge ratio, as well as their fragmentation patterns. For that, the molecules to be analyzed are firstly ionized in the gas phase and then submitted to combinations of electrical and magnetic fields, having their trajectories analyzed. The combination of two mass spectrometers and the use of additional reaction steps can enhance its ability to analyze complex samples, being the fundamentals of tandem mass spectrometry (MS/MS) (McLafferty, 1981).

One of the most relevant challenges when using MS is the ionization of neutral molecules, especially large and complex ones (such as biomolecules) without promoting their degradation or substantial changes. Advancements in ionization sources and methods, however, have further expanded the applicability of MS (Domon & Aebersold, 2006). For example, the use of matrix-assisted laser desorption/ionization (MALDI) is key when proteins, peptides, or polymers are being analyzed (Bonk & Humeny, 2001). This soft ionization technique uses a short laser pulse to ionize the sample, which is mixed with a matrix (e.g., 2,5-dihydroxybenzoic acid) and placed in a target plate. The matrix has the function to absorb the energy from the laser, avoiding sample degradation. The laser leads to the transfer of the molecules to the gas phase and the generation of sample ions. The charged molecules are then accelerated through the use of an electric field and commonly enter a time of flight (TOF) analyzer—a field-free analyzer that measures the time ionized molecules need to travel a fixed distance (El-Aneed et al., 2009). Along with MALDI, electrospray ionization (ESI) is one of the most used ionization methods for MS, as these promote good mass range and great sensitivity (Loo et al., 1989). It is worth notice mentioning that biomolecules of various sizes can be analyzed by the combination of a plethora of ionization and measurement methods and that extremely low concentrations can be targeted (attomole or nanomole), making MS a versatile and relevant technique (Thermo Fisher Scientific, 2023c).

Sample preparation is often cumbersome for MS. When considering proteomics, for example, there is no standard method for analyzing a sample—with the ideal protocol being defined by many different variables such as the type of sample, its complexity, the subcellular location of the target protein, and its concentration. The methods often include several sequential steps, such as lysate preparation, protein digestion, and peptide extraction, enrichment, or cleanup (Bodzon-Kulakowska et al., 2007; Thermo Fisher Scientific, 2023c). If the sample is too complex, the spectrum might be very difficult to analyze. Therefore MS is often coupled to separation techniques, generating the hyphenated techniques (GC-MS, LC-MS, LC-MS/MS, CE-MS, etc.). The main advantages of this coupling are the ability to promote more complex analysis, increased reproducibility, and faster analysis (Awuchi et al., 2022; Chromatography today, 2023).

Although MS and its derivate techniques present extremely low detection limits and are powerful tools for analyzing a variety of samples, equipment prices can range up to $100,000 (Frayling, 2023), besides requiring specialized personnel to operate it and for sample preparation. Furthermore, equipment is often not portable and results commonly require a long time to be collected. Therefore, MS has an invaluable part in precision medicine—allowing the determination of countless biomarkers and the study of complex samples—but the quantification of biomarkers themselves should be more accessible, rapid, decentralized, and require lower costs to be widely applied.

5.3.2 PCR

The combination of PCR, electrophoresis, and fluorescence comprises a powerful basis for the traditional analysis of genetic material for medical purposes. Although it presents multiple applications, some of its highlights lie on the detection of pathogens, allowing the determination of not only the microorganism itself but also its strain—assuring that the most adequate treatment will be provided (Botes et al., 2013; van Belkum, 1994). Although traditional PCR itself is not able to directly determine some specific mutations, recent methods allow doing so, but the combination with separation techniques might be required. When considering mutation identification, PCR-derived techniques are most commonly applied as an amplification technique, providing the desired amplicons for subsequent analysis like single-stranded conformational polymorphism (SSCP), denaturing gradient electrophoresis (DGGE) or sequencing (Frayling et al., 2006).

In short, the PCR technique promotes the amplification of portions of the target gene. For that, primers specifically designed for the amplification of the desired portions are added to the sample along with a polymerase enzyme (usually Taq), allowing the amplification of the desired DNA fragment through thermal cycling. In the beginning of each cycle, temperatures around 94°C–98°C are used to denature the double-stranded DNA, forming single strands. Following, the temperature is lowered to 50°C–65°C, allowing the annealing of the primers to the DNA strands. The third and final step of the cycle comprises the use of temperatures around 72°C for the extension of the genetic material; that is, Taq polymerase synthesizes new DNA strands. This cycle allows the exponential amplification of the desired fragment, being able to identify even concentrations as low as a few hundreds of copies per mL (Arnaout et al., 2020). Usually, this cycle is repeated 20–40 times, taking about 45 minutes to 1 hour for a complete test (Bustin, 2017; Thermo Fisher Scientific, 2023a). Although specific primers are commonly used for the determination of each target, some different approaches are also currently being investigated. For example, Barghouthi et al. have designed a general PCR primer based on the 16S gene that is able to amplify the genetic material of many dozens of microorganisms. Later, the amplicons are sequenced and aligned against a database, allowing the identification of the bacteria. The idea can be expanded to other organisms, including protists and nematodes (Barghouthi, 2011).

When using traditional PCR, the fragments in the resulting solution are separated by size using electrophoresis, allowing the identification of the amplicon (Wittmeier & Hummel, 2022). When the presence of a microorganism is being investigated, it is common for RNA to be the genetic material to be determined. In this case, reverse transcriptase (RT) is used to generate the complementary DNA (cDNA) of the genetic material present in the sample, with the corresponding cDNA then being amplified through PCR. The technique is then called RT-PCR (reverse transcription polymerase chain reaction) and is also commonly applied to detect the presence of mRNAs, pre-mRNAs, noncoding RNAs, or others (Rio, 2014; Vidovic, 1998). When quantitative measurements are required, real-time PCR (qPCR) can be applied. In this case, fluorescent probes are used to follow the amplification in real time, allowing the rapid and precise identification of the target sequence. When used along with RT for the determination of RNA, the technique is named RT-qPCR (Mackay, 2002).

As previously mentioned, traditional PCR and its different approaches are routinely used in medical practice especially to diagnose infections and identify pathogens, but can also be used for determining genetic mutations or variants (Matsuda, 2017; van Belkum, 1994). It is a powerful tool for precision medicine, as genetic traits influence the patients' therapy responses, and the determination of the specific infection type can greatly aid in designing treatments. Furthermore, the determination of mutations can determine personalized treatments or promote preventive actions, as discussed for BRCA-1. Digital PCR, specifically, holds great promise in the area (Wong et al., 2017; Zhang et al., 2022).

More recently, isothermal amplification techniques have caught the attention of the scientific community. In this case, the amplification of the desired gene portion is performed at a single temperature, not requiring the traditional PCR cycling. Regarding the development of biosensors, such techniques are of great interest due to the possibility of simplifying testing by integrating sample amplification and analysis even in remote settings (Abdul Ghani et al., 2022; Das et al., 2022). Some of the most relevant techniques include loop-mediated isothermal amplification (LAMP), rolling circle amplification (RCA), nucleic acid sequence-based amplification (NASBA), and helicase-dependent amplification (HDA) (Gill & Ghaemi, 2008; Zhao et al., 2015). For more detailed information, please refer to Chapter 7, dedicated to describing the recent advancements in detection and amplification techniques.

5.3.3 Next-generation sequencing

Next-generation sequencing (NGS) is a powerful technique for precision medicine, as it allows the sequencing of a whole human genome in periods as short as a day, providing complete information about a patient's genetics. For that, millions of small DNA fragments are sequenced in parallel, with bioinformatic tools being used to match the sequenced fragments to a reference template (Fig. 5.1) (Behjati & Tarpey, 2013). In short, NGS is composed of four main steps: (1) Nucleic acid extraction and isolation: required to obtain the desired genetic material for the analysis (e.g., total RNA, genomic DNA, mRNA, etc.). A yield of nanograms to micrograms is commonly required for the analysis, besides the sample being free of enzyme inhibitors (e.g., ethanol, heparin, etc.) and presenting high quality (e.g., heterogeneous, representative of the source, intact genetic material). (2) Library preparation: the extracted material is then fragmented to appropriate sizes and modified to interact with the NGS platform. Additional processes might be required to improve the outcome, including the purification of the library by nucleic acid size, or the conversion to cDNA if RNA is the starting material; (3) Clonal amplification and sequencing: in this stage, the fragments are amplified to ensure fluorescent signals will be captured by the NGS equipment, with solid-phase amplification being commonly used, and sequenced (e.g., by sequencing by synthesis [SBS]); (4) Data analysis: the final step involves the processing, analysis and interpretation of the acquired data. This is commonly a challenging task especially due to the high cost of the computational infrastructure required, and the complexity of the generated data (AAT Bioquest, 2023; Thermo Fisher Scientific, 2023b). It is important to notice that there are numerous different NGS platforms, and discussing them in detail is beyond the theme of this book. Relevant review papers with further details about the functioning of such devices are found in the literature (Goodwin et al., 2016; Liu et al., 2012; Slatko et al., 2018).

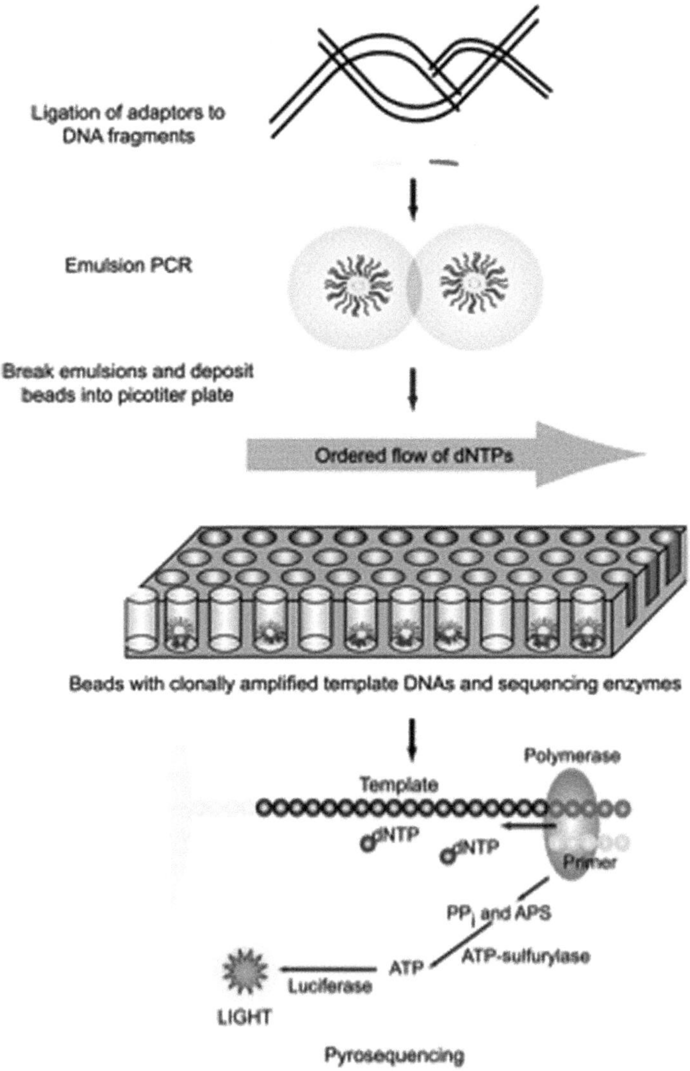

FIGURE 5.1 Representation of the basics of a Roche 454 GS FLX sequencing. After DNA extraction, the template is fragmented, linked to adapters, and clonally amplified. Later, the beads are added to wells containing sequencing enzymes, and the event of nucleotide addition leads to luminescence. APS, adenosine-5′-phosphosulfate; PPi, pyrophosphate. Voelkerding et al. (2009) *Reprinted with permission from Voelkerding, K. V, Dames, S. A., & Durtschi, J. D. (2009). Next-generation sequencing: From basic research to diagnostics.* Clinical Chemistry, 55*(4), 641–658.*

NGS has the potential to revolutionize healthcare through the detailed description of one's genome, allowing the investigation of numerous mutations, including substitutions, deletions, insertions, duplications, copy number changes and chromosome inversions/translocations. The technique is reliable, providing accurate data through multiple readouts, besides being much cheaper than traditional sequencing methods, as Sanger's, if the cost per base is considered (Behjati & Tarpey, 2013). Some of its most important uses in healthcare include the early identification of disorders, assessing predispositions and the determination of pharmacogenetics markers (Gonzalez-Garay, 2014). NGS is also being explored for the identification of tumors and their genetic characterization (Strom, 2016), the determination of microorganisms and their strains (MEDINGER et al., 2010), for a better understanding of RNA expression (Grada & Weinbrecht, 2013), and for investigating Mendelian diseases (Jamuar & Tan, 2015).

Despite all the potential NGS has, its practical clinical applications are currently limited. Many of the genetic anomalies detected by the technique still present unknown effects on one's health, and substantial research is still required for NGS routine practical purposes (Gupta & Verma, 2019; Healio, n.d.; Strom, 2016). However, we expect it to bring significant advantages in the next decades, generating valuable information for precision medicine.

5.3.4 ELISA

ELISA is a widespread analytical technique that can be used for the detection of various targets including proteins, polypeptides, antibodies, and small molecules (Fig. 5.2). Commercial kits are produced by a great variety of manufacturers, with prices reaching a few hundred dollars for a typical 96-test microwell plate (Merck, 2023a, 2023b). Although many different test formats exist (sandwich, competitive, and direct and indirect ELISA), the main principles of the technique are common among them. To discuss the functioning of ELISA, sandwich, the most common format, will be described for the determination of an antigen.

ELISA relies on the attachment of the desired analyte onto a solid surface. In the sandwich format, this attachment is commonly performed through the binding of the analyte to an antibody specific to the target that is placed in the surface of microwells. An enzyme-labeled antibody (usually horseradish peroxidase [HRP], alkaline phosphatase [AP], or Beta-galactosidase) is then added to the wells, forming a second immunocomplex with the analyte. Following, the substrate of the enzyme is added and allowed to react, forming a colored product. The reaction is usually stopped after a few dozens of minutes and the color of the well is registered using a UV–Vis spectrophotometer, with the absorbance being directly correlated to the concentration of the analyte. A very relevant feature when considering ELISA for diagnostics relies on the fact that many different sample types can be used (plasma, urine, culture medium, etc.) and that small sample volumes are required (typically 100 μL).

ELISA is one of the gold standards for biomarker detection, being widely commercially available and presenting good sensitivity, specificity, and

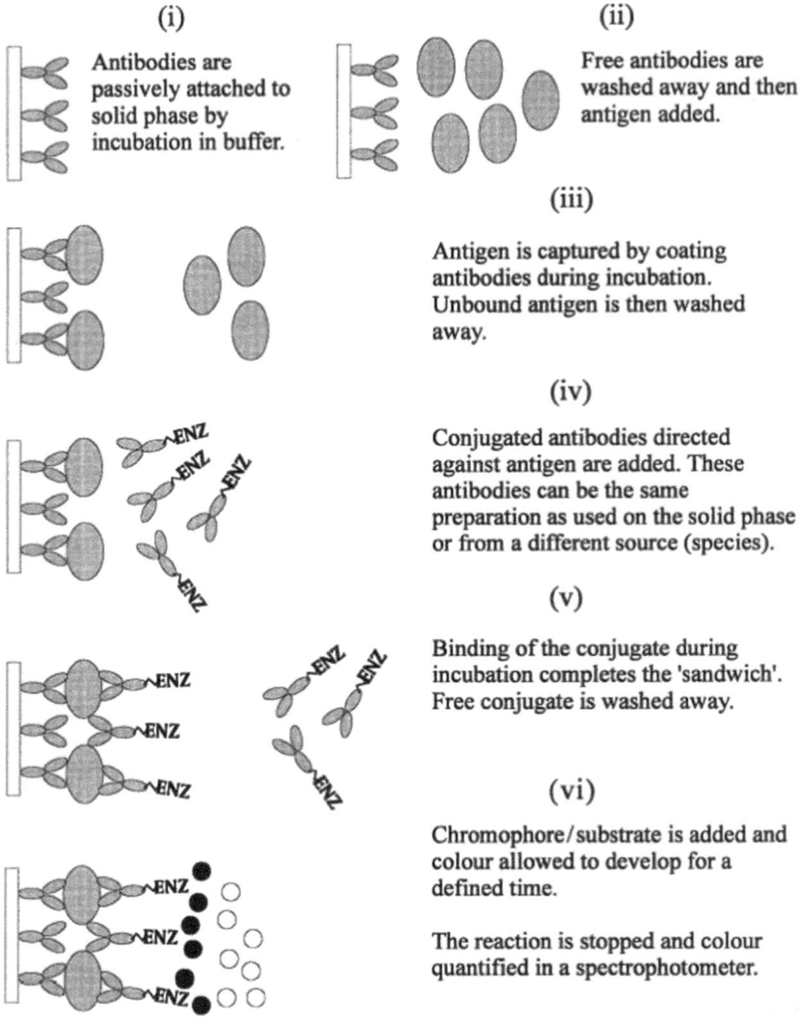

FIGURE 5.2 Scheme of the fundamental steps of a sandwich ELISA. Antibodies attached to a solid phase capture the analyte. Following, an enzyme-labeled antibody is added to the plate, and the color developed when the enzyme substrate is added is used for target determination. *Reprinted with permission from Crowther, J. R. (2009). Systems in ELISA. In J. R. Crowther (Ed.),* The ELISA Guidebook *(2nd ed., pp. 9–42). Humana Press.*

detection limits that can reach concentrations lower than pg/mL (Zhang et al., 2014). However, it also presents significant drawbacks, including requiring the need for the sample to be transported to a laboratory, the use of bulky and expensive equipment, requiring specialized workers, being relatively expensive, and taking long periods to obtain results (around 6 hours, typically) (Leng et al., 2008). Therefore although ELISA is commonly used in clinical laboratories, some of these characteristics hamper its widespread use. Its use in precision medicine relies, mainly, on the determination of protein-based biomarkers (Wang et al., 2019).

5.3.5 Western blot

Western blotting (WB) (also called protein blotting or immunoblotting) is an important technique for the determination of proteins in complex samples, particularly the ones of low abundance (Fig. 5.3). WB is especially useful

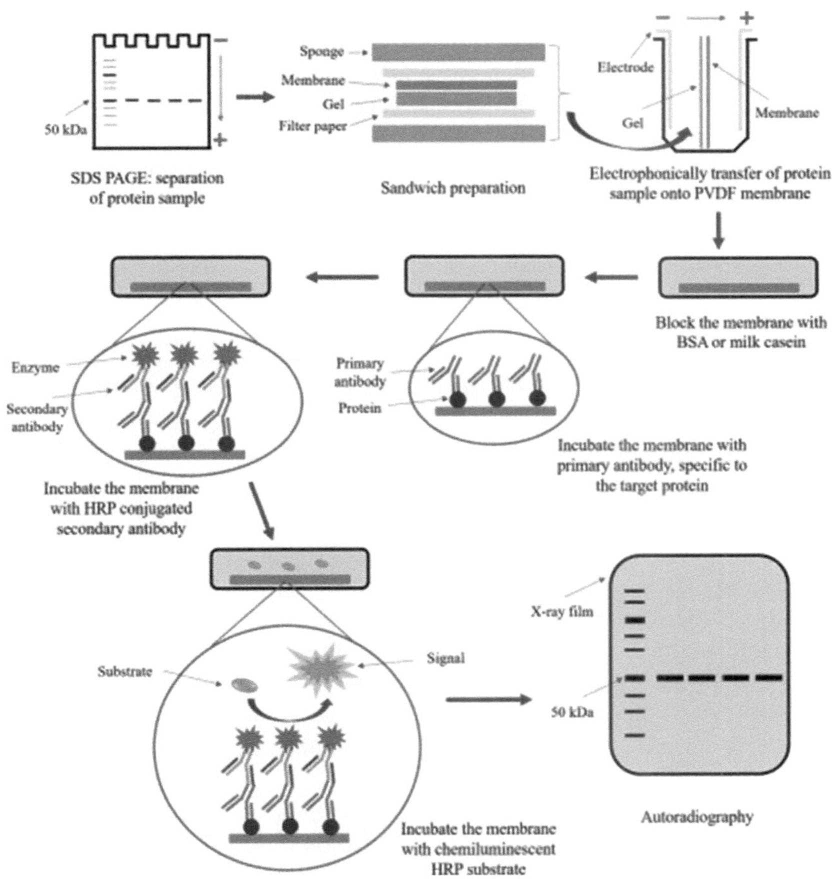

FIGURE 5.3 Basic principles of WB and immunodetection. First, the sample is separated using electrophoresis and the gel pattern is transferred to a membrane. Later, antibodies specific to the desired target are added to the blocked membrane, and secondary enzyme-labeled antibodies are used to display a readable signal (Roy et al., 2019). *Reprinted with permission from Roy, J., Jain, N., Singh, G., Das, B., & Mallick, B. (2019). Small RNA proteome as disease biomarker: An incognito treasure of clinical utility. In B. Mallick (Ed.), AGO-Driven Non-Coding RNAs (pp. 101–136). Elsevier. 2019 Elsevier Inc.*

for determining biomarkers, for the investigation of posttransduction modifications, and the study of protein expression, or phosphorylation, among others (Henderson & Wolf, 1992; Merck, 2023c). It relies on the transfer of the proteins from an electrophoresis gel (usually native or sodium dodecyl sulfate [SDS]-polyacrylamide gels) to an adsorbent membrane (e.g., nitrocellulose, polyvinylidene difluoride [PVDF]), forming a replica of the gel and providing a reliable basis for further experiments. The transfer of the electrophoretic pattern to the membrane can then be performed through different methods, usually based on simple diffusion, vacuum-assisted solvent flow, or electrophoretic elution (Kurien & Scofield, 2006).

Gel electrophoresis is a powerful technique for analyzing biomolecules, including proteins, polypeptides, and nucleic acid fragments. Its principles rely on the differential migration of biomolecules throughout a gel in the presence of electric fields, usually separating the biomolecules by size, shape, and/or electrical properties (e.g., charge). For complex samples, 2D electrophoresis can enhance the separation resolution. In this case, two different electrophoresis methods are combined in an orthogonal fashion. Usually, for bioanalytical ends, isoelectric focusing under denaturing conditions is combined with SDS polyacrylamide gel electrophoresis (Westermeier, 2014).

Now, regarding the detection of specific proteins on immunoblots, immunodetection is commonly the technique of choice. The methods vary, but a traditional approach consists of the use of a primary antibody specific to the desired target in combination with a secondary, enzyme-labeled antibody. Usually, peroxidases or phosphatases are used, allowing the development of a response (colored or chemiluminescent, for example) after the addition of the desired substrate and, therefore, the detection of the target (Egger & Bienz, 1994). The adaptation of western blotting traditional methods permits the simultaneous detection of multiple targets, bringing advantages to healthcare purposes (Berkelman, 2020).

Some of the main highlights of this technique are the availability of the proteins for their interaction with different ligands, the reliability of the transfer, and the storage of the transferred patterns, for example (Henderson & Wolf, 1992). However, the technique presents a relatively high cost, depends on the structure of a laboratory, and requires specialized researchers. Despite that, the technique is commonly used for medical diagnosis. Examples include testing for HIV (Human Immunodeficiency Virus), FIV (Feline Immunodeficiency virus), HBV (Hepatitis B Virus), and more. (Sino Biological, 2023)

5.3.6 Electrochemistry

As thoroughly mentioned in Chapter 3, electrochemistry is one of the main techniques used in the development of biosensors due to its sensitivity, low detection limits, for being able to provide in situ measurements, and more

(Brazaca et al., 2021, 2022; Oliveira et al., 2020). Electrical parameters (e.g., current, potential, impedance) are monitored and related to (bio)chemical events on the surface of the electrode. When the target analyte is electrochemically active (e.g., uric acid, dopamine, vitamin C, 3-nitro-l-tyrosine, etc.), redox reactions can be directly monitored through the use of voltammetric or amperometric techniques, allowing the precise determination of the analyte (Blasques et al., 2022; Carvalho et al., 2023; Sempionatto et al., 2019). On the other hand, when the target is not electrochemically active (e.g., proteins, whole cells, etc.), tags or redox mediators are commonly applied, and the use of other techniques such as electrochemical impedance spectroscopy (EIS) is common (Brazaca, Bramorski, et al., 2016; Brazaca, Janegitz, et al., 2016).

Electrochemical biosensors are very versatile, being able to determine specific proteins, genetic material, peptides, whole cells, and so on. In regular clinical practice, glucometers are one of the most important examples of the use of electrochemistry in medicine. An updated review of the available commercial devices for precision medicine—including the glucometer—and its uses can be found in Chapter 11 of this book. In short, currently, portable equipment for punctual glucose measurements is available commercially at accessible prices, as well as wearable devices that monitor the molecule constantly (Hovorka, 2006; Vashist, 2013). Due to the high volume of collected data, current devices are even able to predict health risk situations, alerting the user for potential glucose peaks or depletion. Such approaches allow the adequation of treatment, including medication type and dosage for specific situations, being a highlight example of precision medicine application (FreeStyle Libre, 2023).

Furthermore, electrochemistry is a powerful technique to be explored for precision medicine especially due to its plethora of methods, which allow the quantification of many different types of molecules rapidly and at potential low costs. Also, its equipment is commonly portable and can be used by untrained personnel, allowing the performance of in situ measurements even in low-resource settings (Bianchi et al., 2020; Blanco et al., 2006; Huang et al., 2007).

5.4 Biosensors in clinical applications

The use of analytical methods is of ultimate importance for clinical applications, especially regarding the diagnosis of diseases and infections. Usually, biological samples such as blood, stool, urine, and others are collected from patients and sent to the laboratory for analysis regarding the presence of pathogens or drugs. For this, equipment and protocols previously mentioned are commonly employed, meaning that each sample needs to undergo special preparation and storage, which increases the cost of each analysis. Moreover, the test results are not in real time, which might hinder the diagnosis processes. For close to real-time results, the patient needs to be admitted to the hospital, which generates prohibitive costs for most of the population. Repetitions of the same tests for confirmation might be needed; in this case, the patient has the burden of performing several hospital visits, for in general, many invasive procedures and sample collection.

Ideally, the monitoring of metabolites, drugs, and other biomarkers in a hospital setting should follow the continuous monitoring already performed in the intensive care units (ICUs) where monitors displaying heart rate, blood pressure, oxygen levels, and so on are used continuously. For continuous and real-time monitoring of metabolites, biosensors must be employed.

As detailed in Chapters 3 and 4, biosensors are autonomous devices able to measure quantitative or semiquantitative analytical information about the target analyte using a biological element of biorecognition (bioreceptor) that is in close contact with the transducer. Biosensors do not require additional reagents or special sample treatment, which differs from other analytical systems. Therefore biosensors are portable analytical devices that can be used on site, generating results in a minute scale. The easy to use and fast response from biosensors makes them an appropriate tool for bedside monitoring, generating results close to real time. Moreover, biosensors employed in clinical setups have the advantage of fast results and low sample volumes, decreasing the invasiveness of the process. For example, disposable hemoglobin and iron biosensors are commonly used to measure red blood cell levels before blood donation to ensure that the hemoglobin will stay at a normal range after the blood donation. Thus depending on the results, the blood donor will be either allowed or deferred to give blood.

The most common example of biosensors for clinical applications is the glucometer. The glucose biosensor was one of the first biosensors developed (Teymourian et al., 2020). It is largely used by the diabetes patient's community to monitor their glucose levels and assess whether insulin must be ministered or not. Briefly, the enzyme glucose oxidase is immobilized on an electrode (transducer) surface, and upon interaction with blood glucose, the enzymatic reaction generates products that can be electrochemically detected and the signal is proportional to the glucose concentration (Min et al., 2021).

The use of biosensors allows the self-monitoring of biomarkers at home, by the patient, decreasing the cost of health care, while increasing the possibility of early diagnosis and prevention of disease. There are several point-of-care biosensors used for at-home self-monitoring, including glucose, lactate, creatinine, hemoglobin, cholesterol, triglycerides, uric acid, and other biosensors. Biosensors for lactate detection, combined with blood pressure levels, are commonly used in clinical settings for the early diagnosis of sepsis (Hotchkiss et al., 2016; Reddy et al., 2018). The advancements in biosensor research have allowed to expand the biomarkers detection to amino acids, and hormones, by employing aptamers, antibodies, and molecularly imprinted polymers (Plaxco & Soh, 2011; Rangel et al., 2020; Sempionatto et al., 2022; Wang et al., 2022). Thus one can expect that a comprehensive biosensor platform will soon be available to analyze the levels of several pathogens at the same time.

5.5 Biosensors in precision medicine

The development of new classes of biosensors allowed the rise of precise or personalized medicine. Precision medicine considers every individual as unique with different metabolisms, different responses to drug treatment, and different genetics (Ashley, 2016; Hodson, 2016). For example, the response to a drug can have different effects depending on the patient, this is usually observed for generalized dosages or treatments that are bonded to the concept of "one size fits all." The treatment of bacterial infections is a clear example of unnecessary overdosage, that creates super bacteria that can worsen the condition and have life-threatening outcomes. In this case, precision medicine is compromised to closely monitor the type and population number of the bacteria so the patient can have a personalized treatment (Brasier et al., 2023; Johnston & Badran, 2022). Another common example of application in precision medicine is the closed-loop insulin pump. There are commercially available glucose sensors that can be placed on the body, so the blood glucose levels are continually measured by a small sensing needle, the information is then communicated to the insulin pump, also adhered to the patient body, which is able to deliver the correct dosage of insulin based on the glucose concentration measured by the sensor. This is called a closed feedback loop (Hovorka, 2011). There are other loops similar to the insulin pump, but to deliver levodopa to Parkinson's disease patients (Teymourian et al., 2022). Personalized medicine is crucial for patients with Parkinson's disease. Parkinson's disease is a neurodegenerative disorder that affects 2%–3% of the population over 65 years old. Neuronal loss in the substantia nigra, causes dopamine deficiency, one of the common symptoms of the disease is uncontrolled muscle spasms or tremors (Grayson, 2016; Poewe et al., 2017). Parkinson's is the second cause of dementia in the United States. The treatment for Parkinson's disease is a medication known as levodopa; however, during the advanced stages of the disease, the management of levodopa treatment is very challenging. Levodopa has a short half-life and narrows the therapeutic window, causing interpatient and intrapatient variations in the dose–response relationship. Thus for Parkinson's patients, the dosage needs to be personalized according to the disease evolution and side effects, the optimal dosage is achieved after several visits to the doctor's office by trial and error. Due to the severity and wide spread of this disease, several efforts were directed at the development of a closed-loop system for levodopa delivery (Teymourian et al., 2022). Biosensors connected to the body can be used to send the blood levodopa levels to a levodopa pump that keeps the medication at the pharmacologic level (Fig. 5.4A).

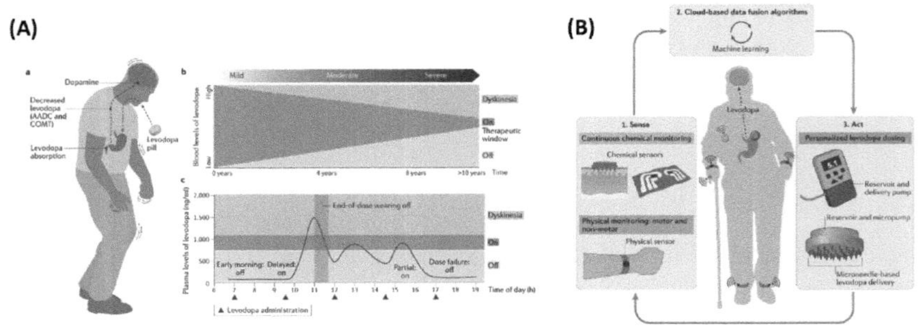

FIGURE 5.4 (A) Levodopa oral absorption path through the upper intestine, where it is metabolized by enzymes, decreasing its bioavailability until crossing the blood–brain barrier and arriving at the substantia nigra to be converted to dopamine (a). Therapeutic window for levodopa, low levels have no clinical effects (off; light gray), and high levels result in dyskinesia (red), the narrowing of the therapeutic range is shown in dark gray (b). Fluctuations in plasma levodopa related to motor symptoms during a day. There is an early off-clinical effect, followed by delayed treatment and a wear-off stage before the next dosage. Unexpected wear-off can also happen (c). (B) Proposed multimodal wearable network of sensors and biosensors for closed loop of levodopa administration. *Reprinted with permission from Teymourian, H., Tehrani, F., Longardner, K., Mahato, K., Podhajny, T., Moon, J. M., Kotagiri, Y. G., Sempionatto, J. R., Litvan, I., & Wang, J., 2022. Closing the loop for patients with Parkinson disease: Where are we?* Nature Reviews Neurology, 18, *497–507.*

Currently, efforts in the field of biosensors have been directed to the development of wearable noninvasive biosensors. Such devices can continuously track biological information using noninvasive biofluids such as sweat, saliva, tears, and interstitial fluid (Kim et al., 2019). Several wearable biosensors have been presented for the monitoring of sweat biomarkers (Bariya et al., 2018; Brothers et al., 2019), used for disease prevention and diagnosis, nutrition, and sports. The challenges and opportunities of working with this wearable biosensor platform will be further discussed in detail in Chapter 9. The development of these platforms will allow the continuous and concurrently monitoring of several molecules. This opens a large number of possible applications for personalized medicine, including the monitoring of illicit drugs for law enforcement; precision nutrition for disease prevention and balanced diet; and more importantly, when combined with machine learning approaches, these wearable platforms will be able to discover new biomarker signatures, where not only one, but a group of molecules will identify the disorder allowing early diagnosis and prevention of several diseases that are currently difficult to diagnosis (Sempionatto et al., 2022). The development of such integration can be applied to drug trials, by measuring several parameters to identify positive side effect reactions. Most impressive, with the advances in the field of wearable biosensors and big data analysis, it will be possible to perform prediction of events that are now considered sudden fatalities such as a heart attack, a stroke, and epilepsy episodes. Once a large amount of data are analyzed, the artificial intelligence system will be able to predict those events based on the early changes of key biomarker levels, thus, a heart attack could be predicted within minutes to hours from happening, saving thousands of lives.

Precision medicine can also be applied to nutrition, there are several chronic conditions that are closely related to diet, such as cancer (Adlercreutz, 2002; Kushi et al., 2012; Mayne et al., 2016), and cardiometabolic disorder, which is characterized by obesity, type II diabetes, elevated lipidic levels, hypertension, and cardiac issues (Aron-Wisnewsky & Clément, 2016; Cani & Van Hul, 2021). Cardiometabolic disease affects almost 40% of the world's population, and cardiac disease and hypertension are the number one killer in America (Gerdts & Regitz-Zagrosek, 2019). What is curious, is that this condition is completely preventable and reversible if diagnosed in its early stages, the decline of the cardiometabolic disorder is related to diet and lifestyle. Wearable biosensors can continuously monitor the user's parameters related to glucose, cholesterol, triglycerides, blood pressure, and heart rate (Sempionatto et al., 2021). When combined with machine learning approaches, the wearable platform can send the user real-time feedback to induce behavioral and lifestyle changes. This could save millions of dollars for the health system. Moreover, the field of precision medicine is quickly advancing with more efforts invested in the continuous monitoring of drug pharmacokinetics via wearable biosensors. The monitoring of the drug's pharmacokinetics in each patient allows the development of personalized treatment with optimum dosage. In this scenario, extensive research needs to be directed to closed-loop feedback systems, such as the insulin pump, for the autonomous "sense and act" approach. Personalized medicine allows a target treatment, with lower drug concentrations and less toxicity. Strategies for the design and validation of biosensors are presented in Chapters 4 and 6. A discussion on the drug delivery system is presented next.

5.6 Strategies for drug delivery systems

Precision medicine tailors to the individual level, treatments, and interventions based on the response of each patient to a drug and/or the development of the disease (Ashley, 2016; Hodson, 2016). To track such responses, continuous monitoring of biomarkers is performed using wearable or point-of-care biosensors. For example, the pharmacokinetics of drugs and medications can be closely monitored to ensure their therapeutic range. The therapeutic range is achieved by the intake of a new dosage of medication (Fig. 5.4B), or, for autonomous systems, by the delivery of the next concentration of drug directly inside the body, in a so-called, closed-loop system (Fig. 5.4B, Fig. 5.5A).

There are several strategies related to biosensors and drug delivery development (Manikkath & Subramony, 2021; Yu et al., 2018). For an efficient

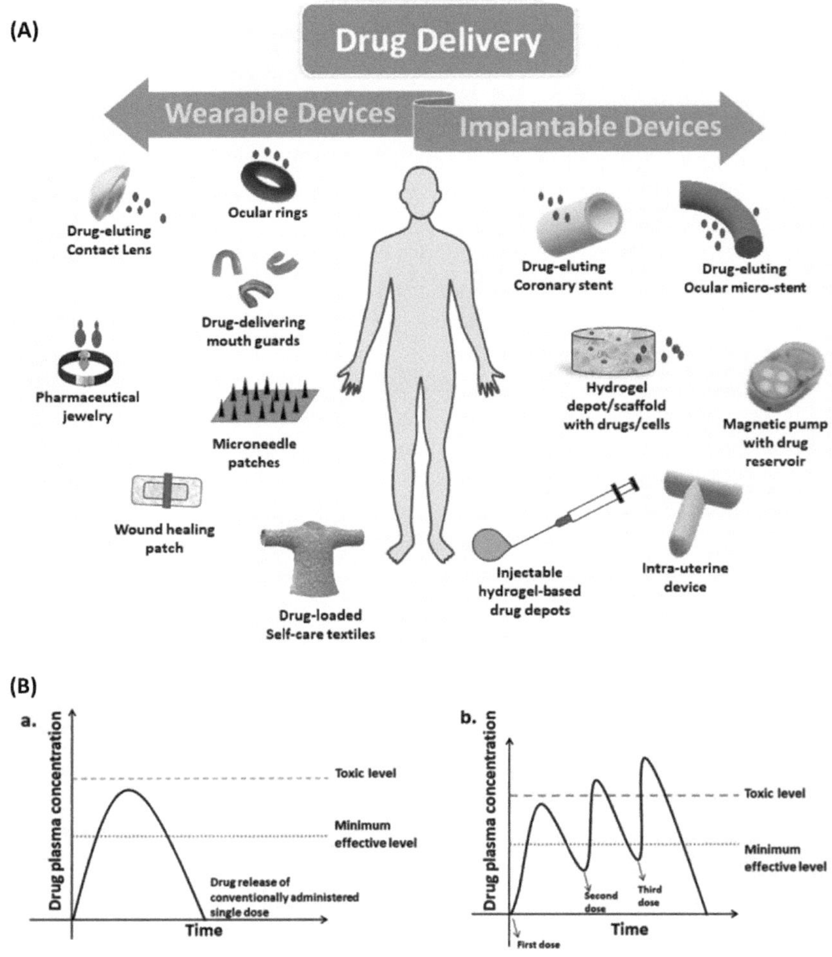

FIGURE 5.5 (A) Examples of wearable and implantable drug delivery systems. (B) Concentration of the drug in plasma after (a) single dose administration and (b) multiple dosages based on the "wear-off" timing of the drug. This interval fluctuates for each individual and must be continuously monitored to achieve the desired therapeutical window and clinical effect. *Reprinted with permission from (A) Kar, A., Ahamad, N., Dewani, M., Awasthi, L., Patil, R., & Banerjee, R., 2022. Wearable and implantable devices for drug delivery: Applications and challenges.* Biomaterials, 283, *121435. and (B) Adepu, S., Ramakrishna, S., 2021. Controlled drug delivery systems: Current status and future directions.* Molecules.

drug delivery system, considerations of the target, the delivery path, and the carrier must be taken for the controlled release of a molecule (Manzari et al., 2021). The target location will dictate the type of delivery system with possibilities that include transdermal, oral, implantable, and other delivery paths (Manzari et al., 2021). The delivery path is important to preserve the drug stability, for example, there are several pH changes in the path using oral delivery. The carrier includes a controlled releasing system using polymers, nanoparticles, cells, liposomes, encapsulation strategies, and so on (Adepu & Ramakrishna, 2021). The carrier will thus dictate the loading and physical properties of the drug (liquid, aerosol, powder, etc.). In general, drug delivery is implemented to ensure optimum pharmacologic range by enhancing the bioavailability while minimizing toxicity and side effects due to overdosage to improve patient compliance. There are numerous strategies applied to the drug delivery system (Fig. 5.5A). The main considerations for different delivery systems are discussed next (Kar et al., 2022).

5.7 Transdermal drug delivery

There are three types of transdermal delivery systems (Prausnitz & Langer, 2008). (1) Topical formulations or adhesive patches that are placed over the skin and the drug migrates to the outermost layer of the skin. This approach is used for systemic delivery to deliver small, lipophilic, low-dose drugs. Examples of such patches include nicotine patches, scopolamine patches used to treat motion sickness, and lidocaine patches for back pain. The main advantage of this system is noninvasiveness, and the delivery of drugs that can be prematurely metabolized if taken orally (Fox et al., 2011). (2) Systems that can enhance transdermal delivery such as iontophoresis. The delivery rate can be easily controlled when using iontophoresis, making it a great candidate for closed-loop systems. Iontophoresis can be performed by applying a low current or voltage between two electrodes (cathode and anode). The drug is loaded in a hydrogel that interfaces the electrodes and skin. Mostly small, charged molecules can be delivered by this technique. Positively charged drugs are loaded on the positive electrode and vice-versa. The drug is delivered by electrostatic repulsion and by the disruption of the skin cell caused by the current/voltage. The amount of drug delivered is proportional to the electrical stimuli; however, delivering higher dosages might cause skin irritation due to the higher current/voltage necessary for the delivery. Iontophoretic systems can also be used to remove molecules from the skin by electroosmotic flow. Iontophoretic systems are used to deliver pilocarpine, used to stimulate sweat for cystic fibrosis diagnosis, and lidocaine for local anesthesia (Dhote et al., 2012; Wu et al., 2020). (3) Transdermal delivery assisted by microneedles that physically disrupt the stratum corneum layer for more effective delivery allowing delivery of higher molecular weight molecules including therapeutic proteins and vaccines. For this, hollow, solid, or dissolvable microneedles can be used; the drug is usually loaded inside, around, or within the microneedles, respectively. Other approaches with transdermal delivery via microneedles comprise using the microneedles for the stratum corneum disruption followed by the application of a patch (Luo et al., 2022; Wermeling et al., 2008).

5.8 Implantable drug delivery

Drugs that cannot be taken orally (due to degradation by the stomach or intestine) and are not suitable for transdermal delivery can be delivered via implantable devices. Just like transdermal delivery, implantable drug delivery devices have the advantage of targeted treatment, requiring lower dosages for therapeutic results. However, the application of such devices is invasive requiring specialized tools, and qualified professionals. The drug can be delivered from the implantable devices either passively or actively. Passive implantable devices are the most used, they rely on the diffusion or dissolution of the drug from the device to the bloodstream. One common example of these types of devices is implantable contraceptives. Passive-release devices can also be biodegradable, dismissing completely the need to recover the implant once the drug is depleted. Nevertheless, regardless of the type of implantable device, the implant design, the drug properties, and the desired concentration must be considered in the system development. Among these considerations, the fabrication technique has the most impact on the performance of such devices. There are several manufacturing techniques including solvent casting, injection molding, compression, extrusion, or 3D printing (Stewart et al., 2018). When using any of these approaches, one needs to be aware of the drug's stability, it must be compatible with the manufacturing process, and the microstructure of the implant material must be reproducible once it dictates the releasing rate of the drug. If an uncontrollable burst occurs, there is the risk of reaching toxic levels of the drug, causing overdosage and life-threatening side effects.

Another challenge in developing such technology is the minimization of surface thrombosis and or fibrous encapsulation (Capuani et al., 2022). When a foreigner object is inserted in the body, a layer of scar tissue cells is formed around the object to protect the body from harm; this tissue formation alters the delivery rate and increases the risks on the device removal process, requiring, eventually, surgery approaches. This natural response of the body is the main challenge for implantable devices (Kar et al., 2022).

Active devices deliver drugs on demand upon a stimulus, such as ultrasound, electrical, and thermal, and are the most suitable for closed-loop systems. Pump systems are among the most used active releasing devices. With the advancements in microfabrication techniques, it was possible to have miniaturized pump systems for implantable drug delivery devices (Domingo-Lopez et al., 2022). An example of an implantable device coupled with a pump for delivery is the insulin pump used for the control of type I diabetes. The "artificial pancreas," as it is known, is coupled with a glucose biosensor that monitors the glucose levels continuously, and in real time, when glucose levels are elevated, the insulin pump is activated and the optimum dosage of insulin is delivered autonomously (Domingo-Lopez et al., 2022). There are several types of pumps currently used for closed-loop systems including osmotic pumps, positive displacement pumps, peristaltic pumps, infusion pumps, and controlled-release micropumps (Kim et al., 2016; Sengupta et al., 2014).

Overall, the use of transdermal or implantable drug delivery systems has advantages and disadvantages; drug delivery systems can target the treatment, decreasing the pharmacologic concentration of the drugs while increasing bioavailability. Moreover, there is a possibility to have personalized dosage by combining the drug delivery system with biosensors for a closed feedback loop, and this approach improves patient compliance and avoids over/under dosages. The challenges of drug delivery systems are finding biocompatible materials that can be in close contact or within the body, the need to remove some implants after drug depletion, and the stability of the drugs in the device (Stewart et al., 2018).

5.9 Prospects and outlook

The use of biosensors for continuous monitoring of disease can promote several advancements in precision medicine; as discussed in this chapter, biosensors enable personalized medicine, where the drug concentration is tailored to the individual level, replacing the idea of "one size fits all" of generalized interventions. Precision medicine takes into consideration unique parameters and metabolism responses, the genetic load, and the patient's lifestyle. When combined with automation, biosensors can integrate closed-loop systems that can improve therapy and decrease toxicity, overdose, and under dosage while ensuring patient compliance. There are numerous possibilities when combining different sensor modalities with data analysis and data mining tools in precision medicine. Multiplexed wearable sensors are devices capable of tracking biomolecular and physical (blood pressure, heart rate, temperature) information concurrently. The combination of variations among these parameters allows us to make specific correlations that are used as new disease biomarkers (Sempionatto et al., 2022).

Further, the discovery of new biomarkers and new biomarker signatures can be used for the prevention and prediction of a health-related event. For example, sensor systems comprising continuous glucose sensors combined with insulin pumps can make recommendations of insulin dosage for patients with type 1 diabetes, the sensors can predict the insulin bolus and perform self-calibration and correction based on previous data set and the patient's lifestyle (Zhu et al., 2022). A similar approach was also performed for other closed-loop systems such as the levodopa pump used by patients with Parkinson's disease that combined levodopa biosensors with sensors able to measure patterns of movement, gait, and posture (Teymourian et al., 2022). These types of sensors can generate a large set of data, requiring special analytical tools such as machine learning algorithms to perform an accurate analysis. Besides increasing the power of diagnostics, the combination of biosensors, physical sensors, and machine learning also improves the (bio) chemical sensor reliability, by making real-time corrections of the parameters with changing variables such as body motion, temperature, pH, and humidity.

Biosensors can replace expensive benchtop equipment that requires trained personnel, by offering fast analysis at a low cost. Integrated biosensor systems can track information that was not previously achieved by traditional techniques. Biosensors can correlate fast dynamic events from our body to symptoms, diseases, or drug side effects. Precision medicine has been largely studied by a variety of researchers and engineers; only combined actions will push the technology forward.

References

Aat Bioquest. (2023). *What are the 4 steps of next generation sequencing (NGS)*? https://www.aatbio.com/resources/faq-frequently-asked-questions/What-are-the-4-steps-of-next-generation-sequencing-NGS.

Abdul Ghani, M. A., Nordin, A. N., Zulhairee, M., Che Mohamad Nor, A., Shihabuddin Ahmad Noorden, M., Muhamad Atan, M. K. F., Ab Rahim, R., & Mohd Zain, Z. (2022). Portable electrochemical biosensors based on microcontrollers for detection of viruses: A review. *Biosensors*, *12*(8), 666. Available from https://doi.org/10.3390/bios12080666.

Adepu, S., & Ramakrishna, S. (2021). Controlled drug delivery systems: Current status and future directions. *Molecules (Basel, Switzerland)*. Available from https://doi.org/10.3390/molecules26195905.

Adlercreutz, H. (2002). Phyto-oestrogens and cancer. *The Lancet Oncology*, *3*, 364–373. Available from https://doi.org/10.1016/S1470-2045(02)00777-5.

Ahmed, S., Zhou, Z., Zhou, J., & Chen, S. Q. (2016). Pharmacogenomics of drug metabolizing enzymes and transporters: Relevance to precision medicine. *Genomics, Proteomics and Bioinformatics*, *14*(5), 298–313. Available from https://doi.org/10.1016/j.gpb.2016.03.008.

Ang, M. Y., Low, T. Y., Lee, P. Y., Wan Mohamad Nazarie, W. F., Guryev, V., & Jamal, R. (2019). Proteogenomics: From next-generation sequencing (NGS) and mass spectrometry-based proteomics to precision medicine. *Clinica Chimica Acta*, *498*, 38–46. Available from https://doi.org/10.1016/j.cca.2019.08.010.

Arnaout, R., Lee, R. A., Lee, G. R., Callahan, C., Yen, C. F., Smith, K. P., Arora, R., & Kirby, J. E. (2020). SARS-CoV-2 testing: The limit of detection matters. *BioRxiv*, *2020*, 06.02.131144.

Aron-Wisnewsky, J., & Clément, K. (2016). The gut microbiome, diet, and links to cardiometabolic and chronic disorders. *Nature Reviews Nephrology*, *12*, 169–181. Available from https://doi.org/10.1038/nrneph.2015.191.

Ashley, E. A. (2016). Towards precision medicine. *Nature Reviews. Genetics*, *17*, 507–522. Available from https://doi.org/10.1038/nrg.2016.86.

Astrazeneca. (2022). *Precision medicine for chronic diseases*. Available from https://www.astrazeneca.com/r-d/precision-medicine.html#asthma-0.

Awuchi, C. G., Twinomuhwezi, H., & Awuchi, C. G. (2022). *Hyphenated techniques. Analytical techniques in biosciences* (pp. 125–145). Elsevier. Available from https://doi.org/10.1016/B978-0-12-822654-4.00015-4.

Barghouthi, S. A. (2011). A universal method for the identification of bacteria based on general PCR primers. *Indian Journal of Microbiology*, *51*(4), 430–444. Available from https://doi.org/10.1007/s12088-011-0122-5.

Bariya, M., Nyein, H. Y. Y., & Javey, A. (2018). Wearable sweat sensors. *National Electronics*. Available from https://doi.org/10.1038/s41928-018-0043-y.

Behjati, S., & Tarpey, P. S. (2013). What is next generation sequencing. *Archives of Disease in Childhood: Education and Practice Edition*, *98*(6), 236–238. Available from https://doi.org/10.1136/archdischild-2013-304340.

Berkelman, T. (2020). Fluorescent western blotting: High sensitivity detection of multiple targets. *Current Protocols in Pharmacology*, *88*(1). Available from https://doi.org/10.1002/cpph.72.

Bianchi, V., Boni, A., Fortunati, S., Giannetto, M., Careri, M., & De Munari, I. (2020). A Wi-Fi cloud-based portable potentiostat for electrochemical biosensors. *IEEE Transactions on Instrumentation and Measurement*, *69*(6), 3232–3240. Available from https://doi.org/10.1109/TIM.2019.2928533.

Blanco, J. R., Ferrero, F. J., Campo, J. C., Anton, J. C., Pingarron, J. M., Reviejo, A. J., & Manso, J. (2006). Design of a low-cost portable potentiostat for amperometric biosensors. *2006 IEEE Instrumentation and Measurement Technology Conference Proceedings*, 690–694. Available from https://doi.org/10.1109/IMTC.2006.328670.

Blasques, R. V., Stefano, J. S., Camargo, J. R., Guterres E Silva, L. R., Brazaca, L. C., & Janegitz, B. C. (2022). Disposable prussian blue-anchored electrochemical sensor for enzymatic and non-enzymatic multi-analyte detection. *Sensors and Actuators Reports*, *4*, 100118. Available from https://doi.org/10.1016/j.snr.2022.100118.

Boccia, S., Pastorino, R., Ricciardi, W., Ádány, R., Barnhoorn, F., Boffetta, P., Cornel, M. C., De Vito, C., Gray, M., Jani, A., Lang, M., Roldan, J., Rosso, A., Sánchez, J. M., Dujin, C. M. V., El, C. G. V., Villari, P., & Zawati, M. H. (2020). How to integrate personalized medicine into prevention? Recommendations from the Personalized Prevention of Chronic Diseases (PRECeDI) Consortium. *Public Health Genomics*, *22*(5–6), 208–214.

Bodzon-Kulakowska, A., Bierczynska-Krzysik, A., Dylag, T., Drabik, A., Suder, P., Noga, M., Jarzebinska, J., & Silberring, J. (2007). Methods for samples preparation in proteomic research. *Journal of Chromatography B*, *849*(1–2), 1–31. Available from https://doi.org/10.1016/j.jchromb.2006.10.040.

Bonk, T., & Humeny, A. (2001). MALDI-TOF-MS analysis of protein and DNA. *The Neuroscientist*, *7*(1), 6–12.

Botes, M., De Kwaadsteniet, M., & Cloete, T. E. (2013). Application of quantitative PCR for the detection of microorganisms in water. *Analytical and Bioanalytical Chemistry*, *405*(1), 91–108. Available from https://doi.org/10.1007/s00216-012-6399-3.

Brasier, N., Ates, H. C., Sempionatto, J. R., Cotta, M. O., Widmer, A. F., Eckstein, J., Goldhahn, J., Roberts, J. A., Gao, W., & Dincer, C. (2023). A three-level model for therapeutic drug monitoring of antimicrobials at the site of infection. *The Lancet Infectious Diseases*, *3099*, 1–9. Available from https://doi.org/10.1016/S1473-3099(23)00215-3.

Brazaca, L. C., Bramorski, C. B., Cancino-Bernardi, J., Janegitz, B. C., & Zucolotto, V. (2016). A genosensor for sickle cell anemia trait determination. *Electroanalysis*, *29*(3), 773–777. Available from https://doi.org/10.1002/elan.201600573.

Brazaca, L. C., Dos Santos, P. L., De Oliveira, P. R., Rocha, D. P., Stefano, J. S., Kalinke, C., Abarza Muñoz, R. A., Bonacin, J. A., Janegitz, B. C., & Carrilho, E. (2021). Biosensing strategies for the electrochemical detection of viruses and viral diseases – A review. *Analytica Chimica Acta*, *1159*, 338384. Available from https://doi.org/10.1016/j.aca.2021.338384.

Brazaca, L. C., Imamura, A. H., Gomes, N. O., Almeida, M. B., Scheidt, D. T., Raymundo-Pereira, P. A., Oliveira, O. N., Janegitz, B. C., Machado, S. A. S., & Carrilho, E. (2022). Electrochemical immunosensors using electrodeposited gold nanostructures for detecting the S proteins from SARS-CoV and SARS-CoV-2. *Analytical and Bioanalytical Chemistry*. Available from https://doi.org/10.1007/s00216-022-03956-1.

Brazaca, L. C., Janegitz, B. C., Cancino-Bernardi, J., & Zucolotto, V. (2016). Transmembrane protein-based electrochemical biosensor for adiponectin hormone quantification. *ChemElectroChem*, *3*(6), 1006–1011. Available from https://doi.org/10.1002/celc.201600099.

Brothers, M. C., Debrosse, M., Grigsby, C. C., Naik, R. R., Hussain, S. M., Heikenfeld, J., & Kim, S. S. (2019). Achievements and challenges for real-time sensing of analytes in sweat within wearable platforms. *Accounts of Chemical Research*, *52*, 297–306. Available from https://doi.org/10.1021/acs.accounts.8b00555.

Bustin, S. A. (2017). How to speed up the polymerase chain reaction. *Biomolecular Detection and Quantification*, *12*, 10–14. Available from https://doi.org/10.1016/j.bdq.2017.05.002.

Cani, P. D., & Van Hul, M. (2021). Do diet and microbes really 'PREDICT' cardiometabolic risks? *Nature Reviews Endocrinology*, *17*, 259–260. Available from https://doi.org/10.1038/s41574-021-00480-7.

Capuani, S., Malgir, G., Chua, C. Y. X., & Grattoni, A. (2022). Advanced strategies to thwart foreign body response to implantable devices. *Bioengineering Translational Medicine*, *7*, e10300. Available from https://doi.org/10.1002/BTM2.10300.

Carvalho, J. H. S., Stefano, J. S., Brazaca, L. C., & Janegitz, B. C. (2023). New conductive ink based on carbon nanotubes and glass varnish for the construction of a disposable electrochemical sensor. *Journal of Electroanalytical Chemistry*, *937*, 117428. Available from https://doi.org/10.1016/j.jelechem.2023.117428.

Center for Research on Science, Technology and Society. (2023). *Precision medicine: What is it and what benefits does it bring?* https://www.ipea.gov.br/cts/en/all-contents/articles/articles/128-precision-medicine

Chromatography today. (2023). *What are hyphenated techniques?* https://www.chromatography-today.com/news/lc-ms/48/breaking-news/what-are-hyphenated-techniques/56946#:~:text=Hyphenated%20techniques%3A%20the%20basics&text=More%20specifically%2C%20it%20refers%20to,and%20capillary%20electrophoresis%20(CE

Crowther, J. R. (2009). Systems in ELISA. In J. R. Crowther (Ed.), *The ELISA Guidebook* (2nd ed., pp. 9–42). Humana Press. Available from https://doi.org/10.1007/978-1-60327-254-4_2.

Das, D., Lin, C.-W., & Chuang, H.-S. (2022). LAMP-based point-of-care biosensors for rapid pathogen detection. *Biosensors*, *12*(12), 1068. Available from https://doi.org/10.3390/bios12121068.

Dhote, V., Bhatnagar, P., Mishra, P. K., Mahajan, S. C., & Mishra, D. K. (2012). Iontophoresis: A potential emergence of a transdermal drug delivery system. *Scientia Pharmaceutica*. Available from https://doi.org/10.3797/scipharm.1108-20.

Domingo-Lopez, D. A., Lattanzi, G., Schreiber, H. J., Wallace, L., Wylie, E. J., O'sullivan, R., Dolan, J., & Duffy, G.P., E. B. (2022). Medical devices, smart drug delivery, wearables and technology for the treatment of diabetes mellitus. *Advanced Drug Delivery Reviews*, *185*, 114280. Available from https://doi.org/10.1016/j.addr.2022.114280.

Domon, B., & Aebersold, R. (2006). Mass spectrometry and protein analysis. *Science (New York, N.Y.)*, *312*(5771), 212–217. Available from https://doi.org/10.1126/science.1124619.

Economopoulou, P., Dimitriadis, G., & Psyrri, A. (2015). Beyond BRCA: New hereditary breast cancer susceptibility genes. *Cancer Treatment Reviews*, *41*(1), 1–8. Available from https://doi.org/10.1016/j.ctrv.2014.10.008.

Egger, D., & Bienz, K. (1994). Protein (western) blotting. *Molecular Biotechnology*, *1*(3), 289–305. Available from https://doi.org/10.1007/BF02921696.

El-Aneed, A., Cohen, A., & Banoub, J. (2009). Mass spectrometry, review of the basics: Electrospray, MALDI, and commonly used mass analyzers. *Applied Spectroscopy Reviews*, *44*(3), 210–230. Available from https://doi.org/10.1080/05704920902717872.

Fernández-Lainez, C., Aguilar-Lemus, J. J., Vela-Amieva, M., & Ibarra-González, I. (2012). Tandem mass spectrometry newborn screening for inborn errors of intermediary metabolism: Abnormal profile interpretation. *Current Medicinal Chemistry*, *19*(26), 4511–4522.

Fox, L. T., Gerber, M., Du Plessis, J., & Hamman, J. H. (2011). Transdermal drug delivery enhancement by compounds of natural origin. *Molecules (Basel, Switzerland)*. Available from https://doi.org/10.3390/molecules161210507.

Frayling, I.M. (2023). *What is a mass spectrometer and how does it work*. https://www.excedr.com/blog/mass-spectrometer-function/#: ~ :text = Cost%20to%20Buy%20Mass%20Spectrometers,under%20%2410%2C000%20to%20nearly%20%24100%2C000

Frayling, I. M., Monk, E., & Butler, R. (2006). *PCR-based methods for mutation detection. Molecular diagnostics* (pp. 65–74). Humana Press. Available from https://doi.org/10.1385/1-59259-928-1:065.

Freestyle Libre. (2023). Trend arrows in depth. https://www.freestyle.abbott/ae-en/discover-freestyle-libre/Understanding-reports-and-data/Trend-arrows-in-depth.html

Garattini, L., & Padula, A. (2019). Precision medicine and monoclonal antibodies: Breach of promise. *? Croatian Medical Journal*, *60*(3), 284–289. Available from https://doi.org/10.3325/cmj.2019.60.284.

Garattini, L., Van De Vooren, K., & Curto, A. (2015). Cost-effectiveness of trastuzumab in metastatic breast cancer: Mainly a matter of price in the EU. *Health Policy (Amsterdam, Netherlands)*, *119*(2), 212–216. Available from https://doi.org/10.1016/j.healthpol.2014.12.002.

Gerdts, E., & Regitz-Zagrosek, V. (2019). Sex differences in cardiometabolic disorders. *Nature Medicine*. Available from https://doi.org/10.1038/s41591-019-0643-8.

Gill, P., & Ghaemi, A. (2008). Nucleic acid isothermal amplification technologies—A review. *Nucleosides, Nucleotides and Nucleic Acids*, *27*(3), 224–243. Available from https://doi.org/10.1080/15257770701845204.

Glish, G. L., & Vachet, R. W. (2003). The basics of mass spectrometry in the twenty-first century. *Nature Reviews. Drug Discovery*, *2*(2), 140–150. Available from https://doi.org/10.1038/nrd1011.

Gonzalez-Garay, M. L. (2014). The road from next-generation sequencing to personalized medicine. *Personalized Medicine*, *11*(5), 523–544. Available from https://doi.org/10.2217/pme.14.34.

Goodwin, S., Mcpherson, J. D., & Mccombie, W. R. (2016). Coming of age: Ten years of next-generation sequencing technologies. *Nature Reviews. Genetics*, *17*(6), 333–351. Available from https://doi.org/10.1038/nrg.2016.49.

Grada, A., & Weinbrecht, K. (2013). Next-generation sequencing: Methodology and application. *Journal of Investigative Dermatology*, *133*(8), 1–4. Available from https://doi.org/10.1038/jid.2013.248.

Grayson, M. (2016). Parkinson's disease. *Nature*, *538*, S1. Available from https://doi.org/10.1038/538S1a, S1.

Gupta, N., & Verma, V. K. (2019). *Next-generation sequencing and its application: Empowering in public health beyond reality. Microbial technology for the welfare of society* (pp. 313–341). Springer, https://doi.org/10.1007/978-981-13-8844-6_15.

Healio. (n.d.). *Whole-genome sequencing*. Retrieved June 28, 2023, from https://www.healio.com/hematology-oncology/learn-genomics/whole-genome-sequencing/whole-genome-whole-exome-sequencing-and-targeted-sequencing-panels.

Henderson, C. J., & Wolf, C. R. (1992). Immunodetection of proteins by western blotting. In M. M. Manson (Ed.), *Immunochemical Protocols* (pp. 221–234). Humana Press. Available from https://doi.org/10.1385/0-89603-204-3:221.

Hodson, R. (2016). Precision medicine. *Nature*, *537*, S49. Available from https://doi.org/10.1038/537S49a, S49.

Hotchkiss, R. S., Moldawer, L. L., Opal, S. M., Reinhart, K., Turnbull, I. R., & Vincent, J.-L. (2016). Sepsis and septic shock. *Nature Reviews Disease. Primers*, *2*, 16045. Available from https://doi.org/10.1038/nrdp.2016.45.

Hovorka, R. (2006). Continuous glucose monitoring and closed-loop systems. *Diabetic Medicine*, *23*(1), 1–12. Available from https://doi.org/10.1111/j.1464-5491.2005.01672.x.

Hovorka, R. (2011). Closed-loop insulin delivery: From bench to clinical practice. *Nature Reviews Endocrinology*. Available from https://doi.org/10.1038/nrendo.2011.32.

Hristova, J., & Svinarov, D. (2022). Enhancing precision medicine through clinical mass spectrometry platform. *Biotechnology & Biotechnological Equipment*, *36*(1), 107–117. Available from https://doi.org/10.1080/13102818.2022.2053342.

Huang, C.-Y., Syu, M.-J., Chang, Y.-S., Chang, C.-H., Chou, T.-C., & Liu, B.-D. (2007). A portable potentiostat for the bilirubin-specific sensor prepared from molecular imprinting. *Biosensors and Bioelectronics*, *22*(8), 1694–1699. Available from https://doi.org/10.1016/j.bios.2006.07.036.

Ilan, Y. (2022). Next-generation personalized medicine: Implementation of variability patterns for overcoming drug resistance in chronic diseases. *Journal of Personalized Medicine*, *12* (8). Available from https://doi.org/10.3390/jpm12081303.

Jamuar, S. S., & Tan, E.-C. (2015). Clinical application of next-generation sequencing for Mendelian diseases. *Human Genomics*, *9*(1), 10. Available from https://doi.org/10.1186/s40246-015-0031-5.

Johnston, C. W., & Badran, A. H. (2022). Natural and engineered precision antibiotics in the context of resistance. *Current Opinion in Chemical Biology*, *69*, 102160. Available from https://doi.org/10.1016/j.cbpa.2022.102160.

Kar, A., Ahamad, N., Dewani, M., Awasthi, L., Patil, R., & Banerjee, R. (2022). Wearable and implantable devices for drug delivery: Applications and challenges. *Biomaterials*, *283*, 121435. Available from https://doi.org/10.1016/j.biomaterials.2022.121435.

Kim, H., Kim, K., & Lee, S. J. (2016). Compact and thermosensitive nature-inspired micropump. *Scientific Reports*, *6*, 36085. Available from https://doi.org/10.1038/srep36085.

Kim, J., Campbell, A. S., De Ávila, B. E.-F., & Wang, J. (2019). Wearable biosensors for healthcare monitoring. *Nature Biotechnology*, *37*, 389–406. Available from https://doi.org/10.1038/s41587-019-0045-y.

Kraus, J. M., Lausser, L., Kuhn, P., Jobst, F., Bock, M., Halanke, C., Hummel, M., Heuschmann, P., & Kestler, H. A. (2018). Big data and precision medicine: Challenges and strategies with healthcare data. *International Journal of Data Science and Analytics*, *6*(3), 241–249. Available from https://doi.org/10.1007/s41060-018-0095-0.

Kurien, B. T., & Scofield, R. H. (2006). Western blotting. *Methods (San Diego, Calif.)*, *38*(4), 283–293. Available from https://doi.org/10.1016/j.ymeth.2005.11.007.

Kushi, L. H., Doyle, C., Mccullough, M., Rock, C. L., Demark-Wahnefried, W., Bandera, E. V., Gapstur, S., Patel, A. V., Andrews, K., & Gansler, T. (2012). American cancer society guidelines on nutrition and physical activity for cancer prevention. *CA: A Cancer Journal for Clinicians*, *62*, 30–67. Available from https://doi.org/10.3322/caac.20140.

Le Tourneau, C., Borcoman, E., & Kamal, M. (2019). Molecular profiling in precision medicine oncology. *Nature Medicine*, *25*(5), 711–712. Available from https://doi.org/10.1038/s41591-019-0442-2.

Leng, S. X., Mcelhaney, J. E., Walston, J. D., Xie, D., Fedarko, N. S., & Kuchel, G. A. (2008). ELISA and multiplex technologies for cytokine measurement in inflammation and aging research. *The Journals of Gerontology Series A: Biological Sciences and Medical Sciences*, *63*(8), 879–884. Available from https://doi.org/10.1093/gerona/63.8.879.

Liu, L., Li, Y., Li, S., Hu, N., He, Y., Pong, R., Lin, D., Lu, L., & Law, M. (2012). Comparison of next-generation sequencing systems. *Journal of Biomedicine and Biotechnology*, *2012*, 1–11. Available from https://doi.org/10.1155/2012/251364.

Loo, J. A., Udseth, H. R., & Smith, R. D. (1989). Peptide and protein analysis by electrospray ionization-mass spectrometry and capillary electrophoresis-mass spectrometry. *Analytical Biochemistry*, *179*(2), 404–412. Available from https://doi.org/10.1016/0003-2697(89)90153-X.

Luo, X., Yu, Q., Liu, Y., Gai, W., Ye, L., Yang, L., & Cui, Y. (2022). Closed-loop diabetes minipatch based on a biosensor and an electroosmotic pump on hollow biodegradable microneedles. *ACS Sensors*, *7*, 1347–1360. Available from https://doi.org/10.1021/acssensors.1c02337.

Mackay, I. M. (2002). Real-time PCR in virology. *Nucleic Acids Research*, *30*(6), 1292–1305. Available from https://doi.org/10.1093/nar/30.6.1292.

Manikkath, J., & Subramony, J. A. (2021). Toward closed-loop drug delivery: Integrating wearable technologies with transdermal drug delivery systems. *Advanced Drug Delivery Reviews*, *179*, 113997. Available from https://doi.org/10.1016/j.addr.2021.113997.

Manzari, M. T., Shamay, Y., Kiguchi, H., Rosen, N., Scaltriti, M., & Heller, D. A. (2021). Targeted drug delivery strategies for precision medicines. *Nature Reviews Materials*, *6*(4), 351–370. Available from https://doi.org/10.1038/s41578-020-00269-6.

Matsuda, K. (2017). *PCR-based detection methods for single-nucleotide polymorphism or mutation*, 45–72. Available from https://doi.org/10.1016/bs.acc.2016.11.002.

Mayne, S. T., Playdon, M. C., & Rock, C. L. (2016). Diet, nutrition, and cancer: past, present and future. *Nature Reviews Clinical Oncology*, *13*, 504–515. Available from https://doi.org/10.1038/nrclinonc.2016.24.

Mclafferty, F. W. (1981). Tandem mass spectrometry. *Science (New York, N.Y.)*, *214*(4518), 280–287. Available from https://doi.org/10.1126/science.7280693.

Medinger, R., Nolte, V., Pandey, R. V., Jost, S., Ottenwälder, B., Schlötterer, C., & Boenigk, J. (2010). Diversity in a hidden world: potential and limitation of next-generation sequencing for surveys of molecular diversity of eukaryotic microorganisms. *Molecular Ecology*, *19*, 32–40. Available from https://doi.org/10.1111/j.1365-294X.2009.04478.x.

Merck. (2023a). *Human CD30 Elisa Kit*. Available from https://www.sigmaaldrich.com/US/en/product/sigma/rab0491.

Merck. (2023b). *Mouse CD40 Elisa Kit*. Available from https://www.sigmaaldrich.com/BR/pt/product/sigma/rab0086.

Merck. (2023c). *Western blot*. Available from https://www.sigmaaldrich.com/BR/pt/applications/protein-biology/western-blotting.

Min, J., Sempionatto, J. R., Teymourian, H., Wang, J., & Gao, W. (2021). Wearable electrochemical biosensors in North America. *Biosensors & Bioelectronics*, *172*, 112750. Available from https://doi.org/10.1016/j.bios.2020.112750.

Naithani, N., Atal, A. T., Tilak, T. V. S. V. G. K., Vasudevan, B., Misra, P., & Sinha, S. (2021). Precision medicine: Uses and challenges. *Medical Journal Armed Forces India*, *77*(3), 258–265. Available from https://doi.org/10.1016/j.mjafi.2021.06.020.

Naithani, N., Sinha, S., Misra, P., Vasudevan, B., & Sahu, R. (2021). Precision medicine: Concept and tools. *Medical Journal Armed Forces India*, *77*(3), 249–257. Available from https://doi.org/10.1016/j.mjafi.2021.06.021.

Oliveira, G. C. M., De, Carvalho, J. H., De, S., Brazaca, L. C., Vieira, N. C. S., & Janegitz, B. C. (2020). Flexible platinum electrodes as electrochemical sensor and immunosensor for Parkinson's disease biomarkers. *Biosensors and Bioelectronics*, *152*, 112016. Available from https://doi.org/10.1016/j.bios.2020.112016.

Paul, A., & Paul, S. (2014). The breast cancer susceptibility genes (BRCA) in breast and ovarian cancers. *Frontiers in Bioscience-Landmark*, *19*(4), 605–618.

Plaxco, K. W., & Soh, H. T. (2011). Switch-based biosensors: A new approach towards real-time, in vivo molecular detection. *Trends in Biotechnology*. Available from https://doi.org/10.1016/j.tibtech.2010.10.005.

Poewe, W., Seppi, K., Tanner, C. M., Halliday, G. M., Brundin, P., Volkmann, J., Schrag, A.-E., & Lang, A. E. (2017). Parkinson disease. *Nature Reviews Disease Primers*, *3*, 17013. Available from https://doi.org/10.1038/nrdp.2017.13.

Prausnitz, M. R., & Langer, R. (2008). Transdermal drug delivery. *Nature Biotechnology*, *26*, 1261–1268. Available from https://doi.org/10.1038/nbt.1504.

Rangel, A. E., Hariri, A. A., Eisenstein, M., & Soh, H. T. (2020). Engineering aptamer switches for multifunctional stimulus-responsive nanosystems. *Advanced Materials*. Available from https://doi.org/10.1002/adma.202003704.

Reddy, B., Hassan, U., Seymour, C., Angus, D. C., Isbell, T. S., White, K., Weir, W., Yeh, L., Vincent, A., & Bashir, R. (2018). Point-of-care sensors for the management of sepsis. *Nature Biomedical Engineering*, *2*, 640–648. Available from https://doi.org/10.1038/s41551-018-0288-9.

Rio, D. C. (2014). Reverse transcription–polymerase chain reaction. *Cold Spring Harbor Protocols*, *2014*(11). Available from https://doi.org/10.1101/pdb.prot080887, pdb.prot080887.

Roy, J., Jain, N., Singh, G., Das, B., & Mallick, B. (2019). Small RNA proteome as disease biomarker: An incognito treasure of clinical utility. In B. Mallick (Ed.), *AGO-driven non-coding RNAs* (pp. 101–136). Elsevier. Available from https://doi.org/10.1016/B978-0-12-815669-8.00005-1.

Sempionatto, J. R., Brazaca, L. C., García-Carmona, L., Bolat, G., Campbell, A. S., Martin, A., Tang, G., Shah, R., Mishra, R. K., Kim, J., Zucolotto, V., Escarpa, A., & Wang, J. (2019). Eyeglasses-based tear biosensing system: Non-invasive detection of alcohol, vitamins and glucose. *Biosensors and Bioelectronics*, *137*, 161–170. Available from https://doi.org/10.1016/j.bios.2019.04.058.

Sempionatto, J. R., Lasalde-Ramírez, J. A., Mahato, K., Wang, J., & Gao, W. (2022). Wearable chemical sensors for biomarker discovery in the omics era. *Nature Reviews Chemistry*, *2022*(6), 1–17. Available from https://doi.org/10.1038/s41570-022-00439-w.

Sempionatto, J. R., Lin, M., Yin, L., De La Paz, E., Pei, K., Sonsa-Ard, T., De Loyola Silva, A. N., Khorshed, A. A., Zhang, F., Tostado, N., Xu, S., & Wang, J. (2021). An epidermal patch for the simultaneous monitoring of haemodynamic and metabolic biomarkers. *Nature Biomedical Engineering*..

Sengupta, S., Patra, D., Ortiz-Rivera, I., Agrawal, A., Shklyaev, S., Dey, K. K., Córdova-Figueroa, U., Mallouk, T. E., & Sen, A. (2014). Self-powered enzyme micropumps. *Nature Chemistry*, *6*, 415–422. Available from https://doi.org/10.1038/nchem.1895.

Sino Biological. (2023). *Western blot applications for medical diagnosis*. https://www.sinobiological.com/category/wb-medical-diagnosis#:~:text=In%20recent%20medical%20field%2C%20Western,)%2C%20HBV%20(Hepatitis%20B%20Virus.

Slatko, B. E., Gardner, A. F., & Ausubel, F. M. (2018). Overview of next-generation sequencing technologies. *Current Protocols in Molecular Biology*, *122*(1). Available from https://doi.org/10.1002/cpmb.59.

Stewart, S., Domínguez-Robles, J., Donnelly, R., & Larrañeta, E. (2018). Implantable polymeric drug delivery devices: Classification, manufacture, materials, and clinical applications. *Polymers (Basel)*, *10*, 1379. Available from https://doi.org/10.3390/polym10121379.

Strom, S. P. (2016). Current practices and guidelines for clinical next-generation sequencing oncology testing. *Cancer Biology & Medicine*, *13*(1), 3–11.

Subramanian, M., Wojtusciszyn, A., Favre, L., Boughorbel, S., Shan, J., Letaief, K. B., Pitteloud, N., & Chouchane, L. (2020). Precision medicine in the era of artificial intelligence: Implications in chronic disease management. *Journal of Translational Medicine*, *18* (1). Available from https://doi.org/10.1186/s12967-020-02658-5.

Teymourian, H., Barfidokht, A., & Wang, J. (2020). Electrochemical glucose sensors in diabetes management: An updated review (2010–2020). *Chemical Society Reviews*. Available from https://doi.org/10.1039/d0cs00304b.

Teymourian, H., Tehrani, F., Longardner, K., Mahato, K., Podhajny, T., Moon, J. M., Kotagiri, Y. G., Sempionatto, J. R., Litvan, I., & Wang, J. (2022). Closing the loop for patients with Parkinson disease: Where are we. *Nature Reviews Neurology*, *18*, 497–507. Available from https://doi.org/10.1038/s41582-022-00674-1.

Thermo Fisher Scientific. (2023a). *PCR cycling parameters—six key considerations for success*. https://www.thermofisher.com/br/en/home/life-science/cloning/cloning-learning-center/invitrogen-school-of-molecular-biology/pcr-education/pcr-reagents-enzymes/pcr-cycling-considerations.html#: ~ :text = The%20number%20of%20cycles%20is,to%20produce%20a%20sufficient%20yield.

Thermo Fisher Scientific. (2023b). *Next-generation sequencing illumina workflow—4 key steps*. https://www.thermofisher.com/br/en/home/life-science/cloning/cloning-learning-center/invitrogen-school-of-molecular-biology/next-generation-sequencing/illumina-workflow.html

Thermo Fisher Scientific. (2023c). *Protein sample preparation for mass spectrometry*. https://www.thermofisher.com/br/pt/home/life-science/protein-biology/protein-biology-learning-center/protein-biology-resource-library/pierce-protein-methods/sample-preparation-mass-spectrometry.html

U.S. Food and Drug Administration. (2017). *FY2015 regulatory science research report: Narrow therapeutic index drugs*. https://www.fda.gov/industry/generic-drug-user-fee-amendments/fy2015-regulatory-science-research-report-narrow-therapeutic-index-drugs

U.S. Food and Drug Administration (FDA). (2023). *Precision medicine*. https://www.fda.gov/medical-devices/in-vitro-diagnostics/precision-medicine#: ~ :text = Precision%20medicine%2C%20sometimes%20known%20as,genes%2C%20environments%2C%20and%20lifestyles

Van Belkum, A. (1994). DNA fingerprinting of medically important microorganisms by use of PCR. *Clinical Microbiology Reviews*, *7*(2), 174–184. Available from https://doi.org/10.1128/CMR.7.2.174.

Vashist, S. (2013). Continuous glucose monitoring systems: A review. *Diagnostics*, *3*(4), 385–412. Available from https://doi.org/10.3390/diagnostics3040385.

Vidovic, M. (1998). Reverse transcription polymerase chain reaction. In (1st ed.). R. Martin (Ed.), *Neuroscience methods* (Vol. 1). CRC Press.

Voelkerding, K. V., Dames, S. A., & Durtschi, J. D. (2009). Next-generation sequencing: From basic research to diagnostics. *Clinical Chemistry*, *55*(4), 641–658. Available from https://doi.org/10.1373/clinchem.2008.112789.

Wang, M., Yang, Y., Min, J., Song, Y., Tu, J., Mukasa, D., Ye, C., Xu, C., Heflin, N., Mccune, J. S., Hsiai, T. K., Li, Z., & Gao, W. (2022). A wearable electrochemical biosensor for the monitoring of metabolites and nutrients. *Nature Biomedical Engineering*, *6*, 1225–1235. Available from https://doi.org/10.1038/s41551-022-00916-z.

Wang, X., Zhong, W., Bu, J., Li, Y., Li, R., Nie, R., Xiao, C., Ma, K., Huang, X., & Li, Y. (2019). Exosomal protein CD82 as a diagnostic biomarker for precision medicine for breast cancer. *Molecular Carcinogenesis, 58*(5), 674–685. Available from https://doi.org/10.1002/mc.22960.

Wermeling, D. P., Banks, S. L., Hudson, D. A., Gill, H. S., Gupta, J., Prausnitz, M. R., & Stinchcomb, A. L. (2008). Microneedles permit transdermal delivery of a skin-impermeant medication to humans. *Proceedings of the National Academy of Sciences., 105*, 2058–2063. Available from https://doi.org/10.1073/pnas.0710355105.

Westermeier, R. (2014). Looking at proteins from two dimensions: a review on five decades of 2D electrophoresis. *Archives of Physiology and Biochemistry, 120*(5), 168–172.

Wheeler, H. E., Maitland, M. L., Dolan, M. E., Cox, N. J., & Ratain, M. J. (2013). Cancer pharmacogenomics: Strategies and challenges. *Nature Reviews Genetics, 14*(1), 23–34. Available from https://doi.org/10.1038/nrg3352.

Wittmeier, P., & Hummel, S. (2022). Agarose gel electrophoresis to assess PCR product yield: Comparison with spectrophotometry, fluorometry and qPCR. *Biotechniques, 72*(4), 155–158. Available from https://doi.org/10.2144/btn-2021-0094.

Wong, Y. K., Tsang, H. F., Xue, V. W., Chan, C. M., Au, T. C., Cho, W. C., Chan, L. W., & Wong, S. C. (2017). Applications of digital PCR in precision medicine. *Expert Review of Precision Medicine and Drug Development, 2*(3), 177–186. Available from https://doi.org/10.1080/23808993.2017.1347482.

Wu, C., Jiang, P., Li, W., Guo, H., Wang, J., Chen, J., Prausnitz, M. R., & Wang, Z. L. (2020). Self-powered iontophoretic transdermal drug delivery system driven and regulated by biomechanical motions. *Advanced Functional Materials, 30*, 1907378. Available from https://doi.org/10.1002/adfm.201907378.

Yu, J., Zhang, Y., Yan, J., Kahkoska, A. R., & Gu, Z. (2018). Advances in bioresponsive closed-loop drug delivery systems. *International Journal of Pharmaceutics, 544*, 350–357. Available from https://doi.org/10.1016/j.ijpharm.2017.11.064.

Zhang, L., Parvin, R., Fan, Q., & Ye, F. (2022). Emerging digital PCR technology in precision medicine. *Biosensors and Bioelectronics, 211*, 114344. Available from https://doi.org/10.1016/j.bios.2022.114344.

Zhang, S., Garcia-D'angeli, A., Brennan, J. P., & Huo, Q. (2014). Predicting detection limits of enzyme-linked immunosorbent assay (ELISA) and bioanalytical techniques in general. *Analyst, 139*, 439–445.

Zhao, Y., Chen, F., Li, Q., Wang, L., & Fan, C. (2015). Isothermal amplification of nucleic acids. *Chemical Reviews, 115*(22), 12491–12545. Available from https://doi.org/10.1021/acs.chemrev.5b00428.

Zhou, S.-F., Di, Y. M., Chan, E., Du, Y.-M., Chow, V. D.-W., Xue, C. C., Lai, X., Wang, J.-C., Li, C. G., Tian, M., & Duan, W. (2008). Clinical pharmacogenetics and potential application in personalized medicine. *Current Drug Metabolism, 9*(8), 738–784. Available from http://www.imm.ki.se/.

Zhu, T., Uduku, C., Li, K., Herrero, P., Oliver, N., & Georgiou, P. (2022). Enhancing self-management in type 1 diabetes with wearables and deep learning. *npj Digital Medcine, 5*, 78. Available from https://doi.org/10.1038/s41746-022-00626-5.

Chapter 6

Recent advancements in bioreceptors and materials for biosensors

Ashutosh Kumar[1] **and Kuldeep Mahato**[2]

[1]*Department of Electrical Engineering, University of Notre Dame, Notre Dame, IN, United States,* [2]*Department of Nanoengineering, University of California, San Diego, CA, United States*

6.1 Introduction

Millions of people are affected by severe forms of communicable and noncommunicable diseases each year due to lack of or inadequate healthcare, which makes the biggest concerns about one's well-being in recent days. Infectious and noninfectious diseases such as diabetes, cancer, malaria, and chronic lung ailments cause approximately 56 million deaths every year. Addressing the challenges in a global context, the close collaboration of government policies, private partners, and healthcare supply manufacturers has constantly worked to deliver cost-effective ways for diagnoses and cures. As a result, various such systems and modules for health monitoring have been leveraged in the recent past. Microbial culture-based diagnosis, chromatography, mass spectrometry, and spectroscopy are among such additions to the effective diagnosis of diseases (Kumar, Roy et al., 2019). However, their sophistication, bulkiness, and need for dedicated space and skilled operators limit the scope of their usage as monitoring devices. In recent advancements, several modules have also been developed based on biosensors for easy, quick, and sensitive detection of biomarkers but also are reported for achieving comparable analytical performance (Chandra & Mahato, 2022). Biosensors are miniaturized analytical devices that include a transducer and a biological recognition element, which function as a device to quantify and/or detect the target analyte even in a small amount of the sample (Kumar et al., 2019). These miniaturized devices are also capable of predicting or diagnosing various lethal diseases in their early stages owing to their excellent analytical performance. For more information regarding the basics of biosensors and their applications in the medical area, please refer to Chapters 3, 4, and 5 of this book.

Biosensors in Precision Medicine. DOI: https://doi.org/10.1016/B978-0-443-15380-8.00007-2

Although biosensors have been widely explored across the globe to push their capabilities toward excellent analytical performance viz. wide detection range, lower detection limit, high selectivity, better sensitivity, extended shelf-life, etc. (Mahato, Kumar, Maurya et al., 2018), the intrinsic nature of the bioreceptor and the sensing surface materials put a barrier toward pushing its limits (Bollella & Katz, 2020). The proteinaceous nature of most bioreceptors (enzymes, antibodies, cell receptors, etc.) majorly adds up the resistance in the electron shuttling during signal transduction, which impacts the electrochemical analytical performance (Bollella & Katz, 2020). In addition, the sensing surfaces and the materials for the fabrication support also play a crucial role in achieving better analytical performance in detecting/monitoring important analytes (Beaver et al., 2021). Hence, nanodimensional, conducting, and flexible materials are recently being employed to achieve durable, conformally tuned, materials for delivering the unobtrusive monitoring of the analyte. In this chapter, we comprehensively discuss various recent advancements regarding bioreceptors and materials used for developing selective and sensitive biosensors. We have briefly described the architecture, the bioreceptor selection criteria, and the formats of the biosensor to give better insights to the readers. Specifically, we describe the type of the emerging bioreceptors including the engineered antibody, enzymes, and molecularly imprinted polymers alongside various other conventionally used ones (Fig. 6.1). Here, the categorization of the biosensors is on the basis of their mode of interaction and origin, that is, enzymatic, or affinity-based or synthetically made molecularly imprinted polymer (MIP) nanomaterials, etc. The materials used for sensing are also described for achieving durable, reliable, and excellent, signal transduction. While the hydrogel-based materials

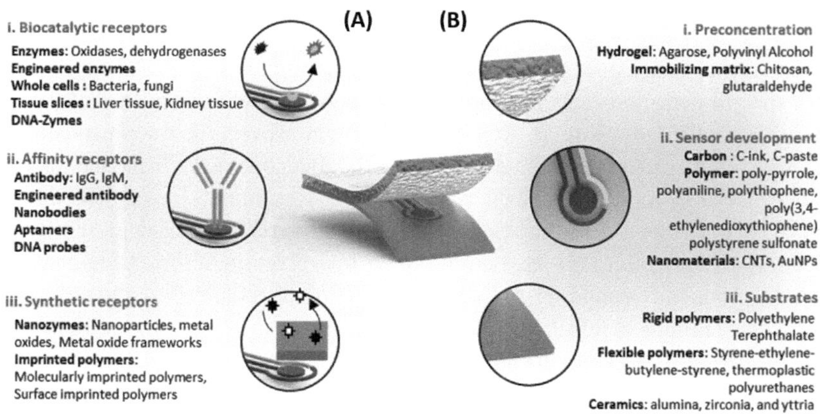

FIGURE 6.1 Biorecognition modalities and materials for biosensor development: (A) bioreceptors: (i) biocatalytic receptors, (ii) bioaffinity-based receptors, and (iii) synthetic receptors; (B) materials: (i) preconcentration, (ii) sensor materials, and (iii) base substrate materials for developing biosensors components.

are mainly used for analyte preconcentration, the cross-linking polymeric matrices such as chitosan, glutaraldehyde, and Nafion provide the efficient immobilization of the bioreceptors and sometimes improve the selectivity by repelling the same charged interfering molecular species from the sensing surfaces.

6.2 Architecture, types, and receptor selection of the biosensor

Biosensing modules can be hand-held, wearable, implantable, and ingestible devices where the desired form factor, sensing mechanism, and applications rule its overall architecture (Morales & Halpern, 2018; Perumal & Hashim, 2014; Soleymani & Li, 2017). For example, biosensors for the quantification of blood biomarkers typically require high sensitivity and analytical approaches capable of quantification, thus the electrochemical modules would be preferred over colorimetry-based biosensors, which usually support qualitative estimations. Mainly, the recognition elements are the specific enzymes, antibodies, nucleic acids, or aptamers under optimal conditions for the detection of particular analytes. In addition, ion-selective membranes are used as a recognition layer for chemical ion sensing.

The trade-off between the sensing module with the form factor design is balanced to develop efficient, sensitive, customizable, and usable biosensors for dedicated applications (Mahato & Wang, 2021). Although the major factors in developing the biosensor are governed by its application, the selection of the sensing mechanism, biorecognition elements, and desirable form factors are the crucial parameters for developing the sensors. For instance, the selection of the enzymes as a biorecognition component is not only meant to introduce selective detection but also to attain operational stability for the biosensors for metabolite detections (Newman & Setford, 2006). Enzymes may provide the continuous monitoring of clinically important metabolites (Tehrani et al., 2022) (viz. glucose, lactate, alcohol, etc.) owing to the constant catalysis and their regenerable properties. On the other hand, the enzyme's globular sheath-like structures prevent the channelization of the electrons from the active centers to the electrode surface (Pravda et al., 2014). Also, its susceptibilities toward the temperature cause the degradation of the conformal native structure which eventually decreases the analytical performance by losing the active site or the binding sites (Prasad et al., 2016).

In addition, antibodies are other biorecognition elements that are commonly used for the detection of biomarkers associated with various diseases. While enzymes are limited to fewer biomarkers (metabolites such as glucose and lactate), the use of an antibody in biosensors offers the detection of a wide range of biomarkers of various diseases, including cancer, neurodegenerative disorders, inflammatory diseases, and diabetes. The antibody-based biosensors offer a sensitive and selective estimation of such biomarkers; however, these have limitations for real-time, continuous assessments due to their nonregenerable interactions with the targets. Thus alternate options have also been explored including the nanozymes and genetically modified enzymes for developing the biosensors.

Unlike the enzymatic biorecognition molecules, bioaffinity-based biorecognition elements rely on biological interactions such as antibody−antigen, and avidin−biotin-based interactions. In the case of biosensors, antibodies, truncated antibodies, nanobodies, etc. have been used as bioreceptors (Dkhar et al., 2022; Li, 2020). Antibodies not only leverage high selectivity and sensitivity but also detect a wide range of targets of clinical importance, which are majorly proteinaceous. Similarly, to preserve the native state of the bioreceptors, stabilizers can be incorporated into the fabrication system, which enables its elongated shelf-life (Bhatnagar et al., 2018; Liébana & Drago, 2016; Tavakoli & Tang, 2017). In a study, the immunosensors were tested with and without the stabilizers, and they found a significant change in the fluctuation of the analytical performances when checked after 60 days (Park et al., 2003); in such cases, the retention of the stability can be attributed to the properties of the immobilized antibody (viz. hydrophobicity, electrostatic potential, van der Waals forces, hydrogen bonding) surface, and the stabilizer matrix. The following section describes the various advances in bioreceptor development and the materials used for enhanced analytical performance and robustness.

6.3 Bioreceptors for biosensors

The biorecognition element or bioreceptor is a primordial part of the biosensors, which is responsible for selectively or specifically interacting with the analyte. Processes related to design, manufacturing, and quality control must be carefully considered. The interaction between the target analyte and the biorecognition element is the crucial event to obtaining the measurable signal generation (Mahato, Baranwal et al., 2018). These interactions of the bioreceptors and the analytes are driven by intermolecular forces (hydrophobic interaction, hydrogen bonds, etc.). For the catalytic-based bioreceptor, the substrate interacts with the bioreceptors (i.e., enzymes), which induces the reorientation of the substrates and the conformational changes to the enzymes at the active site for perfect rearrangements of the substrate followed by the molecular interaction of the substrate to the active site (metallic core) of the enzyme that initiates the catalysis. In affinity-based receptors, hydrophobic interactions between epitope and paratope play an important role. Mimicking the fundamental interactions between, various synthetic bioreceptors have been reported where the nanozymes and the imprinted molecules have particularly found great attention. While the nanozymes mimic the catalytic activities by interacting with the substrates (such as nanoparticles, AuNPs, AgNPs, and metal oxides), the imprinted molecules use the cavities that selectively interact with the target molecules. A biosensor is a device that uses biological recognition components to find and quantify particular analytes. The component known as a transducer transforms the signal produced by the recognition element into a measurable signal, such as an electrical, optical, or mass-based signal. To maximize the sensitivity and specificity of the biosensor, the transducer must be integrated with the recognition component. This can require creating unique microfabricated components. To provide precise and trustworthy results, the biosensor's signal must be processed and examined in real-time. Advanced hardware and software components may be needed for this. To provide a quantitative measurement of the analyte, the signal processor amplifies and processes the signal produced by the transducer. Biosensors must be reliable and consistent across time and with various samples. Integrating these components into a fabricated biosensor device involves several difficulties. To guarantee excellent sensitivity, specificity, and stability, the recognition element must be optimized. A crucial step to maximize the sensitivity and specificity of the biosensor is the integration of the transducer with the recognition element. The selection of the biorecognition element is commonly performed based on the analyte of interest. For instance, if the blood glucose is being targeted, the glucose-cleaving enzymes, viz. glucose oxidase, and glucose dehydrogenase are commonly employed as biorecognition elements (Purohit et al., 2022; Wang, 2008). Similarly, for any protein molecule, antibodies or aptamers are typically used as a biorecognition element (Mahato, Kumar, Srivastava et al., 2018; Purohit, Kumar et al., 2019).

6.3.1 Biocatalytic recognition elements

In the context of biosensors, the device relies on a chemical reaction facilitated by macromolecules, which can be found in their natural environment, previously isolated, or artificially produced, which mainly include enzymes (single or multiple), entire cells like bacteria, fungi, eukaryotic cells, yeast, or cell components like mitochondria or cell walls, and slices of plant or animal tissues. The reaction between one or multiple analytes (substrates) and the enzyme, cells, or tissue in the presence of enzyme(s) results in the production of one or more products and the signal generated in this process is used for quantifying the level of the targets in the samples. The enzyme-based catalytic bioreceptor exploits the intrinsic catalytic behavior of enzymes (Agarwal, 2006). As the enzymes contain active site(s) for selectively catalyzing its substrates, in a biosensor, the isolates of such enzymes are employed to catalyze the metabolite of interest. These metabolites are mainly a signature of the clinical conditions. For instance, high glucose level is related to diabetes, also known as hyperglycemia. Therefore, the immobilization of the glucose-catalyzing enzymes (i.e., glucose oxidase, and glucose dehydrogenase) has been used for developing the biosensors. While enzymes have naturally evolved to induce cellular metabolic activities in their native state, these are commonly exploited for various bioengineering applications, including biosensors. The enzymes are mainly categorized under oxidoreductases, transferases, hydrolases, lyases, ligases, and isomerases (Bell et al., 2021; Wu, 2021). The majority of enzymes applied to biosensors belong to the oxidoreductase family due to their properties of catalyzing the substrate and generating either an electroactive product or by-product, which have been employed for the indirect sensing of the molecules of clinical importance. The catalytic activity is performed by the specific site called the "active site" or "active center," which is highly specific to the substrate, and thus the selectivity is ensured in the biosensing (Morrison, 2020). Using the oxidoreductase family of enzymes, sensors for glucose (Cheng et al., 2022; Jędrzak et al., 2022; Kabir et al., 2022), uric acid (Jain et al., 2019; Verma et al., 2019; Xiao et al., 2022), ascorbate (Barberis et al., 2014; Nakamoto et al., 2010; Sempionatto et al., 2020), lactate (Meng et al., 2021; Regiart et al., 2022; Zhu et al., 2022), etc. have been developed. Using enzymes has great advantages for the regeneration of the biosensing probe and thus facilitates the continuous monitoring of the analytes. This capability of the enzyme has widened the development of biosensors for wearables and continuous monitoring modules; however, their intrinsic limitation has restricted its application in temperature-sensitive environments. Owing to its protein folding, enzymes get denatured when exposed to higher environmental temperatures, which cause a decrease in analytical performance (Peterson et al., 2007).

In some cases, whole cells and microbes have also been used as bioreceptors for detecting clinical biosignatures. Incorporation of these bioreceptors has leveraged the unique advantages of rapid, in situ assessments due to their naturally inbuilt sensor arrays for various sensors such as cellular bioreceptors, enzymes, and ion channels. The major advantages of these types of bioreceptors are their responsiveness in the physiological regimen of the analyte concentrations, which makes them better suitable for the sensing of clinically relevant analytes (Liu et al., 2014). Owing to such advantages, these types of sensing have actively been incorporated in the studies of medical diagnoses, pharmaceutical evaluation, physiological analysis, etc. However, the developments are still facing the intrinsic limitations of satisfactory stability, selectivity, and prolonging the bioreceptor lifespan for prolonged sensing (Bousse, 1996). Various sensors have been developed for bacterial pathogens (Wu, Wang et al., 2021), extracellular pH, metabolites, action potentials, etc. by employing whole cells as bioreceptors; moreover, whole cell-based biosensors also have shown the multiplexed capability for the detection of renal biomarkers for chronic kidney disease targeting urea and uric acid (Köse et al., 2023).

Another category of catalytic bioreceptors consists of tissue slices from animal or plant origins. Tissue slices are the reservoir of the secretory and intracellular enzymes. Employing the tissue slices from specialized organs such as the liver (Rechnitz et al., 1979), kidney, etc.; as bioreceptor offers various advantages. These offer the ability to use the nature-optimized enzyme pathways, enhanced stability of the enzyme in a native environment, multistep process for the detection the large supply of the fresh catalyst and the cofactors. Using organ tissue slices, various biosensors have been reported in the past including arginine, glutamine, and hydrogen peroxide; similarly, plant tissue slices and homogenates have also been used for the detection of the analytes for clinical importance in the recent past, which includes phenolic compounds (Ozcan & Sagiroglu, 2014), catechol (Dinçkaya et al., 1999), and tyrosine (Schubert et al., 1983). However, the challenges associated with these include the regeneration and shelf-life, heterogeneity in the cell populations, and high interference with the signals generated from the coexisting enzymes and the analyte couple during the sensing process.

The deoxy ribozymes (DNAzymes) have also been used for the development of catalysis-based biosensors using ligation, phosphorylation, and cleavage of the DNA as well as RNA. Its activity mostly relies upon the presence of specific molecules of metallic and proteinaceous nature, which facilitate in designing the sensors for detecting the specific biomarker/biochemical (Khan et al., 2021). The major advantages of these recognition elements are their relative stability and sensitivity. Using DNAzymes, various sensors have been developed for the detection of pathogens and biomarkers of clinical importance (Ali et al., 2019; Lee et al., 2019). The general schemes for biocatalysts-based biosensing are shown in Fig. 6.2.

In addition to the traditional catalytic biorecognition elements, such as enzymes, cells, and tissue slices-based biorecognition elements, there are a new generation of recognition elements has also been employed, where the emerging biorecognition elements have been used. In the following sections, such biocatalytic bioreceptors for sensing are discussed.

6.3.1.1 Recombinant enzymes

The catalytic activity of the enzymes is related to the operating pH. Thus, fluctuation in pH-inducing factors (such as substrate, oxygen, and microenvironment) plays a crucial role in its analytical performance (Kumašin et al., 2022). Therefore, pH and temperature-tolerant enzymes are of utmost requirement to develop biosensors for use in extreme environments and sustain a longer shelf-life. In this view, genetically engineered (recombinant) enzymes have found great attention for enhancing pH and temperature resistance. Recombinant enzymes are genetically engineered counterparts of the enzymes. Typically, the active sites of a naturally occurring enzyme are maintained intact, while undesirable protein fragments are removed (Hartley et al., 2019; Zhang et al., 2016). For producing recombinant enzymes, the protein-coding DNA sequences are engineered by recombinant DNA technology. Since the wild-type enzymes are unstable in the nonnative environment, that is, standard laboratory conditions and immobilized onto a surface,

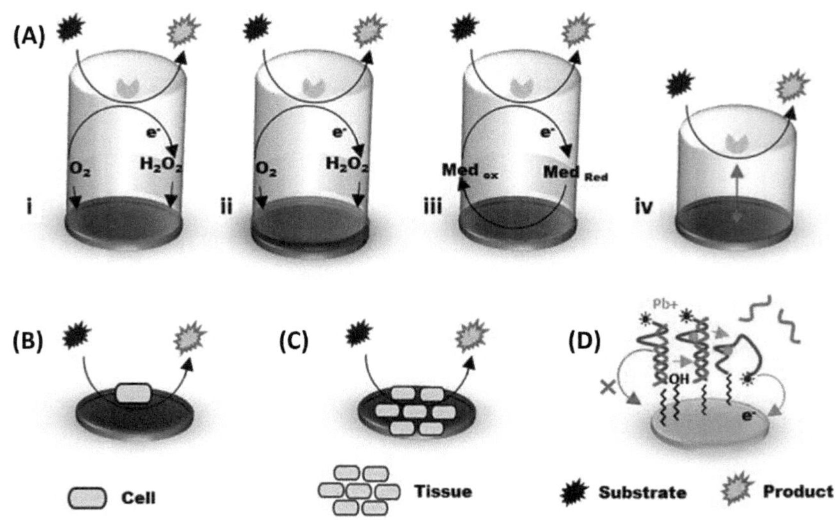

FIGURE 6.2 Schemes for the biosensing based on the biocatalytic bioreceptors. (A) General scheme involving enzyme-based catalysis: (i–ii) first-generation glucose biosensing mechanism, involving the (i) only electrode surface and (ii) electrode surface doped with the mediating molecule. (iii) Second-generation sensing mechanism. (iv) A mediator-less third-generation mechanism where the electron is transferred to the electrode directly from the active center of the enzyme. (B) A general scheme involving cells/microbes. (C) General scheme involving animal/plant tissue. (D) General scheme based on DNA-zyme cleavage.

their long-term stability is impacted, and therefore the analytical performances of biosensors based on them are compromised over time. To incorporate enhanced sensitivity and stability to the biosensors, the enzymatic bioreceptors are engineered, which leverage their prolonged stability (Campàs et al., 2009).

In addition, the engineered enzymes can offer temperature and pH resistance, which makes the enzyme tolerant to electrode modification, and extreme environments, typically offering better analytical performances than the native wild-type enzymes. For example, recombinant forms of horseradish peroxidase have shown better sensitivity toward the peroxides over the wild types (Csöregi et al., 1993). Fig. 6.3 shows the different architecture and electron transfer mechanisms in the wild type and engineered enzyme type. As in the wild types, the orientations of adsorbed enzymes, and glycosylation of the redox protective sheath, lower the electron transfer to the electrode (Csöregi et al., 1993). In another example, a genetically engineered acetylcholinesterase enzyme was used in biosensing to get sensitive detection of dichlorvos. The wild-type enzyme has a deep-seated active site (20 Å) and a narrow entrance from its surface, which hinders the electron shuttling to the electrodes. The replacement of the glutamic acid residue (Glu 69) from the active site with bulky side chains viz. tryptophan or tyrosine has been reported

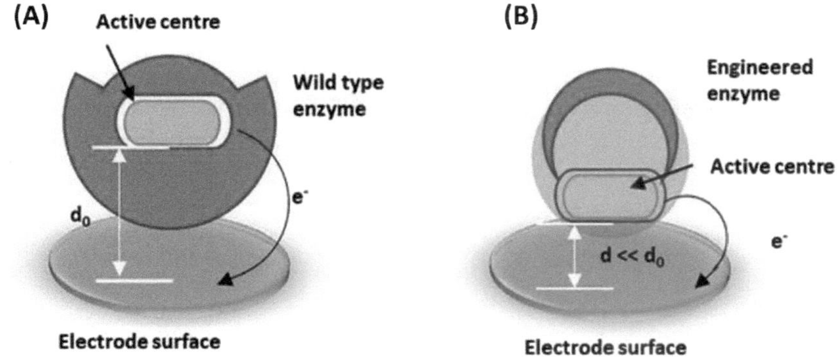

FIGURE 6.3 Architecture of the sensor based on (A) wild type and (B) engineered enzyme.

to significantly increase the inhibition constant, which eventually favors the enhanced sensitivity (Sotiropoulou et al., 2005). Similarly, various other recombinant enzymes have been reported to have shown more excellent catalytic activities than the wild-type counterparts, especially in the biosensing modules, such as Recombinant Mn peroxidase (Izadyar et al., 2019), glucose oxidase (Chen et al., 2002), and lactate oxidase (Hiraka et al., 2020).

6.3.1.2 DNAzymes

Another type of catalytic bioreceptor, the deoxy ribozymes, which are also called DNA enzymes, DNAzymes, or catalytic DNA, catalyze the nucleotides strand in the presence of the specific ions employing the general acid–base mechanism (Breaker, 1997); as the name suggests, the function of the bioreceptor catalyzes using various biochemical reactions viz. by catalyzing the DNA and RNA cleavage, ligation, and the phosphorylation. The DNAzymes are usually obtained by in vitro selection and comprise substrate-binding and catalytic sites. Owing to its better metal binding affinity, specificity, low nonspecific adsorption, and stable nature, they have leveraged the molecular tool for developing various biosensors (Hwang et al., 2016; Yang et al., 2021). For developing the sensor based on DNAzymes, catalytic DNA is required to be immobilized onto the sensor surface. The immobilized probe performs metal-dependent catalysis in the presence of specific metallic ions to generate the signal (Lake et al., 2019). Briefly, the enzyme strand hybridizes with the substrate strand having the sensor tag, which gets cleaved by metal-dependent catalytic activity freeing the cleaved substrate strand upon the application of the melting temperature. For example, Ren et al. have reported a microRNA sensor based on the Mg^{2+}-dependent DNAzyme. The hairpin probe DNA strand with a ribonuclease recognition site for Mg^{2+} ions and G-quadruplex sequence (for hemin binding) is immobilized at the gold. Using the cascade of the complementation process of the microRNA with the hairpins results from the binding complex to the sensing probe. This complex when exposed to the Mg $^{2+}$ ion gets cleaved and releases the G-quadruplex sequence, which further gets complexed with hemin to generate the analytical signal (Ren et al., 2020). DNAzymes are also used in the detection of proteinaceous biomarkers such as alpha-fetoprotein and prostate-specific antigen. For probing these proteins, the immunoassays were coupled with the immobilized metal-assisted DNAzymes to the sensing surface. The metalized detection antibody for selective immune sensing provides the ions for initiating the DNAzyme catalysis, which eventually results from the analytical signal. While the microRNA and protein sensing using the DNAzymes-based sensors are more complex, the ion sensing can be directly employed by modifying the ribozyme site for the probe strand to make it specific to the target ions. When a specific ion binds, the enzyme stand cleaves the substrate strand which upon providing the melting temperature frees the cleaved and tagged substrate strand and eventually produces electrochemical analytical signals. Using the DNAzyme-based approaches, various biomarkers of clinical importance have been detected including pathogens (Alizadeh et al., 2017; Li, Liu et al., 2018; Liu, Zhang et al., 2018), oligonucleotides (Lee et al., 2020), proteins (Kong et al., 2019; Sun et al., 2019), and ions (Liao et al., 2018; Tang et al., 2013; Zhang, Yu et al., 2021).

6.3.2 Affinity-based biorecognition elements

The growing demand for improved diagnosis has pushed the exploration of disease-specific authentic biomarkers (see Chapters 1 and 2 of this book). While many existing biosensing modules rely on enzymes and biocatalytic mechanisms for analyte detection/monitoring, affinity-based biosensing is crucial for the specific detection of proteinaceous biomarkers associated with various diseases. However, other types of biomarkers such as DNA, RNA, and mRNA have also been targeted as these can carry specific information about various intrinsic diseases such as sickle cell anemia and cancer, as well as can be utilized to detect disease-causing agents including bacteria, viruses, and other microbial forms. The generalized schemes associated with the antibody, aptamer, and DNA-probe-based detection mechanisms are shown in Fig. 6.4. While antibody−antigen interplay relies on electrostatic interactions, hydrogen bonds, van der Waals forces, and hydrophobic interactions, the proper orientation and distance between the epitope and paratope play a crucial role. Antibody−antigen (analyte) interaction can be both competitive (Fig. 6.4A) and noncompetitive (Fig. 6.4B and C). The biosensing signal comes when the proper immune complexation of the capture antibody−antigen (analyte)-detection antibody occurs. However, in some formats, the targets generate the signal, where these analytes are mainly the enzymes or chromogenic determinants and are the indicators of clinical conditions. For instance, aptamers bind to the target and fold toward the anchored

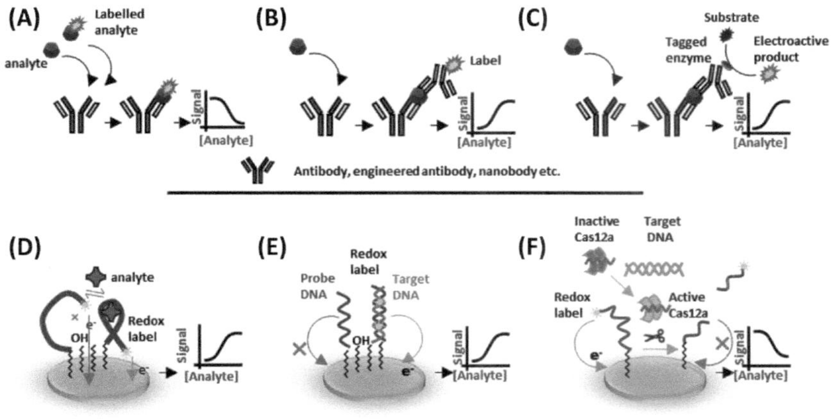

FIGURE 6.4 Schemes for electrochemical biosensing based on the bioaffinity-based receptors: (A) a competitive immune assay, (B) a noncompetitive immune assay with a tagged label, (C) a noncompetitive immune assay with tagged enzyme, (D) aptamer-based assay, (E) hybridization-based assay, and (F) CRISPR-Cas based assay.

electrode surface thereby the labeled tag generates the signal (Fig. 6.4D). Another kind of DNA-based sensor relies on hybridization (Fig. 6.4E) and CRISPR-Cas (Fig. 6.4F)-based mechanisms to generate the analytical signal. In the following subsections, we describe various aspects of affinity-based bioreceptors.

6.3.2.1 Antibody

Antibodies are frequently used for introducing specific detection for a variety of biomarkers viz. proteins (Li, Zhu et al., 2018; Vargas et al., 2020) and peptides (Hloma et al., 2022; Melo et al., 2022), also called antigens, where the antigens are the molecules, which trigger the immune system. Structurally, an antibody comprises four polypeptide chains (two light ones and two heavy ones), which fold to provide the native Y-shaped structure with the help of disulfide and noncovalent interactions. An antibody consists of variable and constant regions. While the variable region provides the antigen-binding sites, the constant region defines the antibody types. Also, the constant region of the heavy chain decides the immunoglobulin types and varied physiological roles. Based on this, the immunoglobulins are categorized under A, D, E, M, and G types (Kumar et al., 2019). Under physiological conditions, the antibody has three functional units including two Fabs and an Fc. For developing the antibody-based sensors, the antibody is immobilized via various bioconjugation-based immobilization techniques exploiting the binding sites such as $-COOH$ and NH_2 moieties present in the peptide chain terminals or the $-R$ groups of the amino acid residues in the chains. However, the biorecognition occurs only at the variable region of the Fab's. Among all types of antibodies, the G type has mostly been used for biosensing applications.

The application of antibodies in the biosensing module increases the target specificity and offers the detection of a wide range of targets for clinical importance. Therefore, antibodies are commonly used in various formats to develop the sensor for disease diagnoses and prognoses (Kumar et al., 2019; Mahato, Kumar, Srivastava et al., 2018). Moreover, the commercial availability of a range of antibodies for clinical biomarkers leverages the major attention for the development of sensitive detection modules based on the antibody. The major advancements in biosensing applications employ the use of antibodies by adopting the various formats of conventional immunoassays to develop extremely precise, point care devices for the detection of clinically important biomarkers such as alkaline phosphatase (Mahato et al., 2021a), insulin (Li, Tian et al., 2018; Martínez-García et al., 2018), cancer biomarkers (Kabay et al., 2022; Wei et al., 2022), and viral determinants (Cardoso et al., 2022; Hryniewicz et al., 2022; Shahdeo et al., 2022). However, their proteinaceous nature, bulky size, and devoid of central metallic atoms lead to their apparent passivation of the surface, which can affect the analytical performances in electrochemical formats of the immunosensors (Bhand & Singh, 2019; Felix & Angnes, 2018; Wild & Davies, 2013). Although various techniques for antibody immobilization have been employed for preventing such passivation including the use of linker molecules, optimized spacers, etc. (Gao et al., 2022), the large molecular size and its proteinaceous nature eventually limit the electron flow toward the sensing surface. To address such limitations, engineered antibodies have been incorporated by tailoring out the nonsignificant determinants. This engineering of proteins leverages the antibodies with lower molecular weight, structural stability, and flexibility while keeping the antigen-binding sites intact. In addition, the cleaved antibodies may offer flanking groups for the immobilizations at the electrode surface (Sharma et al., 2016). As the separation of the heavy and light chains offers an abundance of the thiol linkages, these can be exploited for stable immobilization at the gold-based sensor surface for attaining sensitive detection. Initial attempts for achieving such engineered antibodies were the proteolytic digestion of antibodies to the Fab and Fc. Also, advances in biotechnology have leveraged the techniques to synthesize fully functional antibody determinants (i.e., Fab and Fc fragments) (Zeng et al., 2012). The following section describes various engineered antibody-based receptors that have been successfully implemented in biosensing.

6.3.2.2 *Engineered antibody*

Engineered antibodies have been classified based on their unique properties and their occurrences such as engineered antibodies, nanobodies, bispecific antibodies, single chain variable fragments, and diabodies. To make the sensor probes less insulating, the Fc region of the antibodies is truncated out to obtain the Fab less bulky by retaining the recognition capability for sensitive detection. Since the Fab (consisting of the recognition site Fv) presents a low molecular weight of $\sim$60 Da and corresponds to only a part of the entire antibody, it contributes with less resistance in the electrochemical sensing format. These are mainly produced by slicing the traditional antibodies using site-specific enzyme cleavage. For instance, the action of enzymes known as papain and pepsin cleaves antibodies and generates the Fab and (Fab) 2′ domains, respectively (Zeng et al., 2012). The truncation of the Fc region eventually cuts off the $\sim$50 kD of the peptide chain from the sensing surface. The truncation of the antibody not only provides enhanced stability to the Fab fragments but also facilitates purification, thus lowering the price of fabricating the sensors (Sharma et al., 2016). In addition, bispecific antibodies have also been utilized for biosensing applications. These synthetic antibodies consist of two different specific fabs that can target two different targets (Ma et al., 2021). Using this strategy, Zhang et al. have reported a biosensor for detecting aflatoxin B1 and amantadine in clinical samples (Zhang et al., 2022). Since the hydrophobic interactions between the V_H and V_L chains of antibodies are prone to break in high temperatures and abnormal pH, researchers have coupled both chains using peptide linkers such as (Gly4Ser)3 to produce the single chain fragment variable (scFv) via genetic engineering procedure. These antibody fragments present a molecular mass of only 27 kDa with intact functionality (Ahmad et al., 2012). For instance, Grewal et al. have developed immunosensors based on the scFv probe to demonstrate sensitive detection of *Entamoeba histolytica* antigen (Grewal et al., 2013). The scFv fragments have shown to be stable and functional antibody fragments that have been immobilized to the sensor surface by using the avidin–biotin complexations. The flexibility of coupling multiple scFv together may leverage the formation of multimeric scFvs for specific targets leveraging the sensitive estimation of the biomarkers (Cho et al., 2018; Kim & Park, 2021).

6.3.2.3 Nanobodies

Nanobodies are single-domain antigen-binding fragments ($\sim$12–15 kDa) found in the species of the family *Camelidae., Elasmobranchii, and Chimeras* (Chiu et al., 2019). Comprising only heavy chains and single binding sites, the nanobodies are less bulky than the conventional antibodies, and thus impart less packing/covering to the sensing surface (Guo et al., 2021; Katz, 2021). The advantages of using nanobodies in affinity-based sensing include their thermal stability, high solubility, small size, and refolding capacity (Dkhar et al., 2022). Therefore, besides leveraging sensitive detections, the use of nanobodies also brings more stability to the devices and provides robust and reliable determination of the analytes in variable conditions.

In addition, the nanobodies are resistant to alkaline and acidic pH, being easy to store, which proves the biosensing modules with extended shelf-life (Hosseindokht et al., 2021). Using nanobodies, Guo et al. have developed a transistor-based biosensing module for detecting the severe acute respiratory syndrome coronavirus-2 (SARS-CoV-2) and the Middle East respiratory syndrome coronavirus spike proteins from nasopharyngeal swabs and saliva samples (Guo et al., 2021).

6.3.2.4 Aptamer

Aptamers are single-stranded oligonucleotides (10–100 nucleotides) that can bind to a specific target. Such molecules show affinity toward various targets, including metallic ions, proteins, viruses, bacteria, and small molecules (Chandra, 2016; Villalonga et al., 2020). The unique properties of the aptamers such as high affinity, reproducibility, high chemical stability, and significant conformational changes upon binding with the target offer the development of tools not only for diagnostics but also for therapeutics (Song et al., 2008). The change in the well-defined three-dimensional conformation of the aptamer upon binding with the targets has been employed to generate redox-mediated signals in electrochemical biosensors. To obtain the redox activity, a tag of the redox-active molecule is commonly attached to the flanking end of the aptamer while immobilizing the aptamer to the electrode. The binding process either increases (signal-off) or decreases (signal-on) the distance between the redox tag (signal-on) and the electrode surface, allowing the detection of the target. Employing these mechanisms, several electrochemical aptamer-based sensors have been developed (Nur et al., 2021) for antibiotics (Yadav et al., 2014; Zhu et al., 2013), insulin (Wu et al., 2019), cancer biomarkers (Zhu et al., 2013), viruses (Abrego-Martinez et al., 2022), etc.

6.3.2.5 DNA probes

Another important type of affinity-based bioreceptors is oligonucleotides, especially DNA. In this case, the sensing mechanism is based on the hybridization of the oligonucleotide probe (immobilized onto the sensing platform) with the target oligonucleotide sequence. As the nucleic acids selectively bind to the complementary base pair, the probe is designed using the sequences complementary to the target sequence for sensing. These target sequences may be associated with specific genes and mRNA, related to particular clinical conditions and diseases (Drummond et al., 2003). Nucleic acid detection is conventionally performed by the polymerase chain reaction (PCR)-assisted detection strategies coupled with the nucleic acid hybridization, and next-generation sequencing (NGS) techniques. While the DNA hybridizations are based on the pairing of complementary strands, the NSG offers the precise, high-throughput sensitive estimation of the nucleic acid detection of a clinically important biomarker. For instance, DNA-hybridization-based biosensors have been reported for the detection of the clinically relevant biomarkers associated with invasive aspergillosis and enterotoxigenic *Escherichia coli* pathogen (Bhatnagar et al., 2018; Kashish et al., 2017). Since this interaction of the fragments is ultra-specific to the complementary strands, the sensor is highly specific. For instance, Bhatnagar et al. have developed a hybridization-based sensor for detecting the gene associated with invasive aspergillosis. The sensor is capable of detecting the 1×10^{-14} to 1×10^{-2} M of the target and the detection limit of $0.32 \pm 0.01 \times 10^{-14}$ (Bhatnagar et al., 2018). Similarly, the other genes from the pathogens have also been detected using hybridization-based sensing probes (Kashish et al., 2017; Tripathy et al., 2020). In recent days, three-dimensional (3D) DNA structures have been achieved using DNA origami (Chandrasekaran et al., 2016; Chandrasekaran, 2016; Xavier & Chandrasekaran, 2018; Xiao et al., 2019) technology for the development of DNA-based sensor probe. For this, several in silico tools that predict the DNA sequence and its structure are available. Employing the DNA origami–based strategy Han et al., have reported a microRNA sensor. This work uses the DNA origami probe for the sensitive detection of the cancer biomarker mRNA-21 and shows a wider range of 1 pM–1 nM with a lower detection of 79.8 fM (Han et al., 2019). Long DNA strands can be precisely folded with this method into intricate, nanoscale forms that can be made to attach to particular proteins, tiny chemicals, or cell types with high affinity and selectivity. Due to the ability to create DNA origami structures with a huge surface area, which increases the number of binding sites open to target molecules, this high specificity is possible in DNA origami–based sensors. In addition, the CRISPR-Cas-based system has also found great attention for the detection of DNA-based biomarkers (see Chapter 7). The CRISPR-Cas-based system is originally a defense mechanism of microbes, which they employ to selectively cleave foreign DNA. This system can be incorporated into electrochemical biosensing, where the target DNA sequence activates the Cas proteins, which then cleaves a redox-labeled probe DNA sequence to generate the analytical signals (Kaminski et al., 2021). Using this system several biosensing modules have been reported so far, which include the detection of nucleic acids (Dai et al., 2019) and miRNA (Bruch et al., 2019).

6.3.3 Synthetic recognition elements

In addition, to the bioreceptors, receptors of nonbiological origin have also been used for the development of sensors. These can bring higher stability in a wider range of temperatures and pH (Karadurmus et al., 2022). This section discusses the use of different types of receptors of nonbiological origin for detecting target analytes. Unlike biological receptors, polymeric and synthetic bioreceptors not only enable mass production, improved shelf-life, and lower cost but also leverage minimum batch-to-batch variations. These recognition elements mainly mimic the affinity-based recognition of antibodies, aptamers, and oligonucleotides. Fig. 6.5 shows the general mechanism of the biorecognition based on MIP (Fig. 6.5A).

6.3.3.1 Molecularly imprinted polymers

MIPs are synthesized by templet assisted method, where the target molecules have been used as a templet for making the cavities for specific sites as biorecognition elements (Uzun & Turner, 2016). The synthesis of MIPs was first reported by Dickey in 1949 along with a few key applications including templet-based target molecules. MIP is a polymeric matrix with cavities that offers binding with specific targets. The synthetic process for developing the specific MIP involves the incubation of the template molecules in monomers, polymerization of the monomers, effective entrapment of the analyte molecules, and template extraction. Owing to their superior chemical and thermal stability, and extended shelf-life, MIPs have found great attention in developing the receptors for targeting several biomarkers for acute myocardial infarction, sepsis, stress, etc. (Crapnell et al., 2020). For instance, a biomarker for cardiac infraction cardiac troponins (i.e., T and I) has been reported with *o*-phenylenediamine (Karimian et al., 2013), polyaniline (Phonklam et al., 2020), acrylamide (Moreira et al., 2011), poly-pyrrole (Silva et al., 2016), etc. based MIPs for sensitive detections. Similarly, several targets have been detected using MIP-based recognition elements including proteins (Di Giulio et al., 2020), hormones (Tang et al., 2021), and cancer antigens (Jolly et al., 2016). MIPs are known for their advantages of leveraging the detection of a wide number of clinically relevant molecules including the smaller ones, cortisol, dopamine, hormones, peptides, and proteins. Another major advantage of MIP-based sensors is that they offer the detection of nonelectroactive molecules. The capacity to molecularly imprint makes them exceptionally sensitive and selective. Because of their durability,

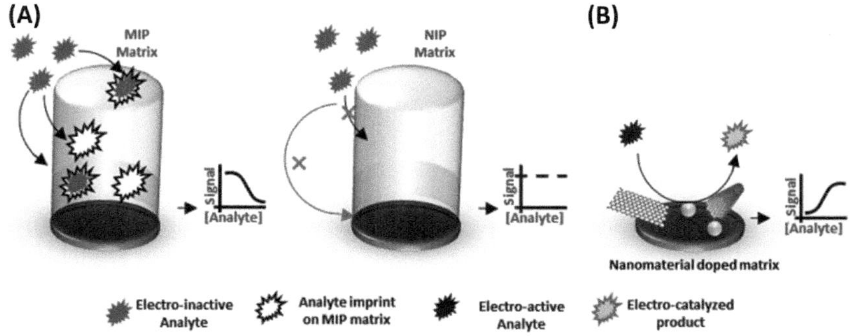

FIGURE 6.5 Synthetic matrix as the recognition matrix. (A) Molecularly imprinted polymers (MIPs) and nonimprinted polymers (NIPs) for affinity-based sensing. **(B)** Nanomaterial-based catalytic biorecognition (nanozymes) elements.

affordability, and scalability, they offer many possibilities in fields as diverse as environmental monitoring, food safety, and medical diagnostics. However, these are limited when it comes to selective detections. As the cavities of MIP are accessible to nonspecific molecules of similar size and structures, the sensors based on it suffer from compromised selectivity. Such cases are more often seen in the sensors of large target molecules such as peptides and proteins, where the interference from the smaller coexisting molecules is prominent. The interaction of the smaller interfering molecules usually competes or sometimes poisons the cavities, which eventually compromises the selectivity.

6.3.3.2 Nanozymes

Nanozymes are nanomaterials that possess intrinsic enzyme-like properties (Mahmudunnabi et al., 2020). These have drawn significant attention due to their unique and distinct capability of catalyzing the molecules when contrasted with their amino acid-based equivalents (Wu, Darland et al., 2021). Nanozymes usually include metallic nanoparticles (Elakkiya & Maduraiveeran, 2019; Rao et al., 2017), nanoclusters (Wang et al., 2022), porous carbon (Peng et al., 2022; Zhang, Chen et al., 2021), nanowires (Guo et al., 2020), and metal-organic frameworks (Chen et al., 2022; Portorreal-Bottier et al., 2022), which achieve catalysis. The major advantages of using such materials are that they typically offer high catalytic activities, robustness in operation, prolonged shelf-life, and can be easily modified (Białas et al., 2022), being great to be used as signal tags and as biorecognition elements. So far, more than 40 nanozymes that are capable of catalyzing biochemical reactions have been documented (Huang et al., 2019). The nanozymes are majorly classified under oxidoreductase and hydrolases families. While oxidoreductase nanozymes are involved in redox-based catalysis, hydrolases are responsible for catalyzing the hydrolysis of the substrate (Huang et al., 2019). Metallic nanoparticles, owing to their unique atomic structure and valence properties, can show oxidoreductase activity, and thus, the gold and silver nanoparticles have frequently been used as peroxidase mimics. Similarly, metal oxide nanoparticles (viz. cobalt, chromium, copper, manganese, ruthenium, etc.), and metal-organic frameworks nanocrystals also show peroxidase-like activity. In addition, nanozymes that mimic the process of oxidase of their counterparts have also been reported. These follow glucose oxidase-like and sulfite oxidase-like mechanisms to catalyze the reactions with target molecules. The gold ion and oxygen-charged couple generate a dioxo-gold intermediate that eventually facilitates the electron transfer in a glucose catalytic reaction (Comotti et al., 2006). The catalytic effect of nanozymes is dependent on various factors, that is, particle shape, particle size, the presence of a passivation layer, and particle morphology, and thus can be tuned by changing the above-stated parameters. A smaller size increases the surface area, enhancing the interaction with the substrate and increasing sensitivity;

however, a larger size has also been reported to have higher activity, possibly due to the greater number of metallic ions. The nanozymes can be used in various detection mechanisms including an optical and electrochemical sensor. So far, nanozymes have been employed in biosensors for the detection of various targets, including neurochemicals (Mahato et al., 2019), glucose (Gong et al., 2019), peroxides, (Purohit, Mahato et al., 2019), vitamin C (Musarraf Hussain et al., 2020), lactic acid (Zaryanov et al., 2017), and drugs (Kumar, Purohit, Mahato, Mandal et al., 2019). In biosensing, Au-based nanozymes have extensively been used for achieving oxidase, catalase, reductase, etc. like activities (Lou-Franco et al., 2020). A report shows the catalytic properties of the AuNPs with various passivating layers, where the bare surface has shown comparable catalysis with the HRP enzyme (Lou-Franco et al., 2020). Similarly, the peroxidase property of the carbon-based nanozymes has also been employed for the detection of ion sensor, glucose (Song, Qu et al., 2010), single nucleotide polymorphism (Song, Wang et al., 2010), cancer cells (Song et al., 2011), cholesterol (Qian et al., 2015), etc. The colorimetric formats have also been adopted for biosensing based on nanozymes, where peroxidase activity has been used by coupling the chromogenic compound such as the tetramethylbenzenes (TMB) (Shi et al., 2011). Using the TMB—nanozyme system, various optical biosensors have been reported for the detection of important analytes including glucose and ascorbic acid (Darabdhara et al., 2017; Wang, Hou et al., 2017).

6.4 Materials for biosensors

Materials play a crucial role in developing sensors. These are not only required for obtaining enhanced analytical performances but also to develop the desired portable modules for various needs such as treatment and disease management (Kumar et al., 2020; Kumar, Roy et al., 2019). To realize such biointegrated sensing devices, chemists and material scientists have closely collaborated to synthesize flexible, conducting, robust, functional materials and enhance the production processes of devices based on them, as these eventually leverage the development of biosensing modules with various form factors (Nemčeková & Labuda, 2021). The materials used in biosensor manufacture are critical and are determined by the individual application as well as the intended performance parameters of the biosensor. Gold nanoparticles and quantum dots, for example, can be employed in biosensors to improve sensitivity and specificity by expanding the surface area for reaction, improving mass transfer, or modifying the optical or electrical properties of the sensor. Because of their capacity to change electrical characteristics in response to changes in analyte concentration, conductive polymers can be utilized as sensing materials in biosensors. Because of their great surface area, mechanical stability, and electrical conductivity, carbon nanotubes and graphene are two further forms of nanomaterials that can be employed in biosensors. Finally, the material used in a biosensor is determined by the application's unique needs, such as sensitivity, specificity, stability, and cost. In addition, the substance must be compatible with both the biological component and the target analyte. The biosensing modules require functional materials for supporting the bioreceptor anchorage, conducting surface for electron transduction, and substrate as the base material (Kumar et al., 2019; Mahato, Kumar et al., 2020). Generally, in biosensors, the analytical performance can be enhanced by utilizing the material that can improve the presence of target molecules around the sensing surfaces, viz. hydrogels, signal-amplifying nanomaterials, and selective bioreceptors of biological or chemical origin. This section discusses the material requirements for developing the sensing modules.

6.4.1 Materials for sensing components

When it comes to the construction of biosensors, the choice of materials to use for the sensing components is an extremely important factor in determining how well the biosensor will function as a whole. Electrodes are often constructed out of conductive materials such as gold, silver, carbon, and platinum. The electrical signals that are produced by the biosensor can be measured due to the path that is provided by these materials. Materials that are capable of transduction are required to change the biological signal into an electrical signal. Enzymes, antibodies, nucleic acids, and even complete cells can fall within this category of substances. The sensing components are held in place by supporting materials, which are often nonconductive materials such as hydrogels, polymers, or ceramics. These materials provide a supportive matrix for the sensing components. The specificity of the biosensor relies heavily on recognition elements, which can take the form of antibodies, enzymes, or nucleic acids, and interact in a highly targeted manner with the analyte of interest. In the end, biocompatible materials such as polymers and ceramics are utilized to ensure that the biosensor does not affect biological systems and may be employed close to biological tissues. This is accomplished by ensuring that the biosensor is constructed out of materials. To achieve high levels of sensitivity, specificity, and stability in the biosensor system, it is very necessary to pick these materials with extreme care.

6.4.2 Materials for signal amplification

In biosensors, the signal is generated with the help of fundamental components including bioreceptors, a sensing surface, and a transducer. Commonly, the signals rely on the electroactive surface itself. However, in some cases, the signal is much fainter than the subtle systemic noise in the devices, and analyzers can interfere with and suppress the actual amplitude (Kumar, Purohit, Maurya et al., 2019). Also, the deep position of the active sites in the bioreceptor molecules makes it hard for the electrons to move from the biorecognition process to the transducing surface (Kumar et al., 2018). Thus, signal-amplifying strategies are employed to deal with such biosensing hurdles (Gao et al., 2011; Huang et al., 2021; Lei & Ju, 2012; Liu, Liu et al., 2018). Apart from bioreceptor engineering, nanomaterial-assisted strategies have found great attention as these provide an easy solution for some system limitations by employing nanomaterial's unique properties viz. larger active surface area for electron transfer and bioreceptor immobilization, catalysis of byproducts, etc. Metal nanoparticles (NPs), semiconductor NPs, and carbon nanostructures are majorly used for this purpose (Medintz, 2009). In addition, these biofunctionalized nanomaterials can serve as carriers (Lang et al., 2020), tracers (Kratz et al., 2021), catalysts (Voiry et al., 2018), electronic conductors (Zhang et al., 2019), and optical emitters (Gupta et al., 2021) to obtain an enhanced detection signal and stabilized recognition probes (Haun et al., 2010). The following subsection describes the materials used for obtaining amplified signals. Signal amplification materials in biosensors are those that boost the sensor's sensitivity or specificity by boosting the signal produced by the target analyte. Examples of such substances include nanoparticles, enzymes, antibodies, and other biomolecules. However, nanozymes are a special kind of nanoparticle that can imitate the catalytic activity of enzymes. They catalyze targeted reactions with the analyte of interest, allowing biosensors to detect and identify targets with greater precision. While some signal amplification materials may boost the signal through factors like increased surface area, mass transfer, or electrical conductivity, nanozymes have a catalytic activity that sets them apart. In addition, nanozymes are typically engineered to perform a single catalytic function, while alternative materials for signal amplification may be capable of a wider variety of tasks.

6.5 Nanocomposites and nanomaterials

To address the limitations present in the sensors, including signal-to-noise ratio, lower detection range, and lower operational stability, the incorporation of nanocomposites has been adopted. Nanocomposites are synthetic materials that have the ability to generate new properties, such as better stability and better charge transfer with the help of constituting materials. This mixing not only provides the collected properties but also can leverage new unique behavior. As the sensing surface is required to be conducted in the electrochemical sensors, nanodimensional and structured materials of various noble metals, graphene, carbon nanotube, etc. have shown great promise (Yang et al., 2020; Zhou et al., 2015). As the electron shuttling in the biosensors is another limitation of having poor sensitivity, various nanomaterials have also been employed for obtaining signal amplification, as these offer high conductivity as well as presenting large surface areas for bioreceptor immobilizations. Nanomaterials such as single-walled-carbon nanotubes (SWCNTs) can electronically connect the enzyme's active center to the conducting electrode surface to facilitate electron shuttling. In a study, the presence of a 50-nm SWCNT in the modified electrode has shown an interfacial electron transfer rate constant of 42 s^{-1}, indicating the possibility of developing an electrochemical biosensor without the use of a charge carrier (Patolsky et al., 2004). In another study, a DNA probe immobilized using SWCNTs offered an amplified signal based on the direct current response of guanine with excellent analytical performance (Wu et al., 2009).

In addition, the active surface area of the sensor transducer plays a crucial role in signal amplification. Because functional graphene sheets have a greater surface area to anchor higher protein amounts, they are an appropriate substrate for collecting proteinaceous bioreceptors and ultimately enhancing the analytical response. In a study, the authors reported an electrochemical immunosensor with a sevenfold increase in the signal compared to the one having no graphene modification and carbon nanospheres labeling, which was used to increase the effective surface area. Similarly, to improve the signal by increasing the surface area, several strategies using nanomaterials have been used, which include electrochemical stripping of the nanoparticle-based aggregates. Also, nanomaterials-based catalysis has been employed for achieving ultrasensitive detection. For instance, palladium nanoparticles (PdNPs) have been used for DNA sensing by Das et al., in this investigation, copolymer-modified electrodes demonstrated very minimal DNA-conjugated PdNPs nonspecific binding. A silane copolymer-modified indium tin oxide (ITO) electrode was covalently linked to amine-terminated capture probes for a sensing electrode. Target DNAs were hybridized with capture probes on the electrode, then detection probe-conjugated Pd NPs. The rapid hydrolysis of $NaBH_4$ on PdNPs produces atomic hydrogens, which are quickly absorbed by PdNPs increasing their electrocatalytic activity and producing amplified analytical responses (Das et al., 2009). The activity level of Fe_3O_4 NPs is 40 times higher than HRPs for the same molar concentration (Gao et al., 2007). Based on this discovery, new immunoassays can be created, in which NPs with antibody modification perform three tasks simultaneously: target capture, isolation, and sensing.

6.6 Conducting polymers

Conducting polymers (CPs) are a wide class of organic polymers characterized by molecular interconnections that contain orbital π-conjugated structures (Aydemir et al., 2016). This one-of-a-kind construction enables them to exhibit properties including a broad range of conductance dependent, on the amount of doping, high redox activities, and the capacity for combined ionic and electronic transport (Ates, 2013). In most cases, undoped CPs will have extremely poor conductivity. To endow conductive nanocomposite with the necessary electrical conductivity, conductive fillers are typically incorporated into polymers during the manufacturing process. The conductivity of CPs can be tailored by controlling the dopants viz. p-type (oxidation) or n-type (reduction) (Pang et al., 2013; Wycisk et al., 2002). In addition, the physicochemical characteristics, including electrical, optical, and mechanical properties of CPs can be altered by the use of doping to increase their functional applications (Hackett et al., 2017; Khokhar et al., 2021). Furthermore, strategies for preparing compatible conducting materials for sensing purposes have also been adopted by mixing the conducting materials with natural biopolymers (Wang et al., 2021). So far, metallic NPs, graphene, carbon nanotubes, etc. have been mixed with biocompatible ion conductive fillers (Suginta et al., 2013). Also, the development of polymer technologies has attracted significant research efforts to design promising biocompatible conducting materials, such as poly(3,4-ethylenedioxythiophene):sulfonated lignin (PEDOT:SL) (Wang, Pan et al., 2019), poly(3,4-ethylenedioxythiophene) polystyrene sulfonate (PEDOT:PSS) (Yang et al., 2015), and polyaniline (PANi) (Wang, Lin et al., 2016). In biosensor fabrication, CPs like polyaniline (PANi) can be employed to create MIPs. As MIPs are very specific, they can be utilized as recognition components in biosensors to detect only the analytes of interest. In conducting polymer biosensors, the binding of the target analyte causes a change in conductivity, which is then used as the basis for the detecting mechanism for MIPs. Due to the doping and undoing of polymer chains, the conductivity of the polymer changes as the target binds to the MIPs.

The pH PANi sensor is a typical type of pH sensor that uses PANi as the conducting polymer. To improve the sensor's conductivity, PANi is doped with a protonic acid like HCl. By adjusting the sample's pH, the polymer's proton concentration will shift, and the resulting change in conductivity will be quantifiable. This sensor can detect pH changes by tracking conductivity shifts, due to its built-in doping/undoing mechanism. The conductivity of a pH PANi sensor is affected by the pH due to the binding of protons to the polymer, which is pH-sensitive. Because of an abundance of protons at low pH, the polymer can be doped to increase its conductivity. In contrast, conductivity drops as the pH is increased because fewer protons are available to dope the polymer. This doping/undoing process enables real-time pH monitoring since variations in conductivity are directly proportional to pH levels. In pH sensing, the doping/undoing process enables the detection of pH changes with great sensitivity and specificity, making this a common use for doping/undoing biosensors. The majority of CPs can either effectively entrap or assimilate biological functional components, or they can supply sites for their connection through the chemical alteration of ligands including carboxyl or amino groups (Dakshayini et al., 2019; Wang et al., 2020). These methodologies have made it possible to effectively couple biorecognition components (such as enzymes, antibodies, and DNA/RNA) to develop biosensor systems.

6.7 Materials for analyte enrichment and immobilization matrices

One of the numerous elements influencing the performance of biosensors is analyte enrichment. Preconcentration is used in biosensors to increase the sensitivity of the device by increasing the amount of analyte (the substance being measured) on the surface. By preconcentrating the analyte, the limit of detection is lowered, and the biosensor can detect smaller amounts of the analyte. An example of an application for preconcentration in biosensors is PCR, a technique used to amplify DNA sequences. PCR requires very small amounts of DNA to start with, so a preconcentration step is necessary to increase the amount of DNA in the sample before amplification. Other established techniques for preconcentration include solid-phase extraction (SPE), liquid-phase microextraction (LPME), and adsorption. In some cases, biosensors use tags, such as antibodies, to specifically bind to the analyte and increase the sensitivity of the sensor. However, these tags can increase the time and cost of the device. Additionally, the use of tags may introduce other problems, such as cross-reactivity (when the tag binds to something other than the analyte) or nonspecific binding (when the tag binds to multiple analytes). Concentrating the analyte directly, without the use of tags, is, therefore, more advantageous because it is faster, cheaper, and eliminates the possibility of cross-reactivity or nonspecific binding. Preconcentration processes can therefore be added to enrich the analytes. This not only improves the analytes but also guards against mechanical aberration and bioreceptor leaking. As yet, several techniques have been used, including time-induced preconcentration and enrichment supported by hydrogels, among others (Al Sulaiman et al., 2018; Jung et al., 2017).

6.8 Hydrogel matrices

Hydrogel-based systems offer significant improvements in analytical performances toward the detection of a wide variety of biomarkers. Owing to the cross-linked network architectures, hydrogels contain high amounts of water (Al Sulaiman et al., 2018). Also, the hydrogels' role as a biomimetic tridimensional (3D) matrix assists in immobilizing bioreceptors (Jung et al., 2017). The application of hydrogels also leverages an aqueous microenvironment favorable for the interaction of bioreceptors with target molecules. The ability of hydrogels to provide better stability over external factors makes them desirable materials for sensor matrices. In a study carried out by Pan and coworkers, the hydrogel was used to develop multiplex sensors for glucose, lactate, and triglycerides with better sensitivity and high selectivity compared to contemporary sensors (Li, Pan et al., 2018) Similarly, hydrogels have also been used in more complex matrices. In this context, Sun et al., have reported a hydrogel-based sensor for amyloid-beta oligomers in blood and cerebrospinal fluid (Sun et al., 2018). Also, hydrogels can be designed to respond to specific biological signals, such as pH changes or the presence of certain biomolecules, and can be used as the sensing element in biosensors. Additionally, hydrogels can also be used as actuators, as they can change shape or release a substance in response to a stimulus. One example of a hydrogel-based biosensor is the use of pH-sensitive hydrogels in the detection of acidity levels in the body. These hydrogels can change color or fluorescence in response to changes in pH, allowing for easy visual detection of pH changes. Another example is the use of enzyme-sensitive hydrogels as biosensors for glucose levels in the blood. These hydrogels can change color or fluorescence in response to the presence of glucose, which can be easily measured using a spectrophotometer or other instrument.

Similarly, hydrogels can also be used as actuators. For example, temperature-sensitive hydrogels can be designed to change shape in response to changes in temperature, which can be used to control the release of drugs or other substances. However, it is important to note that there are also safety and ethical concerns surrounding the use of hydrogels in biosensors and actuators. For example, the long-term biocompatibility and safety of hydrogels implanted in the body must be thoroughly evaluated, and the risk of unintended interactions between hydrogels and other biological systems must be carefully considered. Concisely, hydrogels are a promising material for the development of biosensors and actuators, as they can be designed to respond to specific biological signals and can be used as both sensing and actuating elements. However, it is important to consider safety and ethical concerns in the development and use of hydrogel-based biosensors and actuators.

6.9 Immobilization matrices

The immobilization of bioreceptors onto the substrate can be performed by various methods, including covalent and noncovalent immobilization. If covalent immobilization is employed, the substrate offers functional groups to chemically bind the bioreceptor (e.g., carboxylic acid, amino, thiol, etc.). In a different approach, receptors such as enzymes can be entrapped in conducting porous matrices through the cross-linking of copolymers, which not only prevents receptor leaching but also provides a conductive media for sensitive determinations. For the sensor to be used in biomedical applications, these are required to be biocompatible. To achieve biocompatible sensor probes, materials of biological origins have been employed in sensor fabrications. Most of these polymers (such as chitosan and cellulose) leverage the protection against leaching out the immobilized bioreceptors, thus thin membranes can be used for sensor developments. Also, some of these are known for their soft texture and stretchability, which have been exploited for the development of conformally adjustable bioelectronics for wearable sensing modules. Owing to their excellent biocompatibility, biodegradability, wide availability, long-term sustainability, and functionalization flexibility, biopolymers have been widely accepted for developing sensor components (Hosseini et al., 2021). In addition, polysaccharides, viz. cellulose, (Chen & Hu, 2018), chitin/chitosan (Baranwal et al., 2018), and alginates (Márquez et al., 2017), and proteinaceous materials (viz. silk and gelatine) have been used for immobilizing the bioreceptors as biocompatible components of the soft bioelectronic sensing modules (Chen & Hu, 2018; Wang, Xia et al., 2019; Zhu et al., 2016).

6.9.1 Substrate materials

In the context of biosensors, a substrate is a substance or surface upon which the biosensor's detecting element is positioned. The substrate is vital to the performance and functionality of the biosensor since it must be chemically and physically compatible with the sensing element and must not interfere with the detection of the target analyte. For point care applications of biosensors, such as disease diagnostics and its management, the use of a specific, resilient, compact, and affordable substrate is of great importance. Depending on the purpose and sense method, various types of substrates can be utilized in biosensors. Some instances where gold is preferred since it is a typical substrate material used for biomolecule immobilization in biosensors. Gold surfaces have high chemical stability, are biocompatible, and may be easily functionalized with a variety of biomolecules, including enzymes, antibodies, and DNA. Gold surfaces are also compatible with numerous detection techniques, including surface plasmon resonance (SPR) and electrochemistry. Similarly, optical biosensors frequently employ transparent substrates, including glass and plastic. Important for optical detection techniques such as fluorescence and absorbance, these substrates let light pass through the sensing element. Biosensors designed for usage in hostile settings, such as high temperatures or pH extremes, may require substrates that are robust to these circumstances. For instance, pH-resistant substrates can be composed of materials such as carbon glass or silicon, whereas temperature-resistant substrates can be composed of ceramics or silicon carbide.

Also, the substrate should be chemically and physically inert, and the ideal substrate for biosensors should not interfere with the detection of the target analyte. Silicon, glass, and quartz are examples of inert substrates that are resistant to chemical reactions and physical degradation. In conclusion, the choice of a substrate in biosensors is determined by the application and sensing mechanism, and the substrate must be chemically and physically compatible with the sensing element and not interfere with the detection of the target analyte. The substrate not only provides support to the sensing component but also offers conformability and ease based on the application. To sort these out, substrates based on paper and flexible polymeric materials have opened new ways for developing chemical sensing modules (Shafiee et al., 2015). Owing to their unmatchable properties such as being thin, lightweight, flexible, easily accessible, and abundant, these have traced another parallel path for the development of biosensor including wearable and implantable sensors (Mahato, Purohit et al., 2020). The following section describes various materials used as sensor substrates.

6.9.1.1 *Rigid polymeric and ceramic materials*

Rigid polymeric and ceramic materials play an important role in the fabrication of biosensors. Rigid polymeric materials are often used as the substrate material for biosensors, providing a stable platform and support for the active sensing elements. These materials have good mechanical stability and can be made from various polymers such as polycarbonate, polymethyl methacrylate, and polyethylene terephthalate. On the other hand, ceramic materials, like Alumina, Zirconia, and Titania, are widely used as sensing materials in biosensors due to their biocompatibility, high mechanical stability, and chemical resistance. These materials are suitable for pH sensors, oxygen sensors, and glucose biosensors. They can also be used as a sensing surface for the attachment of biological recognition elements such as enzymes and antibodies. Thus, rigid polymeric and ceramic materials play a crucial role in the fabrication of biosensors, providing the necessary stability and support for effective and reliable sensing. Conventional biosensors are commonly constructed with rigid substrates based on hard polymers and ceramics. These substrate materials are used for protecting the electrode systems printed on them. Polymers like PMMA, PVC, and cellulose acetate film (Leng et al., 2019) have been used as substrate material. As early SPEs were built with carbon inks without elastomeric doping, these were not mechanically resilient; thus, the rigid yet flexible surface was of utmost requirement. SPEs based on rigid substrates have frequently been used for onsite detection.

6.9.1.2 *Flexible materials*

While a rigid surface protects the bioreceptor assembly, it cannot be used for applications that require flexibility or stretchability, such as wearable devices. Thus, currently, flexible and conformally adaptive materials are commonly preferred. Flexible materials can bend as per requirement, leaving no changes in the features, that is aberration or cracks in the circuitry (Han et al., 2017; Wang, Chinnasamy et al., 2016; Xu et al., 2019). The more flexibility that the substrate offers, the better the conformal compatibility enhances interfacial interaction with a sample in wearable settings, which eventually improves the analytical performance of the sensor. Mainly, flexible biosensing substrates are obtained by using synthetic polymers, papers, textiles, and fibers. Synthetic polymers are the most regularly used substrates due to their adaptability and ease of processing, which enable the construction of flexible architectures at lower production costs. For example, polyethylene terephthalate (PET), polyethylene naphthalate (PEN), polyimide (PI), and polydimethylsiloxane (PDMS) have been extensively used for developing sensing modules (Lau et al., 2013; Liao et al., 2015). PET, owing to its unique material properties, such as mechanical endurance, chemical inertness, heat resistance, and competitive prices, has found great application in biosensors (Reddish, 1950). Similarly, PEN is another synthetic polymer, and it exhibits excellent chemical and hydration resistance; heat and thermos oxidation resilience; and ultraviolet resistance, being also widely used in developing optoelectronic biosensors (Murakami et al., 1995). Polyimides (PI) are also commonly used polymers for building wearable modules due to their properties of foldability, molding, high thermal stability, and stretchability, besides possessing excellent dielectric coefficient, high dynamic strength, and architectural stability. PIs show increased conductivity and altered photo-resistivity when decorated with metallic nanoparticles. For making robust base materials in biosensing modules, various biocompatible fillers have been used for stretchable conductive composites including carbon-based materials, metal-based compounds, and CPs. In this context, synthetic polymers such as polydimethylsiloxane (PDMS) (Wolf et al., 2018), Eco-flex (Amjadi et al., 2015), polyurethane (Das & Mahanwar, 2020), poly(styrene-butadiene-styrene) (Ellingford et al., 2018), and poly-(styrene ethylene butylene styrene) (Ayyaru et al., 2012) and natural biopolymers including silk (Wang, Xia et al., 2019), gelatin (Pohanka et al., 2013), alginates (Márquez et al., 2017), cellulose (Chen & Hu, 2018), and chitosan (Baranwal et al., 2018) have been employed extensively to achieve biocompatible elastomeric polymers (Wang, Chen et al., 2017). In the upcoming sections, synthetic polymers and natural polymers that are widely used for biosensor fabrication will be discussed.

6.9.1.2.1 Synthetic polymers

Numerous techniques, including polymerization (addition, condensation, copolymerization), molding (injection, blow, thermoforming), extrusion (film, fibers, profiles), and solution casting, can be used to create synthetic polymers. The choice of strategy will rely on the desired polymer qualities and the specifications for the finished product. These processes differ in complexity, cost, and scalability. For instance, solution casting is the best method for manufacturing thin films of consistent thickness while injection molding is frequently used to produce complex shapes in large quantities. These are also utilized as platforms for brain implants, fuel cells, and in a variety of many other systems through the process of thin film as well as double-layered film fabrication (Liaw et al., 2012). PDMS is a silicone-based elastomer commonly used for the fabrication of microfluidic devices and biomedical equipment, being integrated into catheters, membrane oxygenators, and cartilage grafts, among many others. Microfluidics involves the control of minute liquid quantities and the incorporation of sensors. Microfluidics in vitro have the potential to increase the sensitivity and specificity of the biosensor as it allows for several analytical stages to be performed while maintaining tight control over the fluidic environment. As a result of microfluidics' ability to provide high-throughput analysis and downsizing of biosensors, the technology is now well suited for point-of-care and portable uses. Polymers, such as elastomers, provide a flexible and long-lasting material for constructing microfluidic devices, which greatly improves their utility. Because of their elastomeric nature, they can be imprinted with a master substrate, allowing for the fabrication of microfluidic channels with finely tunable dimensions and fluid flow rates. Because of their low mass and high resistance to damage, polymers are also a good material choice for wearables and transportable biosensors. Due to its low price, chemical stability, heat resistance, and oxygen permeability, the use of PDMS for building biosensors has seen a rapid expansion in recent years (Mata et al., 2005; Qin et al., 2010).

6.9.1.2.2 Paper

Paper is an appealing seminatural platform that offers an affordable substrate for sensor development, bringing important features, especially for resource-limited settings. Paper-based platforms offer a broad range of sensing modalities, including optical, electric, and electrochemical (Mahato et al., 2021b; Mahato, Purohit et al., 2020). Another advantage of paper substrates is that they can be easily molded into composites due to their intrinsic low density, thin architecture, and porous structure. Paper substrates can be used for pumpless microfluidic operations due to their self-wetting phenomena derived from capillary action, allowing the development of devices such as lateral flow assays (Mahato et al., 2021a; Mahato, Purohit et al., 2020). As a microfluidic device, paper is commonly employed in lateral flow biosensors due to its adaptability and low cost. Detecting analytes like proteins and tiny molecules in a portable and user-friendly manner is a major selling point for paper as a microfluidic device in lateral flow biosensors. For these biosensors, paper is used as a substrate for the immobilization of bioreceptors and as a channel for the transport of fluid and analytes. It is common practice to provide a hydrophilic coating to the paper to avoid sample loss and promote uniform fluid flow. Paper-based lateral flow biosensors find widespread application in point-of-care diagnostics, ecological monitoring, and food safety testing. Furthermore, bioactive paper can be obtained by combining paper with various biomaterials to make composite paper materials. For instance, cellulose fibers and nanofibers (CNFs) can successfully be mixed for reinforcing polymeric materials. For example, glyoxylate polyacrylamides (GPAM) and/or polyamide-epichlorohydrin (PAE) give enhanced, long-lasting strength (Pelton, 2009). While the paper substrate provides excellent mechanical properties by tuning the cellulosic component, it shows the limitation of tearing and poor level of stability in water environments. Paper can be used as a substrate in the production of biosensors, which confers a variety of benefits to these devices and increases their desirability for use in a diverse range of contexts. Paper is made from cellulose, which is a raw material that is readily available and inexpensive. This is one of the primary benefits of the papermaking process. Paper is a material that is well suited for use in biosensors due to the fact that it is not only mechanically strong but also flexible and porous. In addition, paper may be easily manufactured through the use of straightforward printing methods, which makes it possible to generate biosensors on a massive scale with only a small investment in labor-intensive production procedures. In addition, paper can be used in applications where biodegradability is a primary concern because it is both biocompatible and biodegradable. This makes paper an attractive material for use in such settings. Paper is already used in a variety of sensors, such as pregnancy tests, COVID testing, glucose monitoring devices, and pH strips. These sensors are not only inexpensive but also quite convenient, which makes them available to a large number of different users. The fact that cellulose is used as the raw material helps to keep production costs low. Additionally, the presence of functional groups (COOH and OH) on the surface of the paper makes it simpler for proteins to adsorb as well as covalently bound onto the surface, which in turn increases the sensor's sensitivity (Hillscher et al., 2021; Mahato et al., 2021b; Mahato, Purohit et al., 2020; Mahato, Srivastava et al., 2017; Parolo & Merkoçi, 2013; Shafiee et al., 2015).

6.9.1.2.3 Textiles and fibers

Textiles can be manufactured as fibers or woven and can be stitched or sculpted into other quasi-forms to offer support and stability for the development of sensitive devices. They represent a broad range of innovative materials that can be used for flexible biosensor applications. Because of their inherent flexibility, sturdiness, and reliability, these materials, which include wool, cotton, and synthetics (such as nylon and polyester), are good candidates for use as substrates. In addition, the incorporation of conductive materials onto textile substrates is feasible using a variety of manufacturing methods. For instance, to create conductive filaments, fabrics have been dipped in or soaked in conductive polymers. Integrating textiles during the polymerization process is another method for creating conductive yarns, which can then be utilized as substrate. The transitioned electrical components on textile substrates have the potential to bridge the gap between potent and conceivably multiplexed electrochemical sensors, while also rewarding the needs of everyday heavy usage, with features such as lightweight, comfortable, integrated capabilities, simple operation, inexpensive, and promoting real-time detection.

The multiple benefits of wearable biosensors make them an attractive option for a diverse set of health, fitness, and environmental monitoring applications. Integrating biosensors into clothing is a major benefit of wearable technology as it creates a wide surface area for sensing and a high level of wearability. This allows for the long-term tracking of physiological signals and environmental factors, which can provide important information about a person's health and lifestyle. One benefit of wearing biosensors is that they may be integrated into microfluidic systems, which utilize fluidics for liquid transportation and manipulation. This allows for the examination of more intricate biological samples, such as blood and saliva, and so provides more in-depth insight into a person's health. As these sensors can deteriorate with time, change color to signify infection, or perform other activities crucial to monitoring patient health, their incorporation into sutures and smart stitching further increases the potential applications of wearable biosensors. Carbon fibers, which are extremely resistive, conductive, and compact, can also be used to create wearable biosensors. High-performance suits can now incorporate these fibers to provide real-time monitoring of the wearer's environment and vitals. Wearable biosensors made from carbon fibers offer a high level of sensitivity and accuracy, allowing for the precise monitoring of a variety of biological and environmental factors. In conclusion, wearable biosensors may drastically improve our health tracking and management. Wearable biosensors allow people to better understand their health and make educated decisions about their lifestyle and habits by offering continuous and real-time monitoring of critical physiological and environmental data. Medical professionals can also benefit from the data collected by wearable biosensors, as this will allow them to make better treatment and care decisions for their patients. With fluidics and sophisticated materials like carbon fibers, wearable biosensors can be integrated into garments, sutures, and smart stitches, expanding their usefulness and utility. Wearable biosensors have a bright future ahead of them, and they will certainly play a larger part in healthcare and wellness in the years to come.

6.10 Conclusions

Biosensors have found great attention for various applications including clinical and health monitoring on a daily basis. While the research activities for developing the biosensors are in full swing, various technical challenges in efficient electron transfer in the biosensing systems for stable and sensitive detection are yet to be addressed to obtain reliable, accurate, and commercially viable biosensing modules. This chapter emphasized the fundamental requirements for developing biosensors in the context of bioreceptors and materials. Although the bioreceptor's consistent regeneration is the most desirable property for long-term continuous monitoring operations., the shelf-life stability becomes more important for disposable single-use strip-based sensors. As the bioreceptors components are highly susceptible in the nonnative (ex vivo and in vitro conditions) environment due to the variable temperature and pH, resistant bioreceptors are in great demand. Conventionally used receptors are biologically active receptors, such as enzymes and antibodies, which show limited resistance to temperature and pH changes. Such vulnerabilities may impact the biosensor's analytical performance and reliability, which are the utmost requirements for various clinical applications. In addition, the wider range of biomarkers in clinical diagnoses and the limited availability of the bioreceptors also lead to the development of synthetic biorecognition elements. Considering the demand, various bioreceptors such as aptamers, nanomaterials-based catalysts, and MIPs have been employed in developing robust sensing modules. Engineered enzymes and modified antibodies have also shown great promise to mitigate such challenges; however, most of the reported biosensors have used in-house engineered variants. While bioreceptors limit the analytical performance of sensors, advanced and functional materials play a crucial role in signal transduction. The functionality of these materials not only assists signal amplification but also leverages the robustness and stability of the sensors. Moreover, obtaining ultra-high sensitivity is required for the detection of biomarkers found in ultra-low concentrations such as metabolite and peptide biomarkers. In such cases, signal amplification is required to enhance the sensitivity by using nanomaterials. The signal amplification in the biosensors is majorly obtained by increasing the active surface area of the sensor or recycling the substrate/mediators. In addition, the use of various functional materials for developing biosensors may increase their durability, robustness, and mechanical stability. However, to enhance their functionality, the vulnerabilities of bioreceptors are required to be sorted out by adopting resistant and systemically favorable variants.

Although attempts have been made to fill the gaps by availing efficient and robust bioreceptors for glucose and lactate detections, these are yet to be explored for many other biomarkers. Another major challenge is systemic limitations, which arise due to the complex nature of the real sample matrices in the clinical/biomedical samples. The coexisting molecules cause surface fouling and interfere with the sensor's performance. The efforts for resolving such challenges have also been made by employing the coating of the polymeric layers over the sensing surface, that is, surface fouling and biofouling, which can be addressed by various charged/uncharged polymeric matrices. However, these do not follow any defined recipe, thus optimization in every case is the only solution. In conclusion, the materials that are employed in the construction of a biosensor play a significant role in defining the overall performance of the device. For accurate detection, the materials utilized to capture the analyte need to have certain features, such as being biocompatible and having a high affinity for the analyte being targeted. Flexible materials provide better durability and simplicity, while materials that enhance the signal assure reliable and sensitive readings. As a result, carefully considering and selecting appropriate materials is necessary to guarantee the success of biosensor research and applications. In this regard, there is a need for biomedical and material scientists to work together more closely on projects to improve the biosensor sector as a whole. Future works will be directed toward the exploration of bioreceptors for stable biorecognition to achieve stability in sensing/monitoring. New research exploring the properties of CPs and nanomaterials would leverage the sensitivity enhancements.

References

Abrego-Martinez, J. C., Jafari, M., Chergui, S., Pavel, C., Che, D., & Siaj, M. (2022). *Biosensors & Bioelectronics, 195*, 113595.

Agarwal, P. K. (2006). *Microbial Cell Factories, 5*, 1–12.

Ahmad, Z. A., Yeap, S. K., Ali, A. M., Ho, W. Y., Alitheen, N. B. M., & Hamid, M. (2012). *Clinical & Developmental Immunology, 2012*.

Al Sulaiman, D., Cadinu, P., Ivanov, A. P., Edel, J. B., & Ladame, S. (2018). *Nano Letters, 18*, 6084–6093.

Ali, M. M., Wolfe, M., Tram, K., Gu, J., Filipe, C. D. M., Li, Y., & Brennan, J. D. (2019). *Angewandte Chemie International Edition, 58*, 9907–9911.

Alizadeh, N., Hallaj, R., & Salimi, A. (2017). *Biosensors & Bioelectronics, 94*, 184–192.

Amjadi, M., Yoon, Y. J., & Park, I. (2015). *Nanotechnology, 26*, 375501.

Ates, M. (2013). *Materials Science Engineering C, 33*, 1853–1859.

Aydemir, N., Malmström, J., & Travas-Sejdic, J. (2016). *Physical Chemistry Chemical Physics: PCCP, 18*, 8264–8277.

Ayyaru, S., Letchoumanane, P., Dharmalingam, S., & Stanislaus, A. R. (2012). *Journal of Power Sources, 217*, 204–208.

Baranwal, A., Kumar, A., Priyadharshini, A., Oggu, G. S., Bhatnagar, I., Srivastava, A., & Chandra, P. (2018). *International Journal of Biological Macromolecules, 110*, 110–123.

Barberis, A., Spissu, Y., Bazzu, G., Fadda, A., Azara, E., Sanna, D., Schirra, M., & Serra, P. A. (2014). *Analytical Chemistry, 86*, 8727–8734.

Beaver, K., Dantanarayana, A., & Minteer, S. D. (2021). *The Journal of Physical Chemistry. B, 125*, 11820–11834.

Bell, E. L., Finnigan, W., France, S. P., Green, A. P., Hayes, M. A., Hepworth, L. J., Lovelock, S. L., Niikura, H., Osuna, S., Romero, E., Ryan, K. S., Turner, N. J., & Flitsch, S. L. (2021). *Nature Reviews Methods Primers, 11*(1), 1–21.

Bhand, S., & Singh, A. C. (2019). *RSC Detection science, 14*, 21–41.

Bhatnagar, I., Mahato, K., Ealla, K. K. R., Asthana, A., & Chandra, P. (2018). *International Journal of Biological Macromolecules, 110*, 449–456.

Białas, K., Moschou, D., Marken, F., & Estrela, P. (2022). *Microchimica Acta, 189*, 1–20.

Bollella, P., & Katz, E. (2020). *Sensors 2020, 20*, 3517.

Bousse, L. (1996). *Sensors Actuators B: Chemical, 34*, 270–275.

Breaker, R. R. (1997). *Nature Biotechnology, 155*(15), 427–431.

Bruch, R., Baaske, J., Chatelle, C., Meirich, M., Madlener, S., Weber, W., Dincer, C., & Urban, G. A. (2019). *Advanced Materials, 31*, 1905311.

Campàs, M., Prieto-Simón, B., & Marty, J. L. (2009). *Seminars in Cell & Developmental Biology, 20*, 3–9.

Cardoso, A. R., Alves, J. F., Frasco, M. F., Piloto, A. M., Serrano, V., Mateus, D., Sebastião, A. I., Matos, A. M., Carmo, A., Cruz, T., Fortunato, E., & Sales, M. G. F. (2022). *Materials Today Bio*, *16*, 100354.

Chandra, P. (2016). *Nanobiosensors for personalized and onsite biomedical diagnosis*. London: IET.

Chandra, P., & Mahato, K. (2022). *Miniaturized biosensing devices: fabrication and applications*. Springer Nature Singapore.

Chandrasekaran, A. R. (2016). *Journal of Chemical Technology and Biotechnology (Oxford, Oxfordshire: 1986)*, *91*, 843–846.

Chandrasekaran, A. R., Anderson, N., Kizer, M., Halvorsen, K., & Wang, X. (2016). *Chembiochem: A European Journal of Chemical Biology*, 1081–1089.

Chen, C., & Hu, L. (2018). *Accounts of Chemical Research*, *51*, 3154–3165.

Chen, L. Q., Zhang, X. E., Xie, W. H., Zhou, Y. F., Zhang, Z. P., & Cass, A. E. G. (2002). *Biosensors & Bioelectronics*, *17*, 851–857.

Chen, S., Xie, Y., Guo, X., & Sun, D. (2022). *Microchemical Journal*, *181*, 107715, devoted to the application of microtechniques in all branches of science.

Cheng, Y., Chen, T., Fu, D., Liu, M., Cheng, Z., Hua, Y., & Liu, J. (2022). *Talanta*, *242*, 123279.

Chiu, M. L., Goulet, D. R., Teplyakov, A., & Gilliland, G. L. (2019). *Antibodies*, *8*, 55.

Cho, I. H., Lee, J., Kim, J., Kang, M. S., Paik, J. K., Ku, S., Cho, H. M., Irudayaraj, J., & Kim, D. H. (2018). *Sensors (Basel)*, 18.

Comotti, M., Della Pina, C., Falletta, E., & Rossi, M. (2006). *Advanced Synthesis Catalysis*, *348*, 313–316.

Crapnell, R. D., Dempsey-Hibbert, N. C., Peeters, M., Tridente, A., & Banks, C. E. (2020). *Talanta Open*, *2*, 100018.

Csöregi, E., Jönsson-Pettersson, G., & Gorton, L. (1993). *Journal of Biotechnology*, *30*, 315–337.

Dai, Y., Somoza, R. A., Wang, L., Welter, J. F., Li, Y., Caplan, A. I., & Liu, C. C. (2019). *Angewandte Chemie International Edition*, *58*, 17399–17405.

Dakshayini, B. S., Reddy, K. R., Mishra, A., Shetti, N. P., Malode, S. J., Basu, S., Naveen, S., & Raghu, A. V. (2019). *Microchemical Journal*, *147*, 7–24, devoted to the application of microtechniques in all branches of science.

Darabdhara, G., Sharma, B., Das, M. R., Boukherroub, R., & Szunerits, S. (2017). *Sensors Actuators B: Chemical*, *238*, 842–851.

Das, A., & Mahanwar, P. (2020). *Advanced Industrial Engineering Polymer Research*, *3*, 93–101.

Das, J., Kim, H., Jo, K., Park, K. H., Jon, S., Lee, K., & Yang, H. (2009). *Chemical Communications*, 6394–6396.

Di Giulio, T., Mazzotta, E., & Malitesta, C. (2020). *Biosensors*, *11*, 3.

Dinçkaya, E., Akyilmaz, E., Telefoncu, A., & Akgöl, S. (1999). *Indian Journal of Biochemistry & Biophysics*, *36*, 36–38.

Dkhar, D. S., Kumari, R., Mahapatra, S., Divya., Kumar, R., Tripathi, T., & Chandra, P. (2022). *International Journal of Biological Macromolecules*, *218*, 225–242.

Drummond, T. G., Hill, M. G., & Barton, J. K. (2003). *Nature Biotechnology*, *21*, 1192–1199.

Elakkiya, R., & Maduraiveeran, G. (2019). *New Journal of Chemistry*, *43*, 14756–14762.

Ellingford, C., Zhang, R., Wemyss, A. M., Bowen, C., McNally, T., Figiel, Ł., & Wan, C. (2018). *ACS Applied Materials & Interfaces*, *10*, 38438–38448.

Felix, F. S., & Angnes, L. (2018). *Biosensors & Bioelectronics*, *102*, 470–478.

Gao, L., Zhuang, J., Nie, L., Zhang, J., Zhang, Y., Gu, N., Wang, T., Feng, J., Yang, D., Perrett, S., & Yan, X. (2007). *National Nanotechnology*, *2*, 577–583.

Gao, S., Guisán, J. M., & Rocha-Martin, J. (2022). *Analytica Chimica Acta*, *1189*, 338907.

Gao, W., Dong, H., Lei, J., Ji, H., & Ju, H. (2011). *Chemical Communications*, *47*, 5220–5222.

Gong, X., Gu, Y., Zhang, F., Liu, Z., Li, Y., Chen, G., & Wang, B. (2019). *Frontiers Materials*, *6*, 3.

Grewal, Y. S., Shiddiky, M. J. A., Gray, S. A., Weigel, K. M., Cangelosi, G. A., & Trau, M. (2013). *Chemical Communications*, *49*, 1551–1553.

Guo, K., Wustoni, S., Koklu, A., Díaz-Galicia, E., Moser, M., Hama, A., Alqahtani, A. A., Ahmad, A. N., Alhamlan, F. S., Shuaib, M., Pain, A., McCulloch, I., Arold, S. T., Grünberg, R., & Inal, S. (2021). *Nature Biomedical Engineering*, *57*(5), 666–677, 2021.

Guo, Q., Zeng, W., Liu, S., & Li, Y. (2020). *Nanotechnology*, *31*, 265501.

Gupta, S. K., Sudarshan, K., & Kadam, R. M. (2021). *Materials Today Communications*, *27*, 102277.

Hackett, A. J., Malmström, J., & Travas-Sejdic, J. (2017). *Progress in Polymer Science*, *70*, 18–33.

Han, S., Liu, W., Yang, S., & Wang, R. (2019). *ACS Omega*, *4*, 11025–11031.

Han, S.-T., Peng, H., Sun, Q., Venkatesh, S., Chung, K.-S., Lau, S. C., Zhou, Y., & Roy, V. A. L. (2017). *Advanced Materials*, *29*, 1700375.

Hartley, C. J., Williams, C. C., Scoble, J. A., Churches, Q. I., North, A., French, N. G., Nebl, T., Coia, G., Warden, A. C., Simpson, G., Frazer, A. R., Jensen, C. N., Turner, N. J., & Scott, C. (2019). *Nature Catalysis 2019*, *211*(2), 1006–1015.

Haun, J. B., Devaraj, N. K., Hilderbrand, S. A., Lee, H., & Weissleder, R. (2010). *National Nanotechnology*, *5*, 660–665.

Hillscher, L. M., Liebich, V. J., Avrutina, O., Biesalski, M., & Kolmar, H. (2021). *ChemTexts*, *7*, 1–22.

Hiraka, K., Kojima, K., Tsugawa, W., Asano, R., Ikebukuro, K., & Sode, K. (2020). *Biosensors & Bioelectronics*, *151*, 111974.

Hloma, P., Uwaya, G. E., & Bisetty, K. (2022). *Biosensors and Bioelectronics: X*, *11*, 100203.

Hosseindokht, M., Bakherad, H., & Zare, H. (2021). *Cancer Cell International*, *21*, 1–9.

Hosseini, E. S., Dervin, S., Ganguly, P., & Dahiya, R. (2021). *ACS Applied Bio Materials*, *4*, 163–194.

Hryniewicz, B. M., Volpe, J., Bach-Toledo, L., Kurpel, K. C., Deller, A. E., Soares, A. L., Nardin, J. M., Marchesi, L. F., Simas, F. F., Oliveira, C. C., Huergo, L., Souto, D. E. P., & Vidotti, M. (2022). *Materials Today Chemistry*, *24*, 100817.

Huang, F., Zhang, Y., Lin, J., & Liu, Y. (2021). *Biosensors*, *11*, 190.

Huang, Y., Ren, J., & Qu, X. (2019). *Chemical Reviews*, *119*, 4357–4412.

Hwang, K., Hosseinzadeh, P., & Lu, Y. (2016). *Inorganica Chimica Acta*, *452*, 12–24.

Izadyar, A., Tran, U., Seok, I., & Hood, E. E. (2019). *ACS Sustainable Chemistry & Engineering*, *7*, 19434–19441.

Jain, S., Verma, S., Singh, S. P., & Sharma, S. N. (2019). *Biosensors & Bioelectronics*, *127*, 135–141.

Jędrzak, A., Kuznowicz, M., Rębiś, T., & Jesionowski, T. (2022). *Bioelectrochemistry (Amsterdam, Netherlands)*, *145*, 108071.

Jolly, P., Tamboli, V., Harniman, R. L., Estrela, P., Allender, C. J., & Bowen, J. L. (2016). *Biosensors & Bioelectronics*, *75*, 188–195.

Jung, I. Y., Kim, J. S., Choi, B. R., Lee, K., & Lee, H. (2017). *Advanced Healthcare Materials*, *6*, 1–19.

Kabay, G., Yin, Y., Singh, C. K., Ahmad, N., Gunasekaran, S., & Mutlu, M. (2022). *Sensors Actuators B: Chemical*, *360*, 131667.

Kabir, M. H., Marquez, E., Djokoto, G., Parker, M., Weinstein, T., Ghann, W., Uddin, J., Ali, M. M., Alam, M. M., Thompson, M., Poyraz, A. S., Msimanga, H. Z., Rahman, M. M., Rulison, M., & Cramer, J. (2022). *ACS Applied Materials Interfaces*, *14*, 24229–24244.

Kaminski, M. M., Abudayyeh, O. O., Gootenberg, J. S., Zhang, F., & Collins, J. J. (2021). *Nature Biomedical. Engeering.*, *57*(5), 643–656.

Karadurmus, L., Bilge, S., Sınağ, A., & Ozkan, S. A. (2022). *TrAC Trends in Analytical Chemistry*, *155*, 116694.

Karimian, N., Vagin, M., Zavar, M. H. A., Chamsaz, M., Turner, A. P. F., & Tiwari, A. (2013). *Biosensors & Bioelectronics*, *50*, 492–498.

Kashish, B., S. Jyoti, A., Mahato, K., Chandra, P., & Prakash, R. (2017). *Electroanalysis*, *29*, 2665–2671.

Katz, H. E. (2021). *Nature Biomedical Engineering.*, *57*(5), 639–640.

Khan, S., Burciu, B., Filipe, C. D. M., Li, Y., Dellinger, K., & Didar, T. F. (2021). *ACS Nano*, *15*(9), 13943–13969.

Khokhar, D., Jadoun, S., Arif, R., & Jabin, S. (2021). *Polymer Technology Materials*, *60*, 465–487.

Kim, J., & Park, M. (2021). *Biosensors*, 11.

Kong, L., Wang, D., Chai, Y., Yuan, Y., & Yuan, R. (2019). *Analytical Chemistry*, *91*, 10289–10294.

Köse, S., Ahan, R. E., Köksaldı, İ. Ç., Olgaç, A., Kasapkara, Ç. S., & Şeker, U. Ö. Ş. (2023). *Biosensors & Bioelectronics*, *223*, 115035.

Kratz, H., Mohtashamdolatshahi, A., Eberbeck, D., Kosch, O., Wiekhorst, F., Taupitz, M., Hamm, B., Stolzenburg, N., & Schnorr, J. (2021). *Nanomaterials*, *11*, 1532.

Kumar, A., Purohit, B., Mahato, K., & Chandra, P. (2019). *Advance engineered nanomaterials in point-of-care immunosensing for biomedical diagnostics. RSC Detection science* (pp. 238–266). Royal Society of Chemistry.

Kumar, A., Purohit, B., Mahato, K., Mahapatra, S., Srivastava, A., & Chandra, P. (2020). *Biointerface engineering: Prospects of medical diagnostics and drug delivery* (pp. 215–231). Springer.

Kumar, A., Purohit, B., Mahato, K., Mandal, R., Srivastava, A., & Chandra, P. (2019). *Electroanalysis*, *31*, 2417–2428.

Kumar, A., Purohit, B., Maurya, P. K., Pandey, L. M., & Chandra, P. (2019). *Electroanalysis*, *31*, 1615–1629.

Kumar, A., Roy, S., Srivastava, A., Naikwade, M. M., Purohit, B., Mahato, K., Naidu, V. G. M., & Chandra, P. (2019). *Nanotherapeutics: A novel and powerful approach in modern healthcare system. Nanotechnology in modern animal biotechnology* (pp. 149–161). Elsevier.

Kumar, A., Sharma, S., Pandey, L. M., & Chandra, P. (2018). *Materials Science for Energy Technologies*, *1*, 38–48.

Kumašin, A., Mahmutović, L., & Hromić-Jahjefendić, A. (2022). *Periodicals of Engineering and Natural Sciences*, *10*, 18–29.

Lake, R. J., Yang, Z., Zhang, J. J., & Lu, Y. (2019). *Accounts of Chemical Research*, *52*, 3275.

Lang, X., Wang, T., Sun, M., Chen, X., & Liu, Y. (2020). *International Journal of Biological Macromolecules*, *154*, 433–445.

Lau, P. H., Takei, K., Wang, C., Ju, Y., Kim, J., Yu, Z., Takahashi, T., Cho, G., & Javey, A. (2013). *Nano Letters*, *13*, 3864–3869.

Lee, S. M., Shin, S., Kim, H. Y., Won, B. Y., Ahn, J. K., Park, K. S., & Park, H. G. (2020). *Biotechnology and Bioprocess Engineering*, *255*(25), 707–714.

Lee, T., Park, S. Y., Jang, H., Kim, G. H., Lee, Y., Park, C., Mohammadniaei, M., Lee, M. H., & Min, J. (2019). *Materials Science and Engineering. C, 99*, 511–519.

Lei, J., & Ju, H. (2012). *Chemical Society Reviews, 41*, 2122–2134.

Leng, T., Pan, K., Zhang, Y., Li, J., Afroj, S., Novoselov, K. S., & Hu, Z. (2019). *ACS Applied Nano Materials, 2*, 6197–6208.

Li, L., Pan, L., Ma, Z., Yan, K., Cheng, W., Shi, Y., & Yu, G. (2018). *Nano Letters, 18*, 3322–3327.

Li, W. (2020). *Nature Biotechnology, 381*(38), 30.

Li, Y., Liu, S., Deng, Q., & Ling, L. (2018). *Journal of Medical Virology, 90*, 699–705.

Li, Y., Tian, L., Liu, L., Khan, M. S., Zhao, G., Fan, D., Cao, W., & Wei, Q. (2018). *Talanta, 179*, 420–425.

Li, Y., Zhu, W., Kang, Q., Yang, L., Zhang, Y., Wang, Y., & Wei, Q. (2018). *ACS Applied Materials & Interfaces, 10*, 38791–38798.

Liao, C., Zhang, M., Yao, M. Y., Hua, T., Li, L., & Yan, F. (2015). *Advanced Materials, 27*, 7493–7527.

Liao, X., Luo, J., Wu, J., Fan, T., Yao, Y., Gao, F., & Qian, Y. (2018). *Journal of Electroanalytical Chemistry, 829*, 129–137.

Liaw, D.-J., Wang, K.-L., Huang, Y.-C., Lee, K.-R., Lai, J.-Y., & Ha, C.-S. (2012). *Progress in Polymer Science, 37*, 907–974.

Liébana, S., & Drago, G. A. (2016). *Essays in Biochemistry, 60*, 59–68.

Liu, M., Zhang, Q., Brennan, J. D., & Li, Y. (2018). *MRS Communications, 8*, 687–694.

Liu, Q., Wu, C., Cai, H., Hu, N., Zhou, J., & Wang, P. (2014). *Chemical Reviews, 114*, 6423–6461.

Liu, Y., Liu, Y., Qiao, L., Liu, Y., & Liu, B. (2018). *Current Opinion Electrochemistry, 12*, 5–12.

Lou-Franco, J., Das, B., Elliott, C., & Cao, C. (2020). *Nano-Micro Letters, 131*(13), 1–36.

Ma, J., Mo, Y., Tang, M., Shen, J., Qi, Y., Zhao, W., Huang, Y., Xu, Y., & Qian, C. (2021). *Frontiers in Immunology, 12*, 1555.

Mahato, K., Baranwal, A., Srivastava, A., Maurya, P. K., & Chandra, P. (2018). *Smart materials for biosensing applications, techno-societal 2016*. Cham: Springer International Publishing.

Mahato, K., Kumar, A., Maurya, P. K., & Chandra, P. (2018). *Biosensors & Bioelectronics, 100*, 411–428.

Mahato, K., Kumar, A., Purohit, B., Mahapatra, S., Srivastava, A., & Chandra, P. (2020). *Nanomaterial functionalization strategies in bio-interface development for modern diagnostic devices. Biointerface engineering: Prospects in medical diagnostics and drug delivery*. Springer.

Mahato, K., Kumar, S., Srivastava, A., Maurya, P. K., Singh, R., & Chandra, P. (2018). *Handbook of immunoassay technologies: Approaches, performances, and applications* (pp. 359–414). Academic Press.

Mahato, K., Purohit, B., Bhardwaj, K., Jaiswal, A., & Chandra, P. (2019). *Biosensors & Bioelectronics, 142*, 111502.

Mahato, K., Purohit, B., Kumar, A., & Chandra, P. (2020). Comprehensive analytical chemistry. https//doi.org/10.1016/bs.coac1.

Mahato, K., Purohit, B., Kumar, A., Srivastava, A., & Chandra, P. (2021a). Immunodiagnostic technologies from laboratory to point-of-care test (pp. 93–110).

Mahato, K., Purohit, B., Kumar, A., Srivastava, A., & Chandra, P. (2021b). Next-generation immunosensing technologies based on nano-bio-engineered paper matrices. In P. Suman, & P. Chandra (Eds.), *Immunodiagnostic technologies from laboratory to point-of-care testing* (pp. 93–110). Springer.

Mahato, K., Srivastava, A., & Chandra, P. (2017). Paper based diagnostics for personalized health care: Emerging technologies and commercial aspects. *Biosensors & Bioelectronics*, *96*, 246–259.

Mahato, K., & Wang, J. (2021). *Sensors Actuators B Chemical*, *344*, 130178.

Mahmudunnabi, R. G., Farhana, F. Z., Kashaninejad, N., Firoz, S. H., Shim, Y. B., & Shiddiky, M. J. A. (2020). *Analyst*, *145*, 4398–4420.

Márquez, A., Jiménez-Jorquera, C., Dom\'\inguez, C., & Muñoz-Berbel, X. (2017). *Biosensors & Bioelectronics*, *97*, 136–142.

Martínez-García, G., Sánchez-Tirado, E., González-Cortés, A., Yáñez-Sedeño, P., & Pingarrón, J. M. (2018). *Microchimica Acta*, *185*, 1–8.

Mata, A., Fleischman, A. J., & Roy, S. (2005). *Biomedical Microdevices*, *7*, 281–293.

Medintz, I. L. (2009). Interfacing biology with nanomaterials. *Materials Today*..

Melo, F. C. C., Alves, R. P., Valle, A. L., Santos, F., de, A. A., Dias, A. C. S., Goulart, I. M. B., Oliveira, E. G. A., Oliveira, G. S., Rodrigues, L. P., & Goulart, L. R. (2022). *Sensors*, *22*, 3681, 2022, Vol. 22, Page 3681.

Meng, L., Turner, A. P. F., & Mak, W. C. (2021). *ACS Applied Materials Interfaces*, *13*, 54456–54465.

Morales, M. A., & Halpern, J. M. (2018). *Bioconjugate Chemistry*, *29*, 3231–3239.

Moreira, F. T. C., Dutra, R. A. F., Noronha, J. P. C., Cunha, A. L., & Sales, M. G. F. (2011). *Biosensors & Bioelectronics*, *28*, 243–250.

Morrison, H. (2020). *Enzyme active sites and their reaction mechanisms. Enzyme active sites and their reaction mechanisms*. Elsevier.

Murakami, S., Nishikawa, Y., Tsuji, M., Kawaguchi, A., Kohjiya, S., & Cakmak, M. (1995). *Polymer*, *36*, 291–297.

Musarraf Hussain, M., Asiri, A. M., & Rahman, M. M. (2020). *Microchemical Journal*, *159*, 105534, devoted to the application of microtechniques in all branches of science.

Nakamoto, K., Kurita, R., & Niwa, O. (2010). *Biosensors & Bioelectronics*, *26*, 1536–1542.

Nemčeková, K., & Labuda, J. (2021). *Materials Science and Engineering C*, 120.

Newman, J. D., & Setford, S. J. (2006). *Molecular Biotechnology*, *323*(32), 249–268.

Nur, Y., Gaffar, S., Hartati, Y. W., & Subroto, T. (2021). *Sensing and Bio-Sensing Research*, *32*, 100416.

Ozcan, H. M., & Sagiroglu, A. (2014). *Artificial Cells, Nanomedicine, and Biotechnology*, *42*, 256–261. Available from http://doi.org/10.3109/21691401.2013.764313.

Pang, H., Piao, Y.-Y., Tan, Y.-Q., Jiang, G.-Y., Wang, J.-H., & Li, Z.-M. (2013). *Materials Letters*, *107*, 150–153.

Park, J. W., Kurosawa, S., Aizawa, H., Wakida, S. I., Yamada, S., & Ishihara, K. (2003). *Sensors Actuators B: Chemical*, *91*, 158–162.

Parolo, C., & Merkoçi, A. (2013). *Chemical Society Reviews*, *42*, 450–457.

Patolsky, F., Weizmann, Y., & Willner, I. (2004). *Angewandte Chemie – International Edition*, *43*, 2113–2117.

Pelton, R. (2009). *TrAC Trends in Analytical Chemistry*, *28*, 925–942.

Peng, C., Miao, L., Qiu, D., & Chen, S. (2022). *Ceramics International*, *48*, 23137–23144.

Perumal, V., & Hashim, U. (2014). *Journal of Applied Biomedical*, *12*, 1–15.

Peterson, M. E., Daniel, R. M., Danson, M. J., & Eisenthal, R. (2007). *The Biochemical Journal*, *402*, 331–337.

Phonklam, K., Wannapob, R., Sriwimol, W., Thavarungkul, P., & Phairatana, T. (2020). *Sensors Actuators B: Chemical*, *308*, 127630.

Pohanka, M., Fusek, J., Adam, V., & Kizek, R. (2013). *International Journal of Electrochemical Science*, *8*, 71–79.

Portorreal-Bottier, A., Gutiérrez-Tarriño, S., Calvente, J. J., Andreu, R., Roldán, E., Oña-Burgos, P., & Olloqui-Sariego, J. L. (2022). *Sensors Actuators B: Chemical*, *368*, 132129.

Prasad, A., Mahato, K., Chandra, P., Srivastava, A., Joshi, S. N., & Maurya, P. K. (2016). *Journal of Molecular Engineering Materials*, *04*, 1640004.

Pravda, L., Berka, K., Svobodová Vařeková, R., Sehnal, D., Banáš, P., Laskowski, R. A., Koča, J., & Otyepka, M. (2014). *BMC Bioinformatics*, *15*, 1–8.

Purohit, B., Kumar, A., Mahato, K., Chandra, P., Purohit, B., Kumar, A., Mahato, K., & Chandra, P. (2022). *Miniaturized Biosensing Devices*, 149–175.

Purohit, B., Kumar, A., Mahato, K., Roy, S., & Chandra, P. (2019). *Cancer cytosensing approaches in miniaturized settings based on advanced nanomaterials and biosensors. Nanotechnology in modern animal biotechnology: Concepts and applications* (pp. 133–147). Elsevier.

Purohit, B., Mahato, K., Kumar, A., & Chandra, P. (2019). *Microchimica Acta*, 186.

Qian, J., Yang, X., Yang, Z., Zhu, G., Mao, H., & Wang, K. (2015). *Journal of Materials Chemistry B*, *3*, 1624–1632.

Qin, D., Xia, Y., & Whitesides, G. M. (2010). *Nature Protocols*, *5*, 491–502.

Rao, H., Zhang, Z., Ge, H., Liu, X., Zou, P., Wang, X., & Wang, Y. (2017). *New Journal of Chemistry*, *41*, 3667–3676.

Rechnitz, G. A., Arnold, M. A., & Meyerhoff, M. E. (1979). *Nature*, *278*, 466–467.

Reddish, W. (1950). *Transactions of the Faraday Society*, *46*, 459–475.

Regiart, M., Ledo, A., Fernandes, E., Messina, G. A., Brett, C. M. A., Bertotti, M., & Barbosa, R. M. (2022). *Biosensors & Bioelectronics*, *199*, 113874.

Ren, R., Bi, Q., Yuan, R., & Xiang, Y. (2020). *Sensors Actuators B: Chemical*, *304*, 127068.

Schubert, F., Wollenberger, U., & Scheller, F. (1983). *Biotechnology Letters*, *5*, 239–242.

Sempionatto, J. R., Khorshed, A. A., Ahmed, A., De Loyola, E., Silva, A. N., Barfidokht, A., Yin, L., Goud, K. Y., Mohamed, M. A., Bailey, E., May, J., Aebischer, C., Chatelle, C., & Wang, J. (2020). *ACS Sensors*, *5*(6), 1804–1813.

Shafiee, H., Asghar, W., Inci, F., Yuksekkaya, M., Jahangir, M., Zhang, M. H., Durmus, N. G., Gurkan, U. A., Kuritzkes, D. R., & Demirci, U. (2015). *Science Reports*, *5*, 1–9.

Shahdeo, D., Roberts, A., Archana, G. J., Shrikrishna, N. S., Mahari, S., Nagamani, K., & Gandhi, S. (2022). *Biosensors & Bioelectronics*, *212*, 114406.

Sharma, S., Byrne, H., & O'Kennedy, R. J. (2016). *Essays in Biochemistry*, *60*, 9–18.

Shi, W., Wang, Q., Long, Y., Cheng, Z., Chen, S., Zheng, H., & Huang, Y. (2011). *Chemistry Communications*, *47*, 6695–6697.

Silva, B. V. M., Rodríguez, B. A. G., Sales, G. F., Sotomayor, M. D. P. T., & Dutra, R. F. (2016). *Biosensors & Bioelectronics*, *77*, 978–985.

Soleymani, L., & Li, F. (2017). *ACS Sensors*, *2*, 458–467.

Song, S., Wang, L., Li, J., Fan, C., & Zhao, J. (2008). *TrAC Trends in Analytical Chemistry*, *27*, 108–117.

Song, Y., Chen, Y., Feng, L., Ren, J., & Qu, X. (2011). *Chemistry Communications*, *47*, 4436–4438.

Song, Y., Qu, K., Zhao, C., Ren, J., & Qu, X. (2010). *Advanced Materials*, *22*, 2206–2210.

Song, Y., Wang, X., Zhao, C., Qu, K., Ren, J., & Qu, X. (2010). *Chemistry – A European Journal*, *16*, 3617–3621.

Sotiropoulou, S., Fournier, D., & Chaniotakis, N. A. (2005). *Biosensors & Bioelectronics*, *20*, 2347–2352.

Suginta, W., Khunkaewla, P., & Schulte, A. (2013). *Chemical Reviews, 113*, 5458–5479.

Sun, D., Lin, X., Lu, J., Wei, P., Luo, Z., Lu, X., Chen, Z., & Zhang, L. (2019). *Biosensors & Bioelectronics, 142*, 111578.

Sun, L. P., Zhong, Y., Gui, J., Wang, X. W., Zhuang, X. R., & Weng, J. (2018). *International Journal of Nanomedicine, 13*, 843–856.

Tang, Q., Yuan, Y., Xiao, X., Guo, P., Hu, J., Ma, D., & Gao, Y. (2013). *Microchimica Acta, 180*, 1059–1064.

Tang, W., Yin, L., Sempionatto, J. R., Moon, J. M., Teymourian, H., & Wang, J. (2021). *Advanced Materials, 33*, 2008465.

Tavakoli, J., & Tang, Y. (2017). *Polymers, 9*, 364.

Tehrani, F., Teymourian, H., Wuerstle, B., Kavner, J., Patel, R., Furmidge, A., Aghavali, R., Hosseini-Toudeshki, H., Brown, C., Zhang, F., Mahato, K., Li, Z., Barfidokht, A., Yin, L., Warren, P., Huang, N., Patel, Z., Mercier, P. P., & Wang, J. (2022). *Nature Biomedical Engineering, 6*, 1214–1224.

Tripathy, S., Shiv, ·, & Singh, G. (2020). *Transactions of the Indian National Academy of Engineering, 52*(5), 205–209.

Uzun, L., & Turner, A. P. F. (2016). *Biosensors & Bioelectronics, 76*, 131–144.

Vargas, E., Povedano, E., Krishnan, S., Teymourian, H., Tehrani, F., Campuzano, S., Dassau, E., & Wang, J. (2020). *Biosensors & Bioelectronics, 167*, 112512.

Verma, S., Choudhary, J., Singh, K. P., Chandra, P., & Singh, S. P. (2019). *International Journal of Biological Macromolecules, 130*, 333–341.

Villalonga, A., Pérez-Calabuig, A. M., & Villalonga, R. (2020). *Analytical and Bioanalytical Chemistry, 412*, 55–72.

Voiry, D., Chhowalla, M., Gogotsi, Y., Kotov, N. A., Li, Y., Penner, R. M., Schaak, R. E., & Weiss, P. S. (2018). Best practices for reporting electrocatalytic performance of nanomaterials. *ACS Nano, 12*(10), 9635.

Wang, C., Xia, K., Zhang, Y., & Kaplan, D. L. (2019). *Accounts of Chemical Research, 52*, 2916–2927.

Wang, C., Yokota, T., & Someya, T. (2021). *Chemical Reviews, 121*, 2109–2146.

Wang, H., Lin, J., & Shen, Z. X. (2016). *Journal of Science Advanced. Materials Devices, 1*, 225–255.

Wang, J. (2008). *Chemical Reviews, 108*, 814–825.

Wang, L., Chen, D., Jiang, K., & Shen, G. (2017). *Chemical Society Reviews, 46*, 6764–6815.

Wang, Q., Pan, X., Lin, C., Lin, D., Ni, Y., Chen, L., Huang, L., Cao, S., & Ma, X. (2019). *Chemical Engineering Journal, 370*, 1039–1047.

Wang, S., Chinnasamy, T., Lifson, M. A., Inci, F., & Demirci, U. (2016). *Trends in Biotechnology, 34*, 909–921.

Wang, X., Hou, C., Qiu, W., Ke, Y., Xu, Q., Liu, X. Y., & Lin, Y. (2017). *ACS Applied Materials & Interfaces, 9*, 684–692.

Wang, Y., Liu, A., Han, Y., & Li, T. (2020). *Polymer International, 69*, 7–17.

Wang, Z., Zhao, H., Chen, K., Zhou, F., Magdassi, S., & Lan, M. (2022). *Biosensors & Bioelectronics, 209*, 114229.

Wei, Z., Cai, X., Cui, W., & Zhang, J. (2022). *Molecules, 27*, 4578.

Wild, D., & Davies, C. (2013). *The immunoassay handbook: Theory and applications of ligand binding, ELISA and related techniques* (pp. 1–26). Elsevier.

Wolf, M. P., Salieb-Beugelaar, G. B., & Hunziker, P. (2018). *Progress in Polymer Science, 83*, 97–134.

Wu, S., Snajdrova, R., Moore, J. C., Baldenius, K., & Bornscheuer, U. T. (2021). *Angewandte Chemie International Edition, 60*, 88–119.

Wu, Y., Darland, D. C., & Zhao, J. X. (2021). *Sensors*, *21*, 5201.

Wu, Y., Midinov, B., & White, R. J. (2019). *ACS Sensors*, *4*, 498–503.

Wu, Y., Wang, C. W., Wang, D., & Wei, N. (2021). *ACS Synthetic Biology*, *10*, 333–344.

Wu, Z., Zhen, Z., Jiang, J.-H., Shen, G.-L., & Yu, R.-Q. (2009). *Journal of the American Chemical Society*, *131*, 12325–12332.

Wycisk, R., Poźniak, R., & Pasternak, A. (2002). *Journal of Electrostatics*, *56*, 55–66.

Xavier, P. L., & Chandrasekaran, A. R. (2018). *Nanotechnology*, 29.

Xiao, J., Luo, Y., Su, L., Lu, J., Han, W., Xu, T., & Zhang, X. (2022). *Analytica Chimica Acta*, *1208*, 339843.

Xiao, M., Lai, W., Man, T., Chang, B., Li, L., Chandrasekaran, A. R., & Pei, H. (2019). *Chemical Reviews*, *119*, 11631–11717.

Xu, M., Obodo, D., & Yadavalli, V. K. (2019). *Biosensors & Bioelectronics*, *124–125*, 96–114.

Yadav, S. K., Agrawal, B., Chandra, P., & Goyal, R. N. (2014). *Biosensors & Bioelectronics*, *55*, 337–342.

Yang, L., Zhang, S., Liu, X., Tang, Y., Zhou, Y., & Wong, D. K. Y. (2020). *Journal of Materials Chemistry B*, *8*, 7880–7893.

Yang, M., Zhang, Y., Zhang, H., & Li, Z. (2015). Characterization of PEDOT: PSS as a biocompatible conductive material. In *10th IEEE international conference on nano/micro engineered and molecular systems* (pp. 149–151).

Yang, Y., Li, W., & Liu, J. (2021). *Analytica Chimica Acta*, *1147*, 124–143.

Zaryanov, N. V., Nikitina, V. N., Karpova, E. V., Karyakina, E. E., & Karyakin, A. A. (2017). *Analytical Chemistry*, *89*, 11198–11202.

Zeng, X., Shen, Z., & Mernaugh, R. (2012). *Analytical and Bioanalytical Chemistry*, *402*, 3027–3038.

Zhang, D., Yu, X., Wu, L., Jin, H., & Wei, M. (2021). *Journal of Electroanalytical Chemistry*, *882*, 114960.

Zhang, K., Cai, H., Lu, M., Wei, D., Yin, J., Ding, N., Lai, W., & Peng, J. (2022). *Food and Agricultural Immunology*, *33*, 403–418, https://doi.org/10.1080/09540105.2022.2080188.

Zhang, L., Chen, F., Ren, G., & Wang, Z. (2021). *Functional Materials Letters*, *15*, https://doi.org/10.1142/S1793604722500047.

Zhang, S., Xu, X., Lin, T., & He, P. (2019). *Journal of Materials Science: Materials in Electronics*, *30*, 13855–13868.

Zhang, Y. H. P., Sun, J., & Ma, Y. (2016). *Journal of Industrial Microbiology & Biotechnology*, *44*, 773–784.

Zhou, S., Yuan, L., Hua, X., Xu, L., & Liu, S. (2015). *Analytica Chimica Acta*, *877*, 19–32.

Zhu, C., Xu, Y., Chen, Q., Zhao, H., Gao, B., & Zhang, T. (2022). *Journal of Colloid and Interface Science*, *617*, 454–462.

Zhu, H., Luo, W., Ciesielski, P. N., Fang, Z., Zhu, J. Y., Henriksson, G., Himmel, M. E., & Hu, L. (2016). *Chemical Reviews*, *116*, 9305–9374.

Zhu, Y., Chandra, P., & Shim, Y. B. (2013). *Analytical Chemistry*, *85*, 1058–1064.

Chapter 7

Recent advancements in detection and amplification techniques

Mylena Lemes Cunha[1], Lucas B. Ayres[2], Carlos D. Garcia[2], Dalila Luciola Zanette[1], Mateus Nóbrega Aoki[1] and Lucas Blanes[1]

[1]Laboratory for Applied Science and Technology in Health, Carlos Chagas Institute, Oswaldo Cruz Foundation (Fiocruz), Curitiba, Paraná, Brazil, [2]Department of Chemistry, Clemson University, Clemson, SC, United States

7.1 Introduction

Efficient and quick technologies capable of detecting disease-related biomarkers are crucial. In recent times, there has been a significant focus on developing reliable biosensing technologies for specific antigens, surface proteins, and nucleic acid biomarkers like circulating tumor-derived DNA and microRNAs. Due to the advancements in efficiency and quick technologies for detecting those biomarkers, researchers can gain a deeper understanding of molecular mechanisms and apply this knowledge to develop early disease diagnosis systems (Zhao et al., 2015). In this context, we examine innovative approaches that utilize biomarkers to establish a robust biosensing platform.

One of the biggest challenges in sensor design is specificity and limit of detection. Complex molecules that are not optic or electrochemically active, or do not have classic biological affinity standards, demand new detection protocols, involving innovative approaches. On the other hand, even if the biomolecule can be easily detected, it is usually present at very low levels; therefore, new amplification protocols must be implemented. The main objective of this chapter is to thoroughly explore and demonstrate the recent technologies employed to amplify the signals and detect biomolecules, with an abundance of examples.

Biosensors in Precision Medicine. DOI: https://doi.org/10.1016/B978-0-443-15380-8.00008-4

7.2 Emerging detection techniques based on nucleic acid detection deployed for point-of-care-based platforms

The following section provides a general overview regarding the basic functionality of new sensing strategies based on nucleic acid detection such as clustered regularly interspaced short palindromic repeats (CRISPR)/Cas systems employed for point-of-care sensing, and their rational design toward the development of analytical assays. Beforehand, it is important to point out that these new strategies are based on the combination of a guided nucleic acid recognition for specific targets (e.g., mRNA, sDNA, etc.) followed by a traditional analytical readout system including fluorescence, colorimetry, and electrochemistry. In this sense, our primary objective is to gain an understanding of the fundamentals of these biomolecular tools, rather than focusing on the mechanisms of commonly studied readout systems in analytical chemistry. We believe that understanding the key aspects of these bioengineering tools along with the analytical chemistry expertise leverages the possibility of developing cost-effective, selective, portable, and innovative diagnostic platforms. The systematic design of these systems is summarized in Fig. 7.1.

In the upcoming sections, we will explore in detail the fundamental aspects of each detection system, including their mechanisms of action and the advantages and limitations of their use. It is worth mentioning that the implementation of these sensing strategies for point-of-care analysis has been supported by several recent research studies (Aggarwal et al., 2023; Gootenberg et al., 2017; Kang et al., 2023; Welch et al., 2022; Zhang et al., 2023; Zhou et al., 2023), demonstrating their effectiveness and potential use in diagnostic settings. Moreover, these detection workflows can be applied before or after nucleic acid amplification (NAA), depending on the instrumental apparatus available as

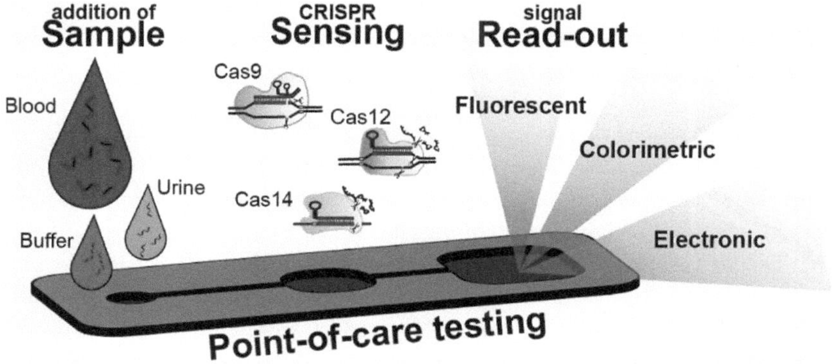

FIGURE 7.1 General workflow used to implement Cas-based detection system in analytical platforms (van Dongen et al., 2020). *Figure reprinted with permission from van Dongen, J. E., Berendsen, J. T. W., Steenbergen, R. D. M., Wolthuis, R. M. F., Eijkel, J. C. T., & Segerink, L. I. (2020). Point-of-care CRISPR/Cas nucleic acid detection: Recent advances, challenges and opportunities. Biosensors and Bioelectronics, 166, 112445. doi:10.1016/j.bios.2020.112445.*

well as the cost and time required for amplification cycles. In general, when deployed after the amplification step, the overall ability of CRISP-based diagnostic assays to recognize specific targets in complex samples is improved, increasing the selectivity and sensitivity of the proposed methodology.

The implementation and operation of these amplification techniques will be discussed further in Section 7.3 of this chapter.

7.2.1 Clustered regularly interspaced short palindromic repeats and associated nucleases (Cas): From gene editing tools to analytical detection systems

In the late 1980s, interspaced repeated DNA sequences were first discovered during research on isozyme conversion of Alkaline phosphatase in *Escherichia coli* (Ishino et al., 1987). At the time, the function of these newly discovered sequences in the bacterial biological system was unknown due to limited DNA sequence data. The following decade saw evidence of similar systems in the archaea group, such as *Haloferax mediterranei* (Mojica et al., 1993), but it was not until the early 2000s that similar array sequences were found in plasmids and bacteriophages, suggesting a possible connection to the bacterial immune system. Simultaneously with these findings, the expression CRISPR was introduced (Makarova et al., 2002), and these sequences were discovered to be associated with CRISPR-related genes (Cas). This strengthened the idea that CRISPR/Cas work in tandem and could be pivotal in the functioning of the adaptive immune defense system. Key aspects of their functionality started to be elucidated in 2005 (Bolotin et al., 2005) with the discovery of a protospacer-adjacent motif (PAM) and the Cas 9. Briefly, it was found that PAMs are short sequence regions adjacent to the targeted protospacer present in the invading DNA, while Cas9 are large proteins associated with nuclease activity. The full elucidation of how these two pillars act together came in 2010 (Garneau et al., 2010) when it was demonstrated that Cas9 can cause double-stranded breaks of invading DNA in specific regions delimited by PAM segments. This breakthrough finding allowed the design of customizable CRISPR/Cas assays for specific applications, levering the knowledge regarding their functionality in different biological systems. As a result of these efforts, the first successful use of CRISPR/Cas as a genome editing tool in eukaryotic cells was demonstrated in 2013 (Cong et al., 2013). Since then, CRISPR has revolutionized the bioengineering field (e.g., personalized therapeutics, studies of gene functions, and bioagriculture), becoming a fast and powerful toolkit for gene editing-related tasks.

The repurposed implementation of these tools in the analytical chemistry field started to be reported with the discovery of a collateral cleavage activity caused by Cas13a (formerly known as C2c2). Shortly, the nuclease activity of Cas13a is only activated if the targeted genetic material (ssRNA, for this specific nuclease) is complementary to the engineered CRISPR sequence. Upon this activation, the Cas13a performs nonspecific collateral cleavage, degrading all the ssRNA present in the environment either in the *cis* or *trans* configuration. In this sense, a probe associated with any genetic material is then released during the indiscriminate cleavage, increasing its concentration in the medium and, therefore, the signal can be recorded. Relying on this mechanism, one of the first demonstrations of a CRISPR/Cas13a-based diagnostic tool for biosensing applications (SHERLOCK) (Gootenberg et al., 2017) was reported in 2017. This work was an important milestone in the development of the new generation of point-of-care platforms empowered by CRISPR. Building on this, different CRISPR proteins such as Cas9, Cas14a/b, Cas13a/b, Cas12, and Cas10 can also be explored as effectors in these platforms, allowing the detection in situ of contagious pathogens such as SARS-CoV-2, Zika virus, and Influenza. An overview of the basic functionality of these CRISPR/Cas assays is demonstrated in Fig. 7.2.

Aiming to facilitate further discussion regarding the functionality of these systems, Table 7.1 summarizes recent CRISPR schemes deployed for point-of-care platforms. Remarkable examples of each sensing strategy will be explored in the following paragraphs.

7.2.2 Fluorescence readout

It is notorious that fluorescent-based probes, such as nanomaterials, fluorophores, and quantum dots, are the most common reporters used in nucleic acid detection in CRISPR/Cas systems. Typically, the fluorogenic reporter is associated with a short sequence of genetic material (ssDNA or ssRNA) depending on the nuclease used in the assay. Once the effector is activated by the presence of the target (which is seen as the pathogen's nucleic acid), indiscriminate cleavage takes place, thereby releasing the reporter and, consequently, increasing the fluorescent signal for detection. It is crucial to emphasize that fluorescence analysis exhibits higher sensitivity and specificity compared to colorimetric assays. This is primarily due to the selective excitation of specific molecules by fluorescence, resulting in a stronger signal when compared to the background. This type of analysis can be performed quickly, in a low-cost portable instrumentation platform, and with minimal sample preparation. Those characteristics make fluorescence well-suited for point-of-care approaches, especially in low-resource-income settings.

A recent application employing this detection method was demonstrated by Chen and collaborators (Chen et al., 2021a) to detect SARS-CoV-2. All the analyses were carried out in an automated plastic cartridge that is integrated into a portable fluorescence detector controlled by a low-cost microcontroller. The analysis takes less than 20 minutes for a high load of viral samples and the results are transmitted via Bluetooth to a smartphone application. A critical aspect of this assay consists of the facilitated elution of the target viral RNA by a pH-mediated electrostatic process. This step takes

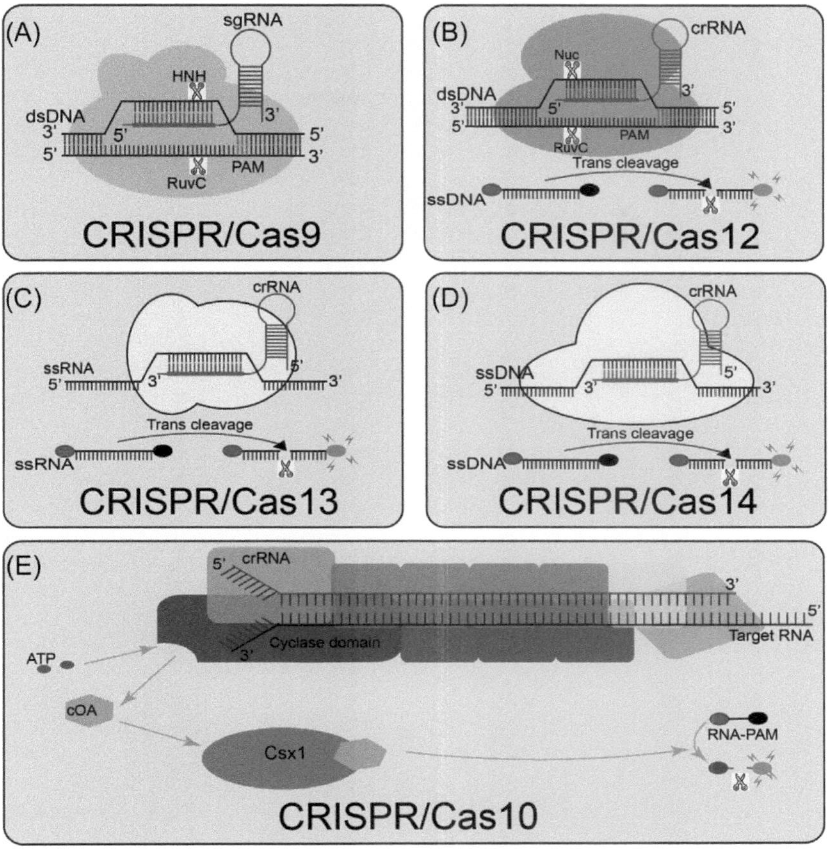

FIGURE 7.2 A synopsis of the actions and catalytic processes of the CRISPR-Cas enzyme includes: (A) Cas9 has the ability to sever both the target and nontarget DNA strands; a brief trinucleotide PAM is also indispensable for the primary DNA attachment; (B) Under the guidance of gDNA, Cas12a can cut dsDNA. The enzyme Cas12a detects the PAM of the initial T-rich spacer and then identifies the target sequence to create PAM distal dsDNA cuts with staggered 5′ and 3′ ends. Cas12 possesses the property of *trans*-cleavage activity. When the sgRNA-guided DNA is joined with Cas12, Cas12 initiates a potent, indiscriminate single-stranded DNA (ssDNA) cleavage activity; (C) By binding to crRNA, Cas13 can trigger its single-stranded RNA (ssRNA) cleavage activity, and it has an extra cleavage activity stimulated by the target RNA; (D) The Cas14 protein is an RNA-guided enzyme that can identify the target ssDNA devoid of restriction sequences and sever it, as well as nonspecifically cut the adjacent ssDNA nuclease molecules; (E) Cas10, a multicomponent and multifunctional immune effector, can be triggered by viral RNA. The viral RNA will activate Cas10 polymerization enzymes to produce approximately 1000 cyclic nucleotides (cOA). cOA activates Csx1, severing the fluorophore connected to the quencher (Li et al., 2022). *Adapted with permission from Li, J., Wang, Y., Wang, B., Lou, J., Ni, P., Jin, Y., Chen, S., Duan, G., & Zhang, R. (2022). Application of CRISPR/Cas systems in the nucleic acid detection of infectious diseases. Diagnostics, 12(10), 2455.*

TABLE 7.1 Recent CRISPR strategies deployed for point-of-care platforms.

Pathogen/ disease	Effector	Reporter	Platform	Redout	Amplification	Analysis time	LOD	Reference
SARS-CoV-2	Cas12-a	Fluorophore-based ssDNA	Automated assay cartridge	F	RT-RPA	<30 min	–	Chen et al. (2021a)
SARS-CoV-2	Cas12-a	FAM-biotin-based ssDNA	Lateral flow strip	C	RT-LAMP	45 min	10 copies /μL	Broughton et al. (2020)
SARS-CoV-2	Cas13a	ssRNA	–	F	RT-PCR	40 min	–	Hou et al. (2020)
SARS-CoV-2	Cas12a	FAM-TTATT-BHQ	Eppendorf	F	RT-LAMP	45 min	5 copies/μL	Wang et al. (2021)
SARS-CoV-2	Cas13a	Redox probe-based ssRNA	Dual electrode	E (SWV)	None	3 h	S gene: 26 copies/ μL	Kashefi-Kheyrabadi et al. (2023)
SARS-CoV-2	Cas13a	Redox probe-based ssRNA	Single electrode	E(DPV)	None	1.5 h	S gene: 8.1 copies/μL	Heo et al. (2022)
Zika virus	Cas9	LacZ-based RNA	Paper	C	NASBA	3 h	fM range	Pardee et al. (2016)
Zika and dengue	Cas13a	Fluorescent probe-based RNA	–	F	RT-RPA	–	2 aM	Gootenberg et al. (2017)
E. coli and *S. aureus*	Cas12a	Redox probe-based ssDNA	Electrode	E (EIS)	PCR	1.5 h	3 nM	Bonini et al. (2021)

Ebola virus	Cas13a	Fluorescent probe-based RNA	Automated microfluidic	F	RT-PCR	5 min	20 pfu/mL	Qin et al. (2019)
M. tuberculosis	dCas9	PC reporter system	–	L	PCR	10 min after PCR	100 copies/μL	Zhang et al. (2017)
Salmonella spp.	Cas13a	Fluorescent probe-based RNA	–	F	PCR	45 min		An et al. (2021)
Influenza A and B	Cas12a	Fluorescent probe-based ssDNA	Lateral flow strip	F	RT-RPA and RT-LAMP	75–85 min	1 PFUS	Park et al. (2021)

Acronyms: F, fluorescence; C, colorimetric; E, electrochemical; L, luminescence; RT, reverse transcription; RPA, recombinase polymerase amplification; PCR, polymerase chain reaction; LAMP, loop-mediated isothermal amplification; NASBA, nucleic acid sequence-based amplification.

place upon manual addition of the sample (unprocessed nasal specimen swab) into the magnetic bead buffer that contains a detergent. The overall schematic of this detection method is shown in Fig. 7.3.

Regarding the analytical performance, the proposed assay correctly identified 11 positive cases in 27 unprocessed clinical swabs, demonstrating a sensitivity of 1 genomic equivalents/μL in a volume of 100 μL. As a drawback, the system still depends on NAA, and it was not well explored with lower virus loads.

Another innovative assay denominated one-pot visual CRISPR (opvCRISPR) was reported by Wang and collaborators (Wang et al., 2021) to detect SARS-CoV-2. Their method also relies on a fluorescent probe, but the amplicon is generated by RT-loop-mediated isothermal amplification (LAMP), enabling the detection of up to a single molecule. The assay consists of using a single microtube and the outcome fluorescence signal, which indicates positive cases, is triggered by a blue light in 45 minutes or less. The functionality of the proposed assay is demonstrated in Fig. 7.4.

As a drawback, the proposed system still depends on RNA extraction, which adds cost, complexity, and the risk of cross-contamination.

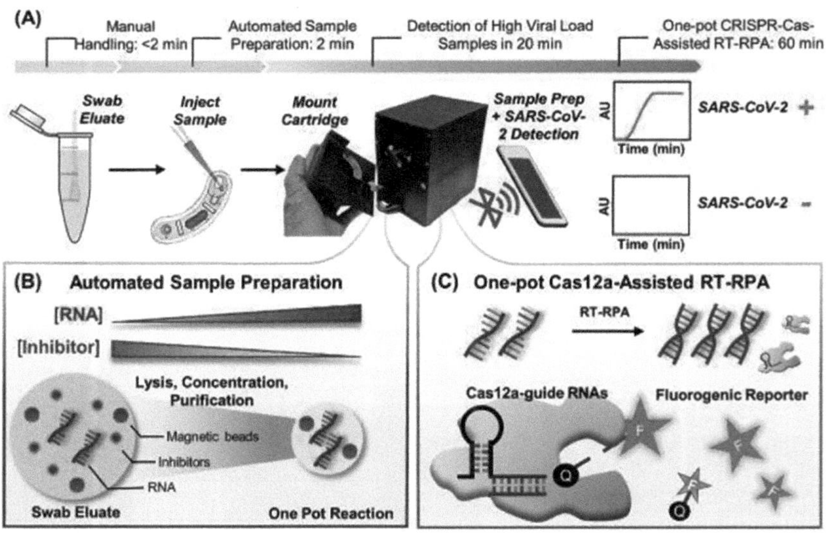

FIGURE 7.3 (A) Employing a handheld droplet magneto fluidic (DM) device, the POC-CRISPR system autonomously identifies the SARS-CoV-2 virus from nasopharyngeal (NP) swab eluates. (B) Within this device, the DM mechanism works on concentrating SARS-CoV-2 RNA, eliminating potential inhibitors, and transporting the RNA into the subsequent CRISPR-Cas reaction mixture. (C) Herein, the SARS-CoV-2 RNA is converted and amplified into detectable DNA fragments via a process known as reverse transcription recombinase polymerase amplification (RT-RPA). These DNA fragments then activate Cas12a-guide RNA complexes, producing fluorescent signals for detection. *Reprinted with permission from Chen, F. E., Lee, P. W., Trick, A. Y., Park, J. S., Chen, L., Shah, K., Mostafa, H., Carroll, K. C., Hsieh, K., & Wang, T. H. (2021a). Point-of-care CRISPR-Cas-assisted SARS-CoV-2 detection in an automated and portable droplet magnetofluidic device. Biosensors & Bioelectronics, 190, 113390. doi:10.1016/j.bios.2021.113390.*

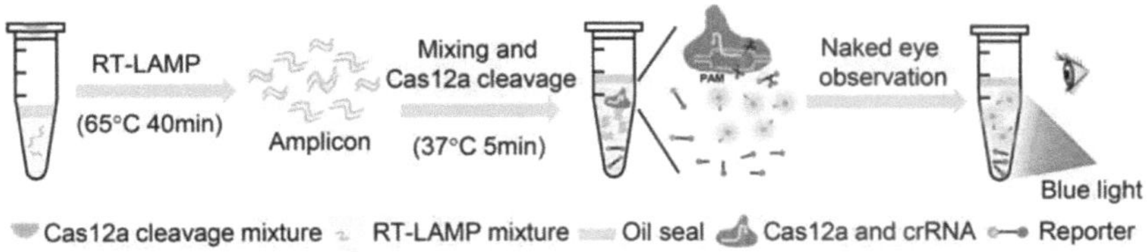

FIGURE 7.4 The one-pot visual reverse transcription loop-mediated isothermal amplification-CRISPR (opvCRISPR) detection method proceeds as follows. For SARS-CoV-2 detection, 20 μL of respiratory swab-extracted RNA templates are mixed with the RT-LAMP solution. This mixture, sealed with 25 μL of oil, is placed at the tube's base, and CRISPR/Cas12a reagents are added within the lid. After a 40-min amplification at 65°C, the tube is shaken to blend with Cas12a agents for cleavage. Activated by the DNA target, Cas12a nuclease divides the dormant fluorescent ssDNA reporter, yielding a fluorescent signal visible under blue light. *Reprinted with permission from Wang, R., Qian, C., Pang, Y., Li, M., Yang, Y., Ma, H., Zhao, M., Qian, F., Yu, H., Liu, Z., et al. (2021). opvCRISPR: One-pot visual RT-LAMP-CRISPR platform for SARS-cov-2 detection.* Biosensors and Bioelectronics, 172, *112766. doi:10.1016/j.bios.2020.112766.*

7.2.3 Colorimetric readout

The color change in response to the presence of an analyte is the basis of colorimetric sensors. These sensors work by using a colorimetric reporter which is linked to a small piece of genetic material. Upon activation of the nuclease enzyme, the reporter is released into the solution and its color change indicates the presence and, if necessary, the concentration of the analyte by utilization of a calibration curve. The simplicity of this detection scheme facilitates their implementation in paper platforms (e.g., lateral flow strips) where a color change, typically a line, will appear in the test region for positive cases. Colorimetric assays are typically more robust and versatile than fluorescence ones due to their ability to operate over a wider range of pH values and temperatures. Moreover, all the reagents can be immobilized on paper, contributing to the portability of these tests, and reducing cross-contamination.

One practical example of a colorimetric test was described for SARS-CoV-2 detection by Broughton and collaborators (Broughton et al., 2020). This assay utilizes RT-LAMP, Cas12 nuclease, and FAM-biotin as reporters. The assay can detect as few as 10 copies of the Sars-CoV-2 genes E and N per microliter of input in approximately 40 minutes for the colorimetric detection mode in the lateral flow assay platform. For this detection mode, the flow strips were designed to capture the labeled nucleic acids, allowing a rapid (within 2 minutes) change in color in the presence of the pathogen's genes. The general functionality of the proposed method is shown in Fig. 7.5.

Moreover, the assay is portable and reduces the risk of cross-contamination as it is performed on paper. The initial line in the strip functions as a control, while the second line (test line) undergoes a color change when Cas12 is activated as a result of the target's presence.

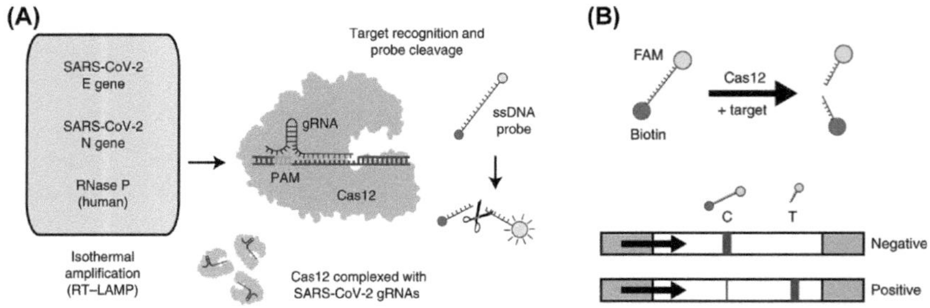

FIGURE 7.5 (A) The suggested method involves recognizing the target and cleaving the probe. (B) Diagram of DETECTR. The uncut FAM-biotinylated reporter molecule advances to the control capture line. When the corresponding target is identified, the Cas-gRNA complex severs the reporter, which moves toward the capture region. *Adapted with permission from Broughton, J. P., Deng, X., Yu, G., Fasching, C. L., Servellita, V., Singh, J., Miao, X., Streithorst, J. A., Granados, A., Sotomayor-Gonzalez, A., et al. (2020). CRISPR–Cas12-based detection of SARS-CoV-2. Nature Biotechnology, 38(7), 870–874. doi:10.1038/s41587-020-0513-4.*

A fast and alternative strategy to detect SARS-CoV-2 was demonstrated by Aoki and collaborators (Aoki et al., 2021). Their assay consists of an amplification step using RT-LAMP with specific primers to selectively detect the E gene and RdRp associated with the pathogen in less than 30 minutes. The amplified target triggers a change in color from pink (negative) to yellow (positive) of the pH-sensitive dye (Phenol red) that can be assessed by the naked eye. Regarding the analytical performance, the proposed test demonstrated a sensitivity between 24 and 240 viral genome gene copies, rendering a clear yellow color for the positive samples. When 62 clinical samples that were positive for SARS-CoV-2 by RT-qPCR were subjected to colorimetric RT-LAMP, 79% of them were also positive, 14.5% were indeterminate, and 6.5% were negative for SARS-CoV-2. Despite the lower positive rate compared to other studies, an orange color was found for the indeterminate samples, and it can be assessed by a spectrophotometer thus levering the positive rate up to 93.5%. It is also important to point out that the sensor is 100% specific for SARS-CoV-2 even in samples ($n = 57$) containing other respiratory disease pathogens, demonstrating the capability of implementing the sensor in point-of-care setups. Mendonça et al. have also developed a lab-on-a-chip device using the same RT-LAMP reaction described earlier (Torezin Mendonça et al., 2022). The hardware consists of a Raspberry Pi computer ($35), a video camera, an Arduino Nano microcontroller, a printed circuit board as a heater, and a 3D-printed housing. The chips were manufactured using a CO_2 laser cutting machine on polymethyl methacrylate (PMMA) and sealed with a PCR optic plastic film. The RT-LAMP SARS-CoV-2 diagnostic was conducted using color image recognition with the OpenCV machine vision software library. These are just two examples of how colorimetric reactions can enable direct visual detection or the use of simplified instrumentation for diagnostics.

7.2.3.1 Toehold switch sensor

Toehold switch detectors are engineered with RNA molecules aiming to detect specific RNA sequences in complex samples. Its functionality is based on the principle of toehold-mediated strand displacement, in which the target RNA binds to its complementary sequence (recognition domain) in the biosensor and triggers a structural change that will expose the signaling domain, generating, therefore, a detectable signal.

Based on this sensing strategy, an innovative method to selectively detect and differentiate strains of the Zika Virus in samples containing similar pathogens such as Dengue was demonstrated by Pardee and collaborators (Pardee et al., 2016). First, the detection of the Zika Virus over Dengue is accomplished by a Toehold Switch RNA detector programmed with a repressed LacZ gene. In the presence of the complementary target Zika RNA, the toehold is activated, leading to conformational changes that will activate gene translation of the enzyme LacZ. Then, this enzyme will convert a yellow substrate (chlorophenol red-β-D-galactopyranoside) into a purple product (chlorophenol red), indicating a positive result for samples containing Zika Virus as demonstrated in Fig. 7.6.

Regarding the differentiation between the lineages (American ZIKV vs African ZIKV), CAS9 was used to cleave DNA from strains containing specific PAM motifs. In this sense, only the uncut genetic material will be amplified and, therefore, utilized to activate the toehold sensor. The proposed assay demonstrated the capability of detecting up to 1 fM of trigger Zika RNA amplicons in approximately 3 hours of diagnostic testing. As a limitation, the assay still depends on the RNA extraction and amplification process, steps that could hinder the full implementation of this detection tool in low-income settings.

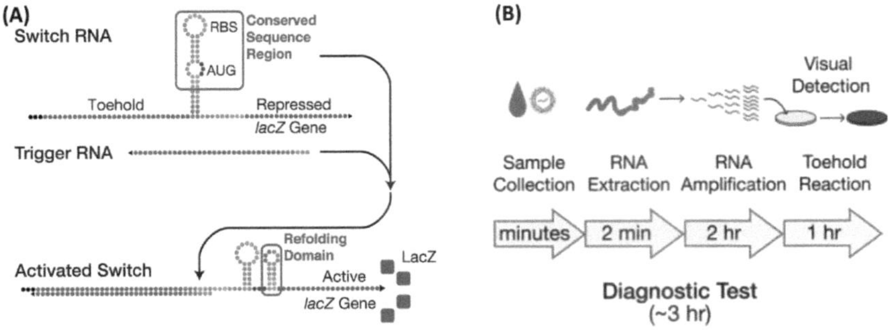

FIGURE 7.6 (A) Toehold activation of LacZ for the biosensor developed for Zika virus. (B) Workflow for the rapid prototyping of paper-based, biomolecular sensors for portable and low-cost diagnostics. *Adapted with permission from Pardee, K., Green, A. A., Takahashi, M. K., Braff, D., Lambert, G., Lee, J. W., Ferrante, T., Ma, D., Donghia, N., Fan, M., et al. (2016). Rapid, low-cost detection of Zika virus using programmable biomolecular components. Cell, 165 (5), 1255–1266. doi:https://doi.org/10.1016/j.cell.2016.04.059.*

7.2.4 Electrochemical readout

The association of redox probes with short genetic materials has allowed the development of electrochemical platforms empowered by CRISPR/Cas systems. Here, the redox reporter is released upon indiscriminate cleavage and the output signal is registered on the electrode's surface by a wide variety of methods such as amperometry, voltammetry, and electrochemical impedance spectroscopy (EIS). The use of different techniques to register the output signal allows the development of versatile and high-sensitivity assays, enabling the detection of trace analyte concentrations. Furthermore, the electrode's surface can be functionalized with functional materials, leveraging the overall analytical performance of these platforms.

7.2.4.1 Square wave voltammetry

As an example of this system, Kashefi-Kheyrabadi and collaborators (Kashefi-Kheyrabadi et al., 2023) have demonstrated an ultrasensitive and amplification-free biosensor to detect SARS-CoV-2 RNA. Their strategy consists of monitoring by square wave voltammetry (SWV) the two probes (methylene blue and ferrocene) that are immobilized with the reporter RNA (reRNA) on the gold-nanostructured electrode. When the target is present, the cleavage capability of Cas13 is triggered, resulting in the cut of the genetic material and, thereby releasing the probe associated with it. As a result, the electrochemical signal is decreased and then associated with the pathogen's concentration as demonstrated in Fig. 7.7.

The proposed assay can detect SARS-CoV-2 genes with a sensitivity of 26.2 and 53.5 copies per microliter of sample in 3 hours of analysis. As a drawback, sample preparation by microfluidic chip and clinical sample treatment are still required.

7.2.4.2 Electrochemical impedance spectroscopy

The implementation of EIS as detection output in CRISPR/Cas12 systems was demonstrated by Bonini and collaborators for bacterial DNA sensing (Bonini et al., 2021). Their strategy consists of accessing modifications on the electrode surface by monitoring the charge transfer resistance values (R_{ct}) in the presence of the redox couple $[Fe(CN)_6]^{3-/4-}$. The surface of the electrode is functionalized with ssDNA, which hinders the exchange of electrons between the redox couple and the sensor. The presence of the target in the solution will trigger the indiscriminate nuclease activity of Cas12, cleaving the ssDNA from the sensor and, consequently, allowing the redox species to reach the electrode surface more easily. As a result, a decrease in the charge transfer resistance can be observed as the diameter of the semicircle also decreases in the EIS spectra. This strategy allows sensing *E. coli* and *Staphylococcus aureus* with a limit of detection (LOD) of 3 nM, with an analysis time of

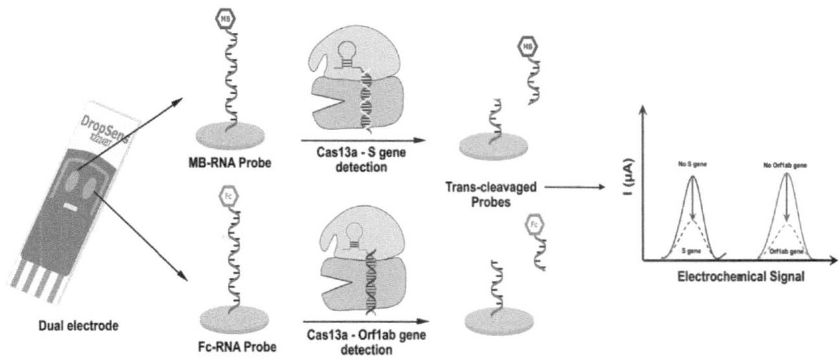

FIGURE 7.7 The functioning concept and essential components of E-CRISPR are explained here. The redox probe conjugated reRNA-modified biosensor encounters the enzymatically activated Cas13a-crRNA-target RNA triplex. When activated, Cas13a severs the RNA, which leads to the release of the redox probe from the reporter RNA (reRNA), ultimately resulting in a reduction in the electrochemical signal. *Reprinted with permission from Kashefi-Kheyrabadi, L., Nguyen, H. V., Go, A., & Lee, M.-H. (2023). Ultrasensitive and amplification-free detection of SARS-CoV-2 RNA using an electrochemical biosensor powered by CRISPR/Cas13a. Bioelectrochemistry (Amsterdam, Netherlands), 150, 108364. doi: https://doi.org/10.1016/j.bioelechem.2023.108364.*

approximately 1 hour and 30 minutes. As a possible drawback, the system still depends on PCR, which increases the analysis time and requires expensive reagents for amplification and purification of the amplicon.

7.3 New amplification techniques

Molecular assays often involve the detection of the target analyte in low concentrations, which can hinder the implementation of biosensor platforms in point-of-care setups. Aiming to overcome this limitation, amplification protocols play a crucial role in these assays due to their ability to improve the sensitivity, selectivity, and stability of the biosensors. Nucleic acid endonucleases, exonucleases, polymerases, and DNAzyme are among the different nuclease tool enzymes that have frequently been used for NAA. It is important to point out that polymerase chain reaction (PCR) is the gold standard method for the amplification step, and it is based on polymerase-dependent NAA (Wang et al., 2022a). The functionality of PCR is based on three cycling steps (denaturation, annealing, and extension) and each one of them takes place at different temperatures aiming to maximize the overall amplification. This method gained notoriety because it requires a minimal amount of sample (in some cases needing only 1 ng of DNA), resulting in amplification of up to 10^6 to 10^9 copies of DNA.

Recently, a variety of new alternatives to PCR assays, including LAMP, target-recycled enzyme-free amplification strategy (TREFAS), strand-displacement amplification (SDA), and rolling circle amplification (RCA), have made their appearance and continue to undergo ongoing optimization. Briefly, the RCA and SDA techniques operate using a similar principle as PCR. On the other hand, in LAMP amplification there is no need for thermocycling reactions due to the displacement caused by the enzyme used in the process, which will be discussed further. Finally, as the name indicates, the TREFAS amplification strategy does not require enzymes, making it more accessible in low-income settings and demonstrating potential for field applications. Examples of biosensors that use these alternative amplification strategies and others will be further discussed in this section.

7.3.1 Strand-displacement amplification

The strand-displacement assay utilizes continuous nicking and polymerization/displacement process catalyzed by the endonuclease and 5′ → 3′ exonuclease-deficient DNA polymerase. The displacement occurs primarily through hybridization dynamics, and the regeneration of nicking sites after each displacement step contributes to repeated cycles of extension, leaving one end of the sequence free for the annealing of primers (Fig. 7.8A), resulting in exponential amplification (Shi et al., 2014).

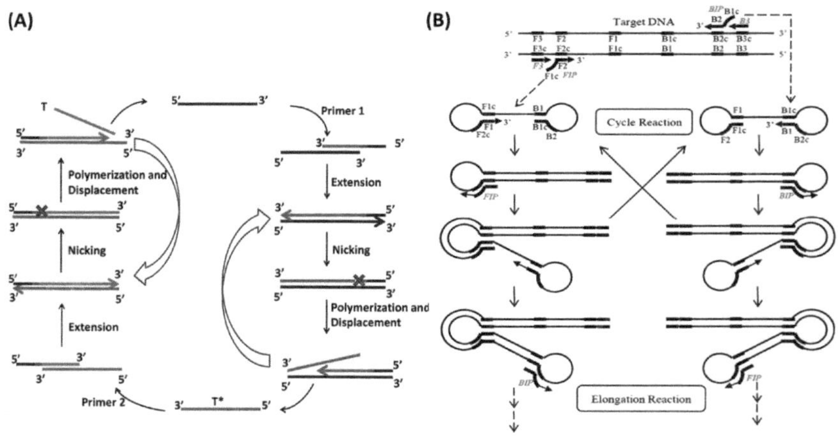

FIGURE 7.8 (A) A diagram of the strand-displacement assay. (B) A depiction of the loop-mediated isothermal amplification (LAMP) assay. In this scheme, the forward inner primer (FIP) aligns with the F2 sequence, having the F1c sequence at the 5′ end. The backward inner primer (BIP) is engineered to complement the B2 sequence, featuring the B1c sequence at the 5′ end. The outer primers are designed at the F3 and B3 regions. It is worth noting that the strand-displacement assay (SDA) activity mentioned in the previous section is visible in this strategy. *(A) Reprinted from Shi, C., Liu, Q., Ma, C., & Zhong, W. (2014). Exponential strand-displacement amplification for detection of microRNAs. Analytical Chemistry, 86(1), 336–339. doi:10.1021/ac4038043. (B) Adapted with permission from Notomi, T., Mori, Y., Tomita, N., & Kanda, H. (2015). Loop-mediated isothermal amplification (LAMP): principle, features, and future prospects. Journal of Microbiology, 53(1), 1–5. doi:10.1007/s12275-015-4656-9.*

A versatile device was employed by Liu et al. (2022a), in which the biosensor aimed to detect small molecules, more specifically, the antibiotic norfloxacin. The device consisted of a nicking endonuclease, a specific antibody, and a pair of antigen-labeled DNA, that worked in an antibody-controlled isothermal chain displacement amplification system (denominated by the authors as ACISDA). This biosensor allows for versatility because of the simple change of the antigen-labeled DNA and specific antibody, allowing other small molecules to be detected. The authors disclosed a linear range of 0.1−500 ng/mL with a detection limit of 0.04 ng/mL for norfloxacin. The SDA method was integrated with assisted CRISPR-Cas12a (SDACC) for the colorimetric detection of the hepatitis B virus (HBV), employed by Gong et al. (2021). The sensitive detection for HBV DNA was achieved by identifying a single-base mismatch located in the DNA sequence, displaying an LOD of 41.8 fM. Another approach coupling CRISPR-Cas12a was demonstrated by Wang et al. (2022b), which developed a synergetic signal amplification system using CRISPR-Cas12a and nicking enzyme-free SDA methodology to detect prostate-specific antigen (PSA). Regarding its analytical performance, an LOD of 0.030 ng mL^{-1} was obtained.

7.3.2 Loop-mediated isothermal amplification

Usually, LAMP uses three pairs of primers (internal, external, and loop) that can recognize up to eight specific locations on the targeted genetic material, contributing to a highly specific assay. The amplification results can go up to 10^9 copies of the amplified DNA within less than an hour. Furthermore, the assay is isothermal due to the use of *Bacillus stearothermophilus* DNA polymerase, commercially named as Bst Polymerase. The strand-displacement activity eliminates the thermal cycling step and the denaturation step, and the optimal temperature for the assay ranges from 60°C to 65 °C. The method consists of the design of a dumbbell-like single-strand DNA structure as a template for cyclic amplification, elongation, and recycling as demonstrated in Fig. 7.8 (Notomi et al., 2015; Soroka et al., 2021). Thus some key advantages are the constant temperature, which eliminates the need for a thermocycler; a higher specificity due to the number of primers used in the reaction; the need for simple reagents (Bst. polymerase, $MgSO_4$, and primers), and enabling rapid turnout time with high quantities of amplicons (Park, 2022).

Recently, LAMP has been used for the implementation of COVID-19 detection devices, as reported by Chen et al. (2021b). The proposed assay was able to identify the presence of SARS-CoV-2 where the SARS-CoV-2 RdRp and *N* genes were analyzed. The authors reported that the complete process was performed in 80 minutes, with a LOD of 20 copies per reaction and a 100% detection specificity, with no apparent cross-reactions to other respiratory pathogens (Chen et al., 2021b). This amplification method has also been employed to detect bacterial genetic material of *Mycoplasma pneumoniae* from nasopharyngeal swab samples, as demonstrated by Xiao et al. (2022). Their assay targeted the community-acquired respiratory distress syndrome (CARDS) toxin gene, achieving a sensitivity of 50 fg per reaction and no cross-contamination with other similar pathogens (Xiao et al., 2022). Li et al. (2019) also used LAMP to detect bacterial genetic material, but targeted the Brucella-specific gene Bscp31 for the detection of infection by *Brucella* spp. The proposed methodology displays 100% analytical specificity and accuracy, allowing a sensitivity of 100 fg of genomic DNA in pure cultures. It is critical to acknowledge that LAMP assays can generate false-positive signals owing to the creation of nonspecific amplicons. These may stem from high DNA primer concentrations causing primer dimerization and mismatched hybridizations. To counter these potential inaccuracies, new strategies are being developed with the goal of improving the reliability and robustness of this technology (Wang et al., 2015; Ku et al., 2022; Garg et al., 2022).

7.3.3 Rolling circle amplification

In this method, primers will extend under the catalysis of polymerase in the 5′-to-3′ direction, which only occurs after the formation of a closed loop from hybridized padlock probes by enzymatic ligation. The hybridization occurs between linear primers and padlock probes. Thus for a successful RCA essay, five elements must be present: a DNA/RNA polymerase, a short DNA or RNA liner single-stranded primer, a circle template (the padlock probe, with 3′-hydroxy and 5′-phosphate), ligase for the specific circularization of the padlock probe, and deoxynucleotide triphosphates (dNTPs)/nucleotide triphosphates (NTPs). In Fig. 7.9, a schematic of the RCA technology illustrates the capability to produce long products rich in tandem sites, which allows for the customization of products with functional nucleic acid sequences. Another unique characteristic is its adaptability, which contributes to the feasibility of the assay to be performed in various clinical samples, from cells and tissues to blood and other bodily fluids (Xu et al., 2021).

The RCA assay is performed in an isothermal condition, at room temperature or 30ºC, which is a great advantage because it removes the need for extra equipment for thermal cycling, affirming its suitability for implantation in point-of-care devices. Furthermore, RCA yields high specificity and sensitivity, allowing the detection of single nucleotide polymorphisms (SNPs), microRNAs, and bacterial and viral nucleic acids, besides other proteins and pathogens (Xu et al., 2021; Goo & Kim, 2016).

Moreover, RCA can have many variations, such as ligation-RCA, based on a padlock probe and a linear single-stranded DNA template DNA, often used to hybridize with target nucleotides such as microRNA. Another variation is the branched RCA, or hyperbranched RCA, which amplifies DNA exponentially, creating double-stranded DNAs (Goo & Kim, 2016).

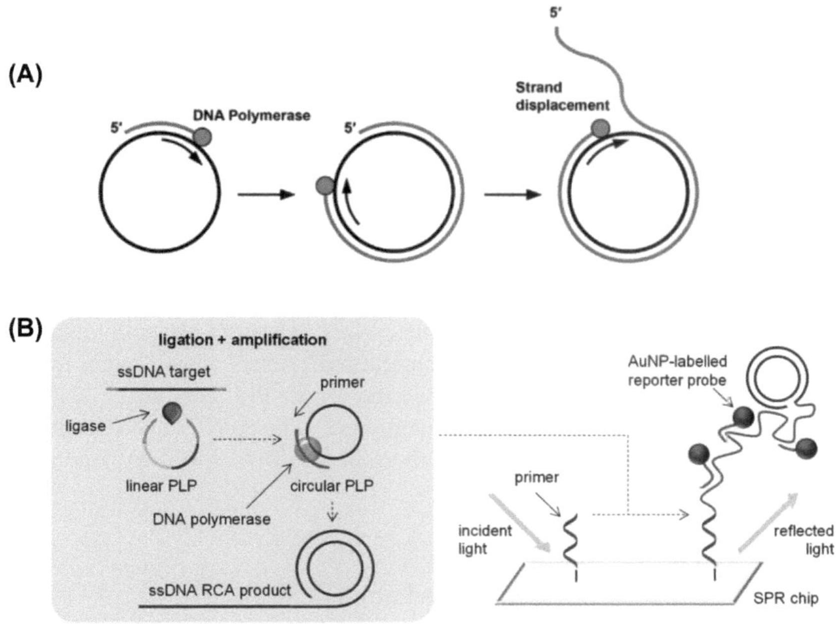

FIGURE 7.9 Schematic representation of RCA amplification and biosensor implementation. (A) The primer DNA is elongated by DNA polymerase. During strand displacement, the DNA polymerase unwinds dsDNA and synthesizes long ssDNA with high specificity. (B) RCA-based SPR biosensor using an immobilized primer and an AuNP-labeled reporter probe (Leonardo et al., 2021). *(A) Image modified from Goo, N.-I., & Kim, D.-E. (2016). Rolling circle amplification as isothermal gene amplification in molecular diagnostics. BioChip Journal, 10(4), 262–271. doi:10.1007/s13206-016-0402-6. (B) Modified with permission from Leonardo, S., Toldrà, A., & Campàs, M. (2021). Biosensors based on isothermal DNA amplification for bacterial detection in food safety and environmental monitoring. Sensors, 21(2), 602.*

As reported by Au-Keller et al. (2022), the technology has been employed for the detection of topoisomerases. The developed biosensor used rolling circle enhanced enzyme activity detection assay for quantitative measurement of topoisomerase 1 (TOP1). The assay consisted of TOP1-mediated cleavage-ligation and, as it occurs, TOP1 converts a DNA oligonucleotide into a closed circle, allowing for a highly sensitive and quantitative assay, which yields about 10^3 tandem repeat rolling circle products. Another example is the biosensor developed by Liu et al. (2022b). The apparatus incorporated technologies such as RCA, CRISPR-Cas9, and catalytic hairpin assembly (CHA) to establish a method for detecting microRNA. The target miRNA cyclized due to the binding with two terminals of dumbbell padlock, triggering the RCA process, which with the assistance of T4 DNA ligase and phi29 DNA polymerase formed a long single-stranded DNA product with a repeated hairpin structure. Double-stranded DNA sequences (dsDNA) were formed with the addition of complementary sequences and can be intensified by the CRISPR/

Cas9 system. Cleavage of the dsDNA happens by the Cas 9 enzyme, and the product is released, resulting in the CHA-based signal amplification. The authors reported high selectivity to the mismatched miRNAs. An additional instance of RCA was shown by Xiang et al. (2013). In the proposed methodology, a biosensor was employed to isothermally target five mutations related to drug resistance in *Mycobacterium.* The device utilizes surface plasmon resonance (SPR) with immobilized primers, surface-anchored RCA, and AuNP-labeled reporter probes (Fig. 7.9B). The biosensor described used different padlock probes (PLPs) containing a target-complementary sequence, a general sequence, and a tag sequence complementary to the specific primers immobilized on distinct channels of the SPR chips. PLPs were circularized by ligation and hybridized to the corresponding immobilized primers when a point mutation recognition occurred. RCA amplified the primer-PLP complexes, and the products of that amplification were hybridized with AuNP-labeled reporter probes for signal amplification. The author reported high specificity and efficiency in mutation discrimination.

7.3.4 Nucleic acid sequence-based amplification

The nucleic acid sequence-based amplification (NASBA) technology is a sensitive transcription system that works as a single-step isothermal assay, with temperature averaging around 41oC. As discussed previously, the isothermal capability excludes the need for a thermocycler. The assay is possible due to the addition of reverse transcriptase/DNA polymerase, RNA polymerase, RNaseH, and T7 promoter-labeled target-specific primer (Fig. 7.10). The presence of the enzyme with RNase activity eliminates the reverse transcription step and contributes to an efficient method for amplifying RNA analytes. The maximum yield of RNA products is not limited by initial primer levels (Deiman et al., 2002). Furthermore, NASBA's sensitivity could be greater than conventional reverse transcription PCR, and it has shown a better turnout for the detection of mRNA in the presence of DNA (Monis & Giglio, 2006).

This technology has been implemented by Ju et al. (2021), where the group integrated a nicking and extension chain reaction system into the NASBA method (NESBA). This strategy allowed for an additional nicking site at the 5′ end of the T7 promoter-labeled target-specific primer, resulting in exponential amplification due to the activities of nicking endonuclease and reverse transcriptase (Fig. 7.10). This technology successfully identified respiratory syncytial virus A (RSV A) genomic RNA (gRNA) with a sensitivity of 1 aM. Later, using a similar biosensor, the research group detected envelope and nucleocapsid genes of SARS-CoV-2 gRNA using the same NESBA method with superb accuracy (100%). The substantial capability of NASBA to be implemented in point-of-care biosensors is evident.

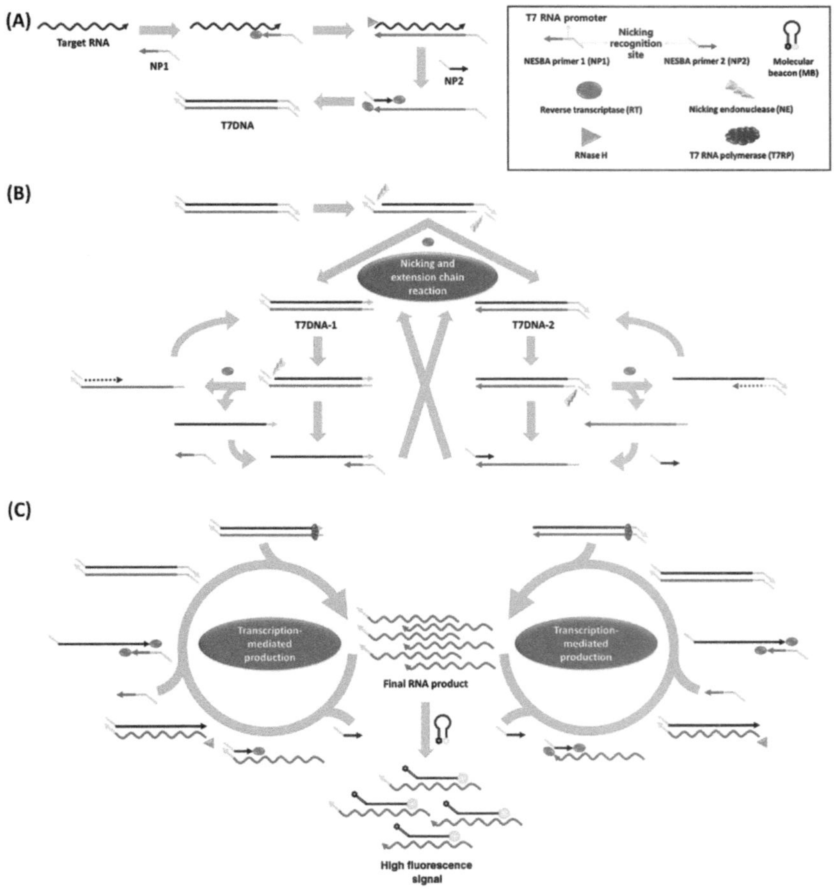

FIGURE 7.10 Diagrammatic depiction of NASBA amplification. RNaseH along with a primer containing a T7 promoter transforms RNA into double-stranded DNA (dsDNA). The T7 RNA polymerase then uses this dsDNA as a template to create numerous antisense RNA copies. (A) T7 DNA formation, (B) nicking and extension chain, (C)transcription-mediated amplification, and generation of a fluorescence signal. *Image modified with permission from Ju, Y., Kim, H. Y., Ahn, J. K., & Park, H. G. (2021). Ultrasensitive version of nucleic acid sequence-based amplification (NASBA) utilizing a nicking and extension chain reaction system. Nanoscale, 13(24), 10785–10791. doi:10.1039/D1NR00564B.*

7.3.5 Helicase-dependent amplification

The technique known as helicase-dependent amplification (HDA) employs helicase to unwind DNA duplexes under isothermal conditions (see Fig. 7.11). Initially, researchers utilized helicase derived from *E. coli*, which exhibited a 3′ to 5′ directionality. In *E. coli*, the helicase activity is stimulated by a protein called methyl-directed mismatch repair (MutL). Thus for the HDA assay to succeed, the presence of MutL protein is necessary, along with a single-

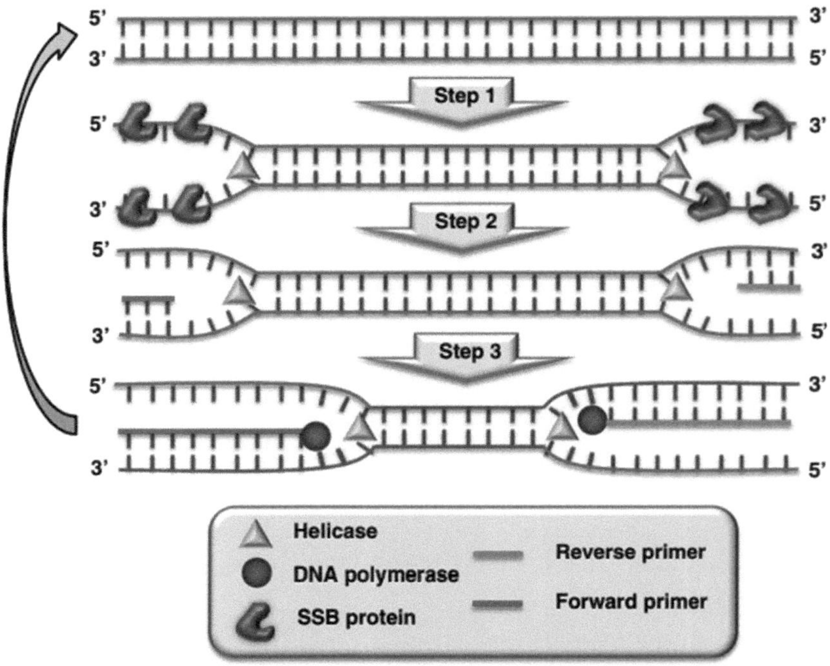

FIGURE 7.11 The schematic diagram illustrates the process of HDA amplification. In Step 1, the helicase unwinds the double-stranded DNA (dsDNA), while the single-stranded DNA-binding protein (SSB) stabilizes the resulting single-stranded DNA (ssDNA). Steps 2 and 3 depict the annealing of primers and the subsequent elongation of primers by DNA polymerase, respectively (Zhao et al., 2015). *The image has been reprinted with permission from Barreda-García, S., Miranda-Castro, R., de-Los-Santos-Álvarez, N., Miranda-Ordieres, A. J., & Lobo-Castañón, M. J. (2018). Helicase-dependent isothermal amplification: a novel tool in the development of molecular-based analytical systems for rapid pathogen detection. Analytical and Bioanalytical Chemistry, 410(3), 679–693 (from NLM). doi:10.1007/s00216-017-0620-3.*

stranded DNA-binding protein (SSB) to stabilize the single-stranded DNA (ssDNA). Once the ssDNA is stabilized, primers bind to the 5′ and 3′ ends of the target sequence, and DNA polymerase extends these annealed primers by incorporating deoxynucleotide triphosphates (dNTPs), resulting in the generation of double-stranded amplification products, similar to conventional PCR. Consequently, this cycle repeats, leading to an exponential amplification of the target dsDNA. An alternative variant of this method is called thermophilic HDA (tHDA), which utilizes a thermostable helicase from *Thermoanaerobacter tengcongensis* in combination with *Bacillus stearothermophilus* DNA polymerase (Bst-DNA polymerase). Unlike the standard HDA, the tHDA system does not require the presence of MutL protein and allows amplification to occur at elevated temperatures (Barreda-García et al., 2018).

This system is widely implemented, even in SARS-CoV-2 detection platforms, as seen in a report by Zasada et al. (2022) where they compared two detection platforms, one using RT-LAMP and the other using RT-HDA. The authors reported no false-positive results for the RT-HDA assay and found that RT-HDA was more sensitive than the RT-LAMP.

7.3.6 Target-recycled enzyme-free amplification strategy

This approach is primarily employed for microRNA detection and involves the hybridization strand displacement of two target sequence-specific hairpin probes (HP1 and HP2), ending in a target recycling (Zhang et al., 2018). As shown in Fig. 7.12, when a target is present, the HP1 hairpin structure unfolds, revealing a region complementary to HP2. The interplay between HP1-micronRNA and HP2 results in the release of microRNA, which is then utilized in the subsequent cycle of hybridization. Consequently, the HP1-HP2 duplex is formed, encapsulating the mold of the target microRNA sequence (Zhang et al., 2020).

This methodology is highly sensitive and precise and provides quicker and more cost-effective alternatives in the amplification phase on biosensor devices, primarily as it does not require enzymes for amplification. An instance of this innovation is a platform conceptualized by Rezaei et al. (2019), where the identification of two Hemophilia A-related microRNAs (miR-1246 and miR-4521) from human serum was carried out. The detection strategy involves the use of gold nanoparticles (AuNPs), heavy metals quantum dots-encapsulated metal-organic frameworks, and CHA. The authors noted that the biosensor could accurately differentiate complementary, 1-

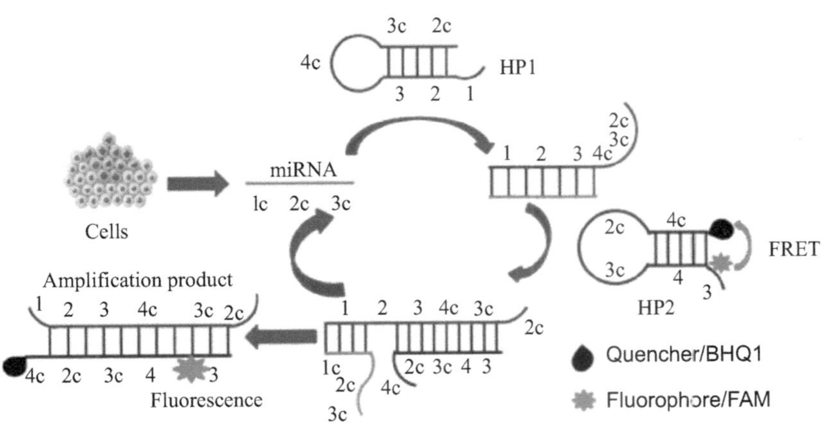

FIGURE 7.12 Schematic representation of TREFAS amplification. In this example, the hairpin HP2 carries a quencher and a fluorophore, deployed for sensing the microRNA. *Image reprinted with permission from Zhang, W., Zhang, P., Zhang, F., Cheng, W., Xu, Y., Zhang, Y., Chen, L., Wang, H., Zhou, Q., & Zhang, X. (2020). Real-time and rapid quantification of microRNAs in cells and tissues using target-recycled enzyme-free amplification strategy. Talanta, 217, 121016. doi:10.1016/j.talanta.2020.121016.*

base mismatched, and noncomplementary sequences and demonstrated a detection range of 1 fM–1 μM with detection thresholds of 0.19 fM and 0.28 fM for miR-1246 and miR-4521, respectively.

7.4 Conclusion

The development of biosensing devices has been mainly driven by the need for sensitive, specific, rapid, and cost-effective detection systems. One of the latest innovations in biosensing technology is the use of the CRISPR-Cas system for detecting DNA and RNA sequences. The CRISPR-Cas system is a highly specific and sensitive tool that has revolutionized gene editing and has also proven to be a robust detection system for analytical purposes.

Several detection methods for biosensor platforms were discussed, including fluorescence readout, colorimetric readout, and electrochemical readout. Fluorescence-based biosensors use a fluorescent molecule that emits light when excited by a specific wavelength of light, providing a sensitive and quantitative detection method. Colorimetric-based biosensors are often used in low-cost, easy-to-use diagnostic tests and rely on a color change in response to the presence of the target analyte. Electrochemical-based biosensors detect the presence of the analyte by measuring changes in electrical properties, such as current or voltage, and can provide real-time monitoring and high sensitivity.

However, the CRISPR-Cas system has shown significant promise in detecting various diseases and can be used in various biosensing devices, including lateral flow assays, electrochemical sensors, and fluorescence-based biosensors. To further enhance the sensitivity of CRISPR-based biosensors, amplification systems such as LAMP and HDA have been used. LAMP and HDA are both isothermal amplification techniques that can rapidly amplify target nucleic acid sequences. LAMP has been shown to be sometimes more sensitive than PCR, and it can detect as low as one copy of DNA. HDA, on the other hand, has shown better specificity than LAMP and can be used to detect multiple targets in a single reaction.

Furthermore, NASBA and SDA are isothermal amplification methods that use specific primers and enzymes to generate an increased number of RNA or DNA molecules. Alternatively, RCA uses a circular DNA template and a DNA polymerase to create long single-stranded DNA copies. It is worth noting that there exist dozens of other isothermal amplification techniques that could be investigated to be used in point-of-care devices.

Moreover, TREFAS allows for nucleic acid detection without the need for enzymes. This characteristic simplifies the assay design and reduces cost. The strategy offers high specificity and sensitivity with the potential to detect targets at low concentrations. These amplification methods have found widespread application and show promising results for implementation in biosensors.

In summary, the field of biosensing devices has been undergoing rapid advancements, and the integration of the CRISPR-Cas system has unlocked fresh possibilities for the sensitive and precise detection of nucleic acids. By incorporating amplification systems like LAMP and HDA, the sensitivity and specificity of CRISPR-based biosensors have been significantly enhanced. These breakthroughs in biosensing technology hold immense potential in revolutionizing disease diagnosis, environmental monitoring, and ensuring food safety, paving the way for transformative approaches in these areas

References

Aggarwal, N., Liang, Y., Foo, J. L., Ling, H., Hwang, I. Y., & Chang, M. W. (2023). FELICX: A robust nucleic acid detection method using flap endonuclease and CRISPR-Cas12. *Biosensors and Bioelectronics*, *222*, 115002. Available from https://doi.org/10.1016/j.bios.2022.115002.

An, B., Zhang, H., Su, X., Guo, Y., Wu, T., Ge, Y., Zhu, F., & Cui, L. (2021). Rapid and sensitive detection of *Salmonella* spp. using CRISPR-Cas13a combined with recombinase polymerase amplification. *Frontiers in Microbiology*, *12*. Available from https://doi.org/10.3389/fmicb.2021.732426.

Aoki, M. N., de Oliveira Coelho, B., Góes, L. G. B., Minoprio, P., Durigon, E. L., Morello, L. G., Marchini, F. K., Riediger, I. N., do Carmo Debur, M., Nakaya, H. I., et al. (2021). Colorimetric RT-LAMP SARS-CoV-2 diagnostic sensitivity relies on color interpretation and viral load. *Scientific Reports*, *11*(1), 9026. Available from https://doi.org/10.1038/s41598-021-88506-y.

Au-Keller, J. G., Au-Mizielinski, K., Au-Petersen, K. V., Au-Stougaard, M., Au-Knudsen, B. R., & Au-Tesauro, C. (2022). Simple and fast rolling circle amplification-based detection of topoisomerase 1 activity in crude biological samples. *JoVE*, *190*, e64484. Available from https://doi.org/10.3791/64484.

Barreda-García, S., Miranda-Castro, R., de-Los-Santos-Álvarez, N., Miranda-Ordieres, A. J., & Lobo-Castañón, M. J. (2018). Helicase-dependent isothermal amplification: a novel tool in the development of molecular-based analytical systems for rapid pathogen detection. *Analytical and Bioanalytical Chemistry*, *410*(3), 679–693. Available from https://doi.org/10.1007/s00216-017-0620-3, From NLM.

Bolotin, A., Quinquis, B., Sorokin, A., & Ehrlich, S. D. (2005). Clustered regularly interspaced short palindrome repeats (CRISPRs) have spacers of extrachromosomal origin. *Microbiology (Reading)*, *151*(Pt 8), 2551–2561. Available from https://doi.org/10.1099/mic.0.28048-0, From NLM.

Bonini, A., Poma, N., Vivaldi, F., Biagini, D., Bottai, D., Tavanti, A., & Di Francesco, F. (2021). A label-free impedance biosensing assay based on CRISPR/Cas12a collateral activity for bacterial DNA detection. *Journal of Pharmaceutical and Biomedical Analysis*, *204*, 114268. Available from https://doi.org/10.1016/j.jpba.2021.114268.

Broughton, J. P., Deng, X., Yu, G., Fasching, C. L., Servellita, V., Singh, J., Miao, X., Streithorst, J. A., Granados, A., Sotomayor-Gonzalez, A., et al. (2020). CRISPR–Cas12-based detection of SARS-CoV-2. *Nature Biotechnology*, *38*(7), 870–874. Available from https://doi.org/10.1038/s41587-020-0513-4.

Chen, F. E., Lee, P. W., Trick, A. Y., Park, J. S., Chen, L., Shah, K., Mostafa, H., Carroll, K. C., Hsieh, K., & Wang, T. H. (2021a). Point-of-care CRISPR-Cas-assisted SARS-CoV-2 detection in an automated and portable droplet magnetofluidic device. *Biosensors & Bioelectronics*, *190*, 113390. Available from https://doi.org/10.1016/j.bios.2021.113390.

Chen, X., Zhou, Q., Li, S., Yan, H., Chang, B., Wang, Y., & Dong, S. (2021b). Rapid and visual detection of SARS-CoV-2 using multiplex reverse transcription loop-mediated isothermal amplification linked with gold nanoparticle-based lateral flow biosensor. *Frontierd in Cellular and Infectction Microbiology*, *11*, 581239. Available from https://doi.org/10.3389/fcimb.2021.581239, From NLM.

Cong, L., Ran, F. A., Cox, D., Lin, S., Barretto, R., Habib, N., Hsu, P. D., Wu, X., Jiang, W., Marraffini, L. A., et al. (2013). Multiplex genome engineering using CRISPR/Cas systems. *Science (New York, N.Y.)*, *339*(6121), 819–823. Available from https://doi.org/10.1126/science.1231143, From NLM.

Deiman, B., van Aarle, P., & Sillekens, P. (2002). Characteristics and applications of nucleic acid sequence-based amplification (NASBA. *Molecular Biotechnology*, *20*(2), 163–179. Available from https://doi.org/10.1385/MB:20:2:163.

Garg, N., Ahmad, F. J., & Kar, S. (2022). Recent advances in loop-mediated isothermal amplification (LAMP) for rapid and efficient detection of pathogens. *Current Research in Microbial Sciences*, *3*, 100120. Available from https://doi.org/10.1016/j.crmicr.2022.100120, From NLM.

Garneau, J. E., Dupuis, M.-È., Villion, M., Romero, D. A., Barrangou, R., Boyaval, P., Fremaux, C., Horvath, P., Magadán, A. H., & Moineau, S. (2010). The CRISPR/Cas bacterial immune system cleaves bacteriophage and plasmid DNA. *Nature*, *468*(7320), 67–71. Available from https://doi.org/10.1038/nature09523.

Gong, S., Zhang, S., Wang, X., Li, J., Pan, W., Li, N., & Tang, B. (2021). Strand displacement amplification assisted CRISPR-Cas12a strategy for colorimetric analysis of viral nucleic acid. *Analytical Chemistry*, *93*(45), 15216–15223. Available from https://doi.org/10.1021/acs.analchem.1c04133.

Goo, N.-I., & Kim, D.-E. (2016). Rolling circle amplification as isothermal gene amplification in molecular diagnostics. *BioChip Journal*, *10*(4), 262–271. Available from https://doi.org/10.1007/s13206-016-0402-6.

Gootenberg, J. S., Abudayyeh, O. O., Lee, J. W., Essletzbichler, P., Dy, A. J., Joung, J., Verdine, V., Donghia, N., Daringer, N. M., Freije, C. A., et al. (2017). Nucleic acid detection with CRISPR-Cas13a/C2c2. *Science (New York, N.Y.)*, *356*(6336), 438–442. Available from https://doi.org/10.1126/science.aam9321.

Heo, W., Lee, K., Park, S., Hyun, K.-A., & Jung, H.-I. (2022). Electrochemical biosensor for nucleic acid amplification-free and sensitive detection of severe acute respiratory syndrome coronavirus 2 (SARS-CoV-2) RNA via CRISPR/Cas13a trans-cleavage reaction. *Biosensors and Bioelectronics*, *201*, 113960. Available from https://doi.org/10.1016/j.bios.2021.113960.

Hou, T., Zeng, W., Yang, M., Chen, W., Ren, L., Ai, J., Wu, J., Liao, Y., Gou, X., Li, Y., et al. (2020). Development and evaluation of a rapid CRISPR-based diagnostic for COVID-19. *PLoS Pathogens*, *16*(8), e1008705. Available from https://doi.org/10.1371/journal.ppat.1008705.

Ishino, Y., Shinagawa, H., Makino, K., Amemura, M., & Nakata, A. (1987). Nucleotide sequence of the iap gene, responsible for alkaline phosphatase isozyme conversion in *Escherichia coli*, and identification of the gene product. *Journal of Bacteriology*, *169*(12), 5429–5433. Available from https://doi.org/10.1128/jb.169.12.5429-5433.1987, From NLM.

Ju, Y., Kim, H. Y., Ahn, J. K., & Park, H. G. (2021). Ultrasensitive version of nucleic acid sequence-based amplification (NASBA) utilizing a nicking and extension chain reaction system. *Nanoscale*, *13*(24), 10785–10791. Available from https://doi.org/10.1039/D1NR00564B, http://doi.org/10.1039/D1NR00564B.

Kang, X., Lei, C., Shi, J., Liu, X., Ren, W., & Liu, C. (2023). A versatile CRISPR/Cas12a-based biosensing platform coupled with a target-protected transcription strategy. *Biosensors and Bioelectronics*, *219*, 114801. Available from https://doi.org/10.1016/j.bios.2022.114801.

Kashefi-Kheyrabadi, L., Nguyen, H. V., Go, A., & Lee, M.-H. (2023). Ultrasensitive and amplification-free detection of SARS-CoV-2 RNA using an electrochemical biosensor powered by CRISPR/Cas13a. *Bioelectrochemistry (Amsterdam, Netherlands)*, *150*, 108364. Available from https://doi.org/10.1016/j.bioelechem.2023.108364.

Ku, J., Chauhan, K., Hwang, S.-H., Jeong, Y.-J., & Kim, D.-E. (2022). Enhanced specificity in loop-mediated isothermal amplification with poly(ethylene glycol)-engrafted graphene oxide for detection of viral genes. *Biosensors*, *12*(8), 661.

Leonardo, S., Toldrà, A., & Campàs, M. (2021). Biosensors based on isothermal DNA amplification for bacterial detection in food safety and environmental monitoring. *Sensors*, *21*(2), 602.

Li, J., Wang, Y., Wang, B., Lou, J., Ni, P., Jin, Y., Chen, S., Duan, G., & Zhang, R. (2022). Application of CRISPR/Cas systems in the nucleic acid detection of infectious diseases. *Diagnostics*, *12*(10), 2455.

Li, S., Liu, Y., Wang, Y., Chen, H., Liu, C., & Wang, Y. (2019). Lateral flow biosensor combined with loop-mediated isothermal amplification for simple, rapid, sensitive, and reliable detection of *Brucella* spp. *Infection and Drug Resistance*, *12*, 2343–2353. Available from https://doi.org/10.2147/idr.S211644, From NLM.

Liu, X., Yang, H., Xu, Z., Liu, R., Zuo, H., Chen, Z., Wang, X., Xia, C., Zhang, Y., Ning, B., et al. (2022a). A novel biosensor based on antibody controlled isothermal strand displacement amplification (ACISDA) system. *Biosensors and Bioelectronics*, *209*, 114185. Available from https://doi.org/10.1016/j.bios.2022.114185.

Liu, X., Zhao, X., Yuan, Y., Cao, Z., Zhu, M., Li, T., & Wu, Z. (2022b). Accurate detection of lung cancer-related microRNA through CRISPR/Cas9-assisted garland rolling circle amplification. *Journal of Thoracic Disease*, *14*(11), 4427–4434. Available from https://doi.org/10.21037/jtd-22-1405, From NLM.

Makarova, K. S., Aravind, L., Grishin, N. V., Rogozin, I. B., & Koonin, E. V. (2002). A DNA repair system specific for thermophilic Archaea and bacteria predicted by genomic context analysis. *Nucleic Acids Research*, *30*(2), 482–496. Available from https://doi.org/10.1093/nar/30.2.482. (acccessed 1/2/2023).

Mojica, F. J., Juez, G., & Rodríguez-Valera, F. (1993). Transcription at different salinities of *Haloferax mediterranei* sequences adjacent to partially modified PstI sites. *Molecular Microbiology*, *9*(3), 613–621. Available from https://doi.org/10.1111/j.1365-2958.1993.tb01721.x, From NLM.

Monis, P. T., & Giglio, S. (2006). Nucleic acid amplification-based techniques for pathogen detection and identification. *Infection, Genetics and Evolution*, *6*(1), 2–12. Available from https://doi.org/10.1016/j.meegid.2005.08.004.

Notomi, T., Mori, Y., Tomita, N., & Kanda, H. (2015). Loop-mediated isothermal amplification (LAMP): principle, features, and future prospects. *Journal of Microbiology*, *53*(1), 1–5. Available from https://doi.org/10.1007/s12275-015-4656-9.

Pardee, K., Green, A. A., Takahashi, M. K., Braff, D., Lambert, G., Lee, J. W., Ferrante, T., Ma, D., Donghia, N., Fan, M., et al. (2016). Rapid, low-cost detection of zika virus using programmable biomolecular components. *Cell*, *165*(5), 1255–1266. Available from https://doi.org/10.1016/j.cell.2016.04.059.

Park, B. J., Park, M. S., Lee, J. M., & Song, Y. J. (2021). Specific detection of influenza A and B viruses by CRISPR-Cas12a-based assay. *Biosensors*, *11*(3), 88.

Park, J.-W. (2022). Principles and applications of loop-mediated isothermal amplification to point-of-care tests. *Biosensors*, *12*(10), 857.

Qin, P., Park, M., Alfson, K. J., Tamhankar, M., Carrion, R., Patterson, J. L., Griffiths, A., He, Q., Yildiz, A., Mathies, R., et al. (2019). Rapid and fully microfluidic ebola virus detection with CRISPR-Cas13a. *ACS Sensors*, *4*(4), 1048–1054. Available from https://doi.org/10.1021/acssensors.9b00239.

Rezaei, H., Motovali-bashi, M., & Radfar, S. (2019). An enzyme-free electrochemical biosensor for simultaneous detection of two hemophilia A biomarkers: Combining target recycling with quantum dots-encapsulated metal-organic frameworks for signal amplification. *Analytica Chimica Acta*, *1092*, 66–74. Available from https://doi.org/10.1016/j.aca.2019.09.037.

Shi, C., Liu, Q., Ma, C., & Zhong, W. (2014). Exponential strand-displacement amplification for detection of MicroRNAs. *Analytical Chemistry*, *86*(1), 336–339. Available from https://doi.org/10.1021/ac4038043.

Soroka, M., Wasowicz, B., & Rymaszewska, A. (2021). Loop-mediated isothermal amplification (LAMP): The better sibling of PCR? *Cells*, *10*(8), 1931.

Torezin Mendonça, G., Cassaboni Stracke, M., de Oliveira Coelho, B., Bruna Soligo Sanchuki, H., Klassen de Oliveira, V., Klerynton Marchini, F., Lucíola Zanette, D., Nóbrega Aoki, M., Ribeiro Viana, E., & Blanes, L. (2022). A new RT-LAMP-on-a-chip instrument for SARS-CoV-2 diagnostics. *Microchemical Journal*, *180*, 107600. Available from https://doi.org/10.1016/j.microc.2022.107600.

van Dongen, J. E., Berendsen, J. T. W., Steenbergen, R. D. M., Wolthuis, R. M. F., Eijkel, J. C. T., & Segerink, L. I. (2020). Point-of-care CRISPR/Cas nucleic acid detection: Recent advances, challenges and opportunities. *Biosensors and Bioelectronics*, *166*, 112445. Available from https://doi.org/10.1016/j.bios.2020.112445.

Wang, D. G., Brewster, J. D., Paul, M., & Tomasula, P. M. (2015). Two methods for increased specificity and sensitivity in loop-mediated isothermal amplification. *Molecules (Basel, Switzerland)*, *20*(4), 6048–6059. Available from https://doi.org/10.3390/molecules20046048, From NLM.

Wang, R., Qian, C., Pang, Y., Li, M., Yang, Y., Ma, H., Zhao, M., Qian, F., Yu, H., Liu, Z., et al. (2021). opvCRISPR: One-pot visual RT-LAMP-CRISPR platform for SARS-cov-2 detection. *Biosensors and Bioelectronics*, *172*, 112766. Available from https://doi.org/10.1016/j.bios.2020.112766.

Wang, W., Liu, J., Wu, L.-A., Ko, C.-N., Wang, X., Lin, C., Liu, J., Ling, L., & Wang, J. (2022b). Nicking enzyme-free strand displacement amplification-assisted CRISPR-Cas-based colorimetric detection of prostate-specific antigen in serum samples. *Analytica Chimica Acta*, *1195*, 339479. Available from https://doi.org/10.1016/j.aca.2022.339479.

Wang, X., Lu, D., Liu, Y., Wang, W., Ren, R., Li, M., Liu, D., Liu, Y., Liu, Y., & Pang, G. (2022a). Electrochemical signal amplification strategies and their use in olfactory and taste evaluation. *Biosensors*, *12*(8), 566.

Welch, N. L., Zhu, M., Hua, C., Weller, J., Mirhashemi, M. E., Nguyen, T. G., Mantena, S., Bauer, M. R., Shaw, B. M., Ackerman, C. M., et al. (2022). Multiplexed CRISPR-based microfluidic platform for clinical testing of respiratory viruses and identification of SARS-CoV-2 variants. *Nature Medicine*, *28*(5), 1083–1094. Available from https://doi.org/10.1038/s41591-022-01734-1.

Xiang, Y., Deng, K., Xia, H., Yao, C., Chen, Q., Zhang, L., Liu, Z., & Fu, W. (2013). Isothermal detection of multiple point mutations by a surface plasmon resonance biosensor with Au nanoparticles enhanced surface-anchored rolling circle amplification. *Biosensors and Bioelectronics*, *49*, 442–449. Available from https://doi.org/10.1016/j.bios.2013.04.044.

Xiao, F., Zhou, J., Sun, C., Huang, X., Zheng, B., Fu, J., Jia, N., Xu, Z., Cui, X., & Wang, Y. (2022). Loop-mediated isothermal amplification coupled with nanoparticle-based biosensor: A rapid and sensitive method to detect mycoplasma pneumoniae. *Frontiers in Cellular and Infection Microbiology*, *12*. Available from https://doi.org/10.3389/fcimb.2022.882855.

Xu, L., Duan, J., Chen, J., Ding, S., & Cheng, W. (2021). Recent advances in rolling circle amplification-based biosensing strategies-A review. *Analytica Chimica Acta, 1148*, 238187. Available from https://doi.org/10.1016/j.aca.2020.12.062.

Zasada, A. A., Mosiej, E., Prygiel, M., Polak, M., Wdowiak, K., Formińska, K., Ziółkowski, R., Żukowski, K., Marchlewicz, K., Nowiński, A., et al. (2022). Detection of SARS-CoV-2 using reverse transcription helicase dependent amplification and reverse transcription loop-mediated amplification combined with lateral flow assay. *Biomedicines, 10*(9), 2329.

Zhang, J., Song, C., Zhu, Y., Gan, H., Fang, X., Peng, Q., Xiong, J., Dong, C., Han, C., & Wang, L. (2023). A novel cascade signal amplification strategy integrating CRISPR/Cas13a and branched hybridization chain reaction for ultra-sensitive and specific SERS detection of disease-related nucleic acids. *Biosensors and Bioelectronics, 219*, 114836. Available from https://doi.org/10.1016/j.bios.2022.114836.

Zhang, J., Zhang, W., & Gu, Y. (2018). Enzyme-free isothermal target-recycled amplification combined with PAGE for direct detection of microRNA-21. *Analytical Biochemistry, 550*, 117–122. Available from https://doi.org/10.1016/j.ab.2018.04.024, From NLM.

Zhang, W., Zhang, P., Zhang, F., Cheng, W., Xu, Y., Zhang, Y., Chen, L., Wang, H., Zhou, Q., & Zhang, X. (2020). Real-time and rapid quantification of microRNAs in cells and tissues using target-recycled enzyme-free amplification strategy. *Talanta, 217*, 121016. Available from https://doi.org/10.1016/j.talanta.2020.121016.

Zhang, Y., Qian, L., Wei, W., Wang, Y., Wang, B., Lin, P., Liu, W., Xu, L., Li, X., Liu, D., et al. (2017). Paired design of dCas9 as a systematic platform for the detection of featured nucleic acid sequences in pathogenic strains. *ACS Synthetic Biology, 6*(2), 211–216. Available from https://doi.org/10.1021/acssynbio.6b00215.

Zhao, Y., Chen, F., Li, Q., Wang, L., & Fan, C. (2015). Isothermal amplification of nucleic acids. *Chemical Reviews, 115*(22), 12491–12545. Available from https://doi.org/10.1021/acs.chemrev.5b00428.

Zhou, B., Ye, Q., Chen, M., Wang, C., Xiang, X., Li, Y., Zhang, J., Zhang, Y., Wang, J., Wu, S., et al. (2023). A label-free AuNP bioprobe-assisted CRISPR/Cas12a colorimetric platform for high-throughput detection of *Staphylococcus aureus* ST398. *Food Control, 145*, 109451. Available from https://doi.org/10.1016/j.foodcont.2022.109451.

Part 3

Trends of biosensors in precision medicine

Chapter 8

Lab-on-a-chip devices

Swapnil Tiwari[1], Diana Bueno Hernandez[2,3], Rupesh Kumar Mishra[4] and Atul Sharma[5]

[1]School of Studies in Chemistry, Pt. Ravishankar Shukla University, Raipur, Chhattisgarh, India, [2]Department of Basic Sciences, Instituto Politecnico Nacional, Unidad Profesional Interdisciplinaria de Biotecnología, Mexico City, Mexico, [3]Instituto Tecnológico de Santo Domingo, Biomedical Engineering, Santo Domingo, Dominican Republic, [4]Bindley Bio-Science Center, Purdue University, West Lafayette, IN, United States, [5]Department of Pharmaceutical Chemistry, SGT College of Pharmacy, SGT University, Gurugram, Haryana, India

8.1 Introduction

Microfluidic systems, popularly known as lab-on-a-chip (LOC), are small platforms and devices that combine various laboratory experiments and processes onto a single chip, for instance, genomic sequencing or cellular reactions (Lim et al., 2010). The field of LOC research encompasses a wide range of applications that consists of designing prototypes for biomarker identification, biosensor diagnostics, and, to a lesser degree, chemical synthesis (Bhatia & Ingber, 2014; Caplin et al., 2015; Esch et al., 2015; Ferrari et al., 2020; Li & Tian, 2018). By reducing the scale of biochemical processes typically carried out in laboratory confinement, these miniaturized devices offer several advantages including economic affordability, rapid analysis, parallelization, and high throughput. The growth of LOC can be attributed to two key technologies: microfluidics and molecular biology (Yetisen et al., 2013). The LOC technology improves the emerging trends in point-of-care (POC) diagnostics through mobility, reduced sample utilization, compactness, and parallelization (Lim et al., 2010). These downscale microfluidic systems are integrated onto a single hand-sized chip. On a single chip that fits in your hand, microfluidic technologies are utilized to fabricate numerous minute channels, each measuring only a few microsizes. The microchannel makes it possible to manage fluids in amounts as little as a few picolitres and to regulate biological processes in incredibly small volumes (DeMello, 2006). These devices encompass more than just an arrangement of micro channels as they must also support various functions. Moreover, for these devices to function as complete diagnostic tools, they may require integrated components and electrical connections and fluidic

Biosensors in Precision Medicine. DOI: https://doi.org/10.1016/B978-0-443-15380-8.00009-6

networks like pumps and sockets and spring loads (Ahn & Choi, 2010; Temiz et al., 2015). The LOC notions aim to consolidate several cellular processes onto a single chip enabling accurate diagnosis of disease utilizing a single drop of blood (Wu et al., 2018). While we are still some distance away from achieving this objective, current technologies are closing the niche by demonstrating a large number of tests on a specialized LOC. Diagnostic procedures are anticipated to alter during the following several decades as a result of developments in LOC technology (Gonzálezgonźgonzález-Guerrero et al., 2016).

In this chapter, the discussion will be focused on POC and self-monitoring health devices applied to precision medicine. This chapter highlights LOC devices that integrate the biorecognition layer inside a microfluidic channel for fast and scalable detection. Microfluidic devices allow the development of multianalyte detection while improving the signal-to-noise ratio, generating highly sensitive devices. Their application and the challenges and opportunities in the field will be discussed.

8.2 Manufacturing technologies: material used in lab-on-chip

A complete LOC must be capable of carrying out typical laboratory tasks, such as processing unprocessed samples, mixing, and reacting to samples and reagents, separating samples, and then detecting analytes of interest. Different conventional functional components, each with a specific function, can be combined on a single chip to achieve the aforementioned sample analysis and processing capabilities. Injector, preparator, transporter, mixer, reactor, separator, detector, controller, and power supply are the major groups into which the significant LOC components can be divided. Fig. 8.1 represents the microfluidic platforms on different substrates.

LOC employs a variety of materials, with the most popular ones being discussed in this section. The most popular production techniques for each material and their advantages and limitations are summarized in Table 8.1.

1. *Glass*: Glass is an excellent candidate for the industrialization of LOC because it is transparent, compatible with micrometer-sized machining, chemically inert, offers a wide range of well-known chemical surface treatments, and has repeatable electrode integration (Berlanda et al., 2021). However, glass LOC manufacturing usually requires a clean room and experts in microfabrication, making it not a common choice in research labs.
2. *Paper*: Paper-based LOC systems are especially interesting for applications that require extremely low prices. The attributes and application of the test determine which type of paper should be used (surface characteristics, surface area, capillary flow rate, pore size, porosity, and thickness). There is not a single kind of paper that can fulfill all application requirements. The porosity, particle retention, and flow rates of different commercial paper grades frequently vary (Yetisen et al., 2013) and may also differ before

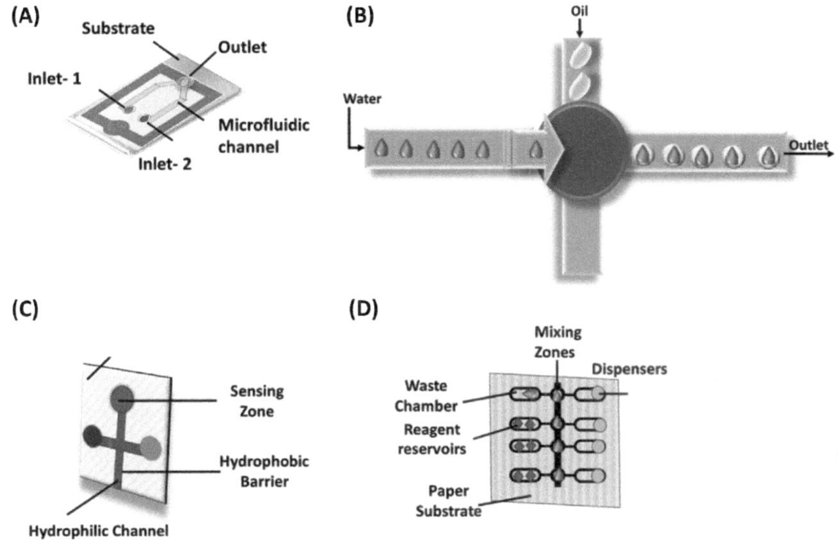

FIGURE 8.1 Schematics showing different types of microfluidic platforms. (A) Polydimethylsiloxane (PDMS)-based microchannels showing a continuous flow channel microfluidic platform. (B) Droplet microfluidic system to create immiscible liquid droplets working as microreactors. (C) A paper-based microfluidic system that works based on capillary action that moves liquid through hydrophilic channels. (D) Digital microfluidic system where liquid droplets move under the influence of electrical potential (illustrated by Qamar and Shamsi (2020)).

and after processing (e.g., when the production of the tests involves the compression of the substrate). Hydrophobic materials (such as photoresist SU-8, wax, or alkyl ketene dimer (AKD)) are utilized to construct hydrophilic channels (Lisowski & Zarzycki). Depending on the hydrophobic agents, the paper porosity can be blocked (after employing SU-8 or PDSM), coated physically (with polystyrene or wax), or chemically changed cellulose fibers (after using AKD) (Abe et al., 2010). Organic solvent washing may significantly decrease the hydrophobicity of paper induced by physical deposition, permitting the production of paper-based microfluidic devices employing organic solvent etching procedures. However, the hydrophobicity of paper generated by chemical alteration cannot be eliminated by organic solvent extraction later (Li et al., 2012). Papers for filters and chromatography are made from superior cotton linter with a cellulose content of at least 98%. The majority of filter sheets are designed for general filtration and have retention levels as low as 2.5 μm. The degree of purity, hardness, and chemical resistance of these filter sheets vary. The use of paper in diagnostics as POC devices is highly promising and has the potential to open up the area of diagnostics and make it more accessible to low-income and resource-limited populations. Contrary to conventional microfluidic devices, paper-based devices can be easily constructed

TABLE 8.1 Comparison of commonly used materials for microfluidic-based POC chips. Fabrication methods, advantages, and limitations are listed (Pandey et al., 2018).

Parameters	glass	Silicon	Polymer	Paper
Fabrication techniques	• Photolithography • Thin-film metallization • Etching	• Bulk or surface micromachining • Buried channel • Nanoimprint lithography • Electron beam irradiation	• Soft lithography • Injection molding • Imprinting • Laser ablation • 3D printing	• Wax and ink-jet printing • Photolithography • Plotting
8.1 Advantages	• Transparent • Inert and stable • Solvent compatible • Hydrophilic	• Mechanically strong • Temperature stable • Very resistant to chemicals	• Transparent • Easy to fabricate • Low cost • High electrical and chemical resistivity	• Flexible and lightweight • Low cost • No need for external pump/valves • Biocompatible • Recyclable (if not chemically modified or printed with ink)
8.2 Limitations	• Brittle • Stiff • High cost	• High cost • Biocompatibility	• Hydrophobic • Shelf life	• Sensitive to humidity and temperature • Difficult to design and integrate into a single chip

without the use of cleanroom facilities, they can be constructed from affordable, easily accessible materials, and they do not need computer-controlled pumps to function. These features make paper-based microfluidics more approachable for researchers from a variety of fields and a perfect platform for the creation of POC diagnostic testing.

3. *Polydimethylsiloxane (PDMS)*: When mixed in appropriate proportion with glass surface, PDMS serves as a transparent and flexible elastomer that finds extensive use in the production of prototypes and LOC devices due to its affordability and ease of manufacturing (Kitsara & Ducree, 2013). PDMS microfluidic devices can be fabricated by soft-lithography methods (Soft Lithography - Xia, 1998). The mixture of PDMS and cross-linking agent (use to cure) is poured into a mold and heated to a high temperature for hardening; upon hardening the mold is separated to obtain a replica of the microchannels. Later, to integrate, the channels for input and output of liquid are punched through a PDMS puncher, and finally, the PDMS with microchannels bond to the glass surface by plasma treatment. The PDMS can mold structures at high resolutions. With some optimization, molding structures of a few nanometers is possible. Furthermore, LOC made of PDMS profits from air permeability for cell culture and research and from the straightforward integration of shock microvalves for rapid flow switching (Nge et al., 2013). PDMS exhibits serious production-related restrictions despite being widely employed for LOC prototyping. It is challenging to incorporate electrodes onto a PDMS chip due to the material's susceptibility to aging and leaching and the fact that PDMS absorbs hydrophobic chemicals (Berthier et al., 2012; Roman et al., 2002). Last but not least, PDMS cannot be used in high-throughput chip production processes like hot embossing or injection molding. PDMS may frequently be chemically treated to solve these problems. The exposed PDMS surface will become hydrophilized after plasma exposure, but this effect is transient for a few minutes (Bodas & Khan-Malek, 2006). This change in surface characteristics is slowed down by the reaction of silanes with silanols created during plasma activation or other processes (Wang et al., 2008).
4. *Thermopolymers*: Elastomeric thermoplastics are widely used in the manufacturing of LOC devices. Developing these bonding materials, such as polycarbonate (PC), polystyrene (PS), poly (methyl methacrylate) (PMMA), and cyclic olefin copolymer, is one of the common steps in designing of LOC-based applications. These polymers offer several other advantages as they exhibit higher chemical inertness compared to PDMS, possess optical clarity, demonstrate resistance to a wide range of chemicals, are compatible with micro-sized lithography, and present more complex and costly deployment. Additionally, their compatibility with microelectrodes is considered as promising candidate for the industrialization of different LOC platforms. This makes them attractive for certain applications like DNA and protein analysis (Kristiansen et al., 2022; Liu & Fan, 2011; Tsao, 2016).

5. *Silicon*: Silicon being the basis of the miniaturization of technologies, microfabrication processes were naturally chosen as a material for the initial design and construction of LOC devices. Due to its cost, lack of optical transparency (apart from IR), need for a clean room, and technical requirements for microfabrication, silicon is currently used for LOC less frequently than it formerly was. The electrical conductivity of silicon also prevents it from being used for LOC procedures that require high voltage (like electrophoresis). However, it appears that it is no longer a viable contender for the industrialization of such devices (Scott & Ali, 2021; Zhang & Xing, 2007). Nonetheless, we believe silicon is still a feasible alternative for the industrialization of several complex LOC applications, such as nucleic acid separation and analysis, LOC polymerase chain reaction (PCR), single cell analysis (or single nucleotide polymorphisms), and so on (Duval et al., 2012; Han et al., 2014; Kumar et al., 2021; Majeed et al., 2012). This presumption considers the high degree of precision achieved in silicon microfabrication and the level of process maturity.

8.3 Advantages of lab-on-a-chip compared to traditional methods

The ultimate objective of LOC technology is to combine one or more laboratory operations or techniques onto a single small chip. When compared to traditional benchtop operations, LOC technology provides several potential advantages. A portable handheld device that can do POC clinical diagnostics, for instance, would be useful at clinics or other remote areas where access to lab facilities and sophisticated equipment is restricted. Because less reagent is used and less chemical waste is generated when standard benchtop processes are scaled down, there are also benefits for cost-effectiveness and the environment. Additionally, by needing smaller samples of blood, urine, saliva, and other bodily fluids, the use of low-volume samples could ease the load on patients. Utilizing LOC technology may result in greater operational efficiency when compared to traditional methods. In theory, integrating microchannels can achieve great parallelization, enabling numerous studies to run concurrently on a single device. Increased ergonomics, diagnostic speed, and sensitivity are further benefits that such little devices may provide. Some of the merits are discussed in detail as follows:

- *Low cost*: Microtechnology has the potential to significantly decrease the cost of analysis, similar to how it has already reduced the computational costs. Through integrated processing and on-chip laboratory capabilities, it can eliminate the requirement of specialized operators making it cheap. The inclusion of various and simultaneous tests on a single chip enables each study to be conducted at a minimal cost thereby minimizing expenses associated with individual analyses.

- *High parallelization*: High parallelization is a notable advantage of LOC technology as it has the capability to combine microfluidic channels. This enables the simultaneous execution of numerous tests on a single device. This level of parallelization empowers doctors to effectively address specific disorders, ideally during the consultation itself, or promptly prescribe the most suitable drugs or therapies for optimal treatment outcomes.
- *Ease of use and compactness*: Easy use and compactness are key advantages of LOC technology. By integrating numerous procedures within a small volume. LOCs enable sophisticated analyses comparable to those conducted in large analytical laboratories. A chip as small as a few centimeter squares, when connected to any readout, can lead to an accurate diagnosis. This significantly reduces the need for complex handling procedures allowing on-site testing to be performed by unskilled individuals.
- *Reduction of human error*: The usage of LOC devices reduces the chance of human mistakes significantly when compared to traditional analytical methods performed in laboratories.
- *Quick responses*: Diffusion of chemicals, flow switch, and temperature gradient and diffusion are all quicker at the downscale sizes. Temperature can be changed in hundreds of milliseconds (allowing for quicker DNA amplification using PCR, for example) or chemicals can be mixed through diffusion in seconds (allowing for faster biochemical processes, for example).
- *Low-volume samples*: Only an aliquot of blood sample is needed for LOC-based analyses, leading to a reduction in analysis costs by minimizing the usage of expensive reagents. Additionally, this technology enables the identification of a wide range of disorders without the necessity of obtaining significant amounts of blood from the patients.
- *Real-time quality control and sensing*: The nanoscale reactivity and controlled environment of the LOC enable easy real-time modulation of chemical reaction settings, improving analyte and receptor interactions and increasing sensitivity.
- *Portable or filed application*: Due to economic feasibility, automechanization, and minimum energy consumption, LOC devices can be effectively employed for outdoor applications such as air and water monitoring without the requirement of a skill individual.
- *Sharing with wider population*: LOC has the potential to reduce diagnostic expenses, costs associated with medical personal training, and infrastructure expenditures. As a consequence, LOC technology will allow the access of present-day medicine more affordable to underdeveloped countries.

8.4 Limitations of lab-on-a-chip compared versus traditional methods

Too far, the most significant impediment to the widespread use and growth of LOC devices has been the development of practical, cheap systems that are effectively developed and constructed. At the micro- and nanoscale, physical phenomena like surface roughness and capillary forces have a significantly greater significance in LOC performance. These strong effects may lead to difficulties that are unexpected in conventional benchtop techniques. Complications like a low signal-to-noise ratio can happen in various situations. As a result, LOC devices could occasionally deliver fewer effective outcomes than traditional methods. Some of the demerits are discussed in detail below:

- *Industrialization*: Most LOC technologies are not yet industrialized. In terms of its primary use, ultra-multiplex diagnosis, we are unsure which fabrication methods will become the norm, such as Clinical Laboratory Improvement Amendments (CLIAs)-based LOC platforms and stereolithography (Rattle et al., 2013). The CLIA-waived LOC systems employ a variety of microfluidic technologies, such as pressure, capillary flow, centrifugal, or electrokinetic effects connected to external detectors for electrochemical or optical examination. Recent advancements in stereolithography, an additive method for printing 3D, have made it possible to manufacture tiny valves, pumps, and other parts that control the liquid flow on LOC devices. The ability to precisely design intricate 3D architectures has become a common reality as research advances.
- *Signal-to-noise ratio*: Miniaturization boosts the signal-to-noise ratio in some applications, and as a consequence, LOC gives better analytical performance than traditional approaches.
- *Ethical consideration and human conduct*: Furthermore, the sequencing capacity of the lab-on-a-DNA chip may allow anybody to sequence someone else's DNA with just a drop of saliva. In the lack of rules, however, real-time processing and the availability of LOC may raise concerns about the general population doing untrained home diagnoses of suspected ailments.
- *LOC necessitates the inclusion of external electronics system*: Despite their compact and powerful nature, it requires precise equipment to function properly, such as electronics or flow control systems. Without a precise method for sample interjection, spitting, and position control. The incorporation of external devices increases the overall size and therefore cost of the system, with certain components, particularly flow control equipment, severely impacting the performance of devices.

8.5 Immobilization of biorecognition element in microfluidic channels

8.5.1 Immobilization strategies

Biorecognition components (enzymes, aptamers, antibodies, etc.) are attached to the immobilization surface using a wide range of immobilization techniques.

When selecting an immobilization approach (e.g., the ability to accurately detect, multiple use of the device, ability to specifically identify, and consistency and reliability), several aspects such as the control surface, matrix, sample compartment, protein characteristics, buffer components, and performance index are considered. In an optimal scenario, it is desirable for the reactants to have unhindered access to active regions where antibody binding or enzymatic conversion is happening. It means that the binding site should be oriented away from the immobilization surface minimizing steric blocking and hindrance and should not be sterically hindered by nearby immobilized proteins (Liébana & Drago, 2016). For a high-performance, repeatable experiment, receptor conformation must remain unaltered upon immobilization to maintain receptor binding properties, permit diffusion of substrates and products, and allow excellent electron transfer. The molecular mechanisms used in bioreceptor immobilization techniques can be divided into physisorption, bioaffinity interaction, covalent binding, and combinations of them (Fig. 8.2).

1. *Adsorption*: The physisorption method is the simplest method for immobilizing a biorecognition component on a surface (i.e., physical adsorption). In physical adsorption, the baroreceptors are immobilized by van der Waals forces, hydrophobic contacts, or hydrogen bonds, whereas enzymes are joined by salt linkages in electrostatic interaction (Fig. 8.2A). The process of noncovalent immobilization is caused by the nature of the forces involved, and it can be stopped by altering the factors that determine how strong the contact is (e.g., pH, salt concentration, solvent polarity, and temperature) (Kim & Herr, 2013; Liébana & Drago,

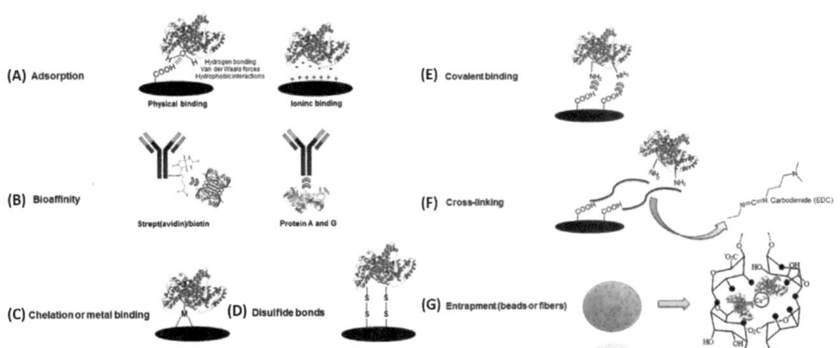

FIGURE 8.2 Schematic representation of physical or reversible immobilization methods: (A) adsorption, physical and electrostatic binding; (B) bioaffinity with strept(avidin)/biotin and Protein A/G; (C) chelation or metal binding; (D) disulfide bonds; (E) covalent binding through primary amines; (F) cross-linking using carbodiimide, and (G) entrapment on beads or fibers and microencapsulation. *Figure reproduced with permission from Liébana, S., & Drago, G. A. (2016). Bioconjugation and stabilisation of biomolecules in biosensors. Essays in Biochemistry, 60, 59.*

2016). The functioning of the biomolecule is often maintained during immobilization by adsorption, which is a simple and straightforward procedure. Weak attachment and unpredictable orientation, which result in desorption and poor reproducibility, are the adsorption mechanism's drawbacks (Nimse et al., 2014).

2. *Bioaffinity*: These bioaffinity interactions, also known as biospecific adsorption (Fig. 8.2B), take advantage of certain binding processes seen in nature. Physisorption cannot compare to the benefits of the bioaffinity association. For example, a directed, highly specific, and considerably stronger protein immobilization is produced through bioaffinity contact. The occurrence of protein leakage can be reduced by employing fixed receptor molecules that offer improved accessibility to binding partners compared to random orientation strategies (Nguyen et al., 2019; Nimse et al., 2014). Additionally, bioaffinity immobilization can be undone through chemical reversal, pH adjustment, or thermal reversal. The natural affinity concept between molecules, such as lectin-sugar, antibody–antigen, and biotin–avidin, has been used to immobilize biomolecules. However, covalent attachment of a costly affinity ligand (such as an antibody or lectin) to the support is typically required. The interactions of streptavidin and biotin, as well as the usage of Protein A or G for antibody immobilization, are the most extensively utilized approaches (Kim & Herr, 2013).
3. *Metal-link binding/chelation*: Fig. 8.2C, the metal salt or hydroxide, principally titanium and zirconium salt, is precipitated and bonded through coordination with nucleophilic groups on the surface (such as cellulose-, chitin-, alginic acid-, and silica-based carriers). Some metal binding sites are empty due to steric restrictions, leaving others available for interactions with biomolecule groups. Although the procedure is simple, there is a lack of repeatability because the adsorption sites are not homogenous and there is significant metal ion leakage.
4. *Sulfur-atom linkage or disulfide bond*: Although a durable covalent link is formed between the support and the bioreceptor, it can be broken by a moderate reaction with a suitable chemical, such as dithiothreitol (DTT) (Fig. 8.2D). Moreover, when employing a thiol-reactive adsorbent having high specificity, the activity yield of disulfide bond-forming procedures is often remarkably high. This is due to the ability to regulate the reactivity of thiols by adjusting pH (Bulaj, 2005).
5. *Grafting or covalent immobilization*: One of the most popular techniques for immobilizing receptors is to use covalent bonding to immobilize them (Fig. 8.2E). In the grafting process, there is a formation of stable bonds between the biomolecule and support that prevents the leaching of the bioreceptor into the solution. To sustain maximum protein activity, it is crucial to make sure that the covalent attachment chemistry

to the support does not damage the active area of the biomolecule. Preventing any interference with the obstruction of necessary amino acid residues regulating the enzyme activity or the antibody recognition area is an essential consideration. This could be difficult to comply with in some circumstances (Prieto-Simon et al., 2008). According to the functional groups present on the target, several different reactions have been devised. Despite the intricacy of the biomolecule's structure, there are just a few functional groups that can be exploited as selective targets for efficiency. Similarly, thiol groups (−SH) that form an integral part of the side chain of cysteine residues are shown to provide good efficacy. In sugars, the Fc region is the main fragment where glycosylation occurs. The diols- moiety are oxidized to aldehydes that involve additional steps depending upon the number of substituents present on carbon atoms bearing hydroxyl groups. However, due to the position of the carbohydrate moiety, the technique typically produces antibody conjugates with high activity. The reductive amination process can be used to directly react sugars with primary amine or hydrazine-based groups. Some of the examples of covalent immobilization are shown in Fig. 8.3A−C.

Using a blocking solution, such as bovine serum albumin (BSA) in phosphate buffer saline (PBS), functional groups that did not attach to the baroreceptors are inactivated (Salva et al., 2021). The usually random orientation of the receptors caused by covalent immobilization, which assures nonreversible attachment of receptors to surfaces, may affect their biological activity and capacity to bind analytes from solution. Covalent immobilization can also take time, especially if low-concentration receptor solutions are applied during the attachment phase (due to their cost). Finally, when proteins need to be immobilized, noncovalent and covalent immobilization can be coupled utilizing a spacer between a surface and receptors (Welch et al., 2017). Alternatively, "random immobilization" describes the placement of receptors on a surface without regard for how they will be oriented. Direct immobilization refers to the attachment of receptors with orientational control; the active site of the receptor is often oriented away from the surface and thus more sterically accessible to species from solution. This results in surfaces with a larger density of receptors.

6. *Physical encapsulation and entrapment*: The capture method involves the confinement of a biomolecule, typically an enzyme with a polymeric network that allows the flow of substrates and products while enclosing the enzyme itself. Unlike the coupling method mentioned earlier, this approach does not involve linking the enzyme to a support or membrane. Enzyme encapsulation can be achieved through various techniques such as micro encapsulation. However, the practical application of these approaches is limited due to the constraints imposed by mass transfer through membranes or gels (Kim & Herr, 2013).

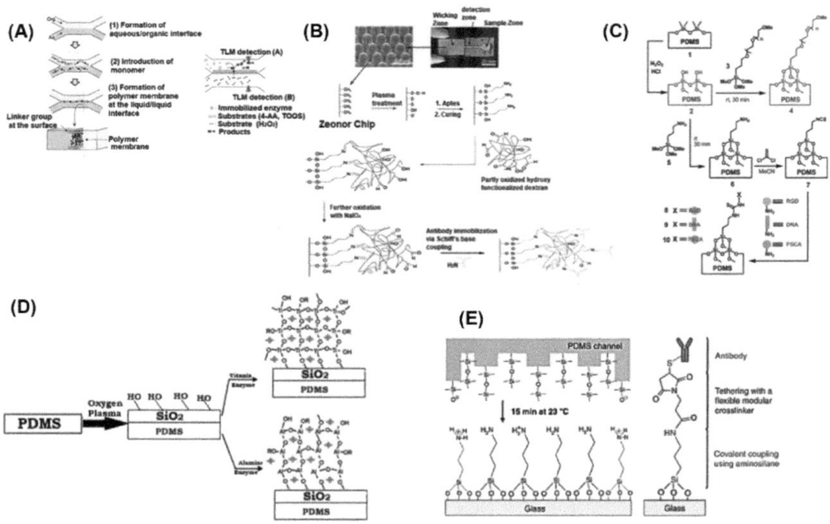

FIGURE 8.3 (A) The in situ polymerization of nylon membrane was performed using an organic/aqueous two-phase flow system made of an X-shaped microchannel chip, and thermal-lens microscopy (TLM) was employed for product and substrate identification. (B) A zero-powered cyclic olefin copolymer immunoassay chip for covalent antibody immobilization with patterned micropillars. First, the cyclic olefin copolymer surface is interacted with oxygen plasma and silanized and functionalized with a dextran matrix. (C) Schiff's base coupling reaction used for antibody immobilization: (1) surface activation through plasma; (2) activated PDMS surface; (3) silanization; (4) silanized PDMS surface; (5) amine-terminated silanization; (6) amine-grafted PDMS microchannels. The activation of amine-grafted microchannels is done with thiophosgene to develop (7) the isothiocyanate-grafted PDMS microchannels as a precursor for (8) the RGD-grafted, (9) DNA-grafted, and (10) PSCA-grafted PDMS microchannels. (D) Process of enzyme-encapsulated sol–gel inside microchannel of PDMS functionalized by oxidation in an oxygen plasma. (E) The mechanism of the bonding occurs through silicon alkoxides ($Si-O^-$) and protonated nitrogens ($-NH + 3H3 +$) at room temperature and under atmospheric pressure. *Figures are reproduced with permission from Deliorman et al. (2020), Hisamoto et al. (1999), Kim and Herr (2013), Sui et al. (2006), and Wu et al. (2004).*

7. *Cross-linking mechanism*: Cross-linking is the process of chemically adding two or more molecules through a covalent bond. Cross-linking agents, sometimes referred to as cross-linkers, are molecules with two or more reactive ends that can chemically connect to a particular functional group of proteins or other biomolecules. Cross-linkers are frequently used to alter drugs, solid surfaces, and nucleic acids. The same chemical processes are used to modify and mark the surfaces of amino acids and nucleic acids (Mishra et al., 2014, 2015; Sharma et al., 2017).

We concentrate on receptor immobilization techniques for static surfaces in this section. In particular, we will give examples of immobilization that takes place directly on the surfaces of microfluidic systems. In microfluidics, the immobilization of receptors (antibodies/enzyme/protein) is frequently carried out directly on the interior of microfluidic structures, such as microchannels (Kim & Herr, 2013). Large regions of immobilization and direct interaction between analytes flowing inside a microchannel and an immobilized receptor are both possible with such an approach. As previously mentioned, microchannels can be made of a variety of substances, including polymers, glass, silicon, and silicon alloys.

Depending on the application covered and the microfabrication process used for the devices, the type of surface to be used to immobilize target receptors must be optimized (Kim & Herr, 2013). The immobilization of receptors depends on a number of variables, including the surface's chemical and physical characteristics, the immobilization technique and chemical conditions (such as the buffer or solution used, pH, and temperature), the characteristics of the molecules that will attach, and the desired position, orientation, and density of the receptors over the surface (Figs. 8.2 and 8.3). It is essential to take into account all of these elements as they can have a notable impact on the sensitivity and limit of detection of the device (Jonkheijm et al., 2008; Kim & Herr, 2013).

8.5.2 2D versus 3D surface immobilization

The surface geometry, which can be either planar (2D) or three-dimensional (3D), has a considerable impact on the immobilization techniques. Microfluidic channels composed of various substrates or surfaces covered with materials like gold, platinum, or graphene are examples of 2D surfaces used for immobilization. The simplest support for immobilizing receptors is a 2D surface; however, the density and number of molecules that can be immobilized on these surfaces are constrained. Glass and silicon are the conventional inorganic materials used to create microfluidic devices. Due to their similar surface chemistry, close techniques can be used to immobilize receptors onto these substrates. Silanization of the surface and receptors attaching to functional groups are two common immobilization techniques (Jonkheijm et al., 2008; Kim & Herr, 2013) (Fig. 8.3B). The absence of functional chemical moieties on the surfaces (PDMS, PMMA, PC, PS, etc.) prevents biological receptors and binding sites from being directly attached. Therefore before immobilizing receptors, these surfaces need to undergo surface treatment to produce (or anchor) functional groups, which help to immobilize receptor molecules and bond with glass/silicon surfaces. Receptors can also be immobilized on a 2D surface using metallic films, and in the case of Au, Pd, and Pt, for example, these films can be functionalized using self-assembled monolayers.

Fig. 8.4 depicts the immobilization of sensors on 3D surfaces (D–F). After 3D geometry has been generated on a substrate or after all of the device's parts have been assembled, it can be functionalized with receptors. Preassembly manufacturing methods involve integrating porous membranes or

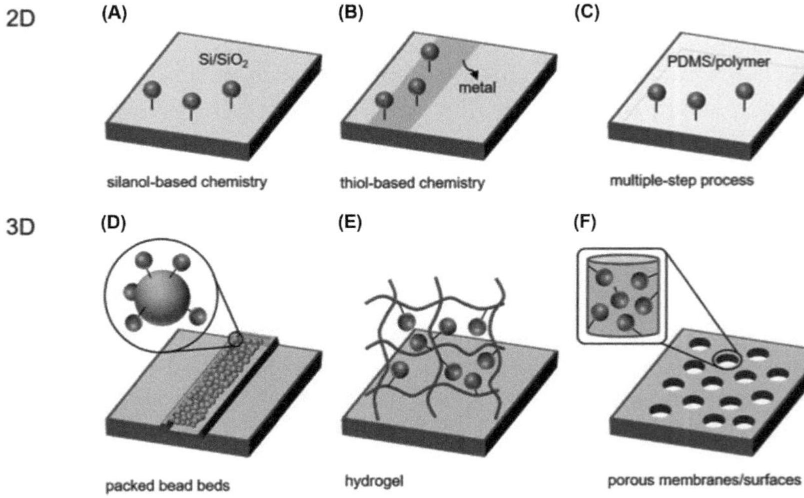

FIGURE 8.4 Immobilization strategies on 2D surfaces and 3D geometries for commonly used receptors in microfluidics such as proteins. (A) Immobilization on silicon or glass, (B) metal, and (C) PDMS or polymer, respectively. (D) Immobilization on beads packed into a 3D structure. Bead materials include polystyrene, silica agarose, and ferromagnetic materials. (E) Immobilization into hydrogel. Commonly used hydrogels include PEGDA, polyacrylamide gel, chitosan, and agarose. (F) Immobilization within a permeable membrane: porous membranes can be fabricated, for example, using polymer monoliths (e.g., ethylene dimethacrylate, acrylamide, 2-hydroxyethyl methacrylate) and porous silicon. Alternatively, porous membranes such as polycarbonate, nitrocellulose, polyvinylidene fluoride, and paper are also commonly used. Small blue dots represent the receptor molecules. *Reproduced with permission from Salva, M. L., Rocca, M., Niemeyer, C. M., & Delamarche, E. (2021). Methods for immobilizing receptors in microfluidic devices: A review. Micro and Nano Engineering,* 11. *Preprint at https://doi.org/10.1016/j.mne.2021.100085.*

patterning microstructures like microposts. The use of paper or nitrocellulose as a substrate, different polymers, and beads can all be used in postassembly manufacturing procedures. From an economic perspective, paper and nitrocellulose are specifically employed to create POC diagnostic devices because they allow for cheap cost and easy-to-use devices, absorption-based immobilization of receptors, and direct viewing of signals (Jane Maxwell et al., 2013). Using copolymerization, covalent bonding, and electrostatic capture, immobilization on polymers like hydrogels is possible. Beads containing receptors can be targeted and packed using certain microstructures or suspended in a solution. In each scenario, the surface makeup of the beads determines the receptor immobilization method. Immobilization techniques for silica beads may resemble those used for silicon and glass surfaces. Numerous methods of immobilization have been devised for agarose beads, including graft polymerization, copolymerization with proteins, and oxidative activation of functional groups. Recently, immobilizing receptors in microfluidic devices has been accomplished using graphene sheets suspended in a solution. 3D surfaces have several advantages over 2D surfaces. First, there are more sites for immobilizing receptors on 3D surfaces than on 2D surfaces because they have a greater surface-to-volume ratio. Additionally, for receptors mounted on 3D surfaces, the diffusion length between reactants is decreased, resulting in increased sensitivity and speed (Salva et al., 2021). Fig. 8.4 depicts techniques used for the immobilization of receptors in 2D and 3D systems.

8.6 Microfluidics integrated point-of-care platforms

8.6.1 Fabric-based microfluidic platforms

"When it comes to a diagnostic and precision medicine," lab-on-fiber technology proves to be a suitable choice due to its affordability and requirement of minimal sample volumes. Recently, various wearable devices utilizing textile threads have emerged. In terms of detection methods, wearable sensors employing these advanced textile threads can be categorized into two classes: (1) optical, which explores colorimetric analysis techniques, with identification of the compounds and quantification performed through alteration and intensity of the colors. Analyses can be performed, for example, through visual comparison or by capturing images with smartphones; (Curto et al., 2012; Xiao et al., 2020; Zhao et al., 2021) and (2) electrochemical analyses using potentiometric, amperometric, impedance, or conductometric techniques. In this case, measurements are taken by miniature electronic systems and/or communicated to mobile devices, particularly smartphones (Kassal et al., 2015; Mostafalu, 2016; Nomura et al., 2018). These two techniques' dominance in the proposed wearable devices can be attributed to their key characteristics for use in analyses and in situ monitoring of compounds in the human body, such as rapid measurements (seconds or minutes), small sample volumes, high sensitivity, better integration with the body and/or textile materials, simplicity of their components, potential for miniaturization, and ease of integration (Agustini et al., 2021).

Threads offer good flexibility, stretchability, and durability as raw materials for fabric microfluidic substrates (Agustini et al., 2021), and maximum surface coverage due to the integration of sensors into clothing, ensuring discretions and wearability. They may be produced at a minimal cost from a wide range of natural or synthetic materials. When compared to conventional microfabrication techniques (such as soft photolithography, screen printing, and stamping), fabric-based microfluidic systems only require straightforward textile processing techniques, like sewing, weaving, and embroidering, which are comparatively quick, affordable, and have a lot of potential for mass production (Neužil et al., 2014). Fabric-based microchannels are made by changing the fabric's wettability and hydrophobicity to regulate the liquid flow. To create channels for the injection and delivery of fluids, hydrophilic threads are sewn into hydrophobic fabrics (Reches et al., 2010). Wearable fabric-based microfluidics have the flexibility to gather perspiration from the skin's surface and wick it through predefined channels to sensing zones because of the capillary effect (Ainla et al., 2017). Hydrophobic and hydrophilic threads can be created by subjecting the materials to a variety of surface treatments such as corona, electrochemical, and UV rays (Islam et al., 2020; Li et al., 2019). Conductive threads are created by altering the material composition, for example, by including different conductive nanomaterials like silver nanoparticles (AgNPs), carbon nanotubes (CNTs), Ag/AgCl, and conductive polymers into the thread (Li et al., 2015; Zhao et al., 2019). Wearable fabric-based microfluidic sensors for physical or chemical analysis can be created by combining multitype threads (Curto et al., 2012; Morris et al., 2009). For example, (Yang et al., 2018). introduced sturdy fabric-based organic electrochemical transistors (OECTs) for glucose monitoring. Nylon fibers covered with a Cr/Au/poly(3,4-ethylenedioxythiophene) polystyrene sulfonate (PEDOT:PSS) (poly(3,4-ethylene dioxythiophene):poly(styrene sulfonate)) conducting layer serves as the transistor's source and drain electrodes. The gate electrode was made of conductive fibers modified with glucose oxidase enzyme. This device is compatible with conventional weaving machines and can be sewn using normal yarns because of the fabric's inherent flexibility. Zhao et al. (2019) demonstrated the fabrication and use of a stretch gold-fiber-based three-electrode electrochemical biosensing device that can meet the criteria for wearable textile glucose monitoring. Based on previous work involving fabricating stretchy gold fibers by dry spinning and electroless-plating (Zhao et al., 2018), additional functionalization of gold fibers was carried out. The functionalization involved the introduction of Prussian blue (PB) and glucose oxidase (GOx) to design the working electrode, while Ag/AgCl modification was implemented to act as the reference electrode, and gold fiber could be used as a counter electrode without further modification (Fig. 8.5). The resulting textile glucose biosensors exhibited the linear range between 0–500 M and 11.7 $\mu A\ mM^{-1}\ cm^2$, respectively. The surface functionalization made possible by the gold fibers facilitated the integration of a three-fiber electrode system into textiles. The combination of the inherent stretchability of the highly conductive gold fiber with the extrinsic helix structure resulted in an electrochemical performance that was strain-insensitive and reproducible from 0 to more than 190% strain. These findings suggest the potential applications of wearable biochemical diagnostics using sweat as a sample medium.

Electrodes can be easily created with personalized geometry and combinations using a computerized embroidery machine to support either custom or commercial electrochemical instrumentation. To sample or detect an analyte of interest, wearable sensing applications allow for the embroidered placement of sensors at precise points on a garment. High-volume manufacturing is also feasible with this technology, lowering the expense of the devices used in in vitro diagnostic testing. Given that most threads are

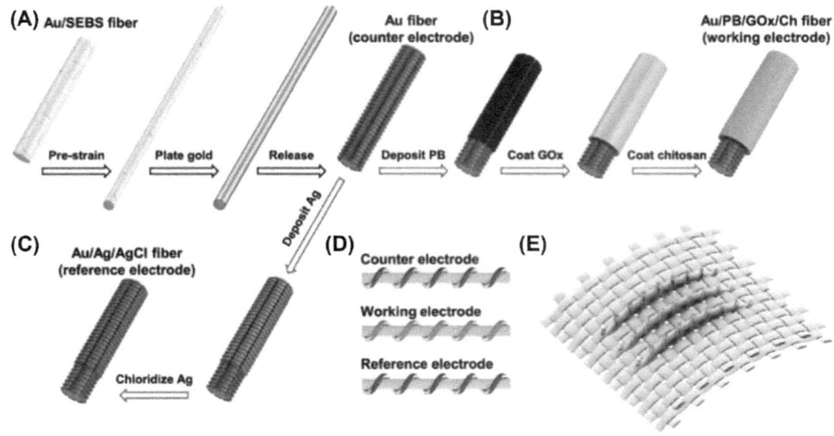

FIGURE 8.5 Schematic depiction of the fabrication process of the stretchable fiber-based glucose sensor: (A – C) illustrations of the fabrication processes of the stretchable Au fiber (counter electrode), the Au/PB/GOx/Ch fiber (working electrode), and the Au/Ag/AgCl fiber (reference electrode), respectively; (D) schematic illustration of the three-electrode glucose sensor; (E) schematic illustration of the glucose sensors integrated into an elastic textile. *Reproduced with permission from Zhao, Y., et al. (2019). Highly stretchable and strain-insensitive fiber-based wearable electrochemical biosensor to monitor glucose in the sweat. Analytical Chemistry,* 91, *34.*

hydrophilic, embroidered sensors can more easily absorb liquids, easing sample loading and enhancing automation. A recently reported approach involves the development of a dual sensing platform using polyester threads, enabling the detection of glucose and lactate (Neužil et al., 2014). The working electrode (WE) and counter electrode (CE) were constructed using thread covered with carbon ink, whereas the reference electrode (RE) was constructed using thread coated with Ag/AgCl ink. To reduce ink oxidation on Ag/AgCl thread, soldering flux was applied to the thread using a flux pen before the ink coating procedure. Furthermore, the biorecognition element can be coated by drop casting and dried under optimum experimental conditions. The performance of the biosensor was tested in response to mechanical deformation for repetitive folding and bending both before and during the test. Both the bent and flat sensors exhibit high accuracy as represented by the low SD (<6%) of multiple measurements. With this set-up, glucose and lactate were detected at up to 40 mM and 50 mM levels, respectively. The linear calibration covers the clinical ranges of glucose and lactate in blood plasma (glucose—3.9–5.6 mM, and lactate—1.0–2.5 mM)

A flexible and wearable band was fabricated as a cheap and portable biosensor for the noninvasive and continuous monitoring of sweat samples (Zhao et al., 2021). This microfluidic band was designed using a straightforward embroidery technique to combine hydrophilic thread and hydrophobic fabric to easily produce microchannels and microreservoirs for on-site sweat collection and monitoring without an additional pumping mechanism. The detection was performed through colorimetric analysis by comparing the results to a reference marker. The assessment of exact RGB value using a smartphone app produced considerably more accurate information. The limit of detection (LOD) and detection ranges for chloride concentration, pH, and glucose concentration, respectively, was 10 mM (Cl^-), 4.0–9.0 pH, and 10 μM (Glu) in sweat. This band addresses the growing demand for mechanical robustness and wearing comfort of flexible microfluidic devices, especially in intense exercise scenarios. This is because the thread and fabric have been well integrated.

8.6.2 Fabrication of microfluidic devices based on paper/μPADs

The concept of microfluidic channels on paper was first introduced by Müller and Clegg in 1949 utilizing filter paper and paraffin (1949) (Müller & Clegg, 1949). However, in 2007 the Whiteside's Group at Harvard University achieved a significant milestone by developing the first functional paper device capable of performing a protein glucose assay (Martinez et al., 2007). Since then, paper has gained immense popularity as a substrate for microfluidic applications like dipstick tests, lateral flow assays (LFAs), and microfluidic analytical devices (PADS) (Akyazi et al., 2018). This is primarily attributed to its biocompatible properties with various substrates, lightweight, flexibility, economic feasibility, hydrophilic characteristics, user friendliness, and availability. Paper is made from raw materials like straw, cotton, jute, bamboo, grass, sisal, linen, hemp, and bamboo (Yetisen et al., 2013). Paper becomes fibrous and porous because of these sources, making paper an ideal substrate—(1) absorbent: to store liquid and permit liquid circulation through capillary action; (2) air permeable: to serve as a filter; (3) has a high surface area-to-volume ratio: to immobilize large volumes of liquid (Akyazi et al., 2018). Two important paper-based substrates of POC testing equipment are cellulose and nitrocellulose. Dipsticks and paper microfludic analytical pads (PADs) are produced using filter and chromatography paper made from cellulose. While LFAs are produced using nitrocellulose (Yetisen et al., 2013).

Two commonly employed methods for fabricating paper-based devices are cutting and hydrophobic/hydrophilic contrast. Cutting involves creating channels on paper using either carbon dioxide laser cutting equipment or a computerized cutter plotter, followed by sealing these channels using adhesive tape. However, this fabrication technique is not widely used due to its reliance on specialized and expensive equipment.

On the other hand, hydrophilic methods involved generating hydrophilic channels on the paper by utilizing hydrophobic patterning agents such as wax, wax-based photoresist, and AKD. Channel patterning can be achieved through three approaches: (1) physically blocking the paper's pores with PDMS or photoresist; (2) physically depositing agents on cellulose fibers using polystyrene or wax; and (3) chemically modifying the paper fibers with AKD (Li et al., 2012). Additional techniques employed in the production of paper-based devices are photolithography, ink-jet printing, PDMS plotting, wax printing, wax dipping, wax screen printing, plasma treatment, ink-jet etching, flexography printing, stamping, wet etching, and cutting (Akyazi et al., 2018; Sachdeva et al., 2020).

The availability of equipment, the cost of materials, the ease of the construction process, and the planned uses of paper-based microfluidic devices all play a role in choosing the best approach. Currently, printing and wax printing exhibit a great promise as most of the favorable techniques, because of quick fabrication process. These methods allow for the production of multiple devices or multizone setups on an A4-sized paper within a span of fewer than 10 minutes, requiring just a single print and heat cycle. It is conceivable that both approaches can be further enhanced to pilot or even large-scale processes. In this envisioned scenario, paper could be initially printed using an ink-jet printer followed by heat treatment in an oven to establish hydrophilic-hydrophobic contrast. Lastly, additional ink-jet printers capable of dispensing reagents for detection could be employed. Although paper-based microfluidic system provides a cost-effective and straightforward platform for multianalyte detection and semiquantitative measurement, complex devices are often needed to multistep procedures like concentration to detection of analyte of interest and recording measurements. Under these conditions, functional components might be added to the devices to regulate fluid flow within paper-based microchannels and to increase their functionality. To address this, Li et al. reported a paper-based reactor capable of facilitating multistep processes as well as a paper-based separator designed for sample filtering (Li et al., 2008). In a similar vein, Whiteside's group (Liu & Crooks, 2011) showcased the development of 3D paper-based microfluidic devices that enable the transport of liquid from a single intake to multiple detection zones through both lateral and vertical fluid distribution. The group also designed and developed a programmable device with "on" buttons that allowed the connection and disconnection of fluid flow between paper-based microchannels. Liu and Crooks utilized the principles of origami to construct 3D paper-based microfluidic devices (Liu & Crooks, 2011). Still, the limits of paper-based POC devices such as sample retention and evaporation are still required to be analyzed and solved, leading to a reduction in the efficacy of the device. LODs are usually high for the optical-based method, which makes it difficult to detect very low analyte concentration and handling of liquid with low-surface tension; liquid penetrates in both hydrophilic and hydrophobic regions (Li et al., 2012).

Besides their applications in precision medicine and analyte detection, paper-based microfluidic devices also known as lateral flow essay have been the topic of active research (Akyazi et al., 2018; Li et al., 2012; Sachdeva et al., 2020). LFAs, such as the at-home pregnancy test, are more complicated instruments that frequently rely on antibodies to offer excellent specificity. A sample pad, conjugate pad, test line, control line, and absorbent pad are among the porous membranes that make up the devices (Fig. 8.6A). The sample is added to the sample pad, wicked through the apparatus, where it interacts with a tagged conjugate antibody and an immobilized capture antibody, and the test is then completed. As the labeled antibody is maintained by the capture antibody through their respective interactions with the antigen, when the antigen (analyte) is present in the sample, a colorful band appears across the strip. The detection abilities of these assays are improved in more complex instances of LFAs using nanoparticles, enzymes, or oligonucleotides (de Puig et al., 2017; Syedmoradi et al., 2017). To identify illnesses, viruses, biomarkers, and poisons, LFAs are being employed in a variety of applications. LFAs have most recently been designed to detect COVID-19 (SARS-CoV-2) (Broughton et al., 2020; Hussein et al., 2020; Nishat et al., 2021). Broughton et al. (2020), successfully developed a quick, user-friendly assay utilizing CRISPR-Cas12 for the detection of COVID-19 (SARS-CoV-2). This assay can perform reverse transcription and isothermal amplification using loop-mediated amplification (RT-LAMP) for extracted RNA from nasopharyngeal or oropharyngeal swabs. A colloidal gold-immunochromatographic assay (GICA) was developed against SARS-CoV-2 by Li et al. (2020) (Fig. 8.6B). The device can detect immunoglobulin M (IgM) and IgG antibodies simultaneously against the SARS-CoV-2 virus in human blood in less than 18 minutes facilitating the detection of various stages of the disease.

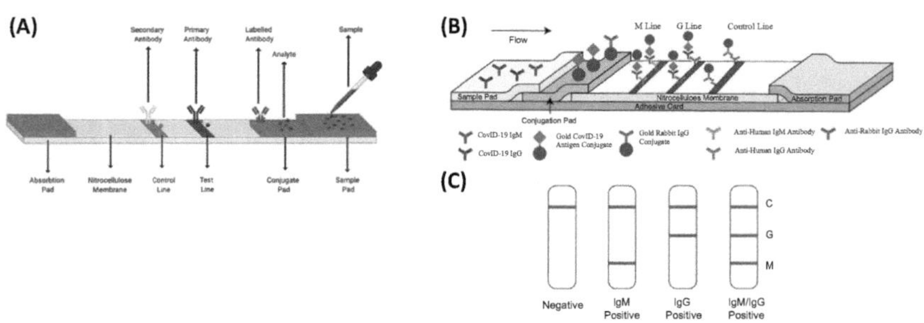

FIGURE 8.6 (A) LFA strip components. A simple LFA test strip consists of a sample pad, a conjugate pad, a test line, a control line, and an absorption pad; a schematic of a SARS-CoV-2 IgM-IgG integrated antibody test. (B) Diagram depicting of the detection device. (C) Schematic of different testing results: C, control line; G, IgG line; M, IgM line. IgG, immunoglobulin G; IgM, immunoglobulin M; SARS-CoV-2, severe acute respiratory syndrome coronavirus 2. *Figures reproduced with permission from Lanceros-Mendez, S., et al. (2021). Microfluidic point-of-care testing: Commercial landscape and future directions. Frontiers in Bioengineering and Biotechnology, 1, 602659. <http://www.frontiersin.org/>; Li, Z., et al. (2020). Development and clinical application of a rapid IgM-IgG combined antibody test for SARS-CoV-2 infection diagnosis. Journal of Medical Virology, 92, 1518–1524.*

8.6.3 Wearable microfluidic platforms

Wearable microfluidic technology is proliferating in a variety of fields, from healthcare to biomedical monitoring systems, allowing for continuous assessment of vital biomarkers for physiological health monitoring and evaluation. Particularly, as the older population increases worldwide, numerous chronic and acute diseases become more significant, and the medical business undergoes a significant change as a result of the need for POC diagnostics and real-time monitoring of long-term health issues. Advantages are continuous sampling, low sample volume requirement, minimal or noninvasive, and low contamination from the skin. Wearable technology has grown steadily to include accessories, integrated garments, body attachments, and body implants. These wearable sensors can monitor your health in real time and include mouthguards, eyeglasses, footwear, ring, necklace, and wristband integrated with sensing technology (Arakawa et al., 2016; Kim et al., 2019; Lee, 2016; Luo & Gao, 2021; Majumder et al., 2017) The human body is completely covered in sweat glands, which is an important and noninvasive biofluid. Sweat can potentially reveal a patient's health status and be used for noninvasive and wearable sensing. For noninvasive personal health monitoring purposes, such as illness state and hydration level, the presence of biomarkers in sweat may be of great importance (like diabetes and cystic fibrosis). Without harming the skin layer, in situ noninvasive sweat monitoring at the epidermis layer further eliminates the need for blood sampling, avoiding potential infections and pain. However, additional testing is necessary to confirm sweat's clinical utility as a biofluid for diagnostics (Heikenfeld, 2016).

With the use of epidermal biosensors, biomarkers in epidermal biofluids (sweat and ISF) may be analyzed in real time, and some systems even could continuously monitor a range of biological and fitness-related applications. These devices depend on ISF or sweat samples being taken at the skin's surface and transported over the biosensor surface (Campbell et al., 2018; Sharma et al., 2022). Reverse iontophoresis can be employed to extract ISF analytes on the skin in a noninvasive manner for monitoring. However, the method's accuracy is influenced by the extraction procedure. Advancements in sweat collection and detection technologies may hasten the practical application of sweat-based diagnostic options with a deep understanding of the chemistry of sweat and the metabolite transport mechanism. The main categories of sweat-based biosensors are textile/plastic and tattoo (epidermal)-based systems and analogs, each with advantages and drawbacks (Steinberg et al., 2016).

Martin et al. demonstrated a microfluidic integrated wearable epidermal electrochemical sensor system for time-dependent and continuous monitoring of sweat compounds like glucose or lactate in athletes (Martín et al., 2017). This soft, skin-mounted gadget intelligently combines with LOC and flexible electronics, smaller in size and capable of wirelessly downsizing for real-time data transmission to a mobile device to enable continuous data collection.

An integrated wearable enzyme-based sweat sensor that takes advantage of near-field communication (NFC) technology to quantify glucose, pH, chloride, and lactate in sweat was demonstrated (Bandodkar et al., 2019). With rising concentration, the lactate sensor's response displays remarkable linearity. The enzymes absorbed on the working electrodes, however, can lose some of their original properties during long-term detection. This is due to some environmental factors like temperature and pH level, which reduce the sensitivity and stability for continuous monitoring. To address the above issue and take advantage of a microfluidic substrate (PDMS), a wearable platform was fabricated as a fully stretchable microfluidic for nonenzymatic glucose monitoring (Bae et al., 2019). The collection and passive, precise conveyance of sweat from the skin to the electrode surface, along with excellent replenishment capabilities, were made possible by embedding elastic cotton fabric as a capillary inside a thin polyurethane nanofiber reinforced PDMS channel. Herein, the working electrode represents an electrocatalytic nanoporous gold electrode and exhibited a high sensitivity to glucose (253.4 A cm^2/mM) and no deterioration under 30% strain. Sweat samples are collected and measured with great accuracy and sensitivity using microfluidic networks made of cloth and PDMS.

Potentiometric sensors consist of ion-selective electrodes (ISEs), which are made of ion-selective materials that interact with ions, to change the voltage signals, and reference electrodes (Ye et al., 2020). The potential of ISE depends on the analyte concentrations, while the potential of the RE is constant. The concentrations of the targeted analytes are calculated using the difference between ISEs and RE. A skin-mounted microfluidic sensor for sensing concentrations of Na^+ and K^+ based on ion-sensitive mechanisms during exercise was reported (Sempionatto et al., 2019). In this study, communication via Bluetooth mode was utilized to transmit the potential signal generated between the working electrode and the reference electrode to a smartphone. By adding ISEs selective for different electrolytes to the microfluidic reservoirs, it is possible to accomplish multiple electrolyte measurements. A plethora of literature has been reported on the development of ion-sensitive field-effect transistors (ISFETs) based on the coupling of field-effect transistors (FETs) and ion-sensitive membranes for electrolyte measurement in a liquid (Cazalé et al., 2016; Nakata et al., 2017). In ISFETs, the gate electrode is changed to a reference electrode that is affected by charged liquid ions or molecules. The size of the complete analytical sensor may be significantly reduced with this approach. For multiplexed sweat electrolyte measurement, a fully lab-on-skin integrated wearable sweat sensing system capable of tracking biochemical information on the skin surface was constructed on a single silicon chip with fully depleted silicon-on-insulator ISFET sensors (Bellando et al., 2018). Real-time experiments show that this ISFET has a high sensitivity and linear response to the number of ions it detects (Na^+ and K^+). ISFETs show considerable promise for detecting numerous electrolytes in large-scale assays due to the tiny size of the sensor (less than 1 cm^2) and ultralow volume of sample needed (less than 200 nL) (Bellando et al., 2018).

SELEX, which stands for systemic evolution of ligands by exponential enrichment, is an in vitro procedure, employed for the selection of aptamers. Aptamers are short sequences of DNA or RNA oligonucleotides that possess the ability to bind to specific targets. These biomolecules are an excellent choice for target analyte detection because of their benefits of greater selectivity, sensitivity, thermal stability, and ease of modification (Sharma et al., 2022; Song et al., 2008). Wang et al. (2022) developed a wearable sensor integrated with aptamer-FETs on a flexible polyimide substrate for cortisol monitoring. The sensor was designed on a thin-film microfluidic device using tape-based fabrication techniques incorporating materials to enable skin-adherent biofluid sampling, routing, and analysis. By monitoring cortisol levels in individuals during a trier social stress test (TSST) and establishing correlations between cortisol in diurnal sweat and saliva samples, the study confirmed the potential utility of cortisol-aptamer-FET sensors for stress detection.

Recently, a new technique that combines aptamer-cross-linked hydrogel with glucoamylase trapping for molecular recognition cascaded enzymatic processes for signal amplification and a PAD for portable readout was demonstrated by Tian et al. (2016). In this design, the hydrogel breaks down when the target is introduced, releasing glucoamylase, which catalyzes the hydrolysis of amylose to yield a significant amount of glucose. By simply folding the PAD, the sample solution containing the glucose product wicks and diffuses parallel to each test zone to conduct homogeneous assays. Using glucose as the substrate, multiple enzymatic and chemical cascade reactions use glucose to produce molecular iodine (I_2), which is responsible for the generation of brown color. A semiquantitative assay is accomplished by the naked eye through color gradient changes based on varying concentrations of the target, and quantitation can be performed using handheld instruments. The effectiveness of the hydrogel-PAD technology was demonstrated by the detection of cocaine in the buffer and urine. More significantly, by adding the matching aptamer to the hydrogel, the hydrogel-PAD system can be expanded to detect a variety of targets. The hydrogel-PAD system described here offers a new platform for the optical detection of a variety of targets that is portable, disposable, and flexible.

8.6.4 Other microfluidic platforms

He et al. (2013) worked on a label-free microfluidic integrated POC device for the electrochemical detection of vasopressin in serum based on aptamer technique. The design and development of the device consist of the process of the immobilization of a bioreceptor (e.g., aptamer) onto preimmobilized

CNTs positioned between lithographically patterned gold electrodes with a distance of 8−10 μm on a silicon surface. This configuration was integrated with a microchannel pattern polydimethylsiloxane (PDMS) base. The sensor response measured in terms of conductivity undergoes change when vasopressin molecules in solution are selectively captured by aptamers immobilized on carbon nanotubes. The binding interaction between the analyte of interest and bioreceptor leads to a decrease in current flowing through sensors as it leads to a reduction of a number of electron carriers on the CNT surface and lowers the local charge field. The downsizing of the device enables the detection of biomarkers from small volumes of bodily fluids on a portable platform. This means that even minute amounts of bodily fluids can be analyzed for the presence of specific biomarkers. The device's portability allows for convenient and rapid testing outside of a laboratory setting. This device achieved an impressive detection limit of 43 pM. Additionally, Shin et al. (2016) developed an aptamer-based electrochemical LOC combined with a microchannel-based platform for in-line detection of secreted protein biomarkers (Fig. 8.7). The sensor was modified with aptamers targeting the cardiac damage biomarker CK-MB. The demonstration of the aptamer-based

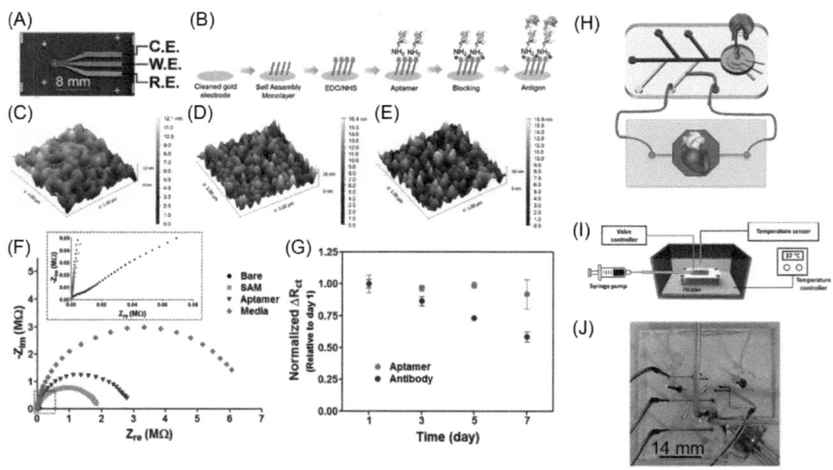

FIGURE 8.7 (A) Image of microfabricated electrode design; (B) step-by-step process flow of immobilization of aptamers on microelectrode surface; AFM of surface morphology studies (C) bare electrode, (D) antibody-, and (E) microelectrode fabricated with aptamer. (F) EIS measurements of microelectrode surface after each step of modification; (G) normalized ΔR_{ct} versus time response showing performance of antibody and aptamer biosensors; (H) heart-on-chip platform; (I) drug injection system that is linked to the microfluidic heart-on-a-chip bioreactor; (J) image of a combination of microelectrode system with microfluidic bioreactor. *AFM*, Atomic force microscopy; *EIS*, Electrochemical impedance spectroscopy. *Figure reproduced with permission from Shin, S. R., et al. (2016). Aptamer-based microfluidic electrochemical biosensor for monitoring cell-secreted trace cardiac biomarkers. Analytical Chemistry, 88, 10019–10027.*

electrochemical immunosensors was also assessed in conjunction with a microfluidic heart-on-a-chip device comprising cardiomyocytes derived from human embryonic stem cells (ESC-CMs). The aptamer-based sensors were employed to record minute variations in secreted CK-MB levels by the cardiac tissues in response to drug, specifically the medication doxorubicin (DOX), in a dose-dependent manner. The changes in CK-MB levels were compared with other tissue and organoid evaluation methods such as beating rates and cellular viability. The study found that the release of CK-MB antigens increased when cardiomyocytes were exposed to doxorubicin, leading to altered beating rates. A live/dead assay was used to assess the vitality of the cardiomyocytes, and it was observed that cell viability decreased over a period in the drug-treated cells, particularly when only the drug was used. These findings supported the hypothesis that doxorubicin can cause cardiotoxicity in an in vitro cardiac model, as detected by microfluidic biosensors. Overall, the developed sensing approach using the aptamer-based electrochemical biosensor and microfluidic organ-on-a-chip device offers flexibility and the potential to detect various cell-secreted biomarkers.

Very recently, the development of a fluorescent biosensor for the detection of amyloid protein oligomer (AO) as an indicator of Alzheimer's disease (AD) in diluted human samples was reported (Wen et al., 2022). In the study mentioned, three-dimensional (3D) Fe_3O_4/MXene nanospheres were synthesized and utilized as quenchers of Forster resonance energy transfer (FRET) for the detection of amyloid-β oligomers (AβO) (Fig. 8.8). The fluorescent reporter group used in this study is carboxyfluorescein (FAM), which emits fluorescence upon excitation at a specific wavelength. An aptamer modified with FAM referred to as FAM-AptAβO, was employed as the fluorescent probe. However, upon the addition of AβO, the quenched fluorescence is restored. Different concentrations of AβO lead to distinct levels of fluorescence restoration. Therefore the concentration of AβO can be detected by monitoring the intensity of the restored fluorescence. To facilitate the detection process, microfluidic chips based on PDMS are utilized as detection platforms (Fig. 8.8B–D). These microfluidic chips offer several advantages, including their small physical size, low sample requirements, and high throughput capabilities. These features make the PDMS-based microfluidic chips well suited for efficient and accurate detection of AβO in the study. The fluorescence intensity and logarithm of the AβO concentration in the 0.10–200 nM range exhibit an excellent linear connection in the Fe_3O_4/MXene-based nanosphere-based biosensor exhibited a detection limit of 50.0 pM with a sample volume of 4.50 μL. Nanoplasmonic biosensors have emerged as viable options because quick and high-sensitivity biosensing technologies are greatly desired for the creation of point-of-care testing (POCT) devices. Gold nanoparticles (AuNPs) are the sensing material used in the majority of nanoplasmonic biosensors. Localized surface plasmon

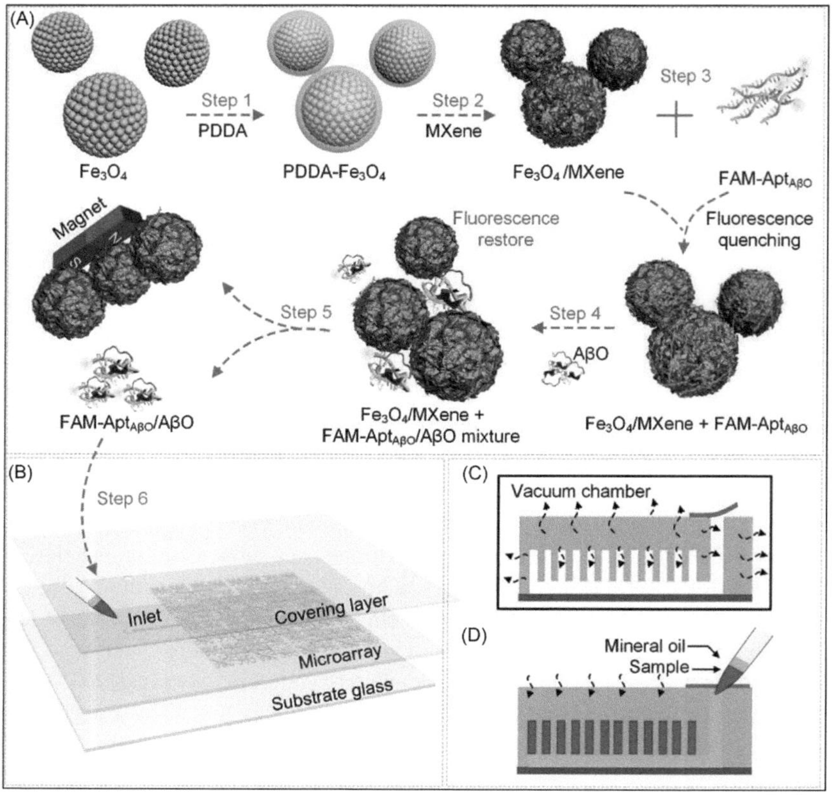

FIGURE 8.8 Schematic illustration of detection processes: (A) FRET-based AβO detection platform; (B) microfluidic chip with a sandwich-like structure; (C) predegassing of the microfluidic chip; (D) loading process of the sample and mineral oil. *Figure reproduced with permission from Wen, X. H. et al. (2022). Fe_3O_4/MXene nanosphere-based microfluidic chip for the accurate diagnosis of Alzheimer's disease. ACS Applied Nano Materials,* 2022, *15925–15933.*

resonance occurs when AuNPs interact with incoming light of a specific frequency. The nanoparticle's extinction spectrum serves to identify particle plasmon resonance (PPR). Even when there is just a little change in the local refractive index (RI) on the nanoparticle surface, the resonance frequency and peak absorbance of the PPR band alter significantly. Based on this phenomenon, a fiber optic particle plasmon resonance (FOPPR) biosensor that used a DNA aptamer as the recognition element for SARS-CoV-2 N-protein in spiked nasopharyngeal swabs was developed by Chang et al. (2022). The sensor meets the acceptable LOD of 106 copies/mL defined by the WHO target product profile (COVID-19) and can detect and quantify the N-protein in 15 minutes with an LOD of 2.8 nM.

8.7 Current challenges and research

It is necessary to note that there are now some important areas of emphasis for LOC research:

1. LOC technologies need to be industrialized to be ready for commercial use. This covers the modification of manufacturing procedures, the design of certain surface treatments, flow control systems, etc. To enable large-scale fabrication, straightforward, inexpensive fabrication techniques are needed. Complex multilayer structures are produced using soft lithography. However, it is a labor-intensive method that involves physical labor. Ink-jet printing and 3D printing are two printing methods that can be used instead of commercially viable microfluidic manufacturing.
2. The advancement lies in the ability to integrate a larger repertoire of biological reactions onto a single chip and parallelize processes, thereby facilitating the detection of a number of pathogens within a single microfluidic cartridge. Fundamental research is required for certain technologies to be applicable, such as DNA reading with nanopores, which has great potential.
3. The mobility and wearability can be further enhanced by improved power sources, better meeting the needs of self-administered personal healthcare. The effectiveness of the NFC technology has been demonstrated for wireless power supply, data transmission, and real-time display. Integration with self-powered sensing devices (based, for example, on triboelectric and piezoelectric effects) will make the system simpler and remove transmission distance restrictions.
4. Intelligent wearable microfluidic sensors are anticipated to give exact assessments of health state as well as intelligent feedback, such as medicine delivery when combined with big data technology. The smart system's foundation of data collecting, storage, and processing eliminates the need for professional staff, enabling its widespread application in households.

For specialized uses like glucose monitoring or pathology diagnosis, LOC have already seen some commercial success. The widespread usage of LOC in hospitals worldwide and eventually in doctor's offices is something we may anticipate in the near future. LOC technologies are anticipated to eventually provide at-home real-time health monitoring. Because it is increasingly obvious that these technologies will alter our daily lives, conventional universities, research institutions, and businesses are spending more and more in LOC.

8.8 Conclusions and perspectives

The engineering challenges of developing next-generation personal healthcare devices, medicinal therapies, and personal electronics can be addressed by integrating existing biosensors and energy technologies into flexible electronics and microfluidics. Work still needs to be done in the fields of soft materials science, sensor technologies, and advanced system design before wearable microfluidic sensors are extensively commercialized; we can now be certain that LOC will transform how we perform diagnostics in the near future based on current research and products that have entered the market. The difficulty for industry research will be to combine as many distinct activities as possible on a single LOC to reduce costs, improve ergonomics, and accelerate diagnostics. Given the absence of unified technologies, it is presently difficult to forecast the most effective materials and technologies for high-throughput diagnostics. Determining suitable solutions will rely on technological potential while also considering economic and industrial factors. Moreover, it is important to evaluate how these solutions align with existing systems such as silicon micromachining to leverage synergies and optimize overall effectiveness.

References

Abe, K., Kotera, K., Suzuki, K., & Citterio, D. (2010). Inkjet-printed paperfluidic immunochemical sensing device. *Analytical and Bioanalytical Chemistry*, *398*, 885–893.

Agustini, D., Caetano, F. R., Quero, R. F., Fracassi da Silva, J. A., Bergamini, M. F., & Marcolino-Junior, L.H., & Pereira de Jesus, D. (2021). Microfluidic devices based on textile threads for analytical applications: state of the art and prospects. *Analytical Methods*, *13* (41), 4830–4857. Available from https://doi.org/10.1039/d1ay01337h.

Ahn, C. H., & Choi, J.-W. (2010). *Microfluidic devices and their applications to lab-on-a-chip. Springer handbook of nanotechnology* (pp. 503–530). Springer. Available from https://doi.org/10.1007/978-3-642-02525-9_18.

Ainla, A., Hamedi, M. M., Guder, F., & Whitesides, G. M. (2017). Electrical textile valves for paper microfluidics. *Advanced Materials*, *29*(38), 1702894. Available from https://doi.org/10.1002/adma.201702894.

Akyazi, T., Basabe-Desmonts, L., & Benito-Lopez, F. (2018). Review on microfluidic paper-based analytical devices towards commercialisation. *Analytica Chimica Acta*, *1001*, 1–17.

Arakawa, T., Kuroki, Y., Nitta, H., Chouhan, P., Toma, K., & Sawada, S. I...., Mitsobayashi, K.. (2016). Mouthguard biosensor with telemetry system for monitoring of saliva glucose: A novel cavitas sensor. *Biosensors & Bioelectronics*, *84*, 106–111. Available from https://doi.org/10.1016/j.bios.2015.12.014.

Bae, C. W., Toi, P. T., Kim, B. Y., Lee, W. I., Lee, H. B., & Hanif, A...., Lee, N.E.. (2019). Fully stretchable capillary microfluidics-integrated nanoporous gold electrochemical sensor for wearable continuous glucose monitoring. *ACS Applied Materials Interfaces*, *11*(16), 14567–14575. Available from https://doi.org/10.1021/acsami.9b00848.

Bandodkar, A. J., Gutruf, P., Choi, J., Lee, K., Sekine, Y., & Reeder, J. T...., Rogers, J.A. (2019). Battery-free, skin-interfaced microfluidic/electronic systems for simultaneous electrochemical, colorimetric, and volumetric analysis of sweat. *Science Advances*, *5*(1), eaav3294.. Available from https://doi.org/10.1126/sciadv.aav3294.

Bellando, F., Zhang, J., Wildhaber, F., Longo, J., Guerin, H., & Ionescu, A. M. (2018). Three-dimensional integrated ultra-low-volume passive microfluidics with ion-sensitive field-effect transistors for multiparameter wearable sweat analyzers. *ACS Nano*, *12*(12), 12646–12656. Available from https://doi.org/10.1021/acsnano.8b07413.

Berlanda, S. F., Breitfeld, M., Dietsche, C. L., & Dittrich, P. S. (2021). Recent advances in microfluidic technology for bioanalysis and diagnostics. *Analytical Chemistry*, *93*, 311–331.

Berthier, E., Young, E. W. K., & Beebe, D. (2012). Engineers are from PDMS-land, biologists are from polystyrenia. *Lab on a Chip*, *12*, 1224–1237.

Bhatia, S. N., & Ingber, D. E. (2014). Microfluidic organs-on-chips. *Nature Biotechnology*, *32*, 760–772.

Bodas, D., & Khan-Malek, C. (2006). Formation of more stable hydrophilic surfaces of PDMS by plasma and chemical treatments. *Microelectronics Engineering*, *83*, 1277–1279.

Broughton, J. P., Deng, X., Yu, G., Fasching, C. L., Servellita, V., & Singh, J..., Chiu, C.Y. (2020). CRISPR–Cas12-based detection of SARS-CoV-2. *Nature Biotechnology*, *38*(7), 870–874. Available from https://doi.org/10.1038/s41587-020-0513-4.

Bulaj, G. (2005). Formation of disulfide bonds in proteins and peptides. *Biotechnology Advances*, *23*, 87–92.

Campbell, A. S., Kim, J., & Wang, J. (2018). Wearable electrochemical alcohol biosensors. *Current Opinion in Electrochemistry*, *10*, 126–135.

Caplin, J. D., Granados, N. G., James, M. R., Montazami, R., & Hashemi, N. (2015). Microfluidic organ-on-a-chip technology for advancement of drug development and toxicology. *Advanced Healthcare Materials*, *4*, 1426–1450.

Cazalé, A., Sant, W., Ginot, F., Launay, J. C., Savourey, G., & Revol-Cavalier, F...., Temple-Boyer, P.. (2016). Physiological stress monitoring using sodium ion potentiometric microsensors for sweat analysis. *Sensors and Actuators. B, Chemical*, *225*, 1–9. Available from https://doi.org/10.1016/j.snb.2015.10.114.

Chang, T.-C., Sun, A. Y., Huang, Y. C., Wang, C. H., Wang, S. C., & Chau, L. K. (2022). Integration of power-free and self-contained microfluidic chip with fiber optic particle plasmon resonance aptasensor for rapid detection of SARS-CoV-2 nucleocapsid protein. *Biosensors*, *12*(10), 785. Available from https://doi.org/10.3390/bios12100785.

COVID-19. *Target product profiles for priority diagnostics to support response to the COVID-19 pandemic v.1.0.* <https://www.who.int/publications/m/item/covid-19-target-product-profiles-for-priority-diagnostics-to-support-response-to-the-covid-19-pandemic-v0.0.1/>.

Curto, V. F., Coyle, S., Byrne, R., Angelov, N., Diamond, D., & Benito-Lopez, F. (2012). Concept and development of an autonomous wearable micro-fluidic platform for real time pH sweat analysis. *Sensors and Actuators B*, *175*, 263–270. Available from https://doi.org/10.1016/j.snb.2012.02.010.

Deliorman, M., Janahi, F. K., Sukumar, P., Glia, A., Alnemari, R., & Fadl, S...., Qasaimeh M.A.. (2020). AFM-compatible microfluidic platform for affinity-based capture and nanomechanical characterization of circulating tumor cells. *Microsystems & Nanoengineering*, *6*(1), 1–15. Available from https://doi.org/10.1038/s41378-020-0131-9.

DeMello, A. J. (2006). Control and detection of chemical reactions in microfluidic systems. *Nature*, *442*, 394–402.

Duval, D., Dante, S., González, A.B., Fernández, L.J. & Lechuga, L.M. (2012). Silicon photonics-based nanobiosensors for lab-on-a-chip integration. In Latin America *o*ptics and *p*hotonics *c*onference, paper LT3B.1. Available from https://doi.org/10.1364/LAOP.2012.LT3B.1.

Esch, E. W., Bahinski, A., & Huh, D. (2015). Organs-on-chips at the frontiers of drug discovery. *Nature Reviews. Drug Discovery*, *14*, 248–260.

Ferrari, E., Palma, C., Vesentini, S., Occhetta, P., & Rasponi, M. (2020). Integrating biosensors in organs-on-chip devices: A perspective on current strategies to monitor microphysiological systems. *Biosensors (Basel)*, *10*.

Gonzálezgonźgonzález-Guerrero, A. B., Maldonado, J., Herranz, S., & Lechuga, L. M. (2016). Trends in photonic lab-on-chip interferometric biosensors for point-of-care diagnostics. *Biosensors*, *8*(48), 8380–8394. Available from https://doi.org/10.1039/c6ay02972h.

Han, J., Tan, M., Sudheendra, L., Weiss, R. H., & Kennedy, I. M. (2014). On-chip detection of a single nucleotide polymorphism without polymerase amplification. *Nano Research*, *7*, 1302.

He, P., Oncescu, V., Lee, S., Choi, I., & Erickson, D. (2013). Label-free electrochemical monitoring of vasopressin in aptamer-based microfluidic biosensors. *Analytica Chimica Acta*, *759*, 74–80.

Heikenfeld, J. (2016). Non-invasive analyte access and sensing through eccrine sweat: Challenges and outlook circa 2016. *Electroanalysis*, *28*, 1242–1249.

Hisamoto, H., et al. (1999). New technology for modern chemistry. *Japan Society of Applied Physics*, *72*, 29.

Hussein, H. A., Hassan, R. Y. A., Chino, M., & Febbraio, F. (2020). Point-of-care diagnostics of COVID-19: From current work to future perspectives. *Sensors*, *20*, 4289.

Islam, G. N., Ali, A., & Collie, S. (2020). Textile sensors for wearable applications: a comprehensive review. *Cellulose*, *27*, 6103–6131. Available from https://doi.org/10.1007/s10570-020-03215-5.

Jane Maxwell, E., Mazzeo, A. D., & Whitesides, G. M. (2013). Paper-based electroanalytical devices for accessible diagnostic testing. *MRS Bulletin/Materials Research Society*, *38*, 309–314.

Jonkheijm, P., Weinrich, D., Schröder, H., Niemeyer, C. M., & Waldmann, H. (2008). Chemical strategies for generating protein biochips. *Angewandte Chemie International Edition*, *47*, 9618–9647.

Kassal, P., Kim, J., Kumar, R., de Araujo, W. R., Steinberg, I. M., Steinberg, M. D., & Wang, J. (2015). Smart bandage with wireless connectivity for uric acid biosensing as an indicator of wound status. *Electrochemistry Communications*, *56*, 6–10. Available from https://doi.org/10.1016/j.elecom.2015.03.018.

Kim, D., & Herr, A. E. (2013). Protein immobilization techniques for microfluidic assays. *Biomicrofluidics*, *7*, 041501.

Kim, J., Campbell, A. S., de Ávila, B. E. F., & Wang, J. (2019). Wearable biosensors for health-care monitoring. *Nature Biotechnology*, *37*, 389–406.

Kitsara, M., & Ducree, J. (2013). Integration of functional materials and surface modification for polymeric microfluidic systems. *Journal of Micromechanics and Microengineering*, *23*(3). Available from https://doi.org/10.1088/0960-1317/23/3/033001.

Kristiansen, P. M., Karpik, A., Werder, J., Guilherme, M., & Grob, M. (2022). Thermoplastic microfluidics. *Methods in Molecular Biology*, *2373*, 39–55.

Kumar, M., Muniswamy, V., Guha, K., Iannacci, J., & Krishnaswamy, N. (2021). Analysis of integrated silicon nitride lab-on-a-chip optofluidic sensor at visible wavelength for absorbance based biosensing applications. *Microsystem Technologies*, *27*, 3499–3506.

Lee, S. I., et al. (2016). Objectively quantifying walking ability in degenerative spinal disorder patients using sensor equipped smart shoes. *Medical Engineering & Physics*, *38*, 442–449.

Li, L., et al. (2015). A nanostructured conductive hydrogels-based biosensor platform for human metabolite detection. *Nano Letters*, *25*, 36.

Li, X., Ballerini, D. R., & Shen, W. (2012). A perspective on paper-based microfluidics: Current status and future trends. *Biomicrofluidics*, *6*.

Li, S., Ma, Z., Cao, Z., Pan, L., & Shi, Y. (2019). Advanced wearable microfluidic sensors for healthcare monitoring. *Small*, *16*(9), 903822. (1 of 15). Available from https://doi.org/10.1002/smll.201903822.

Li, X., & Tian, T. (2018). Recent advances in an organ-on-a-chip: Biomarker analysis and applications. *Analytical Methods*, *10*, 3122–3130.

Li, X., Tian, J., Nguyen, T., & Shen, W. (2008). Paper-based microfluidic devices by plasma treatment. *Analytical Chemistry*, *80*, 9131–9134.

Li, Z., et al. (2020). Development and clinical application of a rapid IgM-IgG combined antibody test for SARS-CoV-2 infection diagnosis. *Journal of Medical Virology*, *92*, 1518–1524.

Lim, Y. C., Kouzani, A. Z., & Duan, W. (2010). Lab-on-a-chip: A component view. *Microsystem Technologies*, *16*, 1995–2015. Available from https://doi.org/10.1007/s00542-010-1141-6, Preprint at.

Lisowski, P., & Zarzycki, P. K. (2013). Microfluidic paper-based analytical devices (lPADs) and micro total analysis systems (lTAS): Development, applications and future trends. *Chromatographia*, *76* (19), 1201–1214. Available from https://doi.org/10.1007/s10337-013-2413-y.

Liu, H., & Crooks, R. M. (2011). Three-dimensional paper microfluidic devices assembled using the principles of origami. *Journal of the American Chemical Society*, *133*, 17564–17566.

Liu, K., & Fan, Z. H. (2011). Thermoplastic microfluidic devices and their applications in protein and DNA analysis. *Analyst*, *136*, 1288.

Liébana, S., & Drago, G. A. (2016). Bioconjugation and stabilisation of biomolecules in biosensors. *Essays in Biochemistry*, *60*, 59.

Luo, H., & Gao, B. (2021). Development of smart wearable sensors for life healthcare. *Engineered Regeneration*, *2*, 163–170.

Majeed, B., et al. (2012). Silicon based system for single-nucleotide-polymorphism detection: Chip fabrication and thermal characterization of polymerase chain reaction microchamber. *Japanese Journal of Applied Physics*, *51*, 04DL01.

Martinez, A. W., Phillips, S. T., Butte, M. J., & Whitesides, G. M. (2007). Patterned paper as a platform for inexpensive, low-volume, portable bioassays. *Angewandte Chemie (International Ed. in English)*, *46*, 1318–1320.

Majumder, S., Mondal, T., & Deen, M. J. (2017). Wearable sensors for remote health monitoring. *Sensors*, *17*(1), 130. Available from https://doi.org/10.3390/s17010130.

Martín, A., et al. (2017). Epidermal microfluidic electrochemical detection system: Enhanced sweat sampling and metabolite detection. *ACS Sensors*, *2*, 1860–1868.

Mishra, G. K., Sharma, A., & Bhand, S. (2015). Ultrasensitive detection of streptomycin using flow injection analysis-electrochemical quartz crystal nanobalance (FIA-EQCN) biosensor. *Biosensors & Bioelectronics*, *67*.

Mishra, G. K., Sharma, A., Deshpande, K., & Bhand, S. (2014). Flow injection analysis biosensor for urea analysis in urine using enzyme thermistor. *Applied Biochemistry and Biotechnology*, *174*.

Morris, D., et al. (2009). Bio-sensing textile based patch with integrated optical detection system for sweat monitoring. *Sensors and Actuators B*, *139*, 231–236.

Mostafalu, P., et al. (2016). A toolkit of thread-based microfluidics, sensors, and electronics for 3D tissue embedding for medical diagnostics. *Microsystems & Nanoengineering*, *2*, 16039.

Müller, R. H., & Clegg, D. L. (1949). Automatic paper chromatography. *Analytical Chemistry*, *21*(9), 1123–1125. Available from https://doi.org/10.1021/ac60033a032.

Nakata, S., Arie, T., Akita, S., & Takei, K. (2017). Wearable, flexible, and multifunctional healthcare device with an ISFET chemical sensor for simultaneous sweat pH and skin temperature monitoring. *ACS Sensors*, *2*, 443–448.

Neužil, P., Campos, C. D. M., Wong, C. C., Soon, J. B. W., Reboud, J., & Manz, A. (2014). From chip-in-a-lab to lab-on-a-chip: towards a single handheld electronic system for multiple application-specific lab-on-a-chip (ASLOC). *Lab on a Chip*, *16*, 2093.

Nge, P. N., Rogers, C. I., & Woolley, A. T. (2013). Advances in microfluidic materials, functions, integration, and applications. *Chemical Reviews, 113*(4), 2550–2583. Available from https://doi.org/10.1021/cr300337x.

Nguyen, H. H., Lee, S. H., Lee, U. J., Fermin, C. D., & Kim, M. (2019). Immobilized enzymes in biosensor applications. *Materials, 12*.

Nimse, S. B., Song, K., Sonawane, M. D., Sayyed, D. R., & Kim, T. (2014). Immobilization techniques for microarray: Challenges and applications. *Sensors, 14*, 22208–22229.

Nishat, S., Jafry, A. T., Martinez, A. W., & Awan, F. R. (2021). Paper-based microfluidics: Simplified fabrication and assay methods. *Sensors and Actuators. B, Chemical, 336*, 129681.

Nomura, K. I., Horii, Y., Kanazawa, S., Kusaka, Y., & Ushijima, H. (2018). Fabrication of a textile-based wearable blood leakage sensor using screen-offset printing. *Sensors, 18*(1), 240. Available from https://doi.org/10.3390/s18010240.

Pandey, C. M., et al. (2018). Microfluidics based point-of-care diagnostics. *Biotechnology Journal, 13*.

Prieto-Simon, B., Campas, M., & Marty, J.-L. (2008). Biomolecule immobilization in biosensor development: Tailored strategies based on affinity interactions. *Protein and Peptide Letters, 15*, 757–763.

de Puig, H., Bosch, I., Gehrke, L., & Hamad-Schifferli, K. (2017). Challenges of the nano–bio interface in lateral flow and dipstick immunoassays. *Trends in Biotechnology, 35*, 1169–1180.

Qamar, A. Z., & Shamsi, M. H. (2020). Desktop fabrication of lab-on-chip devices on flexible substrates: A brief review. *Micromachines, 11*, 126.

Rattle, S., Hofmann, O., Price, C. P., Kricka, L. J., & Wild, D. (2013). Lab-on-a-chip, micro- and nanoscale immunoassay systems, and microarrays. *The Immunoassay Handbook, 175*. Available from https://doi.org/10.1016/B978-0-08-097037-0.00013-0.

Reches, M., Mirica, K. A., Dasgupta, R., Dickey, M. D., Butte, M. J., & Whitesides, G. M. (2010). Thread as a matrix for biomedical assays. *ACS Applied Materials & Interfaces, 2*(6), 1722–1728. Available from https://doi.org/10.1021/am1002266.

Roman, G. T., Hlaus, T., Bass, K. J., Seelhammer, T. G., & Culbertson, C. T. (2002). Sol-gel modified poly(dimethylsiloxane) microfluidic devices with high electroosmotic mobilities and hydrophilic channel wall Characteristics. *Analytical Chemistry, 74*, 1414–1422.

Sachdeva, S., Davis, R. W., & Saha, A. K. (2020). Microfluidic point-of-care testing: Commercial landscape and future directions. *Frontiers in Bioengineering and Biotechnology, 8*, 602659. Available from https://doi.org/10.3389/fbioe.2020.602659, http://www.frontiersin.org.

Salva, M. L., Rocca, M., Niemeyer, C. M., & Delamarche, E. (2021). Methods for immobilizing receptors in microfluidic devices: A review. *Micro and Nano Engineering, 11*. Available from https://doi.org/10.1016/j.mne.2021.100085, Preprint at.

Scott, S. M., & Ali, Z. (2021). Fabrication methods for microfluidic devices: An overview. *Micromachines (Basel), 12*.

Sempionatto, J. R., Martin, A., Garcia-Carmona, L., Barfidokht, A., Kurniawan, J. F., & Moreto, J. R...., Wang, J.. (2019). Skin-worn soft microfluidic potentiometric detection system. *Electroanalysis, 31*(2), 239–245. Available from https://doi.org/10.1002/elan.201800414.

Sharma, A., Istamboulie, G., Hayat, A., Catanante, G., Bhand, S., & Marty, J. L. (2017). Disposable and portable aptamer functionalized impedimetric sensor for detection of kanamycin residue in milk sample. *Sensors and Actuators. B, Chemical, 245*. Available from https://doi.org/10.1016/j.snb.2017.02.002.

Sharma, A., Mishra, R. K., & Marty, J. L. (2022). Aptamer based sensing approaches towards food and biomedical applications. *Biosensors: Fundamentals, Emerging Technologies, and Applications*, 289–312. Available from https://doi.org/10.1201/9781003189435-16.

Shin, S. R., Zhang, Y. S., Kim, D. J., Manbohi, A., Avci, H., & Silvestri, A...., Khademhosseini, A.. (2016). Aptamer-based microfluidic electrochemical biosensor for monitoring cell-secreted trace cardiac biomarkers. *Analytical Chemistry*, *88*(20), 10019–10027. Available from https://doi.org/10.1021/acs.analchem.6b02028.

Soft Lithography – Xia. (1998). Angewandte Chemie International Edition - Wiley Online Library. https://onlinelibrary.wiley.com/doi/10.1002/(SICI)1521–3773(19980316)37:5%3C550::AID-ANIE550%3E3.0.CO;2-G.

Song, S., Wang, L., Li, J., Fan, C., & Zhao, J. (2008). Aptamer-based biosensors. *TrAC Trends in Analytical Chemistry*, *27*, 108–117.

Steinberg, M. D., Kassal, P., & Steinberg, I. M. (2016). System architectures in wearable electrochemical sensors. *Electroanalysis*, *28*, 1149–1169.

Sui, G., Wang, J., Lee, C. C., Lu, W., Lee, S. P., & Leyton, J. V...., Tseng, H. R. (2006). Solution-phase surface modification in intact poly(dimethylsiloxane) microfluidic channels. *Analytical Chemistry*, *78*(15), 5543–5551. Available from https://doi.org/10.1021/ac060605z.

Syedmoradi, L., Daneshpour, M., Alvandipour, M., Gomez, F. A., Hajghassem, H., & Omidfar, K. (2017). Point of care testing: The impact of nanotechnology. *Biosensors & Bioelectronics*, *87*, 373–387. Available from https://doi.org/10.1016/j.bios.2016.08.084.

Temiz, Y., Lovchik, R. D., Kaigala, G. v, & Delamarche, E. (2015). Lab-on-a-chip devices: How to close and plug the lab? *Microelectron Eng*, *132*, 156–175.

Tian, T., Wei, X., Jia, S., Zhang, R., Li, J., Zhu, Z., Zhang, H., Ma, Y., Lin, Z., & Yang, C. J. (2016). Integration of target responsive hydrogel with cascaded enzymatic reactions and microfluidic paper-based analytic devices (μPADs) for point-of-care testing (POCT). *Biosensors and Bioelectronics*, *77*, 537–542.doi. Available from https://doi.org/10.1016/j.bios.2015.09.049.

Tsao, C. W. (2016). Polymer microfluidics: Simple, low-cost fabrication process bridging academic lab research to commercialized production. *Micromachines (Basel)*, *7*.

Wang, B., Zhao, C., Wang, Z., Yang, K. A., Cheng, X., & Liu, W...., Emaminejad, S.. (2022). Applied sciences and engineering wearable aptamer-field-effect transistor sensing system for noninvasive cortisol monitoring. *Science Advances*, *8*(1), 967.

Wang, D., Oleschuk, R. D., & Horton, J. H. (2008). Surface modification of poly(dimethylsiloxane) with a perfluorinated alkoxysilane for selectivity toward fluorous tagged peptides. *Langmuir: the ACS Journal of Surfaces and Colloids*, *24*, 1080–1086.

Welch, N. G., Scoble, J. A., Muir, B. W., & Pigram, P. J. (2017). Orientation and characterization of immobilized antibodies for improved immunoassays (Review). *Biointerphases*, *12*, 02D301.

Wen, X. H., Zhao, X. F., Wang, X. H., Wang, Y., Guo, J. C., & Zhou, H. G...., Lu, H.L.. (2022). Fe3O4/MXene nanosphere-based microfluidic chip for the accurate diagnosis of Alzheimer's disease. *ACS Applied Nano Materials*, *2022*(10), 15925–15933. Available from https://doi.org/10.1021/acsanm.2c04187.

Wu, H., Tian, Y., Liu, B., Lu, H., Wang, X., & Zhai, J....., Wang, H.. (2004). Titania and alumina sol-gel-derived microfluidics enzymatic-reactors for peptide mapping: Design, characterization, and performance. *Journal of Proteome Research*, *3*(6), 1201–1209. Available from https://doi.org/10.1021/pr049889z.

Wu, J., Dong, M., Rigatto, C., Liu, Y., & Lin, F. (2018). Lab-on-chip technology for chronic disease diagnosis. *npj Digital Medicine*, *1*, 1–11.

Xiao, G., He, J., Qiao, Y., Wang, F., Xia, Q., & Wang, X...., Li, C.M.. (2020). Facile and low-cost fabrication of a thread/paper-based wearable system for simultaneous detection of lactate and pH in human sweat. *Advanced Fiber Materials*, *2*, 265–278. Available from https://doi.org/10.1007/s42765-020-00046-8.

Yang, A., Li, Y., Yang, C., Fu, Y., Wang, N., Li, L., & Yan, F. (2018). Fabric organic electrochemical transistors for biosensors. *Advanced Materials*, *30*(23), 1800051. Available from https://doi.org/10.1002/adma.201800051.

Ye, S., Feng, S., Huang, L., & Bian, S. (2020). Recent progress in wearable biosensors: From healthcare monitoring to sports analytics. *Biosensors*, *10*(12), 205. Available from https://doi.org/10.3390/bios10120205.

Yetisen, A. K., Akram, M. S., & Lowe, C. R. (2013). *Paper-based microfluidic point-of-care diagnostic devices. Lab on a Chip*, *13*, 2210–2251.

Zhang, C., & Xing, D. (2007). Miniaturized PCR chips for nucleic acid amplification and analysis: latest advances and future trends. *Nucleic Acids Research*, *35*, 4223.

Zhao, Y., Dong, D., Wang, Y., Gong, S., An, T., Yap, L. W., & Cheng, W. (2018). Highly stretchable fiber-shaped supercapacitors based on ultrathin gold nanowires with double-helix winding design. *ACS Applied Materials & Interfaces*, *10*(49). Available from https://doi.org/10.1021/acsami.8b14769.

Zhao, Y., Zhai, Q., Dong, D., An., T., Gong, S., Shi, Q., & Cheng, W. (2019). Highly stretchable and strain-insensitive fiber-based wearable electrochemical biosensor to monitor glucose in the sweat. *Analytical Chemistry*, *91*(10), 34. Available from https://doi.org/10.1021/acs.analchem.9b00152.

Zhao, Z., Li, Q., Chen, L., Zhao, Y., Gong, J., Li, Z., & Zhang, J. (2021). A thread/fabric-based band as a flexible and wearable microfluidic device for sweat sensing and monitoring. *Lab on a Chip*, *21*(5), 916. Available from https://doi.org/10.1039/D0LC01075H.

Chapter 9

Wearable flexible biosensing devices contributing to modern precision medicine

Itthipon Jeerapan[1,2,3] and Suntisak Khumngern[1,2]
[1]Center of Excellence for Trace Analysis and Biosensor, Prince of Songkla University, Hat Yai, Songkhla, Thailand, [2]Division of Physical Science, Faculty of Science, Prince of Songkla University, Hat Yai, Songkhla, Thailand, [3]Center of Excellence for Innovation in Chemistry, Faculty of Science, Prince of Songkla University, Hat Yai, Songkhla, Thailand

9.1 Introduction

The concept of wearable sensors that are conveniently attached to the human body involves devices generating output signals such as electrochemical signals in the form of current, voltage, or impedance, to determine physical parameters and bio/chemical concentrations (Ghaffari et al., 2021; Promphet et al., 2021). This idea is now expanding to diverse applications, such as sports and biomedical diagnosis (Awotunde et al., 2021; Xuan et al., 2022). Wearable flexible sensors can be a key factor in advancing precision medicine as they allow wearers to track their lifestyle and environment, representing a revolutionary strategy in diagnosing and determining precise therapeutic options by taking into account lifestyle and environmental factors (Liu, Ramli, et al., 2022; Luo & Gao, 2021).

Precision medicine takes into account individual variations in genes, environments, and lifestyles when treating and preventing diseases or abnormal health status (Ashley, 2016; Chen, Zhou, et al., 2022; Ho et al., 2020; Jeong et al., 2019). The concept of precision medicine involves tailoring medical decisions, therapies, practices, or pharmaceutical drugs to small subgroups of patients or even to individual patients, rather than using a one-size-fits-all approach. By using personalized prevention and treatment, therapies are selected based on individual contents, biochemical compositions, and behavioral factors. This approach is a departure from traditional models that treat all patients the same way. Medical professionals can determine the right treatment for an individual based on their body, daily life, and surroundings by using smart wearable sensors (Khoshmanesh et al., 2021; Li, Wei, et al., 2021).

Biosensors in Precision Medicine. DOI: https://doi.org/10.1016/B978-0-443-15380-8.00010-2

Individuals can be divided into subpopulations based on their susceptibility to a particular disease. This is based on the biology or prognosis of diseases or their response to a specific treatment. To obtain a deeper understanding of a specific disease, it is necessary to gain insights into biological, environmental, and behavioral influences, which provide important information on how to prevent or treat it. New wearable sensors can open up ways to avoid illnesses that may impact patients in the future (Siirtola et al., 2018; Zhao, Nyein, et al., 2021). With the use of personalized sensing tools, preventive or therapeutic interventions can be targeted, thereby saving costs and minimizing side effects. Therefore modern healthcare needs many sensing technologies, such as motion detection (Wang et al., 2020; Yang et al., 2018), vital sign detection (Lee & Chung, 2009), biochemical detection (Kim et al., 2019), and environmental detection (Mamun & Yuce, 2020; Tsow et al., 2009). Healthcare fields can benefit from combining personal sensors and big data analytics for one-off prevention and treatment plans (Batko & Ślęzak, 2022; Subrahmanya et al., 2022).

Wearable biosensors offer a modern approach to healthcare and physiological monitoring. These sensors allow for continuous monitoring of vital signs, enabling the detection of changes in the body before symptoms become apparent. Besides blood and urine, alternative biofluids such as sweat (Ghaffari et al., 2021; Sempionatto et al., 2020), tears (Sempionatto et al., 2019; Xu et al., 2022), and interstitial fluids (Abdullah et al., 2022; Kim, Sempionatto, et al., 2018) can be used for analysis. Physiological parameters can be monitored in real time by noninvasive analysis of these fluids, which is relatively easy to obtain and can be performed frequently. As these fluids are close to the body's environment, they also provide a reflection of the body's current state. Wearable devices can be designed to be worn for extended periods and can take the form of shoes, glasses, clothes, socks, and watches.

Comfort is a crucial consideration for wearable sensors, and as such, these sensors should be flexible. Flexible electrochemical sensing systems provide unique opportunities for monitoring comprehensive chemical information (such as glucose and lactate concentrations), as well as tracking physical parameters, such as strain (Chen, Hou, et al., 2022; Niu et al., 2021) and electrocardiograms (Xu et al., 2021; Yoon et al., 2022). As electrochemical sensing systems have developed, unique opportunities for monitoring comprehensive chemical information and tracking physical parameters have emerged. The ideal devices usually consist of a number of components, including sensing interfaces, energy units, and electronics (Jeerapan & Poorahong, 2020). The technological constraint on wearable technology is the bulky, heavy, and rigid nature of batteries, which makes them unsuitable for prolonged and comfortable wear. Several alternatives are being developed by researchers to overcome the challenges associated with powering wearable devices.

In this chapter, we discuss key components of modern on-body tools (including biochemical sensors and wearable energy devices) that can support a closed-loop system (Fig. 9.1). This chapter provides a discussion of wearable biosensors and bioelectronics, with a focus on their basics, materials, processes, devices, and healthcare applications. Rather than aiming to collect a library of all wearable devices, this chapter presents recent representative ideas. The chapter begins by discussing the essential component of the flexible module in wearable sensing devices, which is responsible for detecting metabolites such as glucose and lactate, as well as electrolytes. Next, the chapter highlights wearable energy devices and ideas for the design of flexible energy-storage devices. In addition, this chapter introduces energy-harvesting technologies that offer the potential for a continuous power supply for wearable bioelectronic devices. The integration of microfluidics with flexible wearable bioelectronics is then presented as a means of handling and manipulating low volumes of fluid samples. Finally, the chapter addresses some of the challenges and prospects of flexible sensors.

9.2 How wearables can support precision medicine

Precision medicine can considerably improve patients' quality of life. Currently, monitoring metabolites (such as glucose, uric acid, and lactate) and biomarkers (such as cortisol, carcinoembryonic antigen, and α-fetoprotein) using conventional methods, such as the immunoturbidimetry, the enzyme-linked immunosorbent assay, and the hexokinase-spectrophotometric method. These conventional

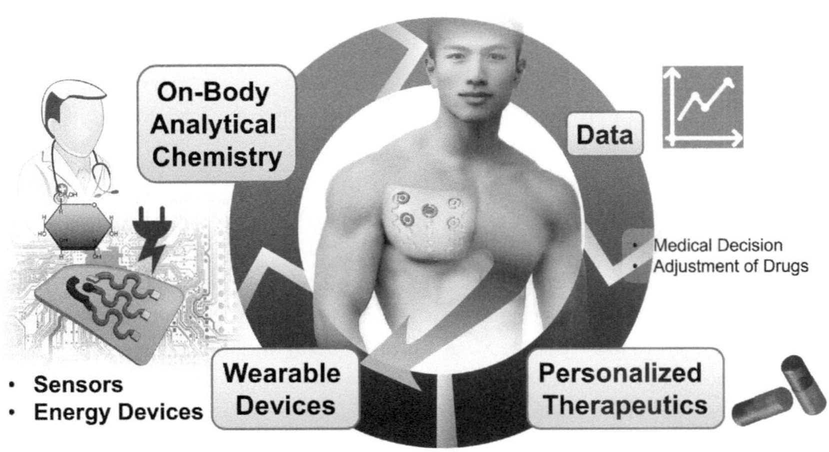

FIGURE 9.1 The general concept of a real-time closed-loop system through wearable sensors. Using wearable sensor technologies along with integrated energy devices facilitates diagnosis and therapeutics to become a personalized manner rather than a traditional benchtop-based diagnosis approach.

ways are time consuming, expensive, and require specialized technicians. Due to disease progression's complexity and dynamic nature, it is challenging to define a prognostic. Biosensors can be highly sensitive and portable. Therefore they are a potential solution (Haleem et al., 2021; Sharifianjazi et al., 2021).

The evolution of wearables from merely entertaining to providing health benefits has been remarkable. In addition, wearables can become an indispensable part of precision medicine initiatives and clinical trials. (Hilty et al., 2021; Tan et al., 2022). The rise of precision medicine has led to an increasing popularity of health tracking and wellness awareness. In the healthcare field, wearables have become essential sources of data for creating personalized treatment plans based on metabolite levels and health indicators (Zheng et al., 2021). This strategy can be used to decide on therapy by conducting a thorough analysis of the patient's status, medical history, and individual needs. Using evidence obtained from wearable sensors, the most appropriate therapy option can be selected. It is also crucial to regularly evaluate the therapy's effectiveness and the patient's response to determine whether to continue, adjust, or change the treatment approach. Wearable devices can also be passive, continuous monitors that do not require much attention from subjects or clinicians. Moreover, wearable sensors can be used in both inpatient and outpatient settings, allowing for continuous, noninvasive, and cost-effective patient monitoring (Dunn et al., 2021; Kwon et al., 2021). It is worth noting that wearables can measure multiple physiological parameters that may correlate with surrogate endpoints, enabling feedback to patients and clinicians.

Wearable sensors can be useful to clinical researchers and decision-makers in several ways. They all fall into two major groups:

9.2.1 Physiological sensors

This type detects signals through electrical (Chen, Qi, et al., 2021; Maddirala et al., 2022), thermal (Butt et al., 2022; Liu et al., 2019), acoustic (Qu et al., 2021; Zhao, Feng, et al., 2021), and optical means (Kazanskiy et al., 2022; Zhou et al., 2021). As a part of the physical evaluation, it is vital to look at physical indicators such as body temperature (Ali et al., 2019), heart rate (Dunn et al., 2021), wrist pulse (Xie et al., 2019), respiration (Dinh et al., 2020), electrophysiological signals (electrocardiogram (ECG or EKG)) (Liu et al., 2021) and electroencephalogram (EEG) (Japaridze et al., 2022; Seok et al., 2021), and body movements (subtle facial movements, swallowing, and large bending movements of arms, legs, and hands) (Ancans et al., 2021; Maity et al., 2021; Moonla et al., 2022).

9.2.2 Biochemical sensors

This type covers devices that transform chemical information into an analytically valuable signal, providing information about sample components and composition. These sensors produce a variety of signals, including electrical, colorimetric, pH, and temperature changes, upon reaction with a target molecule. The resulting signal corresponds to the concentration of the analyte (Ghaffari et al., 2021). This chapter focuses on the principles, design, and applications of electrochemical sensors that utilize an electrochemical transducer.

9.3 Flexible biosensors for on-body detection

Flexible electrochemical sensors convert biological or chemical analytes into electrical signals, which can be correlated with a patient's health status and changes in pattern (Windmiller & Wang, 2013; Yu et al., 2020). These sensors can detect pH levels (Manjakkal et al., 2020; Qin et al., 2019), alcohol (Hauke et al., 2018; Sun et al., 2021; Zhao et al., 2018), glucose (Jin et al., 2022; Zhu et al., 2022), lactate (Komkova et al., 2022), electrolytes (Kim, Yi, et al., 2021) (e.g., sodium, Lim et al., 2021, and potassium, Ozer et al., 2022), etc.

Monitoring the concentration of metabolites such as glucose and lactate has been extensively studied and can assist in identifying abnormalities and promoting self-awareness and behavioral modification (Kim, Campbell, et al., 2018). The most common way to test blood glucose levels is through a fingerstick. By sticking a needle into the fingertip, it is possible to analyze the blood metabolite level by taking a small sample of blood and collecting it into a special test strip or tube for laboratory testing. Nevertheless, the downside is that this approach is invasive and noncontinuous. The provision of noninvasive monitoring solutions makes it possible to monitor blood glucose levels frequently and continuously. This provides actionable feedback that may prevent the development of diabetes in prediabetic individuals (Ma et al., 2022). Biochemical analysis is the primary application of wearables used in clinical research and for ambulatory purposes. Nowadays, significant attention has been paid to noninvasive, wearable sensors for monitoring human health (Hanna et al., 2022; Promphet et al., 2021). Wearable sensors can be attached to a part of the body to conduct noninvasive measurements in real time. For example, it is possible to use a noninvasive clip from GlucoWISE to measure glucose levels (Meta Materials Inc., n.d.), and Epicore patches to measure sweat rate and skin health (Epicore Biosystems, Inc., n.d.). Many wearable biosensors are still in the research and development stage, but they play a crucial role in precision medicine by determining clinically relevant parameters such as glucose concentration. The development of clinically relevant wearable technology in healthcare has been driven by modern lifestyle markets.

9.3.1 Wearable glucose sensors

Glucose is an essential biochemical that supplies energy to the human body and is required for the maintenance of health. Hyperglycemia (high glucose) and hypoglycemia (low glucose) are both dangerous. The abnormal level of glucose can link to a number of symptoms, such as tremors, palpitations, dizziness, and metabolic disorders (Pullano et al., 2022; Zhang et al., 2022). Diabetes is one of the greatest threats to human health involving the inability to produce enough insulin or the ineffective use of insulin, which causes a serious medical condition.

There is a demand for a noninvasive glucose monitoring system for diabetic patients. Wearable sensors that are flexible in nature may be more practical since they can accommodate the movements and activities of daily life while also providing analysis readings. A highly accurate, precise, and flexible wearable sensor is, therefore, required to improve sensor performance.

Glucose biosensors are classified into first, second, and third generations depending on their detection mechanisms (Holade et al., 2017). The core component of these systems is the biocatalyst (such as glucose oxidase (GOx)) to catalyze the oxidation of glucose. For *the first generation*, the production of reaction products (hydrogen peroxide (H_2O_2) or gluconic acid) or the consumption of the reactant (such as oxygen (O_2)) diffuses to the sensing interface and generates an electrical response. In many platforms, this generation is also called indirect detection of glucose through the measurement of H_2O_2 or O_2. With this biosensing design, O_2 can be used as an electron acceptor, and the concentration of glucose can be determined by following either the O_2 consumption or H_2O_2 production. For instance, in the presence of glucose, glucose oxidation (catalyzed by GOx) occurs on the electrode surface that consumes the free O_2 dissolved in the testing biofluid. An increase in glucose concentration can be linked to a decrease in free dissolved O_2. *The second generation* involves redox mediators to mediate electron transfer on the electrode surface. The utilization of redox mediators in second-generation glucose biosensors enables the transfer of electrons between the electrode surface and the active site of GOx, thereby enhancing the accuracy and sensitivity of the biosensor. Moreover, redox mediators aid in ameliorating the effects of O_2 on the glucose oxidation process, leading to reliable measurements. *The ideal third generation* of glucose biosensors, also known as direct electron transfer biosensors, is a type of biosensor that utilizes enzymes, such as GOx, to directly transfer electrons to an electrode surface, without the need for a mediator molecule. Alternatively, a *nonenzymatic* glucose sensor may be introduced as another generation that eliminates biocatalysts by performing direct glucose oxidation.

Enzyme sensors are employed for glucose analysis owing to their high selectivity. However, the sensitivity of these sensors can be limited due to the insulative nature of the enzyme. Nanomaterials exhibiting high conductivity, electron transfer ability, and large surface area are viable alternatives for enhancing the sensitivity of enzyme sensors. Transition metal carbide decorated on metal oxide (such as MXene nanoflakes on ZnO tetrapods (ZnO TPs)) is an example electrode material for fabricating stretchable and wearable glucose sensors (Fig. 9.2A) (Myndrul et al., 2022). ZnO is a metal-

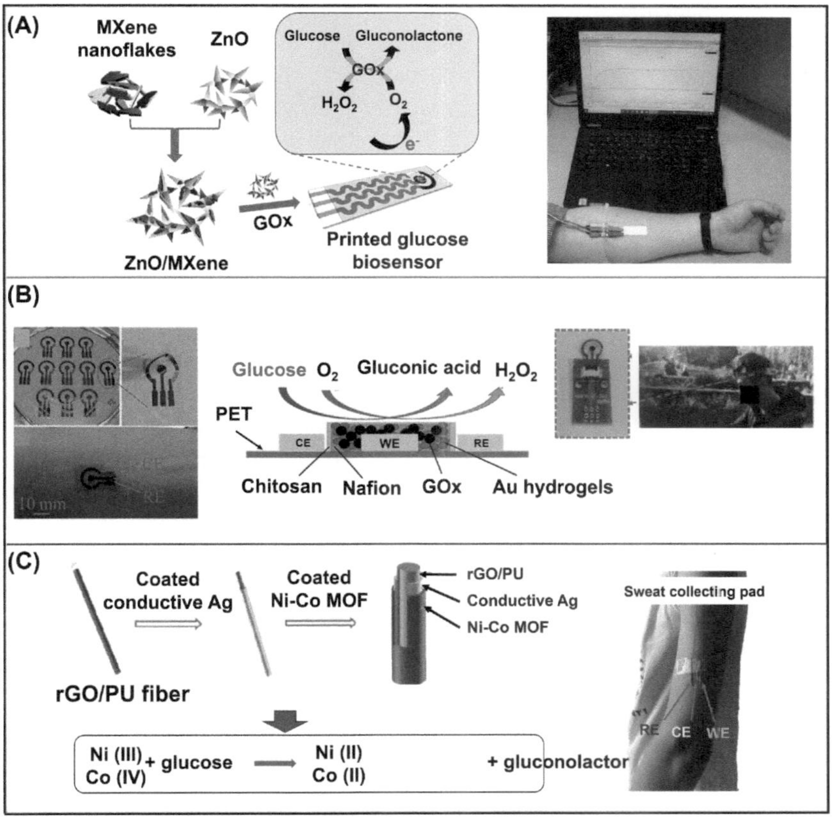

FIGURE 9.2 Wearable glucose sensors. (A) Skin-attachable and stretchable electrodes based on ZnO/MXene nanocomposites for real-time monitoring of postmeal glucose levels in sweat. (B) A flexible wearable glucose biosensor integrating porous Au hydrogel-based electrode for real-time glucose monitoring. (C) A stretchable Ni–Co MOF/Ag/rGO/PU fiber-based wearable sensor for nonenzymatic sweat glucose detection. *(A) Adapted with permission from Myndrul, V., Coy, E., Babayevska, N., Zahorodna, V., Balitskyi, V., Baginskiy, I., Gogotsi, O., Bechelany, M., Giardi, M. T., & Iatsunskyi, I. (2022). MXene nanoflakes decorating ZnO tetrapods for enhanced performance of skin-attachable stretchable enzymatic electrochemical glucose sensor.* Biosensors and Bioelectronics, 207. *https://doi.org/10.1016/j.bios.2022.114141; © 2022, Elsevier. (B) Adapted with permission from Li, R., Wei, X., Xu, J., Chen, J., Li, B., Wu, Z., & Wang, Z. (2021). Smart wearable sensors based on triboelectric nanogenerator for personal healthcare monitoring.* Micromachines, 12*(4). https://doi.org/10.3390/mi12040352; © 2021, American Chemical Society. (C) Adapted with permission from Shu, Y., Su, T., Lu, Q., Shang, Z., Xu, Q., & Hu, X. (2021). Highly stretchable wearable electrochemical sensor based on Ni–Co MOF nanosheet-decorated Ag/rGO/PU fiber for continuous sweat glucose detection.* Analytical Chemistry, 93*(48), 16222–16230, http://pubs.acs.org/journal/ancham. https://doi.org/10.1021/acs.analchem.1c04106; © 2021, American Chemical Society.*

oxide semiconductor that presents excellent properties, including good catalytic activity, nontoxicity, high chemical and mechanical stability, high surface-to-volume ratio, high electron transport capacity, and biocompatibility (Cheng et al., 2021; Ilager et al., 2022). In addition, the nanostructure with large pods and highly porous 3D interconnected networks can improve the active surface area of the electrode sensing. Decorating high-surface-area ZnO TPs with MXene can enhance the conductivity of ZnO TPs, thereby improving the sensitivity of the sensor (Sohan et al., 2021). MXene is a two-dimensional (2D) material transition metal carbide or transition metal nitride that provides high electrical conductivity due to its metal-like conductive properties. The ZnO/MXene composite was deposited on a screen-printed working electrode to fabricate a stretchable electrode. Subsequently, GOx was immobilized on a nanocomposite electrode. By using electrochemical techniques and following the first-generation glucose-sensing mechanism, the reduction current of O_2 can be monitored. Hence, amperometric detection can be employed to monitor the change in current signals at an appropriate potential (such as around -0.2 V or a higher cathodic potential). In general, amperometry measures the current when an electroactive substance loses or gains an electron during electrochemical reactions. Amperometry involves applying a constant potential to induce the oxidation or reduction of specific analytes and then record the resulting current. In this example, O_2 could gain electrons at a reduction potential. The glucose detection on this GOx-based electrode relied on the decrease of the cathodic current of dissolved O_2 reduction. Glucose concentration could be correlated with this cathodic signal. The higher the glucose level, the smaller the cathodic current. The sensor exhibited a linear behavior in glucose concentrations of 50−700 μM, with a limit of detection (LOD) of 21 μM in artificial sweat. The sensor could analyze sweat as an electrolyte, with a stable operation (10 days) and mechanical stability (for 30% elongation), attributed to the thermoplastic polyurethane (TPU)-based ink electrode. The sensor enabled qualitative glucose analysis from sweat without requiring sample pretreatment. Glucose levels measured by the stretchable sensor on the skin showed trends similar to those of a conventional blood glucose meter.

A thin substrate that is stretchable and stable is highly desirable since the body structure is nonplanar. Using rigid electrodes in electrochemical sensing limits the ability of the devices to be applied to soft skin tissues. Another example of a highly flexible wearable sensor uses a porous noble metal (such as Au) hydrogel-enzyme electrode for detecting biomarkers (e.g., glucose) in sweat (Fig. 9.2B) (Li, Hao, et al., 2021). This first-generation biosensor is based on the amperometric detection of produced H_2O_2 from an enzymatic reaction. First, the immobilized oxidase catalyzes the oxidation of the analyte to produce H_2O_2, which can be monitored via an electrochemical sensor. As a result, the H_2O_2 oxidation current can be higher when increasing the concentration of glucose. It is, therefore, necessary to measure the H_2O_2 concentration to measure the glucose target concentration. The catalyst (e.g., porous Au) plays a crucial role here because the detection of H_2O_2 can be difficult without a catalyst. High overvoltages can cause interference in H_2O_2 detection. For example, an overvoltage can also oxidize ascorbic acid (a common component of biological samples), leading to false positive results. Various inorganic catalysts have been developed for H_2O_2 detection, including Au nanomaterials (Dang et al., 2020), Pt nanomaterials (Kucherenko et al., 2019), and Prussian blue (PB) (Khumngern et al., 2021). In addition to improved sensitivity, these catalysts operate at lower overvoltages, which reduces the risk of interference from other compounds in the sample. The use of PB is discussed in Section 9.3.2 "*Wearable lactate sensors.*"

The flexible electrode illustrated in Fig. 9.2B was fabricated using soft-microelectromechanical systems (soft-MEMS) technologies with soft polymers (such as polyethylene terephthalate (PET)) as a substrate, which offer high precision and mass production capability. Soft-MEMS employs a process known as photolithography, which involves the transfer of a design from a mask onto a substrate using a radiation source, such as visible UV light or X-ray (Pinto et al., 2018). PET exhibits high flexibility as well as biocompatibility, making it suitable for use in on-body biosensors. Au hydrogels have also gained attention due to their biocompatibility. The term "hydrogel" generally refers to a three-dimensional network of materials that can absorb significant quantities of water or biological fluids, such as sweat. In this instance, an Au hydrogel was created via the dopamine-induced assembly of Au nanoparticles. Dopamine serves as an effective destabilizing agent in the controlled assembly of Au nanoparticles, leading to the formation of a branched network. The porous structure and large surface area of the hydrogel permit rapid analyte diffusion (e.g., glucose) and offer numerous active sites, thereby enhancing the biosensor's sensitivity (Du et al., 2020; Hu et al., 2021). In addition, an interconnected porous network structure of Au hydrogel enhances the flexibility of the sensing electrode. Au aerogels were fabricated to evaluate the porous structure of Au hydrogel, exhibiting a specific surface area of $\sim 60\ m^2/g$. By combining an Au hydrogel-based platform with GOx immobilization, the wearable biosensing yielded a linear response in concentrations of glucose ranging from 0.2 to 5 mM glucose with a limit detection (LOD) of $\sim 18\ \mu M$ (in 0.1 M phosphate buffer saline (PBS) solution, pH 7.4). This concentration range allows biosensors to be used as wearable glucose monitoring in human sweat, which is in the range of $10\ \mu M-1.11$ mM (Khor et al., 2022; Qiao et al., 2022). Despite 90 degrees of bending angles and many bending cycles, the sensing unit's usability remained desirable. By applying the sensing unit to healthy volunteers' skin, the calibration of glucose in sweat was obtained in the range of up to 2 mM with a LOD of 18 μM (pH 4.4). In addition, the sensitivity was slightly decreased compared to the calibration of glucose in PBS solution. This was due to the change in the activity of GOx in different values of pH. A real-time glucose monitoring system showed ~580 μM glucose in the volunteer, which was consistent with the glucose levels found by high-performance anion exchange chromatography (~605 μM).

Enzyme-based biosensors have excellent catalytic efficiency, specificity, and sensitivity and can operate under mild conditions. Nonetheless, wearable enzymatic sensors may be impacted by several environmental factors, including pH, temperature, chemicals, and humidity (Hwang et al., 2018; Wei et al., 2020). Materials must be engineered to optimize their performance in biosensors by modifying their shape, size, composition, adsorption capacity, and electron transfer properties. Various metals and metal oxides are viable options for glucose-sensing applications (Dong et al., 2021; Naikoo et al., 2021). To construct nonenzymatic sensors, noble metal nanomaterials (Ag, Au, Pd, and Pt) can be promising choices due to their high glucose electro-oxidation activity (He et al., 2023; Sehit & Altintas, 2020). Nevertheless, it is also important to consider the cost of these metals and the influence of other molecules (such as drugs and other metabolized species) when developing sensors.

When designing electronic assemblies to fit around curvilinear human skin with millions of pores, breathability is an important factor to consider. Fiber-based clothing has been preferred by humans because of its softness, flexibility, and breathability, making it an ideal platform for wearable electronics. The soft and flexible nature of fiber assemblies makes them a good fit for wearable systems. Due to their Young's modulus (MPa to kPa ranges), fiber-based devices can deform under external forces. Packing density and fiber orientation are typically defined structurally. They also have large specific surface areas and are porosity-adjustable. It is essential to develop flexible systems that support mobility and high electrical performance, along with mechanical and environmental stability to realize fiber-based wearable biosensing electronics.

For soft devices, textile materials offer several advantages, including high flexibility, wear resistance, mechanical strength, and promising sensing performance. Different approaches can be used to prepare this class of biosensors. Approaches can involve printing or coating textile materials with conductive nanomaterials or inks. Additionally, other technologies such as weaving, knitting, and embroidery can be used to integrate conductive fibers and yarns into textiles. Conductive fiber serves as a fundamental component of textiles for further integration with weaving, knitting, and embroidery techniques, allowing the fabrication of electronic textiles. Flexible and stretchable electrodes can be made of fiber, which are flexible. An example of a nonenzymatic wearable glucose sensor is made of conductive fiber of carbon nanomaterials mixed with a soft polymer with a catalytic composite fiber (such as reduced graphene oxide/polyurethane (rGO/PU)) (Fig. 9.2C) (Shu et al., 2021). This flexible sensor was fabricated by modifying carbon nanomaterial/elastomer fiber with metal-organic framework nanosheets (rGO/PU with Ni–Co MOF). Carbon nanomaterial/elastomer fibers were used to fabricate electrodes with flexibility and stretchability through the application of wet spinning technology. To enhance the conductivity of the modified fiber (Ni–Co MOF/Ag/rGO/PU), a layer of conducting Ag was coated on the nanomaterial/elastomer fiber. In addition, MOF was subsequently coated to provide a large specific surface area and catalytic activity toward glucose oxidation (Barbee et al., 2022; Shu et al., 2022). The fiber electrode's effective catalytic behavior allowed it to monitor glucose levels by detecting an increase in oxidation current. Using amperometric detection, the resulting electrode provided a linear range of $10 - 660$ μM glucose with a LOD of ~ 3 μM in 0.10 M NaOH; this electrode provided a narrower range ($20 - 280$ μM glucose) in neutral PBS solution. The use of this metal-based fiber was possibly limited in acidic conditions because of the dissolution of metals at oxidation voltage. The better performance in the basic solution was due to the high catalytic ability of glucose oxidation, catalyzed by the high amount of metal peroxide species. The fiber electrode provided good electrochemical performance under stretching (up to 100% strain), bending (up to 90 degrees), temperature changes (24–38°C), and storage (7 days). As a result of these characteristics, the sensor could be used as a wearable device on human skin with exceptional mechanical stability. A wearable sensor was constructed by assembling a fiber working electrode, Ag/AgCl/nanomaterial/elastomer reference electrode, and Pt wire counter electrode onto a waterproof and stretchable polydimethylsiloxane (PDMS) substrate using a needle thread. The sensor patch was then applied to the volunteer's arm, with a sweat-absorbent cloth covering the working area. The patch continuously measured the volunteer's sweat during exercises. The detected sweat glucose levels at different time points for one day ranged from 50 to 180 μM. In addition, the sweat glucose level was about 50 μM before a meal and increased to about 175 μM after a meal for 1 hour, which were correlated to commercial sweat and blood glucose meters. Those results confirmed that the nonenzymatic fiber could be used for reliable and real-time sweat glucose monitoring devices.

As we learn more about glucose electrochemical biosensors, we can extend our understanding to other metabolite biosensors (such as lactate) because enzyme-based or nonenzymatic electrochemical reactions are the basis of many sensors. Researchers can gain insight into the factors that affect sensor sensitivity, selectivity, and stability by studying the design and performance of glucose biosensors, and then apply this knowledge to develop biosensors for other analytes.

9.3.2 Wearable lactate sensors

Lactate, a biomarker of tissue oxygenation, can be used to assess human stamina (Jia et al., 2013). Lactate levels in biofluids are important to determine health status (Okorie & Dellinger, 2011). The concentration of lactate in sweat is associated with various health conditions, including heart failure, metabolic disorders, and respiratory and renal failure (Van Hoovels et al., 2021). Prolonged exercise can cause sweat lactate levels to increase up to 25 mM, which can lead to muscle dysfunction and cell acidosis (Mathew et al., 2021; Xuan et al., 2021). Thus monitoring sweat lactate levels is essential, particularly during high-intensity physical activities, to ensure optimal health protection.

An example of a flexible lactate enzymatic sensor utilizes metal nanoparticles, carbon nanomaterials, and a catalyst to modify flexible screen-printed carbon electrodes (SPCEs), along with lactate oxidase (LOx) immobilization (Fig. 9.3A) (Jiang et al., 2022). The lactate-biosensing membrane was designed for real-time lactate monitoring during exercise. PB is an important component applied in many oxidase-based biosensors due to the catalytic behavior toward H_2O_2 reduction. The high-spin Fe^{2+} atoms in the reduced form of PB allow high catalytic activity. A divalent iron ion Fe^{2+} center in PB can enhance the H_2O_2 catalysis (Itaya et al., 1984). The use of PB can be

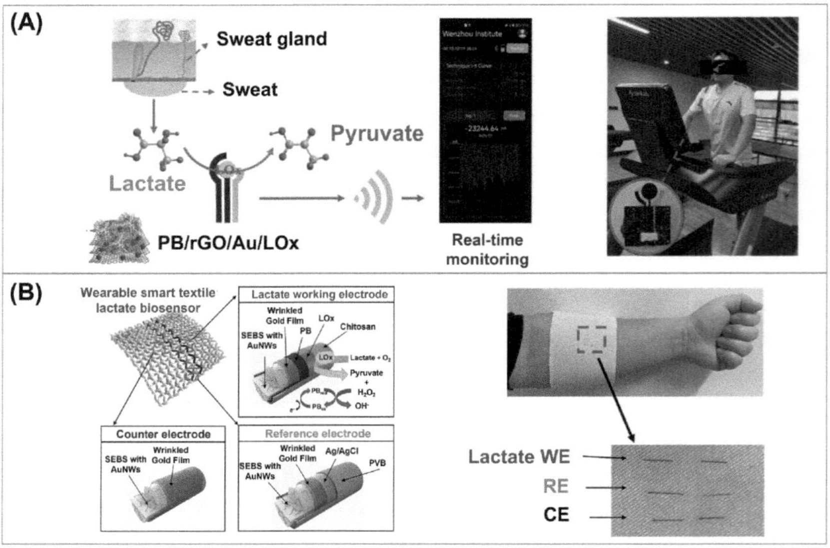

FIGURE 9.3 Wearable lactate sensors. (A) A wearable electrochemical lactate biosensor based on urchin-like Au nanoparticles/PB/RGO sensing membrane. (B) A stretchable gold fiber-based textile electrochemical lactate biosensor. *(A) Adapted with permission from Jiang, D., Xu, C., Zhang, Q., Ye, Y., Cai, Y., Li, K., Li, Y., Huang, X., & Wang, Y. (2022). In situ preparation of lactate-sensing membrane for the noninvasive and wearable analysis of sweat.* Biosensors and Bioelectronics, 210. *https://doi.org/10.1016/j.bios.2022.114303; © 2022, American Chemical Society. (B) Adapted with permission from Wang, R., Zhai, Q., An, T., Gong, S., & Cheng, W. (2021). Stretchable gold fiber-based wearable textile electrochemical biosensor for lactate monitoring in sweat.* Talanta, 222 *(2021). https://doi.org/10.1016/j.talanta.2020.121484; © 2021, Elsevier.*

an alternative catalyst to Pt electrodes or peroxidase enzymes (Karyakin et al., 1994; Zhang et al., 2014). Therefore lactate can be detected by following the change of cathodic reduction currents. In this work, the metallic contacting sites from nanoparticles enhanced the electron transfer process from the active centers of the enzyme. By using amperometric detection, this fabricated wearable lactate biosensor exhibited a wide linear range from 1 μM to 25 mM of lactate and LOD of 0.8 μM. Compared to PBS solution (pH 6.5), responses of the wearable lactate biosensor in artificial sweat were found in the range of 10–25 mM. This range could sufficiently be applied in real sweat lactate monitoring. In addition, the response obtained in spiked artificial sweat (pH 6.5) was similar to the standard curve measured in PBS solution. This biosensor presented good stability to pH variations of sweat from 4 to 7 due to the structure of PB that was capable of high performance in weak acidic sweat (Khumngern et al., 2021). A biosensor was also able to bend 200 times, indicating high flexibility. Biosensors were tested for 20 days and showed acceptable stability with 88% of their initial sensitivity. A flexible lactate biosensor was attached to volunteers' foreheads for real-time epidermal measurement using a portable device built. Lactate levels of volunteers were measured during running at different speeds. The first volunteer running at 7 km/h had lactate levels of approximately 11, 12, and 14 mM at 8, 9, and 10 minutes, respectively. The second and third volunteers, running at 8 km/h and 10 km/h, respectively, showed lactate levels ranging from 12 to 19 mM. This lactate biosensor opens the possibility of real-time monitoring of lactate during exercise in sports and biomedicine.

An example of a stretchable enzymatic textile-based sensor based on gold fiber electrodes for real-time lactate monitoring was demonstrated (Fig. 9.3B) (Wang et al., 2021). This wearable textile biosensor was assembled by weaving working, counter, and reference electrodes-based elastomeric gold fibers into a piece of fabric from a T-shirt. The working electrode was modified with LOx /chitosan on a PB-based electrode to create a lactate biosensor. Chitosan, a linear polysaccharide, was utilized as a coating to immobilize the enzyme and maintain its bioactivity. The reference electrode was coated with polyvinyl butyral (PVB)/Ag/AgCl on a gold fiber. PVB acted as an ion reservoir to maintain chloride concentration, thereby ensuring potential stability. By employing the amperometric technique, the linearity of the lactate biosensor in PBS solution (pH 7.0) was up to 30 mM with a sensitivity of ~19 μA/mM cm^2 (0–5 mM) and ~3 μA/mM cm^2 (10–30 mM), and an LOD of ~0.1 mM. The textile-based lactate sensor provided a highly stable signal even for a long duration (around 2 hours) in PBS solution and was durable for 6-day storage at 4°C. This lactate textile biosensor exhibited a linear range up to 30 mM with a sensitivity of ~15 μA/mM cm^2 for 0–5 mM lactate and a lower sensitivity (~3 μA/mM cm^2) for a high level of lactate (5–30 mM lactate) by testing in artificial sweat. The lower sensitivity observed in artificial sweat compared to PBS solution might be because the electrode surface absorbed multiple chemical species, resulting in poorer performance. However, wearable lactate textiles performed well in both PBS buffer and artificial sweat, even under 100% strain. The wearable biosensor demonstrated an 88% retention of its initial current after multiple stretches of up to 30%. To analyze sweat on a volunteer's forearm, the biosensor was attached to the skin. The measured sweat lactate levels (7.5–10 mM) were found to be lower than those obtained from high-performance liquid chromatography (HPLC) analysis (17.5–24 mM). The observed difference in lactate concentrations may be attributed to a concentration gradient between the skin surface and the sensor surface, which are separated by a 1 mm thickness of textile cloth.

9.3.3 Wearable electrolyte sensors

In addition to metabolite molecules, the measurement of electrolytes in biofluids (such as sweat) provides valuable information for monitoring health conditions (Kukkar et al., 2022). Electrolytes, including sodium, potassium, and chloride, are among the most measured analytes in sweat. Sodium ion (Na^+) is one of the most important markers for many physiological conditions, that is, general health vitality, heat stress, and various diseases such as hyponatremia and cystic fibrosis (Zhao et al., 2022). Therefore on-demand monitoring of Na^+ in sweat is an interesting topic to be approached. To detect such electrolytes (ions), an ion-selective electrode (ISE) can be employed using an ionophore-based potentiometric sensor. It can provide selective detection toward target analytes. An ionophore is a chemical species that captures ions reversibly. Potentiometric sensors utilizing ionophores or ISEs use host–guest chemistry to maximize selectivity to specific ions (such as Na^+). In addition to their versatility, low cost, and low power consumption, ISEs have high selectivity. Several ISEs have been reported, including solid crystalline electrodes (Jasiński et al., 2015), glass electrodes (Enseleit et al., 2021), ceramic electrodes (Razmi et al., 2008), and liquid membrane electrodes (Mou et al., 2021). Many wearable ISEs today have an organic polymer membrane as a sensing layer. An electrode is constructed by confining a component of the sensing layer (including an ionophore) within a polymer membrane. For instance, polyvinyl alcohol (PVC) is one of the most widely used polymers for ISE membranes because of its low cost, good mechanical properties, and plasticization ability. Potentiometric sensors are commonly used for electrolyte sensing applications. Using two electrodes for this sensor, one is a reference electrode whose potential is known and stable, and the other is an indicator electrode that interacts with the solution being analyzed. To determine the analyte concentration, potentiometric sensors primarily track changes in the potential difference (electromotive force or EMF) between the working and reference electrodes at various analyte concentrations under the conditions of zero electrical current flow (Zhai et al., 2022).

An example of a stretchable wearable electrolyte sensor was designed by screen-printing TPU elastomer and exfoliated graphene flakes (TPU/EGFs) ink on a mechanically stretchable TPU substrate. TPU was used as an electrode substrate because of its stretchable and flexible properties. Furthermore, a flexible ink was created by incorporating EGF into a soft polymer, such as TPU, to improve flexibility. A conductive and printable graphene ink was also developed for real-time monitoring of sweat Na^+ levels (Fig. 9.4A) (Park et al., 2021). Stretchable conductor ink was prepared via a fluid dynamics-induced exfoliation and mixing process by incorporating EGFs into a viscoelastic TPU. As a result of the flexibility and stretchability of TPU, the prepared ink would offer more stretchable. Another viscoelastic polymer could also be used such as PDMS (Mikkonen et al., 2020) and elastomeric fluorine copolymer (DAI-EL G801) (Matsuhisa et al., 2015). The screen-printed TPU/EGF conductor showed high conductivity and mechanical properties, including stretchability and flexibility, which are crucial for wearable sensors. The prepared 50-TPU/EGF (50:50 weight ratios) film exhibited a high conductivity of ~400 S/m and a failure strain of ~52%. A 50-TPU/EGF conductor withstood 10,000 cycles of mechanical bending (up to 180 bending angles) without significant changes in resistance. In comparison, the 50-TPU/graphite conductor had lower conductivity (80 S/m) and tensile strain (5.5%), and its resistance changed significantly during bending and 1000 cycles of fatigue testing. The TPU/EGF conductor showed improvements in both electrical and mechanical properties, as demonstrated by this comparison. Moreover, the serpentine-patterned 50-TPU/EGF conductor shows a maximum strain of over 300%. Serpentine patterns outperform other patterns in terms of stretchability, such as wavy, zigzag, fractal, and island-bridge patterns. This was due to the suitable width and connecting angle (θ) of the serpentine structure. The sensor for the quantitative analysis of Na^+ concentrations in human sweat was prepared by printing 50-TPU/EGF conductors on TPU substrates. Then, electrodes were coated with Ag/AgCl ink (to fabricate a reference electrode) and a Na^+-selective membrane cocktail (to fabricate a Na^+ sensor). Na ionophore X was used in this work to selectively bind with Na^+. Na ionophore X or *p*-tert-butylcalixarene-tetraacetic acid tetraethyl ester is generally used in the preparation of ISE due to its selectivity to Na^+ (Bühlmann et al., 1998). Then, EMF signals corresponding to Na^+ concentrations could be measured. The performance of the wearable Na^+ sensor provided good linearity in the range of 10^{-1} to 10^{-4} M with a slope of 58.3 mV dec^{-1}. The LOD was as low as 2.5 μM. This detection range is suitable for monitoring Na^+ concentration in the typical range of human sweat, which is between 10 and 100 mM (Mohan et al., 2020). Moreover, the sensitivities of different calibration curves under 300% and the fatigue test of 10,000 cycles showed no significant difference. The Na^+ sensor showed excellent electrochemical stability even under high strain. Its sensitivity remained unchanged when subjected to normal, bent, and stretched conditions. A wearable Na^+ sensor for real-time monitoring could be demonstrated by integrating the sensor

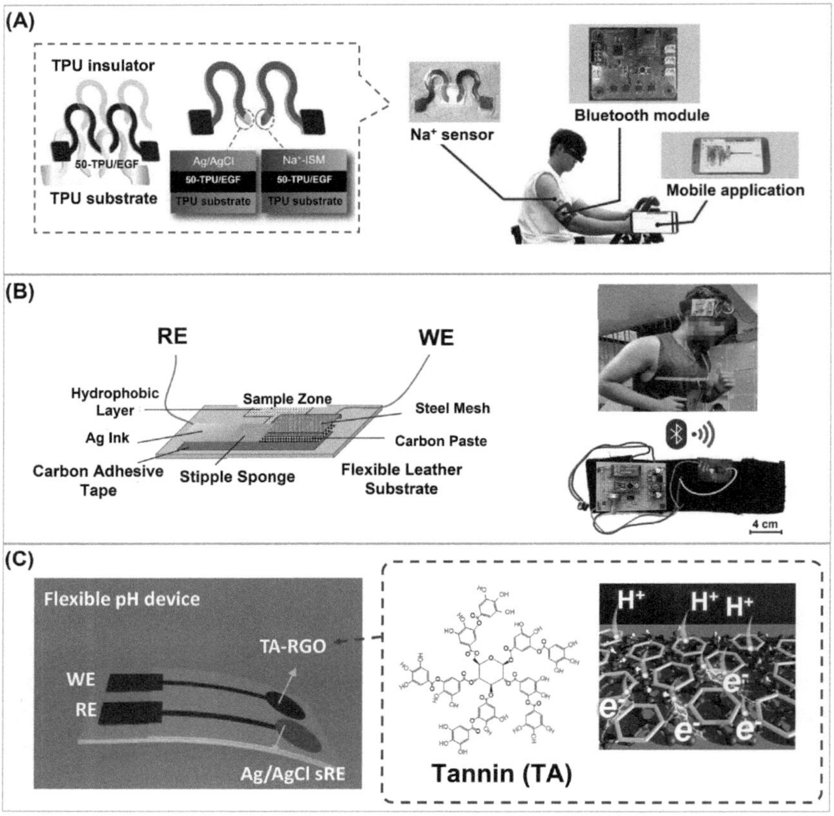

FIGURE 9.4 Wearable electrolyte sensors. (A) A stretchable Na^+ sensor based on highly conductive and stretchable graphene ink electrode for real-time monitoring sweat during stretching exercise. (B) A textile platform-based wearable potentiometric Na^+ sensor in sweat using $Na_{0.44}MnO_2$ as the sensing material. (C) A self-adhesive and pH-sensitive tannin-reduced graphene oxide-based electrode for flexible wearable pH sensor in sweat. *(A) Adapted with permission from Park, H. J., Jeong, J. M., Son, S. G., Kim, S. J., Lee, M., Kim, H. J., Jeong, J., Hwang, S. Y., Park, J., Eom, Y., & Choi, B. G. (2021). Fluid-dynamics-processed highly stretchable, conductive, and printable graphene inks for real-time monitoring sweat during stretching exercise.* Advanced Functional Materials, 31*(21). http://onlinelibrary.wiley.com/journal/10.1002/ (ISSN) 1616–3028. https://doi.org/10.1002/adfm.202011059; © 2021, Wiley Online Library. (B) Adapted with permission from Ghoorchian, A., Kamalabadi, M., Moradi, M., Madrakian, T., Afkhami, A., Bagheri, H., Ahmadi, M., Khoshsafar, H. (2022). Wearable potentiometric sensor based on Na0.44MnO2 for non-invasive monitoring of sodium ions in sweat.* Analytical Chemistry, 94*(4), 2263–2270. http://pubs.acs.org/journal/ancham. https://doi.org/10.1021/acs.analchem.1c04960; © 2022, American Chemical Society. (C) Adapted under the terms and conditions of CC-BY 4.0, from Lin, K., Xie, J., Bao, Y., Ma, Y., Chen, L., Wang, H., Xu, L., Tang, Y., Liu, Z., Sun, Z., Gan, S., & Niu, L. (2022). Self-adhesive and printable tannin–graphene supramolecular aggregates for wearable potentiometric pH sensing.* Electrochemistry Communications, 137. *https://doi.org/10.1016/j.elecom.2022.107261; © 2022, Kanglong Lin et al., published by Elsevier.*

and printed circuit board (PCB) as a band-type wearable sensor that could be displayed on a smartphone. A PCB-based potentiometric system was used to wirelessly transmit data to a smartphone during two types of exercise: constant stationary exercise and physical stretching activities (16% strain). The real-time sweat Na^+ levels did not differ between the two experiments, confirming that the wearable sweat sensor effectively monitored real-time sweat during physical stretching activities. In addition, the calibration curve of the Na^+ sensor remained unchanged before and after the on-body test.

Another example of a sodium sensor uses a carbon paste electrode based on $Na_{0.44}MnO_2$ as a sensing material (Fig. 9.4B) (Ghoorchian et al., 2022). Instead of traditional ionophore-based ion-selective membranes, alternative ion-selective sensing materials with ion-exchange properties can be used such as $Na_{0.44}MnO_2$ (Sauvage, Baudrin, et al., 2007) and zeolites (Jendrlin et al., 2021). This way relies on the similar principle of typical glass electrodes, which are ISEs used to measure cations with high sensitivity. Chemically active, the glass membrane can selectively bind hydronium, sodium, or other simple cations, depending on its composition. Researchers have developed specific materials that can be used to make potentiometric sensors. For instance, zeolite Na-ZSM-5 (DUBBE & MOOS, 2008) and $LiMn_2O_4$ (Lyu et al., 2022) have been explored as selective sensing elements for Na^+ and Li^+ ion detection, respectively. $Na_{0.44}MnO_2$ is also a promising material for sodium-ion sensors because it can accommodate cations reversibly. The structure of $Na_{0.44}MnO_2$ exhibits good crystallization, good electrical conductivity, and stability. An orthorhombic lattice structure has several potential sites available for Na^+ ion storage which enables it to reversibly accommodate Na^+(Liu, Jia, et al., 2022; Sauvage, Laffont, et al., 2007). Therefore $Na_{0.44}MnO_2$ can be applied to potentiometrics to selectively detect Na^+ without the use of traditional ionophores (Zhai et al., 2022).

To achieve a wearable potentiometric sensor, the sensing interface could be constructed on a flexible leather substrate, as demonstrated in this example. Subsequently, the potentiometric sensor could be integrated with a microcontroller that employs wireless Bluetooth communication technology to produce a headband. An integrated headband textile-based sensor could be applied for real-time monitoring of the Na^+ during exercise activity. The off-body test provided a linear response for Na^+ calibration in the range of $\sim 0.2 - 25$ mM with a Nernstian response (58 mV dec^{-1}). On-body testing revealed that a subject had a sweat Na^+ concentration of $\sim 10 - 18$ mM, which falls within the healthy range for individuals (<40 mM) and is consistent with inductively coupled plasma-optical emission spectrometry results ($\sim$20 mM). These findings demonstrate the potential for noninvasive analysis of human sweat using potentiometry.

The pH of the sweat is another important diagnostic parameter because it provides information about health conditions like dehydration, metabolic alkalosis, skin illnesses, cystic fibrosis, and epidermis conditions during wound healing (Escobedo et al., 2021; Pusta et al., 2022). Therefore it is essential to monitor the pH of sweat. A wearable potentiometric pH sensor based on pH-sensitive molecules (such as tannin (TA)) is another example of a wearable electrolyte sensor (Fig. 9.4C) (Lin et al., 2022). Strong adhesion of sensing materials on the electrode substrate is essential when designing for wearable applications. A small amount of binder, for example, Nafion and polyvinylidene fluoride, is required when mixing with electrode materials. However, it may raise the question of biocompatibility and the possibility of detachment. The pH-responsive TA molecule has been recognized as a potential alternative due to the abundant presence of phenolic hydroxyl groups, which are pH-responsive sites. Moreover, a self-adhesive TA-rGO supramolecular aggregate was created by self-assembling TA on rGO surfaces through $\pi-\pi$ conjugation interactions, which resulted in needle-like TA supramolecular aggregates. The high electrical conductivity and electron transfer of rGO promoted electron transfer between TA and the electrode (Karami et al., 2021). Interestingly, TA-rGO could be attached to various substrates owing to its strong adhesion; for example, the TA-rGO film stuck to a glass slide effectively. In addition, TA-rGO could adhere to the skin firmly. The potentiometric response of TA-rGO toward H^+ was based on a classic proton-coupled electron transfer (PCET) process. The linear response was obtained in a range of about 1–10 pH values. Moreover, TA-rGO exhibited high stability in buffer solution (20 hours) with a drift rate of ~0.52 mV/h, which was better than that of polyaniline (PANI) (2 mV/h) (Kaempgen & Roth, 2006) and WO_3 (2.5 mV/h) (Salazar et al., 2016). Moreover, TA-rGO exhibited good long-term stability up to 32 days. These could confirm the strong adhesion of TA-rGO. A flexible pH sensor was created on a PET substrate by employing TA-rGO ink printing and drop-casting a reference membrane solution onto the working and reference electrodes, respectively. This method resulted in consistent calibration responses even during bending states, indicating a high degree of flexibility. As a result, this feature makes it a viable option for practical on-body pH monitoring. During the volunteer's run, real-time pH monitoring revealed gradual increases in the signal, which stabilized after 15 minutes due to continuous sweat production. The pH level observed was approximately 7, which was consistent with the pH test strips. The calibration curves of pH response before and after sweat testing were also comparable, indicating that the wearable pH sensor was reliable. Stability is a crucial factor to consider for sensors, particularly for wearable biosensors. These types of sensors face additional challenges in terms of reliability, stability, and drift due to variations in operating environments, sensor handling, and motion effects resulting from daily life activities. These challenges are greater in comparison to traditional lab-based sensors.

Wearable biosensors are expected to modernize and become more convenient for users, with potential applications ranging from wristbands to textiles. In some wearables, the use of disposable components may be necessary to prevent fouling. However, noninvasive biosensors will need to undergo extensive evaluation and clinical assessment before being widely adopted by the medical community. In the coming years, wearable sensors can transform people's lives by allowing for real-time monitoring of physiological information through noninvasive or minimally invasive sampling.

9.4 Electrical energy for wearable flexible devices

9.4.1 Wearable energy-storage devices for supporting flexible biosensing systems

Regular monitoring of biological signals is critical for timely detection and treatment. However, wearable medical devices powered by bulky batteries can reduce portability, wearability, and comfort due to increased weight and volume (Jeerapan & Poorahong, 2020). Batteries (Ali et al., 2021) and supercapacitors (SCs) (Guo et al., 2021; Zhao, Xia, et al., 2021) can power flexible and wearable electronics. However, developers need to overcome limitations such as bulkiness, low energy density, and poor wearability in energy-storage devices.

Energy-storage modules, such as batteries and SCs, are limited by their rigid and bulky structure as well as safety concerns. Most rechargeable batteries commercially available use strong acid or base electrolytes, which can be hazardous. As an example of a biocompatible energy-storage component, sweat-activated batteries (SABs) were developed (Fig. 9.5A) (Huang et al., 2022). Safe to the skin, and environmentally friendly SAB consisted of an air-permeable tape, a graphene-coated absorbent paper cathode, a thin Mg foil anode, a thin cotton layer with KCl powder electrolyte, and releasing the film. The releasing layer was applied as a good air breathing for O_2 flow. SABs could be used to power ultra-thin, flexible, and garment-based microelectronics based on Mg/O_2-based reaction In this example, sweat was collected in the cotton layer, creating an aqueous electrolytic environment, and magnesium foil can react with O_2 and water to form $Mg(OH)_2$. Basically, the Mg anode in this example was oxidized and produced electrons, whereas the graphene cathode could reduce O_2 and receive electrons. Therefore an electron transport circuit could be formed between the anode and cathode. Generally, a battery has a positive side called the cathode and a negative side called the anode. The negative side is responsible for the flow of electrons that will be transferred from the negative side to the positive side via an external electrical circuit. In a redox reaction, high-energy reactants are transformed into lower-energy products when a battery is connected to an external load, thus producing electricity. The SAB showed excellent performance including good electricity output with a high capacity of ~ 14 mAh using KCl as the electrolyte and a P_{max} of ~ 3.2 mW/cm^2. In addition, it also established excellent flexibility during the bending state from 90 degrees

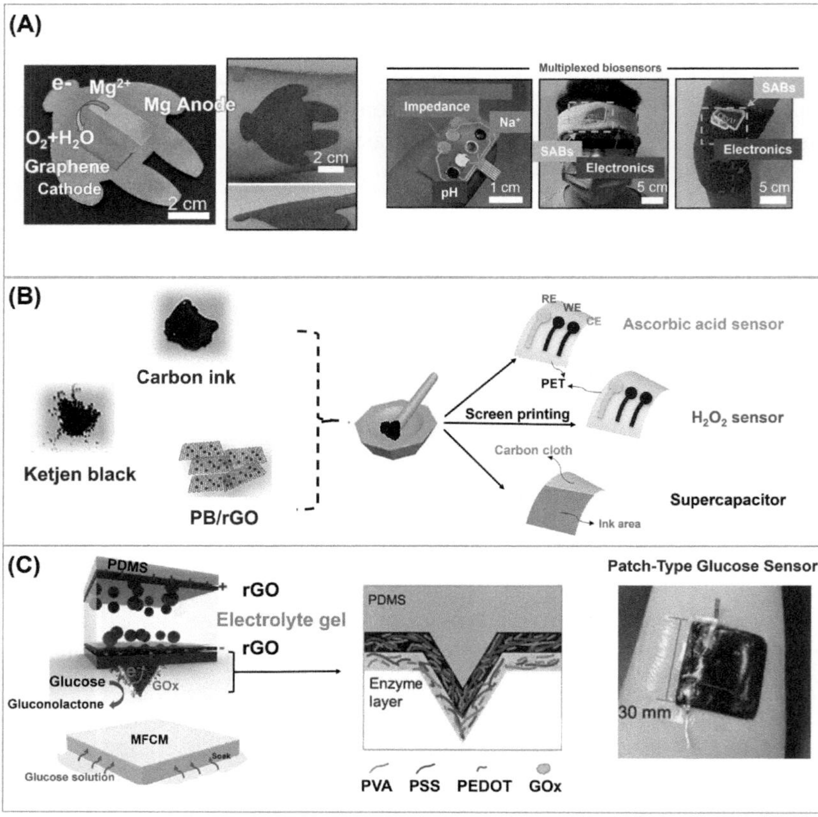

FIGURE 9.5 Flexible energy-storage modules. (A) A thin and flexible electronic system powered by SABs for a multiplexed biosensor patch. (B) A flexible wearable electrochemical biosensor and a supercapacitor based on PB/rGO ink. (C) A self-charging solid-state supercapacitor for a wearable microneedle-type sensor for glucose detection in the interstitial fluids. *(A) Adapted under the terms and conditions of CC-BY 4.0, from Huang, X., Liu, Y., Zhou, J., Nejad, S. K., Wong, T. H., Huang, Y., Li, H., Yiu, C. K., Park, W., Li, J., Su, J., Zhao, L., Yao, K., Wu, M., Gao, Z., Li, D., Li, J., Shi, R., & Yu, X. (2022). Garment embedded sweat-activated batteries in wearable electronics for continuous sweat monitoring.* Npj Flexible Electronics, 6*(1). nature.com/npjflexelectron/. https://doi.org/10.1038/s41528-022-00144-0; © 2022, Springer Nature. (B) Adapted with permission from Ma, J., Cui, Z., Du, Y., Xu, Q., Deng, Q., & Zhu, N. (2021). Multifunctional prussian blue/graphene ink for flexible biosensors and supercapacitors.* Electrochimica Acta, 387. *https://doi.org/10.1016/j.electacta.2021.138496; © 2021, Elsevier. (C) Adapted with permission from Kil, H. J., Kim, S. R., & Park, J. W. (2022). A self-charging supercapacitor for a patch-type glucose sensor.* ACS Applied Materials and Interfaces, 14*(3), 3838–3848. http://pubs.acs.org/journal/aamick. https://doi.org/10.1021/acsami.1c21394; © 2022, American Chemical Society.*

to 180 degrees (open-circuit voltage (OCV) stabilized $\sim$1.52 V) and repeated the bending test at 90 degrees for 600 bending cycles (OCV stabilized between 1.45 and 1.59 V). Multiplexed biosensors and garment-embedded microelectronic systems coupled with wireless transmission were used to continuously detect Na^+ levels, pH values, and skin impedance during exercise. Skin impedance was used to indicate skin moisture for dehydration prevention. The OCV responses were used to monitor the Na^+ concentrations and pH values. Au electrodes were used as a substrate for Na^+ and pH sensors. Na^+ sensor was fabricated by coating layers of poly (3,4-ethylenedioxythiophene):poly(styrene sulfonate) (PEDOT:PSS) and the Na^+ sensitive polymer membrane. pH-responsive PANI-modified gold electrode was used for the pH sensor. PANI is a conductive polymer, which had the pH sensing ability to monitor pH with near-Nernstian response ($\sim$58 mV/pH) based on its transition from emeraldine salt to emeraldine base (Guinovart et al., 2014). The impedance sensor consisted of two Au electrodes. The Na^+ and pH sensors responded well within the range of 10–100 mM Na^+ and 2–8 pH values. In a practical application, 2 SAB cells were used to power the sweat sensing system, enabling continuous support for the microelectronics system for up to 3 hours. The multiplexed sensor patch allowed for real-time sweat monitoring during jogging.

In wearable devices, rGO can be used for sensing and energy storage due to its excellent electron transfer, high conductivity, and large surface area. PB and graphene can be used for biosensors and SCs (Song et al., 2021; Zakaria et al., 2022), with PB acting as an artificial peroxidase that reduces H_2O_2. PB was incorporated into rGO to create flexible sensing arrays and SCs, which is another example of an applicable technology for wearable energy-storage devices (Fig. 9.5B) (Ma et al., 2021). A flexible SC electrode was made from PB/rGO ink, carbon ink, and ketjen black ink that was printed onto a carbon cloth substrate. PB/rGO ink-based SCs exhibited a high capacitance of 105 F/g at 0.5 A/g. In addition, a hybrid full cell prepared by pairing two PB/rGO ink-based electrodes was also constructed. The hybrid full cell presented a specific capacitance of $\sim$15 F/g at a current density of 0.25 A/g with a maximum energy density of $\sim$3 Wh/kg at a power density of 150 W/kg.

The flexible sensor for H_2O_2 was printed using screen-printing techniques on a PET substrate. The sensor patterns consisted of working, counter, and reference electrodes. Working and counter electrodes were prepared using PB/rGO ink, while the reference electrode was prepared using Ag/AgCl ink. The developed sensor and SCs demonstrated excellent flexibility, withstanding over 1000 bending cycles, highlighting their potential for wearable sensors. Furthermore, the flexible SC electrode showed good electrochemical and mechanical stability (50-degree bending). The sensor, based on PB/rGO ink, was used to detect H_2O_2 through amperometry. The current signals increased with the rise in H_2O_2 concentration during the applied oxidation potential of H_2O_2. The results revealed a good response toward H_2O_2 with a LOD of $\sim 2\ \mu M$. Therefore this electrode could be applied to enzymatic biosensors, such as lactate and glucose. Moreover, ascorbic acid or vitamin C sensor could also be evaluated due to the excellent electron transfer and electrocatalysis ability of PB/rGO ink. The ascorbic acid is vital for preventing related diseases like scurvy or numerous infections. The sensor provided a wide linear range of up to 50 mM with a LOD of $\sim 100\ \mu M$. This example of the ascorbic acid sensor also exhibited good selectivity toward alanine, NaCl, and H_2O_2.

Wearable electronic devices should have self-charging systems that allow them to operate without an external power supply (Kim, Khan, et al., 2021; Liu, Wang, et al., 2022). An example of an all-in-one self-charging system of a self-powered solid-state supercapacitor (SPSC) coupled with s microneedle-type glucose sensor was developed (Fig. 9.5C) (Kil et al., 2022). For monitoring biomarkers in real time, microneedle arrays are useful tools for directly accessing the interstitial fluid (ISF). Metabolites such as glucose and lactate can be found in ISF. With sharp microscopic structures, it is possible to puncture the stratum corneum of the skin and place it in the viable epidermis for analyzing a target (Takeuchi et al., 2022; Zhang, Chen, et al., 2019). In this energy device, electrons were created by the oxidation of glucose with the help of GOx. The SPSC functions as a biofuel cell (BFC) based on glucose and has fast-charging properties, making it ideal for continuous energy storage and supply. This makes it useful for monitoring glucose. The microneedle consists of biocompatible PDMS with conductive polymer PEDOT:PSS used as the electrode material. This could measure glucose levels in ISF using a microneedle that minimizes damage to the skin. The SPSC was made of flexible PDMS thin film to create a lightweight and wearable energy-storage device that was suitable for skin attachment. The SPSC was made by coated PDMS thin film and the flat side of the microneedle structure with layers of rGO and the electrolyte gel followed by the assembly between both layers of electrolyte gel. To obtain glucose-based BFC, GOx was coated on the surface of the microneedle of the SPSC. The energy storage or charging of the SPSC occurs when electrons are generated electrochemically by glucose oxidation catalyzed by GOx. The developed SPSC showed a high capacity ($\sim$ 1600 F/g or 800 mF/cm^3) and a power density of 0.62 mW/cm^2 through glucose sensing of 11 mM glucose concentration. In addition, the voltage of the SPSC increased as the glucose concentration was varied, making it applicable to glucose biosensors. Thus the correlation between glucose concentration and the voltage of the self-charging SC was assessed. In experiments conducted on Micropig Franz Cell Membrane (MFCM) artificial skin, which absorbed different concentrations of glucose, the microneedle-type glucose sensor could detect a wide range of glucose concentrations ranging from 10 μM to 11 mM.

9.4.2 Wearable energy-harvesting devices and wearable self-powered biosensors

Renewable and sustainable power supplies are essential for continuous operation and accurate measurements of wearable biosensors. Energy-harvesting devices are "green" power generators that use the energy available in the environment, such as electromagnetic energy (Lee & Roh, 2021; Miao et al., 2022; Zhang, Cao, et al., 2019) and temperature gradients (Kim, Cho, et al., 2021; Pan et al., 2022), to power electronics. There are several energy sources available to power wearable bioelectronics, including mechanical (Chen, Tang, et al., 2021; Mariello et al., 2021), thermal (Sattar & Yeo, 2022), radiant (Tetik, 2022), and biochemical (Jeerapan, et al., 2020; Lv et al., 2021). In this section, we emphasize the last type. A focus will be placed on energy-harvesting devices that utilize sweat-containing biochemicals as a source of energy.

Chemical energy stored in biomolecules can be converted into electrical energy by enzyme-based BFCs. Self-powered devices can collect energy from the surrounding environment without an external source of energy. Several BFCs have been developed for harvesting energy from the human body by the use of chemical energy produced by various biofuels such as fructose, lactate, and glucose, which can then be converted into electrical energy via enzyme catalysis of electrochemical reactions (Serag et al., 2022; Wu, Zhang, et al., 2021). Hence, BFC harvesters can operate continuously and power self-powered biosensors that generate a signal proportional to the amount of fuel analyte in the sample.

In one example of self-powered sensors, stretchable textile-based BFCs used extracted energy from sweat to detect metabolites (such as lactate and glucose) (Fig. 9.6A) (Jeerapan et al., 2016). These devices were fabricated using screen-printing of customized stress-enduring inks on stretchable fabrics and several surfaces of wearable accessories (such as straps, socks, underwear, and flexible watches). The anode was printed using carbon nanocomposite (such as carbon nanotubes (CNTs))-based ink with the addition of a PU elastomer. The addition of the elastomer ensured the ink's stretchability. This anode was then functionalized with a redox mediator (such as 1,4-naphthoquinone (NQ)) and oxidase enzymes (GOx or LOx). The cathode was printed using an Ag/AgCl-based ink composite with CNTs nanofiller and elastomer. The Ag_2O/Ag cathode was obtained by anodization under alkaline conditions. The use of nanomaterial-modified electrodes, possessing a large electrochemically active surface area, facilitated the electron flow. The flexibility of printable bioelectronics devices, including ink formulations and serpentine electrodes, allowed them to withstand mechanical deformations. During the enzymatic reaction, the metabolite, such as glucose, is oxidized in the anode compartment, and the acquired electrons are transmitted to the conductive material. The electrons flow through the external load circuit to the cathode, where they are reduced by the electron acceptor, such as Ag_2O, generating electricity.

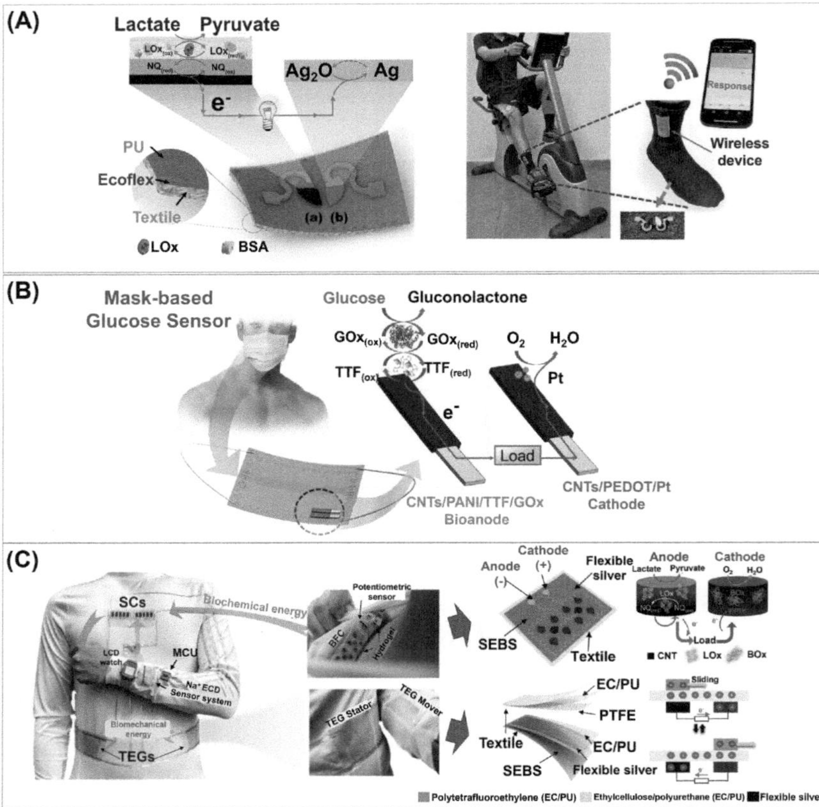

FIGURE 9.6 Energy-harvesting devices and self-powered biosensors. (A) A stretchable textile-based printed biochemical energy harvesters and self-powered lactate and glucose sensor. (B) A self-powered and energy-harvesting device on a mask for monitoring glucose. (C) A biochemical and biomechanical energy harvester using a textile-based multimodular bioenergy microgrid system for sustainable wearable systems. *(A) Adapted with permission from Jeerapan, I., Sempionatto, J. R., Pavinatto, A., You, J. M., & Wang, J. (2016). Stretchable biofuel cells as wearable textile-based self-powered sensors.* Journal of Materials Chemistry A, *4(47), 18342–18353. http://pubs.rsc.org/en/journals/journalissues/ta. https://doi.org/10.1039/C6TA08358G; © 2016, Royal Society of Chemistry. (B) Adapted under the terms and conditions of CC-BY 4.0, from Jeerapan, I., Sangsudcha, W., & Phokhonwong, P. (2022). Wearable energy devices on mask-based printed electrodes for self-powered glucose biosensors.* Sensing and Bio-Sensing Research, 38. *https://doi.org/10.1016/j.sbsr.2022.100525; © 2022, Elsevier. (C) Adapted under the terms and conditions of CC-BY 4.0, from Yin, L., Kim, K. N., Lv, J., Tehrani, F., Lin, M., Lin, Z., Moon, J. M., Ma, J., Yu, J., Xu, S., & Wang, J. (2021). A self-sustainable wearable multimodular E-textile bioenergy microgrid system.* Nature Communications, 12*(1). http://www.nature.com/ncomms/index.html. https://doi.org/10.1038/s41467-021-21701-7; © 2021, Springer Nature.*

A self-powered sensor in this work was operated based on an enzyme-based BFC anode that catalyzes the oxidation of biofuels (i.e., glucose or lactate) and generates electrons. Glucose and lactate BFCs generated the P_{max} values of 160 and 250 μW/cm^2 with OCVs of 0.44 and 0.46 V, respectively. The autonomous mode of self-generated voltage output displays a real-time signal of lactate levels over a range up to 20 mM lactate and a LOD of 0.3 mM. To enable on-body applications, an LOx-based BFC was printed within a sock for comfortable wearing on the foot. This sock-based biosensor allows for real-time monitoring of lactate levels. The signal output was self-generated and was observed after a volunteer performed cycling and secreted lactate-containing sweat. Textile-based self-powered sensors hold significant potential for the next generation of smart clothes, which could be used for health and performance monitoring.

A flexible self-powered device on a mask is another example of a wearable energy-harvesting device that can be applied for continuous monitoring of a metabolite (Fig. 9.6B) (Jeerapan et al., 2022). The BFC platform was fabricated based on the CNTs/PANI/tetrathiafulvalene (TTF)/GOx bioanode and CNTs/PEDOT/Pt cathode. Incorporating conductive materials such as PEDOT, PANI, and CNTs enhanced the electrode's capacitive characteristics, enabling charge storage due to their conductivity and electric double-layer properties. The bioanode productivity of electron transfer was improved by utilizing a TTF mediator, which has been frequently used in biosensing applications to facilitate low-potential bioelectrocatalytic conversion by flavoproteins such as GOx (Kowalewska & Kulesza, 2009). The self-powered system was completed by electrons that are produced by glucose oxidation at the bioanode flowing to the cathode. As a result, the biodevice can function as a BFC, SC, and self-powered biosensor. The flexible mask-based biodevice exhibited an OCV of 0.37 V with a P_{max} of 14 μW/cm^2 at 0.25 V. For the applications of glucose biosensors, by employing GOx, the self-powered glucose biosensor provided a self-generated current response for glucose concentrations up to 10 mM with a LOD of 0.22 mM.

A single component of an energy module may have a limitation in energy availability. Solar cells (Hashemi et al., 2020), triboelectric generators (TEGs) (Jeong et al., 2021), and enzymatic or microbial BFCs (Li et al., 2022) have recently been developed into flexible and wearable electronic devices. Hybrid harvesting systems with multiple inputs have become popular because of their efficiency in scavenging energy. An example of a hybrid harvesting system combined sweat-based BFCs and TEGs for biochemical and biomechanical energy harvesters (Fig. 9.6C) (Yin et al., 2021). The electric generator uses an approach that generates electric charges, separates the charges with opposite signs, and then uses the electrical potential (generated by the charges) to drive electrons. For energy harvesting, TEG utilized biomechanical energy to immediately begin the system by transforming instant motion-induced charge generation. BFCs also harvest biochemical energy through lactate sweat via electroenzymatic reactions.

In this example, modules were printed on a shirt to collect energy from body movements between the forearms and the torso as well as from sweat produced above the chest. In addition, the hydrophobic, waterproof, and stretchable polystyrene–polyethylene–polybutylene–polystyrene (SEBS)-based layer could enhance flexibility, smoothing the textile layer, and insulating and securing the surface and electrode due to the formation of a dense, hydrophobic, waterproof layer upon drying. Flexible and stretchable silver was printed as conductive interconnections to connect all modules in the microgrid system and current collectors in SCs. As a result, the concept of a wearable e-textile system with synergistic energy harvesters could continuously supply liquid crystal displays (LCDs) or a sweat Na^+ sensor-electrochromic display (ECD) system during exercise on a cycling machine.

The TEG module was incorporated with textile material to create a flexible e-textile. This particular TEG module consisted of a flexible negative mover, composed of a material called polytetrafluoroethylene (PTFE), and a positive stator composed of ethylcellulose (EC)-PU. The TEG module was tested by simulating the swinging motion of the arms during walking and running, resulting in a peak voltage of 160 V. The peak current increased as the frequency increased, ranging from 45 to 130 μA. When powering a 100 μF capacitor at a frequency of 1.5 Hz, a current of approximately 6 μA was obtained. The TEG module demonstrated mechanical durability and flexibility, as it was able to withstand folding, crumpling, sliding motion, and washing (20 minutes with tap water).

The wearable BFCs were also screen-printed onto the TEG fabric materials, and CNT-based pellets were attached to interdigitated "island-bridge" interconnections. The island-bridge structure is a design that connects functional electrodes, which are rigid islands, with stretchable bridges (interconnections), to provide stretchability (Fig. 9.7) (Ko et al., 2008; Li et al., 2013; Silva et al., 2020). In the example of these island-bridge BFCs, the electrodes were separated into several islands that were connected by serpentine-based printed silver interconnections. LOx and an NQ mediator were incorporated in a CNT-based bioanode. The oxidation of lactate by the LOx immobilized on the bioanode served as the basis for bioenergy harvesting. The CNT-based biocathode was immobilized with bilirubin oxidase (BOx) and protoporphyrin IX to facilitate catalytic O_2 reduction. Protoporphyrin IX was employed to promote electron transfer. The BFC drove the system by generating electrons by lactate oxidation at bioanode and gaining electrons by O_2 reduction at biocathode. The BFC modules generated a power of 22 W per module in the presence of 15 mM lactate, when discharged at 0.5 V. Furthermore, the power per module increased within a range of 10–25 μW (from 5 to 25 mM lactate). The BFC exhibited desirable mechanical durability, including bending, 7-day storage, and 20-minute washing with tap water. To harvest energy from the human body, the BFC was coated with PVA hydrogel. This hydrogel served as a stable electrolyte to maintain

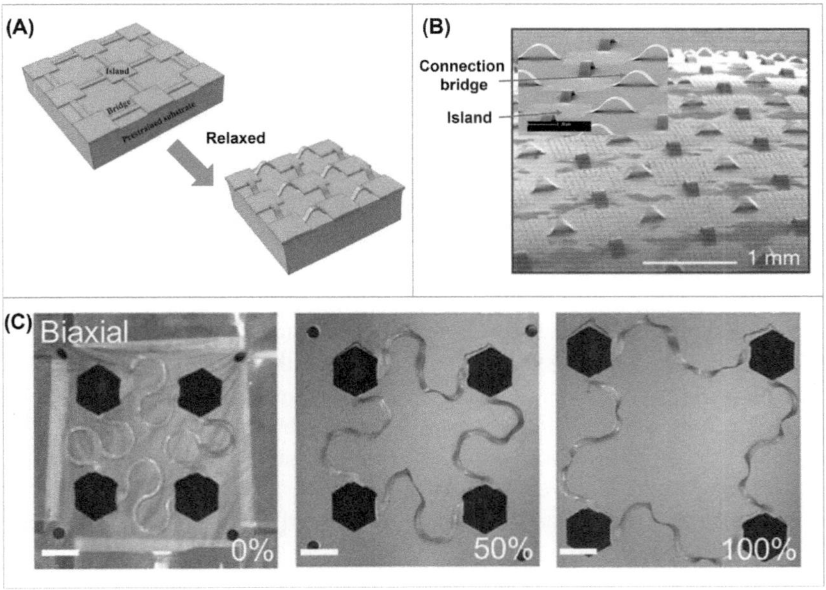

FIGURE 9.7 (A) The island-bridge structure. (B) SEM images of the silicon island-bridge structure on a PDMS substrate. (C) All printed Ag-eutectic gallium indium island-bridge wearable electrochemical device under biaxial mechanical deformations at 0%, 50%, and 100%. *(A) Adapted with permission from Li, R., Li, M., Su, Y., Song, J., & Ni, X. (2013). An analytical mechanics model for the island-bridge structure of stretchable electronics.* Soft Matter, 9*(35), 8476–8482. https://doi.org/10.1039/c3sm51476e; © 2013, Royal Society of Chemistry. (B) Adapted with permission from Ko, H. C., Stoykovich, M. P., Song, J., Malyarchuk, V., Choi, W. M., Yu, C. J., Geddes, J. B., Xiao, J., Wang, S., Huang, Y., & Rogers, J. A. A hemispherical electronic eye camera based on compressible silicon optoelectronics.* Nature, 454*(7205), 748–753. http://www.nature.com/nature/index.html. https://doi.org/10.1038/nature07113; © 2008, Royal Society of Chemistry. (C) Adapted under the terms and conditions of CC-BY 4.0, from Silva, C. A., Lv, J., Yin, L., Jeerapan, I., Innocenzi, G., Soto, F., Ha, Y.-G., & Wang, J. Liquid metal based island-bridge architectures for all printed stretchable electrochemical devices.* Advanced Functional Materials, 30*(30). https://doi.org/10.1002/adfm.202002041; © 2020, John Wiley and Sons.*

the ionic connection between the anode and cathode pellets. It was also a reservoir to absorb and temporarily store the sweat.

To create the textile-based SC energy-storage module, a conductive carbon nanomaterial of CNT and a conductive polymer of PEDOT:PSS hybrid capacitor was chosen. In addition, four interdigitated CNT-PEDOT:PSS electrode segments were coated by a solidified, transparent, flexible, sulfuric acid-crosslinked PVA gel electrolyte. The high specific surface area CNTs' double-layer capacitance and the pseudocapacitance from the PEDOT:PSS served as the hybrid capacitor's energy-storage mechanisms. The printed SC's capacitance could be estimated to be around 10 mF/cm^2. For the mechanical stability of the module, the SC did not exhibit any discernible differences between the 1000 cycles of bending deformation and the initial state.

The hybrid energy harvester module utilized two energy harvesters for in vitro and on-body tests, providing instant startup and prolonged energy harvesting during an activity session. An autonomous wearable system could benefit from this, ensuring reliable and long-lasting operation. A garment-based Na^{+} sensor integrated with a high-power wearable and flexible ECD pixeled display was created as an example for applications using a printed PEDOT:PSS-based ECD system that turns its color from light blue to dark blue in the presence of Na^{+}. The microgrid system had a quick boot time of three minutes. It was also capable of periodically refreshing the results for the 30-minute duration of its operation. These features allowed the wearable microgrid system to be operated sustainably for an extended period of time. For a low-power demand system, a wristwatch with an LCD was selected as a representative utilization. A voltage of a minimum of 2.5 V was required for the LCD to show sufficient contrast. By harnessing energy from human activities, the simultaneous BFC-TEG energy-harvesting system enabled quick booting (2 minutes) and sustained operation for up to 30 minutes. The integration of hybrid bioenergy microgrids on textiles presented a promising approach for creating sustainable wearable devices.

Wearable devices will be adopted by the population, particularly as the population ages and lifestyles change. To ensure a seamless user experience, wearable modules must be integrated. In addition to energy harvesting, other factors must be considered to develop optimal wearable devices with a long life cycle. Overcoming the theoretical limits of wearable energy systems will be a significant challenge for the development of next-generation wearable electronics. Hybridization of different energy harvesters can benefit wearable devices both in terms of electrochemical energy performance and design, enabling long-term operation. However, hybrid energy harvesters produce variable energy outputs, and effective power management tactics are highly necessary to control generated electricity efficiently.

9.5 Integration of microfluidics with flexible bioelectronics

Microfluidics serves as a tool for handling and manipulating low volumes of fluid samples, including sweat collection and transport to the detection zone. Wearable microfluidic devices enable continuous and noninvasive sweat absorption without dilution, while providing consistent flow (Jeerapan et al., 2022; Tabasum et al., 2022). Osmotic pumping for microfluidics can be realized using hydrogels (Shay et al., 2017). The fluid (such as sweat) could be drawn passively through a membrane by using the osmotic properties of hydrogels. This phenomenon is attributed to hydrogels with a high osmotic strength, allowing for sweat extraction with zero electrical power. However, many wearable sweat-sensing devices need frequent replacement of the sweat fluid due to saturation and dilution effects, which can compromise long-term operation. To address this issue, paper-based microfluidics have been developed, comprising a hydrophilic cellulose network that facilitates fluid movement through capillary action from an inlet, through a porous medium, to a desired detection zone or outlet. Such channels are useful for sweat transport and management, and by designing the end part of a paper channel appropriately, it can serve as an evaporation area to maintain a constant flow (Noviana et al., 2021).

An example of a microfluidic design for wearable applications is the continuous monitoring of lactate. This sensing system employs an osmotic approach, integrating a hydrogel for sweat extraction, a paper-based microfluidic channel, and a screen-printed electrochemical sensor (Fig. 9.8A) (Saha et al., 2022). This platform consisted of a low-power and flexible wireless potentiostat system. A hydrogel with a paper-based microfluidic channel could be used to extract sweat from the skin by osmosis, capillary wicking, and evaporation. This approach did not require electricity to extract sweat, thus allowing autonomous sampling. The system was developed by placing the lactate screen-printed electrode (SPE) sensor between the hydrogel and the circular section of the paper channel. The working electrode (WE) of the SPE contained carbon ink with electroactive PB, graphene nanoparticles, LOx, and Nafion. This functional electrode could effectively and specifically detect lactate (see the PB-based lactate biosensor mentioned earlier). Then, the circular end of the paper channel was inserted into the circular space on the PI sheet, which was the area of sweat sampling. Finally, a thin film was applied on top of the hydrogel to retain the electrode. The sensing system showed a linear response up to 15 mM lactate with LOD of $\sim$400 nM in PBS. In addition, the electrode stored at 4°C exhibited good stability for 4 days. An electrochemical analog front end was used to measure amperometric lactate data. The data could be accessed on a smart mobile device by using Bluetooth low energy (BLE) technology. This full system consumed a power of $\sim$0.7 mW. The wearable system, powered by a tiny 12 mAh LiPo battery, could be used for up to $\sim$63 hours, regarding power consumption. The performance for real-time sweat lactate determination on forearm human subjects was evaluated by monitoring lactate concentration at different states. The sweat lactate concentrations showed a maximum of $\sim$2 mM during the resting state trials (blood lactate: $\sim$1.4 mM), $\sim$6 mM during medium-intensity-exercise trials (blood lactate: $\sim$3 mM), and $\sim$13 mM during high-intensity-exercise trials (blood lactate: $\sim$12 mM). Lactate concentrations were correlated at the beginning of the study, but excessive sweating could dilute lactate concentrations.

Another example of a stretchable and self-powered sensor integrated with microfluidics was demonstrated for determining glucose levels in sweat during cycling workouts (Fig. 9.8B) (Bae et al., 2022). This wearable sensor had two components: a microfluidic device embedded in cotton thread and an enzymatic glucose sensor. A noble metal (such as nanoporous Au) anode was used as the basis of the BFC. This porous anode oxidized glucose to

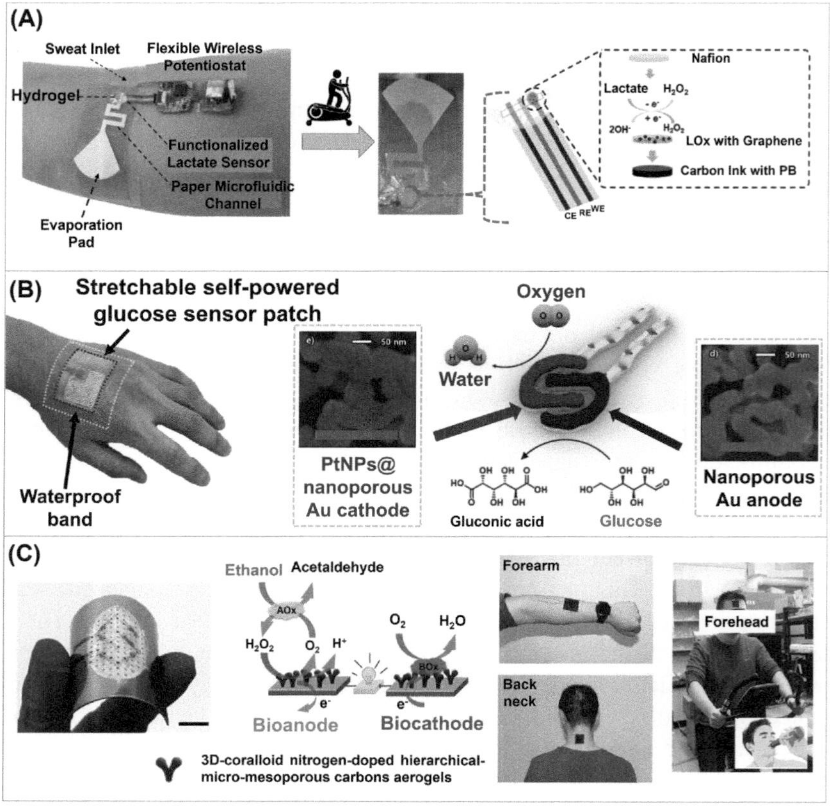

FIGURE 9.8 Flexible wearable microfluidic-integrated bioelectronics. (A) An osmotic system with a lactate biosensor, integrating a flexible wireless potentiostat. (B) A stretchable cotton thread-embedded microfluidic-integrated nonenzymatic fuel cell-based sensor patch for self-powered wearable glucose monitoring. (C) A flexible wearable microfluidic ethanol BFC. *(A) Adapted with permission from Saha, T., Songkakul, T., Knisely, C. T., Yokus, M. A., Daniele, M. A., Dickey, M. D., Bozkurt, A., & Velev, O. D. (2022). Wireless wearable electrochemical sensing polatform with zero-power osmotic sweat extraction for continuous lactate monitoring.* ACS Sensors, *7(7), 2037–2048. http://pubs.acs.org/journal/ascefj. https://doi.org/10.1021/acssensors.2c00830; © 2022, American Chemical Society. (B) Adapted with permission from Bae, C. W., Chinnamani, M. V., Lee, E. H., & Lee, N. E. Stretchable non-enzymatic fuel cell-based sensor patch integrated with thread-embedded microfluidics for self-powered wearable glucose monitoring.* Advanced Materials Interfaces, *9(20). http://onlinelibrary.wiley.com/journal/10.1002/(ISSN)2196–7350. https://doi.org/10.1002/admi.202200492; © 2022, Wiley Online Library. (C) Adapted with permission from Sun, M., Gu, Y., Pei, X., Wang, J., Liu, J., Ma, C., Bai, J., & Zhou, M. (2021). A flexible and wearable epidermal ethanol biofuel cell for on-body and real-time bioenergy harvesting from human sweat.* Nano Energy, 86. *https://doi.org/10.1016/j.nanoen.2021.106061; © 2021, Elsevier.*

gluconic acid and the PtNPs-nanoporous Au cathode reduced O_2 to H_2O. A stretchable cotton thread was embedded into the PDMS layer with a channel pattern for passive control of the constant sweat flow in the sensing chamber. The study demonstrated self-powered glucose detection using this nonenzymatic device. The changes in power and OCV were used to determine glucose concentration, with a linearity of up to 10 mM. The stretchable fuel cell electrode exhibited good endurance against mechanical deformation, including 30% elongation as well as 1000 stretching cycles with 30% elongation strain. Glucose sensor patches were affixed to the back of a volunteer's hand or forehead to conduct an on-body experiment. The volunteer engaged in a 20-minute bike ride, which resulted in the secretion of sweat. Both the output from the OCV signal and the blood glucose concentration (measured by a commercial glucometer) were found to be in concurrence. Furthermore, the study revealed that continuous glucose detection from sweat was feasible through the use of a stretchable fuel cell-based glucose sensor patch.

In addition to glucose or lactate BFCs, an example of a flexible and wearable epidermal microfluidic exogenous substance ethanol/O_2 BFC was demonstrated to harvest the electrical power from alcohol drinkers' respiration (Fig. 9.8C) (Sun et al., 2021). The device consisted of an epidermal microfluidic module for sweat collection and a BFC module for noninvasive and real-time sweat bioenergy harvesting. A flexible and wearable device was fabricated on polyimide (PI) and PET flexible polymeric substrates for on-skin bioenergy generation. The microfluidic module contained five layers, including a patterned adhesive layer, a patterned spacer, and a microfluidic reservoir for skin-on in situ fresh sweat collection, microfluidic channels for sweat transfer, and a microfluidic chamber for sweat storage and excretion. The BFC module consisted of three layers that are PI layer for BFC attachment, enzymatic SPE arrays for bioenergy generation, and the encapsulation layer for BFC encapsulation. In addition, the enzymatic electrode used porous carbon (that is, 3D-coralloid nitrogen-doped hierarchical-micromesoporous carbons aerogels) as its substrate. The 3D porous architecture of the carbon enables a high specific surface area, which allows for a high mass and compact loading of enzymes. This architecture also facilitates continuous diffusion of biofuel, accelerating electron transfer and enhancing bioenergy generation in the BFC. To fabricate the BFC, alcohol oxidase and BOx can be immobilized on porous carbon-modified SPE arrays to obtain the bioanode and biocathode, respectively. Bioenergy harvesting occurs during the consumption of sweat alcohol that generates the electron via enzymatic alcohol oxidation and dissolves O_2 that gains the electron through O_2 reduction. The ethanol/O_2 BFC-based electrode showed a P_{ax} of 1.9 $\mu W/cm^2$ at 0.39 V. The wearable epidermal device exhibited excellent flexibility performance, with no significant differences in P_{max} values observed after bending. To evaluate wearable bioenergy harvesting, different skin regions such as the forearm, back neck, and forehead were tested after consuming a bottle of alcoholic beverage and cycling at a constant level. The epidermal ethanol BFC worn on the forearm exhibited the highest P_{max} ($\sim 1\ \mu W/cm^2$). Higher doses of alcohol resulted in higher power density because more alcohol was eliminated through sweat. As a result, a self-powered sweat ethanol biosensor could be applied to a self-powered sweat ethanol biosensor.

9.6 Conclusion and prospects

This chapter describes wearable biosensing systems and their potential to improve human quality of life in medicine and biomedical engineering. Wearable technology is a breakthrough and has attracted interest due to its potential to revolutionize the way we approach medicine. Wearable sensors have the ability to continuously monitor health both inside and outside care facilities, making them a useful tool for precision medicine. By using smart sensing devices, we can build connected networks with insights about each patient throughout the care delivery cycle, allowing health specialists to see responses and modulate treatment plans. Analyzing data and developing diagnostic tools are key to making precision medicines a reality. Wearable biosensors and technology are particularly useful for measuring biomarkers in biological fluids, especially for patients suffering from a variety of diseases. Further developing wearable biosensors requires identifying potential biomarkers, selecting a suitable transducer, and attaching a suitable biorecognition element.

The rise of wearable technology has stimulated researchers to design personalized, flexible treatment regimens for patients. The strategy to use wearables enables adjustments to be made in reaction to alterations in the patient's health condition. Wearable technology further motivates patients to actively participate in their healthcare routines and improves communication with their doctors.

The widespread use of modern technology has enabled a greater number of people to access effective medical diagnosis and treatment. Medical conditions, including diabetes, can be managed and treated using new sensing devices that exhibit sensitivity, specificity, reliability, and biocompatibility. It is also important to integrate different sensors and functions for comprehensive diagnosis and treatment. An example of this is the development of a smart bandage that can promote healing while simultaneously monitoring the healing process. This type of bandage is particularly beneficial for managing chronic wounds resulting from conditions such as diabetes or burns. A pH sensor can be incorporated into the bandage to evaluate the wound's pH level. Once the pH level surpasses the predetermined threshold, the drug delivery component is activated, releasing antibiotics or antiinflammatory drugs. The integration of multiple sensors and functions for comprehensive diagnosis and treatment is becoming more widespread. An example of this is the utilization of neural networks and aggregated data from consumer wearables with physiological sensors, allowing for early diagnosis of medical conditions such as hypertension, atrial fibrillation, and sleep apnea (Konstantinidis et al., 2022; Oliver & Flores-Mangas, 2006; Wu, Chen, et al., 2021). Collecting and analyzing significant amounts of data from various sources can provide a comprehensive understanding of a patient's overall health status. Personalized healthcare can be optimized through the development of emerging technologies, such as mobile and wearable sensing devices. To address the challenges in healthcare resulting from human suffering and increasing healthcare costs, a paradigm shift from generic medicine to personalized and preventive medicine is important. The realization of the full potential of wearable and biomarker data will necessitate collaboration across multiple domains, including artificial intelligence, big data, data fusion, data analysis, and data imputation.

References

Abdullah, H., Phairatana, T., & Jeerapan, I. (2022). Tackling the challenges of developing microneedle-based electrochemical sensors. *Microchimica Acta*, *189*(11). Available from https://doi.org/10.1007/s00604-022-05510-3, https://www.springer.com/journal/604.

Ali, A. E., Jeoti, V., & Stojanović, G. M. (2021). Fabric based printed-distributed battery for wearable e-textiles: a review. *Science and Technology of Advanced Materials*, *22*(1), 772–793. Available from https://doi.org/10.1080/14686996.2021.1962203, http://www.tandfonline.com/loi/tsta20#.V0KNDU1f3cs.

Ali, S., Khan, S., & Bermak, A. (2019). Inkjet-printed human body temperature sensor for wearable electronics. *IEEE Access*, *7*, 163981–163987. Available from https://doi.org/10.1109/ACCESS.2019.2949335, http://ieeexplore.ieee.org/xpl/RecentIssue.jsp?punumber = 6287639.

Ancans, A., Greitans, M., Cacurs, R., Banga, B., & Rozentals, A. (2021). Wearable sensor clothing for body movement measurement during physical activities in healthcare. *Sensors*, *21*(6). Available from https://doi.org/10.3390/s21062068.

Ashley, E. A. (2016). Towards precision medicine. *Nature Reviews. Genetics*, *17*(9), 507–522. Available from https://doi.org/10.1038/nrg.2016.86, http://www.nature.com/reviews/genetics.

Awotunde, J. B., Folorunso, S. O., Bhoi, A. K., Adebayo, P. O., & Ijaz, M. F. (2021). *Disease diagnosis system for IoT-based wearable body sensors with machine learning algorithm intelligent systems reference library* (pp. 201–222). Nigeria: Springer Science and Business Media Deutschland GmbH. Available from http://www.springer.com/series/8578?detailsPage = titles, 10.1007/978-981-16-2972-3_10.

Bae, C. W., Chinnamani, M. V., Lee, E. H., & Lee, N. E. (2022). Stretchable non-enzymatic fuel cell-based sensor patch integrated with thread-embedded microfluidics for self-powered wearable glucose monitoring. *Advanced Materials Interfaces*, *9*(20). Available from https://doi.org/10.1002/admi.202200492, http://onlinelibrary.wiley.com/journal/10.1002/(ISSN)2196-7350.

Barbee, B., Muchharla, B., Adedeji, A., Karoui, A., Kumar Sadasivuni, K., Sha, M. S., Abdullah, A. M., Slaughter, G., & Kumar, B. (2022). Cu and Ni Co-sputtered heteroatomic thin film for enhanced nonenzymatic glucose detection. *Scientific Reports*, *12*(1). Available from https://doi.org/10.1038/s41598-022-11563-4, http://www.nature.com/srep/index.html.

Batko, K., & Ślęzak, A. (2022). The use of big data analytics in healthcare. *Journal of Big Data*, *9*(1). Available from https://doi.org/10.1186/s40537-021-00553-4, http://journalofbigdata.springeropen.com/.

Bühlmann, P., Pretsch, E., & Bakker, E. (1998). Carrier-based ion-selective electrodes and bulk optodes. 2. Ionophores for potentiometric and optical sensors. *Chemical Reviews*, *98*(4), 1593–1688. Available from https://doi.org/10.1021/cr970113+.

Butt, M. A., Kazanskiy, N. L., & Khonina, S. N. (2022). Revolution in flexible wearable electronics for temperature and pressure monitoring—A review. *Electronics*, *11*(5). Available from https://doi.org/10.3390/electronics11050716.

Chen, B., Tang, W., & Wang, Z. L. (2021). Advanced 3D printing-based triboelectric nanogenerator for mechanical energy harvesting and self-powered sensing. *Materials Today*, *50*, 224–238. Available from https://doi.org/10.1016/j.mattod.2021.05.017, http://www.journals.elsevier.com/materials-today/.

Chen, S., Qi, J., Fan, S., Qiao, Z., Yeo, J. C., & Lim, C. T. (2021). Flexible wearable sensors for cardiovascular health monitoring. *Advanced Healthcare Materials*, *10*(17). Available from https://doi.org/10.1002/adhm.202100116, http://onlinelibrary.wiley.com/journal/10.1002/. (ISSN) 2192–2659.

Chen, M., Zhou, Y., Lang, J., Li, L., & Zhang, Y. (2022). Triboelectric nanogenerator and artificial intelligence to promote precision medicine for cancer. *Nano Energy*, *92*. Available from https://doi.org/10.1016/j.nanoen.2021.106783.

Chen, X., Hou, Z., Li, G., Yu, W., Xue, Y., Niu, G., Xin, M., Yang, L., Meng, C., & Guo, S. (2022). A laser-scribed wearable strain sensing system powered by an integrated rechargeable thin-film zinc-air battery for a long-time continuous healthcare monitoring. *Nano Energy*, *101*. Available from https://doi.org/10.1016/j.nanoen.2022.107606.

Cheng, Z., Song, H., Zhang, X., Cheng, X., Xu, Y., Zhao, H., Gao, S., & Huo, L. (2021). Morphology control of ZnO by adjusting the solvent and non-enzymatic nitrite ions electrochemical sensor constructed with stir bar-shaped ZnO/Nafion nanocomposite. *Sensors and Actuators B: Chemical*, *346*. Available from https://doi.org/10.1016/j.snb.2021.130525.

Dang, W., Sun, Y., Jiao, H., Xu, L., & Lin, M. (2020). AuNPs-NH2/Cu-MOF modified glassy carbon electrode as enzyme-free electrochemical sensor detecting H2O2. *Journal of Electroanalytical Chemistry*, *856*. Available from https://doi.org/10.1016/j.jelechem.2019.113592.

Dinh, T., Nguyen, T., Phan, H. P., Nguyen, N. T., Dao, D. V., & Bell, J. (2020). Stretchable respiration sensors: Advanced designs and multifunctional platforms for wearable physiological monitoring. *Biosensors and Bioelectronics*, *166*. Available from https://doi.org/10.1016/j.bios.2020.112460, http://www.elsevier.com/locate/bios.

Dong, Q., Ryu, H., & Lei, Y. (2021). Metal oxide based non-enzymatic electrochemical sensors for glucose detection. *Electrochimica Acta*, *370*. Available from https://doi.org/10.1016/j.electacta.2021.137744.

Du, R., Joswig, J. O., Hübner, R., Zhou, L., Wei, W., Hu, Y., & Eychmüller, A. (2020). Freeze–Thaw-promoted fabrication of clean and hierarchically structured noble-metal aerogels for electrocatalysis and photoelectrocatalysis. *Angewandte Chemie - International Edition*, *59*(21), 8293–8300. Available from https://doi.org/10.1002/anie.201916484, http://onlinelibrary.wiley.com/journal/10.1002/. (ISSN) 1521–3773.

DUBBE, A., & MOOS, R. (2008). Potentiometric hydrocarbon gas sensing characteristics of sodium ion conducting zeolite ZSM-5. *Sensors and Actuators B: Chemical*, *130*(1), 546–550. Available from https://doi.org/10.1016/j.snb.2007.09.067.

Dunn, J., Kidzinski, L., Runge, R., Witt, D., Hicks, J. L., Schüssler-Fiorenza Rose, S. M., Li, X., Bahmani, A., Delp, S. L., Hastie, T., & Snyder, M. P. (2021). Wearable sensors enable personalized predictions of clinical laboratory measurements. *Nature Medicine*, *27*(6), 1105–1112. Available from https://doi.org/10.1038/s41591-021-01339-0, http://www.nature.com/nm/index.html.

Enseleit, U., Feller, C., Partsch, U., & Vonau, W. (2021). Iodide determination with chalcogenide glass electrodes. *Journal of Solid State Electrochemistry*, *25*(8–9), 2293–2300. Available from https://doi.org/10.1007/s10008-021-05000-1.

Epicore Biosystems, Inc. (2022). 9 16 Gx Sweat Patch Gx Sweat Patch https://www.epicorebiosystems.com/.

Escobedo, P., Ramos-Lorente, C. E., Martínez-Olmos, A., Carvajal, M. A., Ortega-Muñoz, M., Orbe-Payá, Id, Hernández-Mateo, F., Santoyo-González, F., Capitán-Vallvey, L. F., Palma, A. J., & Erenas, M. M. (2021). Wireless wearable wristband for continuous sweat pH monitoring. *Sensors and Actuators, B: Chemical* (327). Available from https://doi.org/10.1016/j.snb.2020.128948, https://www.journals.elsevier.com/sensors-and-actuators-b-chemical.

Ghaffari, R., Rogers, J. A., & Ray, T. R. (2021). Recent progress, challenges, and opportunities for wearable biochemical sensors for sweat analysis. *Sensors and Actuators, B: Chemical* (332). Available from https://doi.org/10.1016/j.snb.2021.129447, https://www.journals.elsevier.com/sensors-and-actuators-b-chemical.

Ghoorchian, A., Kamalabadi, M., Moradi, M., Madrakian, T., Afkhami, A., Bagheri, H., Ahmadi, M., & Khoshsafar, H. (2022). Wearable potentiometric sensor based on Na0.44MnO2 for non-invasive monitoring of sodium ions in sweat. *Analytical Chemistry*, *94*(4), 2263–2270. Available from https://doi.org/10.1021/acs.analchem.1c04960, http://pubs.acs.org/journal/ancham.

Guinovart, T., Valdés-Ramírez, G., Windmiller, J. R., Andrade, F. J., & Wang, J. (2014). Bandage-based wearable potentiometric sensor for monitoring wound pH. *Electroanalysis*, *26*(6), 1345–1353. Available from https://doi.org/10.1002/elan.201300558.

Guo, H., Jiang, Z., Ren, D., Li, S., Wang, J., Cai, X., Zhang, D., Guo, Q., Xiao, J., & Yang, J. (2021). High-performance flexible micro-supercapacitors printed on textiles for powering wearable electronics. *ChemElectroChem*, *8*(9), 1574–1579. Available from https://doi.org/10.1002/celc.202100100.

Haleem, A., Javaid, M., Singh, R. P., Suman, R., & Rab, S. (2021). Biosensors applications in medical field: A brief review. *Sensors International*, *2*. Available from https://doi.org/10.1016/j.sintl.2021.100100.

Hanna, J., Tawk, Y., Azar, S., Ramadan, A. H., Dia, B., Shamieh, E., Zoghbi, S., Kanj, R., Costantine, J., & Eid, A. A. (2022). Wearable flexible body matched electromagnetic sensors for personalized non-invasive glucose monitoring. *Scientific Reports*, *12*(1). Available from https://doi.org/10.1038/s41598-022-19251-z, http://www.nature.com/srep/index.html.

Hashemi, S. A., Ramakrishna, S., & Aberle, A. G. (2020). Recent progress in flexible-wearable solar cells for self-powered electronic devices. *Energy and Environmental Science*, *13*(3), 685–743. Available from https://doi.org/10.1039/c9ee03046h, http://pubs.rsc.org/en/journals/journal/ee.

Hauke, A., Simmers, P., Ojha, Y. R., Cameron, B. D., Ballweg, R., Zhang, T., Twine, N., Brothers, M., Gomez, E., & Heikenfeld, J. (2018). Complete validation of a continuous and blood-correlated sweat biosensing device with integrated sweat stimulation. *Lab on a Chip*, *18*(24), 3750–3759. Available from https://doi.org/10.1039/c8lc01082j.

He, C., Asif, M., Liu, Q., Xiao, F., Liu, H., & Xia, B. Y. (2023). Noble metal construction for electrochemical nonenzymatic glucose detection. *Advanced Materials Technologies*, *8*(1). Available from https://doi.org/10.1002/admt.202200272.

Hilty, D. M., Armstrong, C. M., Luxton, D. D., Gentry, M. T., & Krupinski, E. A. (2021). A scoping review of sensors, wearables, and remote monitoring for behavioral health: Uses, outcomes, clinical competencies, and research directions. *Journal of Technology in Behavioral Science*, *6*(2), 278–313. Available from https://doi.org/10.1007/s41347-021-00199-2, https://www.springer.com/journal/41347.

Ho, D., Quake, S. R., McCabe, E. R. B., Chng, W. J., Chow, E. K., Ding, X., Gelb, B. D., Ginsburg, G. S., Hassenstab, J., Ho, C. M., Mobley, W. C., Nolan, G. P., Rosen, S. T., Tan, P., Yen, Y., & Zarrinpar, A. (2020). Enabling technologies for personalized and precision medicine. *Trends in Biotechnology*, *38*(5), 497–518. Available from https://doi.org/10.1016/j.tibtech.2019.12.021, http://www.elsevier.com/locate/tibtech.

Holade, Y., Tingry, S., Servat, K., Napporn, T. W., Cornu, D., & Kokoh, K. B. (2017). Nanostructured inorganic materials at work in electrochemical sensing and biofuel cells. *Catalysts*, *7*(1). Available from https://doi.org/10.3390/catal7010031, http://www.mdpi.com/2073-4344/7/1/31/pdf.

Hu, X. B., Qin, Y., Fan, W. T., Liu, Y. L., Huang, W. H., & Three-Dimensional, A. (2021). Electrochemical biosensor integrated with hydrogel enables real-time monitoring of cells under their in vivo-like microenvironment. *Analytical Chemistry*, *93*(22), 7917–7924. Available from https://doi.org/10.1021/acs.analchem.1c00621, http://pubs.acs.org/journal/ancham.

Huang, X., Liu, Y., Zhou, J., Nejad, S. K., Wong, T. H., Huang, Y., Li, H., Yiu, C. K., Park, W., Li, J., Su, J., Zhao, L., Yao, K., Wu, M., Gao, Z., Li, D., Li, J., Shi, R., & Yu, X. (2022). Garment embedded sweat-activated batteries in wearable electronics for continuous sweat monitoring. *Npj Flexible Electronics*, *6*(1). Available from https://doi.org/10.1038/s41528-022-00144-0, nature.com/npjflexelectron/.

Hwang, D. W., Lee, S., Seo, M., & Chung, T. D. (2018). Recent advances in electrochemical non-enzymatic glucose sensors – A review. *Analytica Chimica Acta*, *1033*, 1–34. Available from https://doi.org/10.1016/j.aca.2018.05.051, http://www.journals.elsevier.com/analytica-chimica-acta/.

Jeerapan, I., Moonla, C., Thavarungkul., & Kanatharana, P. (2022). *Lab on a body for biomedical electrochemical sensing applications: The next generation of microfluidic devices progress in molecular biology and translational science* (2022, pp. 249–279). Academic Press. Available from 10.1016/bs.pmbts.2021.07.025.

Ilager, D., Malode, S. J., Kulkarni, R. M., & Shetti, N. P. (2022). Electrochemical sensor based on Ca-doped ZnO nanostructured carbon matrix for algicide dichlone. *Journal of Hazardous Materials Advances*, *7*. Available from https://doi.org/10.1016/j.hazadv.2022.100132.

Itaya, K., Shoji, N., & Uchida, I. (1984). Catalysis of the reduction of molecular oxygen to water at prussian blue modified electrodes. *Journal of the American Chemical Society*, *106*(12), 3423–3429. Available from https://doi.org/10.1021/ja00324a007.

Japaridze, G., Loeckx, D., Buckinx, T., Armand Larsen, S., Proost, R., Jansen, K., MacMullin, P., Paiva, N., Kasradze, S., Rotenberg, A., Lagae, L., & Beniczky, S. (2022). Automated detection of absence seizures using a wearable electroencephalographic device: a phase 3 validation study and feasibility of automated behavioral testing. *Epilepsia*. Available from https://doi.org/10.1111/epi.17200, http://onlinelibrary.wiley.com/journal/10.1111/(ISSN)1528-1167.

Jasiński, A., Guziński, M., Lisak, G., Bobacka, J., & Bocheńska, M. (2015). Solid-contact lead (II) ion-selective electrodes for potentiometric determination of lead(II) in presence of high concentrations of Na(I), Cu(II), Cd(II), Zn(II), Ca(II) and Mg(II). *Sensors and Actuators, B: Chemical*, *218*, 25–30. Available from https://doi.org/10.1016/j.snb.2015.04.089.

Jeerapan, I., & Poorahong, S. (2020). Review—flexible and stretchable electrochemical sensing systems: Materials, energy sources, and integrations. *Journal of the Electrochemical Society*, *167*(3). Available from https://doi.org/10.1149/1945-7111/ab7117.

Jeerapan, I., Sempionatto, J. R., & Wang, J. (2020). On-body bioelectronics: Wearable biofuel cells for bioenergy harvesting and self-powered biosensing. *Advanced Functional Materials*, *30*(29). Available from https://doi.org/10.1002/adfm.201906243.

Jeerapan, I., Sangsudcha, W., & Phokhonwong, P. (2022). Wearable energy devices on mask-based printed electrodes for self-powered glucose biosensors. *Sensing and Bio-Sensing Research*, *38*. Available from https://doi.org/10.1016/j.sbsr.2022.100525.

Jeerapan, I., Sempionatto, J. R., Pavinatto, A., You, J. M., & Wang, J. (2016). Stretchable biofuel cells as wearable textile-based self-powered sensors. *Journal of Materials Chemistry A*, *4*(47), 18342–18353. Available from https://doi.org/10.1039/C6TA08358G, http://pubs.rsc.org/en/journals/journalissues/ta.

Jendrlin, M., Khumngern, S., Numnuam, A., Thavarungkul, P., Kanatharana, P., Kirsanov, D., Zholobenko, V. L., Mendecki, L., & Radu, A. (2021). Ion sensing pencil: Draw your own sensor. *Sensors and Actuators B: Chemical*, *337*. Available from https://doi.org/10.1016/j.snb.2021.129751.

Jeong, I. C., Bychkov, D., & Searson, P. C. (2019). Wearable devices for precision medicine and health state monitoring. *IEEE Transactions on Biomedical Engineering*, *66*(5), 1242–1258. Available from https://doi.org/10.1109/TBME.2018.2871638, http://ieeexplore.ieee.org/xpl/RecentIssue.jsp?reload = true&punumber = 10.

Jeong, J., Jeon, S., Ma, X., Kwon, Y. W., Shin, D. M., & Hong, S. W. (2021). Sustainable A and flexible microbrush-faced triboelectric generator for portable/wearable applications. *Advanced Materials*, *33*(39), 1521–4095. Available from https://doi.org/10.1002/adma.202102530, http://onlinelibrary.wiley.com/journal/10.1002/. (ISSN).

Jia, W., Bandodkar, A. J., Valdés-Ramírez, G., Windmiller, J. R., Yang, Z., Ramírez, J., Chan, G., & Wang, J. (2013). Electrochemical tattoo biosensors for real-time noninvasive lactate monitoring in human perspiration. *Analytical Chemistry*, *85*(14), 6553–6560. Available from https://doi.org/10.1021/ac401573r.

Jiang, D., Xu, C., Zhang, Q., Ye, Y., Cai, Y., Li, K., Li, Y., Huang, X., & Wang, Y. (2022). In situ preparation of lactate-sensing membrane for the noninvasive and wearable analysis of sweat. *Biosensors and Bioelectronics*, *210*. Available from https://doi.org/10.1016/j.bios.2022.114303.

Jin, X., Li, G., Xu, T., Su, L., Yan, D., & Zhang, X. (2022). Fully integrated flexible biosensor for wearable continuous glucose monitoring. *Biosensors and Bioelectronics*, *196*. Available from https://doi.org/10.1016/j.bios.2021.113760.

Kaempgen, M., & Roth, S. (2006). Transparent and flexible carbon nanotube/polyaniline pH sensors. *Journal of Electroanalytical Chemistry*, *586*(1), 72–76. Available from https://doi.org/10.1016/j.jelechem.2005.09.009.

Karami, Z., Youssefi, M., Raeissi, K., & Zhiani, M. (2021). An efficient textile-based electrode utilizing silver nanoparticles/reduced graphene oxide/cotton fabric composite for high-performance wearable supercapacitors. *Electrochimica Acta*, *368*. Available from https://doi.org/10.1016/j.electacta.2020.137647.

Karyakin, A. A., Gitelmacher, O. V., & Karyakina, E. E. (1994). A high-sensitive glucose amperometric biosensor based on prussian blue modified electrodes. *Analytical Letters*, *27* (15), 2861–2869. Available from https://doi.org/10.1080/00032719408000297.

Kazanskiy, N. L., Butt, M. A., & Khonina, S. N. (2022). Recent advances in wearable optical sensor automation powered by battery versus skin-like battery-free devices for personal healthcare—A review. *Nanomaterials*, *12*(3). Available from https://doi.org/10.3390/nano12030334, https://www.mdpi.com/2079-4991/12/3/334/pdf.

Khor, S. M., Choi, J., Won, P., & Ko, S. H. (2022). Challenges and strategies in developing an enzymatic wearable sweat glucose biosensor as a practical point-of-care monitoring tool for type II diabetes. *Nanomaterials*, *12*(2). Available from https://doi.org/10.3390/nano12020221, https://www.mdpi.com/2079-4991/12/2/221/pdf.

Khoshmanesh, F., Thurgood, P., Pirogova, E., Nahavandi, S., & Baratchi, S. (2021). Wearable sensors: At the frontier of personalised health monitoring, smart prosthetics and assistive technologies. *Biosensors and Bioelectronics*, *176*. Available from https://doi.org/10.1016/j.bios.2020.112946.

Khumngern, S., Jirakunakorn, R., Thavarungkul, P., Kanatharana, P., & Numnuam, A. (2021). A highly sensitive flow injection amperometric glucose biosensor using a gold nanoparticles/polytyramine/prussian blue modified screen-printed carbon electrode. *Bioelectrochemistry (Amsterdam, Netherlands)*, *138*. Available from https://doi.org/10.1016/j.bioelechem.2020.107718.

Kil, H. J., Kim, S. R., & Park, J. W. (2022). A self-charging supercapacitor for a patch-type glucose sensor. *ACS Applied Materials and Interfaces*, *14*(3), 3838–3848. Available from https://doi.org/10.1021/acsami.1c21394, http://pubs.acs.org/journal/aamick.

Kim, J., Campbell, A. S., de Ávila, B. E. F., & Wang, J. (2019). Wearable biosensors for healthcare monitoring. *Nature Biotechnology*, *37*(4), 389–406. Available from https://doi.org/10.1038/s41587-019-0045-y, http://www.nature.com/nbt/index.html.

Kim, J., Campbell, A. S., & Wang, J. (2018). Wearable non-invasive epidermal glucose sensors: A review. *Talanta*, *177*, 163–170. Available from https://doi.org/10.1016/j.talanta.2017.08.077, https://www.journals.elsevier.com/talanta.

Kim, J., Khan, S., Wu, P., Park, S., Park, H., Yu, C., & Kim, W. (2021). Self-charging wearables for continuous health monitoring. *Nano Energy*, *79*. Available from https://doi.org/10.1016/j.nanoen.2020.105419.

Kim, M. H., Cho, C. H., Kim, J. S., Nam, T. U., Kim, W. S., Il Lee, T., & Oh, J. Y. (2021). Thermoelectric energy harvesting electronic skin (e-skin) patch with reconfigurable carbon nanotube clays. *Nano Energy* (87). Available from https://doi.org/10.1016/j.nanoen.2021.106156, http://www.journals.elsevier.com/nano-energy/.

Kim, T., Yi, Q., Hoang, E., & Esfandyarpour, R. (2021). A 3D printed wearable bioelectronic patch for multi-sensing and in situ sweat electrolyte monitoring. *Advanced Materials Technologies*, *6*(4). Available from https://doi.org/10.1002/admt.202001021.

Kim, J., Sempionatto, J. R., Imani, S., Hartel, M. C., Barfidokht, A., Tang, G., Campbell, A. S., Mercier, P. P., & Wang, J. (2018). Simultaneous monitoring of sweat and interstitial fluid using a single wearable biosensor platform. *Advanced Science*, *5*(10), 2198–3844. Available from https://doi.org/10.1002/advs.201800880, http://onlinelibrary.wiley.com/journal/10.1002/. (ISSN).

Ko, H. C., Stoykovich, M. P., Song, J., Malyarchuk, V., Choi, W. M., Yu, C. J., Geddes, J. B., Xiao, J., Wang, S., Huang, Y., & Rogers, J. A. (2008). A hemispherical electronic eye camera based on compressible silicon optoelectronics. *Nature*, *454*(7205), 748–753. Available from https://doi.org/10.1038/nature07113, http://www.nature.com/nature/index.html.

Komkova, M. A., Eliseev, A. A., Poyarkov, A. A., Daboss, E. V., Evdokimov, P. V., Eliseev, A. A., & Karyakin, A. A. (2022). Simultaneous monitoring of sweat lactate content and sweat secretion rate by wearable remote biosensors. *Biosensors and Bioelectronics*, *202*. Available from https://doi.org/10.1016/j.bios.2022.113970, http://www.elsevier.com/locate/bios.

Konstantinidis, D., Iliakis, P., Tatakis, F., Thomopoulos, K., Dimitriadis, K., Tousoulis, D., & Tsioufis, K. (2022). Wearable blood pressure measurement devices and new approaches in hypertension management: the digital era. *Journal of Human Hypertension*, *36*(11), 945–951. Available from https://doi.org/10.1038/s41371-022-00675-z.

Kowalewska, B., & Kulesza, P. J. (2009). Application of tetrathiafulvalene-modified carbon nanotubes to preparation of integrated mediating system for bioelectrocatalytic oxidation of glucose. *Electroanalysis*. ***21** (3–5)*, 351–359. Available from https://doi.org/10.1002/elan.200804396Poland, http://www3.interscience.wiley.com/cgi-bin/fulltext/121561592/PDFSTART.

Kucherenko, I. S., Soldatkin, O. O., Kucherenko, D. Y., Soldatkina, O. V., & Dzyadevych, S. V. (2019). Advances in nanomaterial application in enzyme-based electrochemical biosensors: A review. *Nanoscale Advances*, *1*(12), 4560–4577. Available from https://doi.org/10.1039/c9na00491b, pubs.rsc.org/en/journals/journalissues/na?_ga2.190536939.1555337663.1552312502-1364180372.1550481316#!issueidna001002&typecurrent&issnonline2516-0230.

Kukkar, D., Zhang, D., Jeon, B. H., & Kim, K.-H. (2022). Recent advances in wearable biosensors for non-invasive monitoring of specific metabolites and electrolytes associated with chronic kidney disease: Performance evaluation and future challenges. *TrAC Trends in Analytical Chemistry*, *150*. Available from https://doi.org/10.1016/j.trac.2022.116570.

Kwon, S., Kim, H., & Yeo, W.-H. (2021). Recent advances in wearable sensors and portable electronics for sleep monitoring. *iScience*, *24*(5). Available from https://doi.org/10.1016/j.isci.2021.102461.

Lee, H., & Roh, J. S. (2021). Charging device for wearable electromagnetic energy-harvesting textiles. *Fashion and Textiles*, *8*(1). Available from https://doi.org/10.1186/s40691-020-00233-6, http://www.springer.com/materials/journal/40691.

Lee, Y. D., & Chung, W. Y. (2009). Wireless sensor network based wearable smart shirt for ubiquitous health and activity monitoring. *Sensors and Actuators, B: Chemical*, *140*(2), 390–395. Available from https://doi.org/10.1016/j.snb.2009.04.040.

Li, G., Hao, J., Li, W., Ma, F., Ma, T., Gao, W., Yu, Y., & Wen, D. (2021). Integrating highly porous and flexible Au hydrogels with Soft-MEMS technologies for high-performance wearable biosensing. *Analytical Chemistry*, *93*(42), 14068–14075. Available from https://doi.org/10.1021/acs.analchem.1c01581, http://pubs.acs.org/journal/ancham.

Li, R., Wei, X., Xu, J., Chen, J., Li, B., Wu, Z., & Wang, Z. (2021). Smart wearable sensors based on triboelectric nanogenerator for personal healthcare monitoring. *Micromachines*, *12*(4). Available from https://doi.org/10.3390/mi12040352.

Li, R., Li, M., Su, Y., Song, J., & Ni, X. (2013). An analytical mechanics model for the island-bridge structure of stretchable electronics. *Soft Matter*, *9*(35), 8476–8482. Available from https://doi.org/10.1039/c3sm51476e.

Li, X., Wu, D., Feng, Q., Zhang, Y., Lv, P., & Wei, Q. (2022). Flexible bioelectrode via in-situ growth of MOF/enzyme on electrospun nanofibers for stretchable enzymatic biofuel cell. *Chemical Engineering Journal*, *440*. Available from https://doi.org/10.1016/j.cej.2022.135719.

Lim, H. R., Lee, Y., Jones, K. A., Kwon, Y. T., Kwon, S., Mahmood, M., Lee, S. M., & Yeo, W. H. (2021). All-in-one, wireless, fully flexible sodium sensor system with integrated Au/CNT/Au nanocomposites. *Sensors and Actuators, B: Chemical* (331). Available from https://doi.org/10.1016/j.snb.2020.129416, https://www.journals.elsevier.com/sensors-and-actuators-b-chemical.

Lin, K., Xie, J., Bao, Y., Ma, Y., Chen, L., Wang, H., Xu, L., Tang, Y., Liu, Z., Sun, Z., Gan, S., & Niu, L. (2022). Self-adhesive and printable tannin–graphene supramolecular aggregates for wearable potentiometric pH sensing. *Electrochemistry Communications*, *137*. Available from https://doi.org/10.1016/j.elecom.2022.107261.

Liu, C., Jia, L., Li, J., Wu, Y., & Zhang, B. (2022). Fast and stable potentiometric Na + ion response of Na0.44MnO2 nanorod-based pseudocapacitive electrode via a short time activation. *Materials Letters*, *308*. Available from https://doi.org/10.1016/j.matlet.2021.131102.

Liu, Q., Tai, H., Yuan, Z., Zhou, Y., Su, Y., & Jiang, Y. (2019). A high-performances flexible temperature sensor composed of polyethyleneimine/reduced graphene oxide bilayer for real-time monitoring. *Advanced Materials Technologies*, *4*(3). Available from https://doi.org/10.1002/admt.201800594.

Liu, T., Yang, Y., Fan, W., & Wu, C. (2021). Few-shot learning for cardiac arrhythmia detection based on electrocardiogram data from wearable devices. *Digital Signal Processing, 116*. Available from https://doi.org/10.1016/j.dsp.2021.103094.

Liu, R., Ramli, A.A., Zhang, H., Henricson, E., & Liu, X. (2022). Lecture notes in computer science (including subseries Lecture notes in artificial intelligence and Lecture notes in bioinformatics). Springer Science and Business Media Deutschland GmbH United States. An overview of human activity recognition using wearable sensors: Healthcare and artificial intelligence, pp. 1–14. https://www.springer.com/series/558 12993. 10.1007/978-3-030-96068-1_1 16113349. 1 2022/01.

Liu, R., Wang, Z. L., Fukuda, K., & Someya, T. (2022). Flexible self-charging power sources. *Nature Reviews Materials, 7*(11), 870–886. Available from https://doi.org/10.1038/s41578-022-00441-0, https://www.nature.com/natrevmats/.

Luo, H., & Gao, B. (2021). Development of smart wearable sensors for life healthcare. *Engineered Regeneration, 2*, 163–170. Available from https://doi.org/10.1016/j.engreg.2021.10.001, https://www.sciencedirect.com/journal/engineered-regeneration.

Lv, J., Chen, J., & Lee, P. S. (2021). Sustainable wearable energy storage devices self-charged by human-body bioenergy. *SusMat, 1*(2), 285–302. Available from https://doi.org/10.1002/sus2.14.

Lyu, Y., Han, T., Zhong, L., Tang, Y., Xu, L., Ma, Y., Bao, Y., Gan, S., Bobacka, J., & Niu, L. (2022). Coulometric ion sensing with Li + -selective LiMn2O4 electrodes. *Electrochemistry Communications, 139*. Available from https://doi.org/10.1016/j.elecom.2022.107302.

Ma, J., Cui, Z., Du, Y., Xu, Q., Deng, Q., & Zhu, N. (2021). Multifunctional prussian blue/graphene ink for flexible biosensors and supercapacitors. *Electrochimica Acta, 387*. Available from https://doi.org/10.1016/j.electacta.2021.138496.

Ma, R., Shao, R., An, X., Zhang, Q., & Sun, S. (2022). Recent advancements in noninvasive glucose monitoring and closed-loop management systems for diabetes. *Journal of Materials Chemistry B, 10*(29), 5537–5555. Available from https://doi.org/10.1039/d2tb00749e, http://pubs.rsc.org/en/journals/journal/tb.

Maddirala, G., Searle, T., Wang, X., Alici, G., & Sencadas, V. (2022). Multifunctional skin-compliant wearable sensors for monitoring human condition applications. *Applied Materials Today, 26*. Available from https://doi.org/10.1016/j.apmt.2021.101361.

Maity, D., Rajavel, K., & Kumar, R. T. Rajendra (2021). MWCNT enabled smart textiles based flexible and wearable sensor for human motion and humidity monitoring. *Cellulose, 28*(4), 2505–2520. Available from https://doi.org/10.1007/s10570-020-03617-5, http://www.springer.com/journal/10570.

Mamun, M. A. A., & Yuce, M. R. (2020). Recent progress in nanomaterial enabled chemical sensors for wearable environmental monitoring applications. *Advanced Functional Materials, 30*(51). Available from https://doi.org/10.1002/adfm.202005703, http://onlinelibrary.wiley.com/journal/10.1002/. (ISSN)1616-3028.

Manjakkal, L., Dervin, S., & Dahiya, R. (2020). Flexible potentiometric pH sensors for wearable systems. *RSC Advances, 10*(15), 8594–8617. Available from https://doi.org/10.1039/d0ra00016g, http://pubs.rsc.org/en/journals/journal/ra.

Mariello, M., Qualtieri, A., Mele, G., & De Vittorio, M. (2021). Metal-free multilayer hybrid PENG based on soft electrospun/-sprayed membranes with cardanol additive for harvesting energy from surgical face masks. *ACS Applied Materials and Interfaces, 13*(17), 20606–20621. Available from https://doi.org/10.1021/acsami.1c01740, http://pubs.acs.org/journal/aamick.

Mathew, M., Radhakrishnan, S., Vaidyanathan, A., Chakraborty, B., & Rout, C. S. (2021). Flexible and wearable electrochemical biosensors based on two-dimensional materials: Recent developments. *Analytical and Bioanalytical Chemistry*, *413*(3), 727–762. Available from https://doi.org/10.1007/s00216-020-03002-y, https://link.springer.com/journal/216/volumes-and-issues.

Matsuhisa, N., Kaltenbrunner, M., Yokota, T., Jinno, H., Kuribara, K., Sekitani, T., & Someya, T. (2015). Printable elastic conductors with a high conductivity for electronic textile applications. *Nature Communications* (6). Available from https://doi.org/10.1038/ncomms8461, http://www.nature.com/ncomms/index.html.

Meta Materials Inc. (2022) 9 16 GlucoWISE®. https://gluco-wise.com/.

Miao, G., Fang, S., Wang, S., & Zhou, S. (2022). A low-frequency rotational electromagnetic energy harvester using a magnetic plucking mechanism. *Applied Energy*, *305*. Available from https://doi.org/10.1016/j.apenergy.2021.117838.

Mikkonen, R., Puistola, P., Jönkkäri, I., & Mäntysalo, M. (2020). Inkjet printable polydimethylsiloxane for all-inkjet-printed multilayered soft electrical applications. *ACS Applied Materials and Interfaces*, *12*(10), 11990–11997. Available from https://doi.org/10.1021/acsami.9b19632, http://pubs.acs.org/journal/aamick.

Mohan, A. M. V., Rajendran, V., Mishra, R. K., & Jayaraman, M. (2020). Recent advances and perspectives in sweat based wearable electrochemical sensors. *TrAC - Trends in Analytical Chemistry*, *131*. Available from https://doi.org/10.1016/j.trac.2020.116024, http://www.elsevier.com/locate/trac.

Moonla, C., Lee, D. H., Rokaya, D., Rasitanon, N., Kathayat, G., Lee, W.-Y., Kim, J., & Jeerapan, I. (2022). Review—lab-in-a-mouth and advanced point-of-care sensing systems: Detecting bioinformation from the oral cavity and saliva. *ECS Sensors Plus*, *1*(2). Available from https://doi.org/10.1149/2754-2726/ac7533.

Mou, L., Xia, Y., & Jiang, X. (2021). Epidermal sensor for potentiometric analysis of metabolite and electrolyte. *Analytical Chemistry*, *93*(33), 11525–11531. Available from https://doi.org/10.1021/acs.analchem.1c01940, http://pubs.acs.org/journal/ancham.

Myndrul, V., Coy, E., Babayevska, N., Zahorodna, V., Balitskyi, V., Baginskiy, I., Gogotsi, O., Bechelany, M., Giardi, M. T., & Iatsunskyi, I. (2022). MXene nanoflakes decorating ZnO tetrapods for enhanced performance of skin-attachable stretchable enzymatic electrochemical glucose sensor. *Biosensors and Bioelectronics*, *207*. Available from https://doi.org/10.1016/j.bios.2022.114141.

Naikoo, G. A., Salim, H., Hassan, I. U., Awan, T., Arshad, F., Pedram, M. Z., Ahmed, W., & Qurashi, A. (2021). Recent advances in non-enzymatic glucose sensors based on metal and metal oxide nanostructures for diabetes management – A review. *Frontiers in Chemistry* (9). Available from https://doi.org/10.3389/fchem.2021.748957, http://journal.frontiersin.org/journal/chemistry.

Niu, S., Chang, X., Zhu, Z., Qin, Z., Li, J., Jiang, Y., Wang, D., Yang, C., Gao, Y., & Sun, S. (2021). Low-temperature wearable strain sensor based on a silver nanowires/graphene composite with a near-zero temperature coefficient of resistance. *ACS Applied Materials and Interfaces*, *13*(46), 55307–55318. Available from https://doi.org/10.1021/acsami.1c14671, http://pubs.acs.org/journal/aamick.

Noviana, E., Ozer, T., Carrell, C. S., Link, J. S., McMahon, C., Jang, I., & Henry, C. S. (2021). Microfluidic paper-based analytical devices: From design to applications. *Chemical Reviews*, *121*(19), 11835–11885. Available from https://doi.org/10.1021/acs.chemrev.0c01335, http://pubs.acs.org/journal/chreay.

Okorie, O. N., & Dellinger, P. (2011). Lactate: Biomarker and potential therapeutic target. *Critical Care Clinics*, *27*(2), 299–326. Available from https://doi.org/10.1016/j.ccc.2010.12.013.

Oliver N., & Flores-Mangas F. (2006). HealthGear: A real-time wearable system for monitoring and analyzing physiological signals. In *International workshop on wearable and implantable body sensor networks (BSN'06), IEEE 2006*. doi:10.1109/BSN.2006.27.

Ozer, T., Agir, I., & Henry, C. S. (2022). Low-cost internet of things (IoT)-enabled a wireless wearable device for detecting potassium ions at the point of care. *Sensors and Actuators B: Chemical*, *365*. Available from https://doi.org/10.1016/j.snb.2022.131961.

Pan, Y., Song, Y., Jiang, Q., Jia, Y., Liu, P., Song, H., & Liu, G. (2022). Solvent treatment of wet-spinning PEDOT:PSS fiber towards wearable thermoelectric energy harvesting. *Synthetic Metals*, *283*. Available from https://doi.org/10.1016/j.synthmet.2021.116969.

Park, H. J., Jeong, J. M., Son, S. G., Kim, S. J., Lee, M., Kim, H. J., Jeong, J., Hwang, S. Y., Park, J., Eom, Y., & Choi, B. G. (2021). Fluid-dynamics-processed highly stretchable, conductive, and printable graphene inks for real-time monitoring sweat during stretching exercise. *Advanced Functional Materials*, *31*(21), 1616–3028. Available from https://doi.org/10.1002/adfm.202011059, http://onlinelibrary.wiley.com/journal/10.1002/. (ISSN).

Pinto, Alexandra M. F. R., Oliveira, V. ânia B., & Falcão, Daniela S. (2018). *Miniaturization of direct alcohol fuel cells: Microfabrication techniques and microfluidic architectures* (pp. 245–264). Academic Press. Available from https://www.sciencedirect.com/science/article/pii/B9780128118498000073, 10.1016/B978-0-12-811849-8.00007-3.

Promphet, N., Ummartyotin, S., Ngeontae, W., Puthongkham, P., & Rodthongkum, N. (2021). Non-invasive wearable chemical sensors in real-life applications. *Analytica Chimica Acta*, *1179*. Available from https://doi.org/10.1016/j.aca.2021.338643.

Pullano, S. A., Greco, M., Bianco, M. G., Foti, D., Brunetti, A., & Fiorillo, A. S. (2022). Glucose biosensors in clinical practice: Principles, limits and perspectives of currently used devices. *Theranostics*, *12*(2), 493–511. Available from https://doi.org/10.7150/thno.64035, https://www.thno.org/v12p0493.

Pusta, A., Tertiş, M., Cristea, C., & Mirel, S. (2022). Wearable sensors for the detection of biomarkers for wound infection. *Biosensors*, *12*(1). Available from https://doi.org/10.3390/bios12010001.

Qiao, Y., Qiao, L., Chen, Z., Liu, B., Gao, L., & Zhang, L. (2022). Wearable sensor for continuous sweat biomarker monitoring. *Chemosensors*, *10*(7). Available from https://doi.org/10.3390/chemosensors10070273.

Qin, M., Guo, H., Dai, Z., Yan, X., & Ning, X. (2019). Advances in flexible and wearable pH sensors for wound healing monitoring. *Journal of Semiconductors*, *40*(11). Available from https://doi.org/10.1088/1674-4926/40/11/111607.

Qu, M., Chen, X., Yang, D., Li, D., Zhu, K., Guo, X., & Xie, J. (2021). Monitoring of physiological sounds with wearable device based on piezoelectric MEMS acoustic sensor. *Journal of Micromechanics and Microengineering*, *32*(1). Available from https://doi.org/10.1088/1361-6439/ac371e.

Razmi, H., Heidari, H., & Habibi, Es (2008). pH-sensing properties of PbO2 thin film electrodeposited on carbon ceramic electrode. *Journal of Solid State Electrochemistry*, *12*(12), 1579–1587. Available from https://doi.org/10.1007/s10008-008-0523-5.

Saha, T., Songkakul, T., Knisely, C. T., Yokus, M. A., Daniele, M. A., Dickey, M. D., Bozkurt, A., & Velev, O. D. (2022). Wireless wearable electrochemical sensing polatform with zero-power osmotic sweat extraction for continuous lactate monitoring. *ACS Sensors*, *7*(7), 2037–2048. Available from https://doi.org/10.1021/acssensors.2c00830, http://pubs.acs.org/journal/ascefj.

Salazar, P., Garcia-Garcia, F. J., Yubero, F., Gil-Rostra, J., & González-Elipe, A. R. (2016). Characterization and application of a new pH sensor based on magnetron sputtered porous WO3 thin films deposited at oblique angles. *Electrochimica Acta, 193*, 24–31. Available from https://doi.org/10.1016/j.electacta.2016.02.040, http://www.journals.elsevier.com/electrochimica-acta/.

Sattar, M., & Yeo, W. H. (2022). Recent advances in materials for wearable thermoelectric generators and biosensing devices. *Materials, 15*(12). Available from https://doi.org/10.3390/ma15124315, https://www.mdpi.com/1996-1944/15/12/4315/pdf?version = 1655537910.

Sauvage, F., Baudrin, E., & Tarascon, J.-M. (2007). Study of the potentiometric response towards sodium ions of Na0.44 – xMnO2 for the development of selective sodium ion sensors. *Sensors and Actuators B: Chemical, 120*(2), 638–644. Available from https://doi.org/10.1016/j.snb.2006.03.024.

Sauvage, F., Laffont, L., Tarascon, J.-M., & Baudrin, E. (2007). Study of the insertion/deinsertion mechanism of sodium into Na 0.44 MnO 2. *Inorganic Chemistry, 46*(8), 3289–3294. Available from https://doi.org/10.1021/ic0700250.

Sehit, E., & Altintas, Z. (2020). Significance of nanomaterials in electrochemical glucose sensors: An updated review (2016-2020). *Biosensors and Bioelectronics, 159*. Available from https://doi.org/10.1016/j.bios.2020.112165.

Sempionatto, J. R., Jeerapan, I., Krishnan, S., & Wang, J. (2020). Wearable chemical sensors: Emerging systems for on-body analytical chemistry. *Analytical Chemistry, 92*(1), 378–396. Available from https://doi.org/10.1021/acs.analchem.9b04668.

Sempionatto, J. R., Brazaca, L. C., García-Carmona, L., Bolat, G., Campbell, A. S., Martin, A., Tang, G., Shah, R., Mishra, R. K., Kim, J., Zucolotto, V., Escarpa, A., & Wang, J. (2019). Eyeglasses-based tear biosensing system: Non-invasive detection of alcohol, vitamins and glucose. *Biosensors and Bioelectronics, 137*, 161–170. Available from https://doi.org/10.1016/j.bios.2019.04.058, http://www.elsevier.com/locate/bios.

Seok, D., Lee, S., Kim, M., Cho, J., & Kim, C. (2021). Motion artifact removal techniques for wearable EEG and PPG sensor systems. *Frontiers in Electronics, 2*. Available from https://doi.org/10.3389/felec.2021.685513.

Serag, E., El-Maghraby, A., & El Nemr, A. (2022). Recent developments in the application of carbon-based nanomaterials in implantable and wearable enzyme-biofuel cells. *Carbon Letters, 32*(2), 395–412. Available from https://doi.org/10.1007/s42823-021-00299-2, https://www.springer.com/journal/42823.

Sharifianjazi, F., Jafari Rad, A., Bakhtiari, A., Niazvand, F., Esmaeilkhanian, A., Bazli, L., Abniki, M., Irani, M., & Moghanian, A. (2021). Biosensors and nanotechnology for cancer diagnosis (lung and bronchus, breast, prostate, and colon): a systematic review. *Biomedical Materials, 17*(1). Available from https://doi.org/10.1088/1748-605x/ac41fd.

Shay, T., Dickey, M. D., & Velev, O. D. (2017). Hydrogel-enabled osmotic pumping for microfluidics: Towards wearable human-device interfaces. *Lab on a Chip, 17*(4), 710–716. Available from https://doi.org/10.1039/C6LC01486K, http://pubs.rsc.org/en/journals/journal/lc.

Shu, Y., Shang, Z., Su, T., Zhang, S., Lu, Q., Xu, Q., & Hu, X. (2022). A highly flexible Ni-Co MOF nanosheet coated Au/PDMS film based wearable electrochemical sensor for continuous human sweat glucose monitoring. *Analyst, 147*(7), 1440–1448. Available from https://doi.org/10.1039/d1an02214h, http://pubs.rsc.org/en/journals/journal/an.

Shu, Y., Su, T., Lu, Q., Shang, Z., Xu, Q., & Hu, X. (2021). Highly stretchable wearable electrochemical sensor based on Ni–Co MOF nanosheet-decorated Ag/rGO/PU fiber for continuous sweat glucose detection. *Analytical Chemistry, 93*(48), 16222–16230. Available from https://doi.org/10.1021/acs.analchem.1c04106, http://pubs.acs.org/journal/ancham.

Siirtola, P., Koskimäki, H., Mönttinen, H., & Röning, J. (2018). Using sleep time data from wearable sensors for early detection of migraine attacks. *Sensors*, *18*(5). Available from https://doi.org/10.3390/s18051374.

Silva, C. A., Lv, J., Yin, L., Jeerapan, I., Innocenzi, G., Soto, F., Ha, Y. -G., & Wang, J. (2020). Liquid metal based island-bridge architectures for all printed stretchable electrochemical devices. *Advanced Functional Materials*, *30*(30). Available from https://doi.org/10.1002/adfm.202002041.

Sohan, A., Banoth, P., Aleksandrova, M., Nirmala Grace, A., & Kollu, P. (2021). Review on MXene synthesis, properties, and recent research exploring electrode architecture for supercapacitor applications. *International Journal of Energy Research*, *45*(14), 19746–19771. Available from https://doi.org/10.1002/er.7068, http://onlinelibrary.wiley.com/journal/10.1002/(ISSN)1099-114X.

Song, X., Song, S., Wang, D., & Zhang, H. (2021). Prussian blue analogs and their derived nanomaterials for electrochemical energy storage and electrocatalysis. *Small Methods*, *5*(4). Available from https://doi.org/10.1002/smtd.202001000.

Subrahmanya, S. V. G., Shetty, D. K., Patil, V., Hameed, B. M. Z., Paul, R., Smriti, K., Naik, N., & Somani, B. K. (2022). The role of data science in healthcare advancements: applications, benefits, and future prospects. *Irish Journal of Medical Science*, *191*(4), 1473–1483. Available from https://doi.org/10.1007/s11845-021-02730-z, http://www.springerlink.com/content/0021-1265.

Sun, M., Gu, Y., Pei, X., Wang, J., Liu, J., Ma, C., Bai, J., & Zhou, M. (2021). A flexible and wearable epidermal ethanol biofuel cell for on-body and real-time bioenergy harvesting from human sweat. *Nano Energy*, *86*. Available from https://doi.org/10.1016/j.nanoen.2021.106061.

Tabasum, H., Gill, N., Mishra, R., & Lone, S. (2022). Wearable microfluidic-based e-skin sweat sensors. *RSC Advances*, *12*(14), 8691–8707. Available from https://doi.org/10.1039/d1ra07888g, http://pubs.rsc.org/en/journals/journal/ra.

Takeuchi, K., Takama, N., Sharma, K., Paul, O., Ruther, P., Suga, T., & Kim, B. (2022). Microfluidic chip connected to porous microneedle array for continuous ISF sampling. *Drug Delivery and Translational Research*, *12*(2), 435–443. Available from https://doi.org/10.1007/s13346-021-01050-0, http://www.springer.com/biomed/pharmaceutical + science/journal/13346.

Tan, M., Xu, Y., Gao, Z., Yuan, T., Liu, Q., Yang, R., Zhang, B., & Peng, L. (2022). Recent advances in intelligent wearable medical devices integrating biosensing and drug delivery. *Advanced Materials*, *34*(27). Available from https://doi.org/10.1002/adma.202108491.

Tetik, E. (2022). Flexible metamaterial design based on wearable material for microwave energy harvesting. *Bulletin of Materials Science*, *45*(4). Available from https://doi.org/10.1007/s12034-022-02760-y, http://www.ias.ac.in/matersci/.

Tsow, F., Forzani, E., Rai, A., Wang, R., Tao, N. J., Tsui, R., Mastroianni, S., Knobbe, C., & Gandolfi, J. (2009). A wearable and wireless sensor system for real-time monitoring of toxic environmental volatile organic compounds. *IEEE Sensors Journal*, *9*(12), 1734–1740. Available from https://doi.org/10.1109/JSEN.2009.2030747.

Van Hoovels, K., Xuan, X., Cuartero, M., Gijssel, M., Swarén, M., & Crespo, G. A. (2021). Can wearable sweat lactate sensors contribute to sports physiology? *ACS Sensors*, *6*(10), 3496–3508. Available from https://doi.org/10.1021/acssensors.1c01403, http://pubs.acs.org/journal/ascefj.

Wang, J., Lu, C., & Zhang, K. (2020). Textile-based strain sensor for human motion detection. *Energy and Environmental Materials*, *3*(1), 80–100. Available from https://doi.org/10.1002/eem2.12041, onlinelibrary.wiley.com/journal/25750356.

Wang, R., Zhai, Q., An, T., Gong, S., & Cheng, W. (2021). Stretchable gold fiber-based wearable textile electrochemical biosensor for lactate monitoring in sweat. *Talanta, 222*. Available from https://doi.org/10.1016/j.talanta.2020.121484.

Wei, M., Qiao, Y., Zhao, H., Liang, J., Li, T., Luo, Y., Lu, S., Shi, X., Lu, W., & Sun, X. (2020). Electrochemical non-enzymatic glucose sensors: recent progress and perspectives. *Chemical Communications, 56*(93), 14553–14569. Available from https://doi.org/10.1039/d0cc05650b, http://pubs.rsc.org/en/journals/journal/cc.

Windmiller, J. R., & Wang, J. (2013). Wearable electrochemical sensors and biosensors: A review. *Electroanalysis, 25*(1), 29–46. Available from https://doi.org/10.1002/elan.201200349.

Wu, H., Zhang, Y., Kjøniksen, A. -L., Zhou, X., & Zhou, X. (2021). Wearable biofuel cells: Advances from fabrication to application. *Advanced Functional Materials, 31*(48). Available from https://doi.org/10.1002/adfm.202103976.

Wu, S., Chen, M., Wei, K., & Liu, G. (2021). Sleep apnea screening based on photoplethysmography data from wearable bracelets using an information-based similarity approach. *Computer Methods and Programs in Biomedicine, 211*. Available from https://doi.org/10.1016/j.cmpb.2021.106442.

Xuan, X., Chen, C., Pérez-Ràfols, C., Swarén, M., Wedholm, L., Cuartero, M., & Crespo, G. A. (2022). A wearable biosensor for sweat lactate as a proxy for sport performance monitoring. *Analysis & Sensing*. Available from https://doi.org/10.1002/anse.202200047.

Xie, L., Chen, P., Chen, S., Yu, K., & Sun, H. (2019). Low-cost and highly sensitive wearable sensor based on napkin for health monitoring. *Sensors, 19*(15). Available from https://doi.org/10.3390/s19153427.

Xu, J., Tao, X., Liu, X., & Yang, L. (2022). Wearable eye patch biosensor for noninvasive and simultaneous detection of multiple biomarkers in human tears. *Analytical Chemistry, 94*(24), 8659–8667. Available from https://doi.org/10.1021/acs.analchem.2c00614, http://pubs.acs.org/journal/ancham.

Xu, X., Chen, Y., He, P., Wang, S., Ling, K., Liu, L., Lei, P., Huang, X., Zhao, H., Cao, J., & Yang, J. (2021). Wearable CNT/Ti3C2Tx MXene/PDMS composite strain sensor with enhanced stability for real-time human healthcare monitoring. *Nano Research, 14*(8), 2875–2883. Available from https://doi.org/10.1007/s12274-021-3536-3, http://www.springer.com/materials/nanotechnology/journal/12274.

Xuan, X., Pérez-Ràfols, C., Chen, C., Cuartero, M., & Crespo, G. A. (2021). Lactate biosensing for reliable on-body sweat analysis. *ACS Sensors, 6*(7), 2763–2771. Available from https://doi.org/10.1021/acssensors.1c01009, http://pubs.acs.org/journal/ascefj.

Yang, Z., Pang, Y., Han, X. L., Yang, Y., Yang, Y., Ling, J., Jian, M., Zhang, Y., & Ren, T. L. (2018). Graphene textile strain sensor with negative resistance variation for human motion detection. *ACS Nano, 12*(9), 9134–9141. Available from https://doi.org/10.1021/acsnano.8b03391, http://pubs.acs.org/journal/ancac3.

Yin, L., Kim, K. N., Lv, J., Tehrani, F., Lin, M., Lin, Z., Moon, J. M., Ma, J., Yu, J., Xu, S., & Wang, J. (2021). A self-sustainable wearable multi-modular E-textile bioenergy microgrid system. *Nature Communications, 12*(1). Available from https://doi.org/10.1038/s41467-021-21701-7, http://www.nature.com/ncomms/index.html.

Yoon, S., Yoon, H., Zahed, M. A., Park, C., Kim, D., & Park, J. Y. (2022). Multifunctional hybrid skin patch for wearable smart healthcare applications. *Biosensors and Bioelectronics, 196*. Available from https://doi.org/10.1016/j.bios.2021.113685, http://www.elsevier.com/locate/bios.

Yu, Y., Nyein, H. Y. Y., Gao, W., & Javey, A. (2020). Flexible electrochemical bioelectronics: The rise of in situ bioanalysis. *Advanced Materials, 32*(15). Available from https://doi.org/10.1002/adma.201902083.

Zakaria, M. R., Omar, M. F., Zainol Abidin, M. S., Md Akil, H., & Abdullah, M. M. A. B. (2022). Recent progress in the three-dimensional structure of graphene-carbon nanotubes hybrid and their supercapacitor and high-performance battery applications. *Composites Part A: Applied Science and Manufacturing, 154*. Available from https://doi.org/10.1016/j.compositesa.2021.106756, https://www.journals.elsevier.com/composites-part-a-applied-science-and-manufacturing.

Zhai, J., Luo, B., Li, A., Dong, H., Jin, X., & Wang, X. (2022). Unlocking all-solid ion selective electrodes: Prospects in crop detection. *Sensors, 22*(15). Available from https://doi.org/10.3390/s22155541.

Zhang, J., Mai, X., Hong, X., Chen, Y., & Li, X. (2022). Optical fiber SPR biosensor with a solid-phase enzymatic reaction device for glucose detection. *Sensors and Actuators B: Chemical, 366*. Available from https://doi.org/10.1016/j.snb.2022.131984.

Zhang, W., Ma, D., & Du, J. (2014). Prussian blue nanoparticles as peroxidase mimetics for sensitive colorimetric detection of hydrogen peroxide and glucose. *Talanta, 120*, 362–367. Available from https://doi.org/10.1016/j.talanta.2013.12.028, https://www.journals.elsevier.com/talanta.

Zhang, X., Chen, G., Bian, F., Cai, L., & Zhao, Y. (2019). Encoded microneedle arrays for detection of skin interstitial fluid biomarkers. *Advanced Materials, 31*(37). Available from https://doi.org/10.1002/adma.201902825.

Zhang, Y., Cao, J., Zhu, H., & Lei, Y. (2019). Design, modeling and experimental verification of circular Halbach electromagnetic energy harvesting from bearing motion. *Energy Conversion and Management, 180*, 811–821. Available from https://doi.org/10.1016/j.enconman.2018.11.037, https://www.journals.elsevier.com/energy-conversion-and-management.

Zhao, H., Su, R., Teng, L., Tian, Q., Han, F., Li, H., Cao, Z., Xie, R., Li, G., Liu, X., & Liu, Z. (2022). Recent advances in flexible and wearable sensors for monitoring chemical molecules. *Nanoscale, 14*(5), 1653–1669. Available from https://doi.org/10.1039/d1nr06244a, http://pubs.rsc.org/en/journals/journal/nr.

Zhao, J., Nyein, H. Y. Y., Hou, L., Lin, Y., Bariya, M., Ahn, C. H., Ji, W., Fan, Z., & Javey, A. (2021). A wearable nutrition tracker. *Advanced Materials, 33*(1), 1521–4095. Available from https://doi.org/10.1002/adma.202006444, http://onlinelibrary.wiley.com/journal/10.1002/. (ISSN).

Zhao, T., Li, J., Zeng, H., Fu, Y., He, H., Xing, L., Zhang, Y., & Xue, X. (2018). Self-powered wearable sensing-textiles for real-time detecting environmental atmosphere and body motion based on surface-triboelectric coupling effect. *Nanotechnology, 29*(40). Available from https://doi.org/10.1088/1361-6528/aad3fc.

Zhao, Z., Feng, X., Chen, X., Dai, L. L., & Xu, Y. (2021). A wearable mechano-acoustic sensor based on electrochemical redox reaction for continuous cardiorespiratory monitoring. *Applied Physics Letters, 118*(2). Available from https://doi.org/10.1063/5.0029108.

Zhao, Z., Xia, K., Hou, Y., Zhang, Q., Ye, Z., & Lu, J. (2021). Designing flexible, smart and self-sustainable supercapacitors for portable/wearable electronics: From conductive polymers. *Chemical Society Reviews, 50*(22), 12702–12743. Available from https://doi.org/10.1039/d1cs00800e, http://pubs.rsc.org/en/journals/journal/cs.

Zheng, Y., Tang, N., Omar, R., Hu, Z., Duong, T., Wang, J., Wu, W., & Haick, H. (2021). Smart materials enabled with artificial intelligence for healthcare wearables. *Advanced Functional Materials, 31*(51). Available from https://doi.org/10.1002/adfm.202105482.

Zhou, Z., Shu, T., Sun, Y., Si, H., Peng, P., Su, L., & Zhang, X. (2021). Luminescent wearable biosensors based on gold nanocluster networks for "turn-on" detection of Uric acid, glucose and alcohol in sweat. *Biosensors and Bioelectronics*, *192*. Available from https://doi.org/10.1016/j.bios.2021.113530.

Zhu, B., Li, X., Zhou, L., & Su, B. (2022). An overview of wearable and implantable electrochemical glucose sensors. *Electroanalysis*, *34*(2), 237–245. Available from https://doi.org/10.1002/elan.202100273, http://onlinelibrary.wiley.com/journal/10.1002/. (ISSN) 1521-4109.

Chapter 10

Improving quality of wearable biosensor data through artificial intelligence

Tiong Hoo Lim[1], Amirul Fahmiin Abdullah[1] and Syazana Abdullah Lim[2]

[1]Electrical and Electronic Engineering, Faculty of Engineering Universiti Teknologi Brunei, Brunei Darussalam, [2]Center for Research on Agri-Food Science and Technology, Universiti Teknologi Brunei, Brunei Darussalam

10.1 Introduction

Biosensors are quantitative and qualitative devices that use biomolecules as bioreceptors for detecting, in a specific manner, a determined target. They can be used for environmental screening, drug monitoring, and medical diagnosis, to name a few. Their basic set-up comprises a biosensor (e.g., DNA, antibodies) and a transducer (e.g., electrochemical, optical). Upon the interaction of analytes and bioreceptors, a signal is generated that can be analyzed locally or sent wirelessly to a data center using the Internet. This data can then be converted into meaningful information for end users. Owing to their characteristics, biosensors offer several advantages, including portability, simplicity, and low cost in comparison with conventional analytical methods that are often expensive, bulky, and require trained personnel for operation (Lim & Ahmed, 2016). Biosensors are becoming increasingly important in healthcare sectors due to health challenges caused by chronic diseases attributed to our modern lifestyles and the need for immediate treatment or action. Screening for biomarkers enables early diagnosis and appropriate treatment as preventive measures could not only lessen the financial burden on the healthcare system but most importantly save one's life (Haleem et al., 2021).

With the advancement of recent technology, notably nanotechnology, improved bioreceptors, and signal enhancement strategies (further detailed in Chapters 6 and 7), biosensors have been experiencing major progress in terms of performance. Unfortunately, the successful commercialization of biosensors has been hindered due to several attributes such as poor shelf-life,

Biosensors in Precision Medicine. DOI: https://doi.org/10.1016/B978-0-443-15380-8.00011-4

stability, and signal-noise ratio. Most importantly, the accuracy and the meaning of the data generated by biosensors are essential parameters that can influence the enhancement of some analytical features, including complex sample analysis, multiplexed biomarkers determination, eliminating interferents, and increasing precision in diagnostics. Thus in search of viable solutions to overcome this hurdle, researchers are actively looking for alternative and creative strategies in the field of artificial intelligence (AI). AI has been employed as a rapid and automated diagnostic tool to analyze data, for instance, images and signal output produced from sensors. AI has been demonstrated to be successful in medical settings, particularly in the diagnosis of a variety of disorders and diseases (Attallah, 2022). The incorporation of AI into biosensing technology can bridge the gap between data generation and analysis for the improvement of accuracy (Jin et al., 2020). Currently, machine learning (ML), a subdivision of AI, has been gaining popularity among researchers to classify data accordingly to create automated disease diagnostic tools. ML is a method that leverages data for algorithms to automatically "learn" in making predictions or decisions to improve performance without explicitly being programmed. For instance, despite electrochemical sensors being one of the most common types of biosensors (Lim & Ahmed, 2016), their signal outputs are commonly not reproducible and unstable in a complex matrix due to interferants and other variables such as pH and temperature. Furthermore, electrodes are prone to biofouling with time. Thus the analysis of one-dimensional data may be insufficient for the precise quantification of the analyte in real samples. This presents an excellent opportunity for merging ML with electrochemical biosensors to assess ML's capability for multidimensional data classification and enhance the accuracy and reliability of sensors. Up to now, ML has been used for diagnosing diseases from scans and X-rays. It can provide disease prediction in a more efficient manner than medical practitioners in less time.

State-of-the-art technologies have resulted in the progressive miniaturization of biosensors leading to further development into wearable and portable devices. The trend of wearable devices is booming with their impact being significant in healthcare applications allowing for continuous and remote monitoring of skin temperature, heart, and respiratory rates, and other physiological parameters. Taking advantage of advanced telecommunication technologies, instantaneous and multidirectional massive data transfer from a group of biosensors has opened up the avenue for remote monitoring of these physiological parameters in real time (Wong et al., 2020). As a result, wearable devices connected through wireless networks reduce healthcare burden in terms of cost and manpower with improved efficiency in disease management. Despite the excellent opportunities, data transmitted through wireless networks experience various problems due to issues such as interferences and hardware problems leading to reduced accuracy and reliability of the devices and quality of biosensor data.

This chapter discusses on the existence of anomalies in reading wearable biosensor data and the application of AI to address this issue. The first part explains the various types of errors found in transmitting accurate data through wireless biosensor networks and will be followed up with a discussion on the strategies used to mitigate these unwanted errors using ML.

10.2 Wireless biosensor networks

Wireless biosensor networks (WBNs) serve as a crucial means to transmit data from wearable biosensors to hospital data centers. These networks consist of biosensor units that capture, process, and transfer data related to biological responses. The sensed signals are transmitted across the networks to a hospital data center. With their low power consumption, intelligent processing capability, and adaptive wireless communication module, WBNs play a vital role in medical parameter monitoring, typically involving tens to several biosensor units (Haleem et al., 2021). The collected information is processed and transmitted over a low data rate wireless link to a data collection unit or sink at regular intervals. Subsequently, the aggregated data is sent to a health center or hospital data center via the Internet for further processing as shown in Fig. 10.1. Once the data has been processed, they are stored in the electronic patient record (EPR) systems. A medical practitioner can then retrieve and analyze the data anytime and anywhere to diagnose the patients and proceed to make critical decisions about the medications or procedures to be taken. (Haleem et al., 2021).

However, the functionality of WBNs can be significantly affected by anomalies, which are deviations from the expected behavior of the network. These anomalies can occur in various aspects of WBNs, including application functionality, data transmission, nodes, networks, and the transmission channel (Haleem et al., 2021). In the context of medical devices and WBNs, these anomalies become even more critical as they can impact

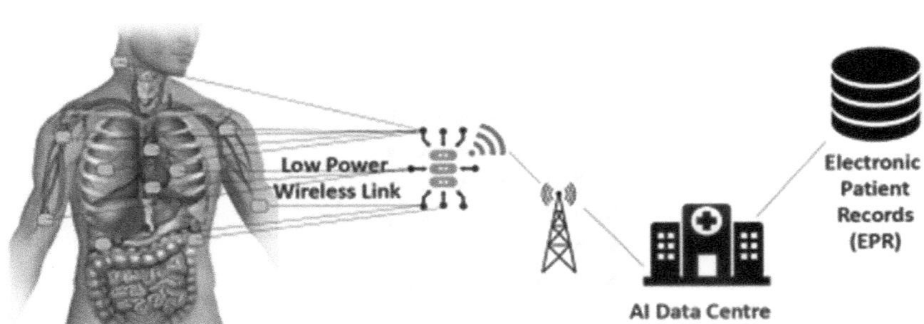

FIGURE 10.1 Biosensing data communication. Transmission of biosensing data to a data center for processing and stored in the electronic patient records database.

decision-making related to medication intake and medical treatment. Therefore ensuring the quality of biosensor data is of utmost importance to avoid potentially life-threatening consequences resulting from poor data accuracy.

10.2.1 Wireless biosensor network architecture

In the context of wireless biosensor network architecture, wearable or biosensor units are typically strategically attached to predefined positions on the human body to facilitate the transmission of information between these units (Zou & Chakrabarty, 2007). These biosensor units serve as interfaces to medical devices, allowing for the monitoring and collection of patient information. The collected data is then processed and organized into packets before being transmitted to a central data collection point, often referred to as the "sink," using either a single-hop or multihop network (Zou & Chakrabarty, 2007) (Fig. 10.2).

In a single-hop network, the data packets are directly transmitted from the source biosensor unit to the sink. This approach offers a higher level of data delivery reliability as the receiving unit can immediately confirm the successful reception of each transmitted packet (Brownfield et al., 2006). However, if the data collection point is far from the source unit, the transmitting biosensor unit needs to increase its transmission range to ensure that the receiver can collect the information. This increased transmission range results in higher battery consumption and reduced battery life for the biosensor unit (Raghunathan et al., 2002).

On the other hand, in a multihop network, nodes can communicate with each other by relaying data packets as long as they are within their respective radio transmission ranges. This allows for lower transmission power and reduced energy consumption in individual nodes (Zou & Chakrabarty, 2007).

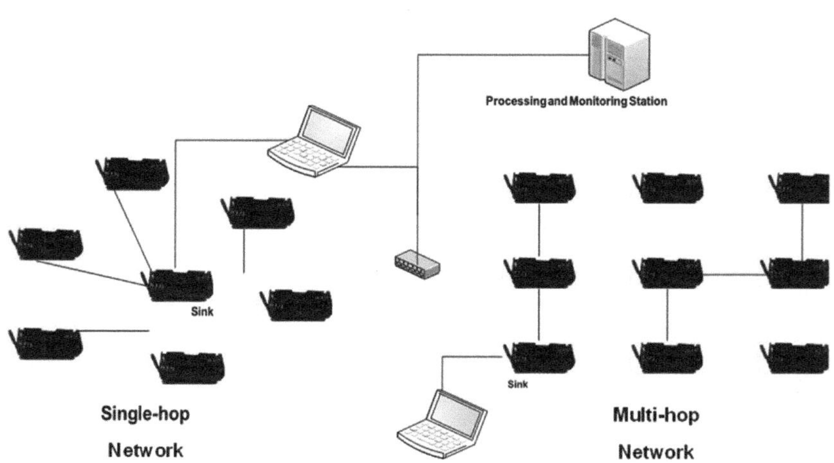

FIGURE 10.2 Wireless biosensor network architecture. Wireless biosensor network architecture differentiating the single-hop (left) and multihop network (right).

However, successful transmission relies on the cooperation and proper functioning of multiple intermediate nodes along the data transmission path.

For example, consider a scenario where a biosensor unit located far from the sink wants to transmit data. In a single-hop network, the unit would need to increase its transmission power to reach the distant sink directly. This would consume more battery power and decrease the unit's operational lifetime. In contrast, in a multihop network, the unit can transmit the data to a nearby node within its transmission range. This node, in turn, can relay the data to another node that is closer to the sink. By utilizing multiple hops, the transmission power can be kept low, reducing energy consumption in individual nodes and potentially extending the overall network lifetime.

10.3 Data errors in wireless biosensor networks

Biosensors can be unreliable as stated in existing literature (Langendoen et al., 2006; Tolle et al., 2005). This may be attributed to data being missing for reasons such as corrupted data, unavailable data during a certain time window, or hardware interference issues occurring during data collection. These missing data need to be addressed especially when continuous monitoring of illness is required. Hence, it is imperative to first understand the fault characteristics of biosensors to solve missing data or errors found in biosensor data.

10.3.1 Classification of anomalies in faults

Biosensors are commonly utilized in an environment where anomalies can occur in both hardware components and communication protocols. To optimize energy consumption and reduce computational costs, application and communication protocols often employ noncomplex algorithms (Langendoen et al., 2006). This allows the devices to operate unattended for extended periods of time, but it can also lead to reduced sensitivity of the sensors when exposed to extreme conditions. As a result, the data generated by the sensors may contain errors such as corruption, missing information, or inaccuracies due to faults (Langendoen et al., 2006).

Hardware faults are typically induced internally, impacting the components and system software of the biosensors due to unreliable hardware and limited resources (Langendoen et al., 2006). On the other hand, interaction faults are externally generated errors that affect the communication protocols. For example, radio interference from other medical devices in the channel can disrupt the links (Langendoen et al., 2006). Additionally, failures and errors can persist in the software, network, communication, or data domains (Langendoen et al., 2006). These errors can often be detected and addressed through the application of AI and data mining techniques.

The subsequent sections examine the four types of anomalies that can arise in real-world biosensor network deployments as a result of these faults.

10.3.1.1 Biosensor node anomalies

Biosensor node anomalies occur when the nodes fail to perform their designated tasks. These anomalies can be attributed to either hardware (Szewczyk et al., 2004) or software faults (Werner-Allen et al., 2006). Hardware faults can manifest as internal malfunctions within biosensor components. These faults can include sensor element failures, power supply issues, or component malfunctions (Langendoen et al., 2006). Unreliable hardware and limited resources can contribute to these hardware faults. For instance, a faulty temperature sensor may produce inaccurate temperature readings, affecting environmental monitoring applications. Typically, they are often attributed to low-cost manufacturing, which renders the nodes susceptible to damage from impacts or heat. Improper handling during deployment can lead to damage to the node's protective enclosure and disconnection or damage to the sensors, resulting in false alarms or missed events (Kahn et al., 1999). For instance, Szewczyk et al. (2004) reported node failure due to a short circuit when the sensor node's circuit board came into direct contact with water. Some nodes may deplete their power due to limited onboard energy resources (Tolle et al., 2005). Limited computational/storage constraints further contribute to node anomalies. Exceeding these limits can result in incomplete tasks, leading to nondeterministic behavior and segmentation faults in the software. Software failures can occur within the biosensor network's operating system or application software. These failures can be attributed to software bugs, glitches, or errors in the firmware controlling the biosensors (Langendoen et al., 2006). Incorrect data processing or interpretation resulting from software failures can lead to misleading or erroneous sensor readings. As these nodes are often deployed in inaccessible locations, maintenance, software updates, and bug fixing are challenging and infrequently performed, exacerbating the challenges in addressing these anomalies.

10.3.1.2 Network anomalies

Network anomaly errors encompass various issues related to the network infrastructure supporting biosensor networks such as sensor nodes, gateways, and transmission radio unavailability. These errors can include packet loss, network congestion, or communication delays. In challenging environments, signal attenuation or interference, or even malicious attacks can cause packet loss and hinder the reliable transmission of sensor data (Jian et al., 2021). Attack-related anomalies are prevalent due to the exploitable nature of existing routing protocols and isolated node deployments (Karlof & Wagner, 2003). Routing protocols in multihop networks introduce challenges such as packet losses due to collisions and hidden terminal effects. Packet losses can occur in routing protocols due to the broadcast nature of communication, particularly when two nodes in proximity transmit simultaneously. Such collisions can lead to the loss of collided packets at the receiver, which is known as the hidden terminal effect (Woo & Culler, 2001). To address this issue, senders must be able to detect these occurrences and initiate retransmission mechanisms to ensure the successful forwarding of data packets to the intended destination.

The dynamic nature of radio communications and network connections further compounds these anomalies, leading to constant changes in routing tables and increased energy consumption. Multihop networks introduce significant volatility in both radio communication and network connectivity, thereby rendering it impractical to ensure an identical packet delivery rate as witnessed during controlled laboratory testing (Szewczyk et al., 2004). According to Ingelrest et al. (2010), the presence of unstable links among nodes can result in continuous fluctuations in the routing table. This frequent updating of the routing table may require additional radio communication, which not only occupies the shared radio channel but also consumes extra energy resources within the nodes. Effective detection and intelligent handling of unstable links are vital to minimize the impact of network anomalies.

10.3.1.3 Signal anomalies

Signal anomalies arise from factors that affect the wireless transmission medium, including signal attenuation, channel interference, and environmental noise (Zhou et al., 2004). Irregularities and interferences caused by nonisotropic characteristics of the propagation media and heterogeneous properties of nodes can result in message loss during wireless propagation. These irregularities and interferences can vary depending on the radio module used. Various physical phenomena, such as metallic objects, other radio-emitting devices, and hardware calibration errors (e.g., antenna gain), can interrupt communication and cause fluctuations in the radio transmission range due to battery depletion (Tolle et al., 2005). Physical objects, such as buildings or trees, can reflect signals, causing network interference. These faults are transient in nature, as the presence of interferences fluctuates nondeterministically (Chipara et al., 2010). Additionally, other wireless devices operating at the same frequency as wireless sensor networks can introduce noise and increase the packet error rate, resulting in transient faults (Gungor & Hancke, 2009). Robust techniques for detecting and mitigating these signal anomalies are essential to ensure reliable communication (Zhou et al., 2004).

10.3.1.4 Data anomalies

Data anomalies pertain to errors in the information received at the sink. Inconsistencies in measurements are considered anomalies and must be detected and eliminated. Processing faults (Elnahrawy & Nath, 2003), hardware faults (Ni et al., 2009), and malicious attacks (Wang et al., 2006) contribute to data anomalies. The cumulative impact of node, network, and signal anomalies can lead to corrupted or missing data, compromising the accuracy of sensor readings. Systematic and random errors are identified as two primary sources of data anomalies (Elnahrawy & Nath, 2003). Systematic errors stem from changes in operating conditions and calibration errors, while random errors result from external noise and inaccuracies in measurements, and computational techniques. These errors can impact the accuracy of sensor readings. For instance, Tolle et al. (2005) showed that as the battery voltage dropped below 2.4 V, the sensor data deviated from the normal range and displayed inconsistencies. Prolonged node operation can give rise to systematic errors that impact components and running applications, including sensor calibration or synchronization timing drift (Ni et al., 2009). Additionally, intentional data anomalies can be introduced when an attacker gains access to a sensor node and injects false or retransmitted packets into the network (Wang et al., 2006). This type of attack not only drains the node's battery but also leads to network congestion (Raymond & Midkiff, 2008).

10.3.2 Relationship between anomalies

The relationship between anomalies in biosensor networks is complex and interconnected, with one anomaly often triggering or influencing another. Understanding these relationships and causality is crucial for comprehending the overall impact and dynamics of anomalies within the network. The occurrence of these faults can have a cascading effect on each other, as depicted in Fig. 10.3. A fault in one component can trigger a partial failure within the system (Tanenbaum & Steen, 2002). The impact of this partial failure on other components depends on the nature of the faults (Avižienis et al., 2004). For instance, in biosensor node anomalies, if a node fails due to a hardware fault, it may lead to a gap in the network topology, causing network partitions or isolated regions. This can subsequently lead to routing failures, packet losses, or increased communication overhead, resulting in network anomalies. A faulty node may exhibit erratic behavior in transmitting or receiving signals, leading to signal attenuation, interference, or intermittent connectivity issues (Huo et al., 2009). The occurrence of node and network anomalies can directly impact data anomalies. Node anomalies, such as hardware faults or limited computational resources, can cause corrupted or missing data, leading to data anomalies. Similarly, network anomalies, such as packet losses or malicious attacks, can compromise the integrity and accuracy of the data received at the sink. The cumulative effect of these anomalies can result in inconsistent measurements, misleading analysis, or unreliable decision-making based on the collected data. Tanenbaum and Steen (2002) stress the importance of constructing a system capable of automatically detecting and recovering from partial failures without compromising overall performance.

It is important to note that the causality between anomalies is not unidirectional. Anomalies can also have feedback effects, where one anomaly exacerbates the occurrence or impact of another anomaly, creating a cycle of

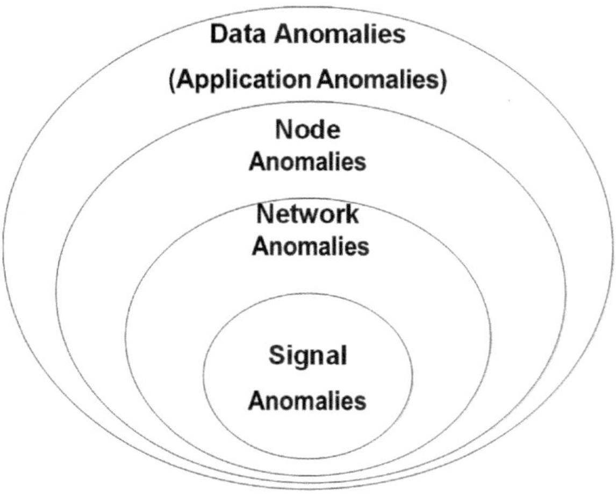

FIGURE 10.3 Causal relationship between anomalies. The intertwining relationship between anomalies in WBNs where each anomaly potentially triggering or contributing to the occurrence of others.

mutual influence. For example, network anomalies, such as packet losses or routing failures, can increase the energy consumption of the nodes due to retransmissions or route rediscovery. This increased energy consumption can further deplete the node's power resources, leading to node anomalies such as power exhaustion or performance degradation.

In summary, the relationship between anomalies in WBNs is intertwined, with each anomaly potentially triggering or contributing to the occurrence of others. Node anomalies, network anomalies, signal anomalies, and data anomalies are interdependent, and their interactions can have a cascading effect on the overall performance, reliability, and data quality within the network. Understanding these relationships is crucial for designing effective anomaly detection, mitigation strategies, and system resilience in WBNs.

10.3.3 Mitigating anomalies for reliable operation

Anomalies in biosensor networks can significantly impact their accuracy and reliability. Understanding the different types of anomalies and their effects on data quality is essential for developing robust biosensor networks. To ensure reliable and accurate operation, biosensor networks need mechanisms to detect and mitigate node, network, signal, and data anomalies. Techniques such as fault-tolerant protocols, anomaly detection algorithms, calibration procedures, and secure communication mechanisms can be employed to address these anomalies and enhance the reliability and accuracy of biosensor data (Jian et al., 2011; Langendoen et al., 2006). Leveraging AI and ML techniques can help overcome these anomalies, enhancing the accuracy of biosensor networks and enabling their effective application in various domains. Both can be adopted in WBNs especially at the application level to ensure the extraction and utilization of useful data in medical applications for disease prediction and diagnosis. Before feature extraction and data classification, the data need to be preprocessed and cleaned by removing anomalies using existing anomaly detection algorithms and replacing missing data. The subsequent section discusses how ML can be employed to eliminate anomalous data.

10.4 Artificial intelligence for biosensor data classification

In the field of wireless biosensor research, the integration of AI and ML techniques has emerged as a promising solution to address the challenges posed by anomalies in WBNs (Haleem et al., 2021). AI and ML algorithms can effectively detect and mitigate data anomalies, contributing to enhanced accuracy and reliability of biosensor data. For example, anomaly detection algorithms can be applied to identify anomalies in the data collected by biosensors. These algorithms analyze patterns and deviations from normal behavior to flag data points that are likely to be erroneous or inconsistent (Langendoen et al., 2006; Tolle et al., 2005). By leveraging AI and ML, it can reduce the need for manual inspection and increase the efficiency of data-cleaning processes. Moreover, AI-based methods can assist in filling missing data gaps caused by anomalies. Missing data can be imputed or predicted based on the available information (Haleem et al., 2021). This allows for more complete datasets, enabling accurate analysis and decision-making. One real-life example of AI application in anomaly detection and data imputation is a study by Tolle et al. (2005). The researchers reported that the data obtained from biosensors exhibited abnormalities when the battery voltage fell below a certain threshold. By employing ML algorithms, they were able to identify and correct these anomalies, ensuring that the data remained accurate and reliable.

In the literature, there are numerous studies that have been conducted on the application of AI and ML in the medical field, most commonly in noncommunicable diseases such as diabetes, cardiovascular diseases, breast cancer, and so on. For example, Vijayan and Anjali (2015) tackled the issue of diabetes mellitus by employing a blend of multiple algorithms. Their approach resulted in a model capable of classifying individuals prone to diabetes mellitus with an accuracy rate of up to 80.7%. In another study by Wei et al. (2018), the highest accuracy of 77.9% was achieved among five distinct individual algorithms, namely Neural Network, Support Vector Machine, Decision tree, Logistic regression, and Naïve Bayes.

Additionally, with the aid of computer vision and convolutional neural networks (CNNs), images from magnetic resonance imaging, radiography, and ophthalmoscopy and others can also be used as training data for the model to learn to identify patterns of diseases within the image itself. This can be seen in the studies published by Lu et al. (2019) and Angara et al. (2019) where CNN was used to detect breast cancer from a collection of classified images. In addition to modeling algorithms for classification, research done on model improvement has also further contributed to this area of study; for example, the work of Banik et al. (2019) where their proposed fused CNN trained faster than a CNN-recursive neural network (CNN-RNN) model when used for classifying images of white blood cell and Srivastava et al. (n.d.), where their novel pixel optimization technique in visualizing diabetic retinopathy indicators managed to rediscover critical features of the disease. Here, artificial neural networks (ANNs) played a major role in forming the deep learning branch of AI study from which CNN is derived and further reliable models that sprouted through the past decade. Both classical and deep learning algorithms are still under vast amounts of research in this field.

Although the main objectives of these studies from Ang et al. (2010) and Liu et al. (2012) were to produce a classification model for the intended specific diseases, data preprocessing beforehand is also an important factor that the authors have acknowledged when it comes to achieving better results. Data preprocessing, which includes "cleaning" the dataset and dimensionality reduction, is the main area of improvement that helps in optimizing the classification results before model training in the previously mentioned research studies. As discussed in the previous section, data received at the data center can be subjected to data anomalies. Data cleaning is needed to remove the data anomalies and replace the missing data accordingly based on the history of sensory data learned. Fig. 10.4 shows a general overview of the ML process flow for training a classification model for a healthcare dataset. Dataset procurement and the preprocessing to form a clean, high-quality dataset are arguably the most time-intensive phases of the process. The anomalous data need to be identified, removed, and replaced to ensure the ML model built is correct. The train and test set split allows for thorough testing of the algorithm before the evaluation of performance is done. Training is then repeated by tweaking several hyperparameters or making use of different techniques until adequate results are obtained.

10.4.1 Data set preprocessing

As shown in Fig. 10.4, preprocessing is the first step that needs to be performed in ML to remove data anomalies. It uses data mining or analytics to transform raw data into a more understandable and useful format. Real-world data collected from biosensors contains missing data, outliers, or errors. Inconsistencies of multiple data being collected from different and nonstandardized methods or sensors from similar locations are commonplace. To preprocess data, the following approaches are employable.

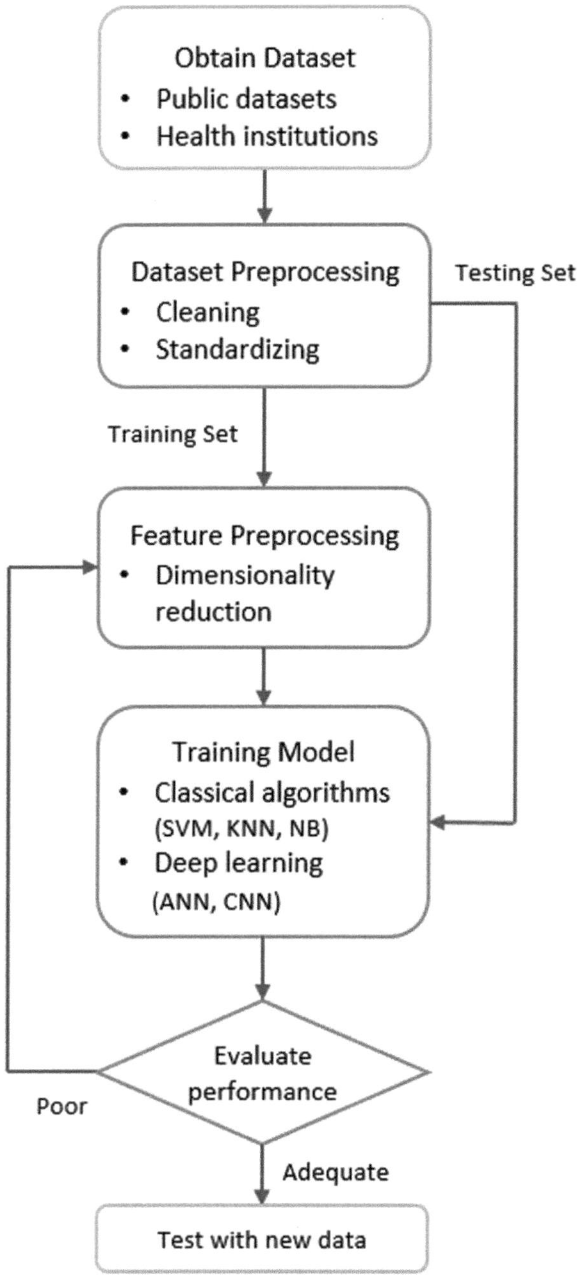

FIGURE 10.4 Machine learning. A general overview of common machine learning process flow.

10.4.1.1 Data cleaning

This is also known as data scrubbing where missing values are filled and anomalies are removed to resolve inconsistencies. It can be used to detect and remove errors and inconsistencies present in the data. Data errors or inconsistencies occur due to faults in the data collection process or transmissions. Missing data due to transmission error or interferences cannot be overlooked in a data set as many AI algorithms cannot accept missing values (except k-nearest neighbor and Naïve Bayes). Missing data can be overcome by filling up the values manually, using the forward or backward value, calculating the attribute mean or median, or applying regression techniques to predict the value. Outliers can be detected using clustering algorithms and removed and replaced using binning and then smoothing.

10.4.1.2 Data integration

This step involves taking clean data from multiple sources. However, duplicate attributes or tuples may occur as a result of repeating measurements or taking multiple data from various sources, resulting in inconsistencies. Data redundancy or inconsistency can be reduced by careful integration of data from multiple sources.

10.4.1.3 Data transformation

Normalization and aggregation of data according to the data set requirements.

10.4.1.4 Data reduction

Reducing the number of records or the number of attributes or dimensions without changing the results of the original data. This can help to improve the classification speed. Once the raw data has been processed and transformed, it is stored in the EPRs.

10.4.2 Feature importance

Although large volumes of features can be obtained from EPRs, not all of them are considered useful and may be redundant due to the vast amount of data collected from different devices (Cios & William Moore, 2002). Redundant features exist where one feature is correlated to another that does not add any relevant information to other features in the overall dataset. In cases where there are only a few redundant features, inclusion does not hurt the classification model's performance. It is only when these types of features outnumber the relevant features that it increases processing time for both training and classification processes. This is detrimental for algorithms built on the premise of time-saving requirements.

According to Dutta et al. (2019), a comprehensive analysis of feature importance was conducted on the publicly available Pima Indians dataset.

The researchers employed the random forest algorithm and achieved the most favorable outcomes. Their findings revealed that out of the eight given features, five played a crucial role in accurately classifying individuals as either diabetics or nondiabetics.

In a separate study by Balakrishnan et al. (2008), the authors utilized a classification algorithm called support vector machine (SVM) in conjunction with a feature extraction technique. They employed this approach on the same dataset, evaluating the prediction accuracy after eliminating the least important feature at each iteration. The results indicated a notable increase in accuracy by 1.88% when approximately 37.5% of the features were removed from consideration. Similarly, Akan and Kılıç (2012) have succeeded in reducing 17 features of their own diabetic dataset to four features using principal component analysis (PCA), outputting an increase in accuracy score when fed to a decision tree (DT) model. These studies have shown the importance of dimensionality reduction and how the importance of each feature contributes toward improved classification performance.

10.4.3 Feature selection methods

Determining feature importance is the key component toward improving the original dataset. It is crucial that the methodology for selection is also accurate and as reliable as that of an experienced practitioner. As it is unfeasible for thousands of features to be handpicked for every dataset imaginable, AI researchers have introduced the use of ML to aid in the process of feature selection.

There are two types of feature selection methods that have been introduced into data mining and preprocessing, namely the filter and the wrapper methods (Visalakshi & Radha, 2015). In filter methods, statistical techniques are used to perform the rank ordering method on the full feature set such as the Pearson correlation technique and Chi-square test (DeepaLakshmi & Velmurugan, 2016). Its simplicity allows for fast computational time at a low cost and enables the quick removal of redundant features before any classification processes. Fig. 10.5 displays the process flow for a classification process using the filter method as its preprocessing technique.

In wrapper methods, the inclusion of a mining algorithm, one set of any common estimators (such as logistic regression, SVM, gradient boosting, and random forest), is necessary to generate the suitable subfeatures via evaluating the performance of the algorithm. This process is done before any

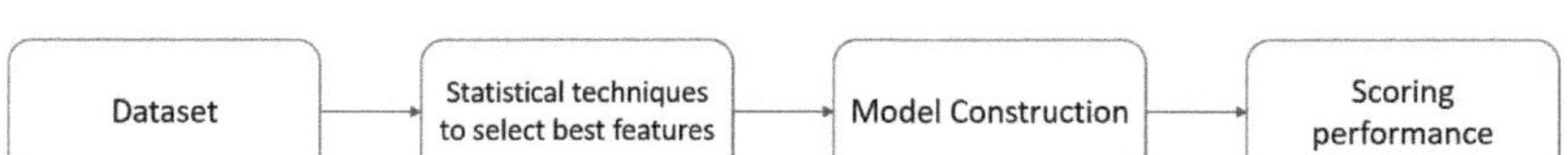

FIGURE 10.5 Preprocessing technique. Filter method process flow.

construction or training of the actual classification model, thus, increasing the computational cost and risk of overfitting (Bai et al., 2014). Fig. 10.6 shows the process flow for a classification process using the wrapper method as its preprocessing technique.

The low count of starting feature pool reduces the effectiveness of filter methods as it is normally used in scaling to high-dimensional datasets, as well as the drawback of interaction separation with the classifying algorithm, which could lead to poor performance as compared to wrapper (Beniwal & Arora, 2012).

The arguments stated to become the justification for wrapper feature selection methods to be the main focus of this study to gain insight into the inner workings of the methods that could potentially contribute better toward the classification of diseases in the chosen low-dimensional medical datasets. The feature selection methods to be discussed in the proceeding subsections are sequential feature selection (SFS) and recursive feature elimination (RFE).

10.4.3.1 Sequential feature selection

The SFS method makes use of common estimators' performances to determine the most optimum feature subset that gives the best result. The four common ones chosen for this research are logistic regression (LR), SVM, gradient boosting (GB), and random forest (RF). It goes through a process of picking the most important features that are added iteratively or conversely and the least important features iteratively removed from the original dataset to finally leave a subset of the most optimum group of features that are used in the following model training and classification. The subset's main notable characteristic is its size, which ideally is only a fraction of the original starting size.

SFS encompasses two variations known as sequential forward selection (SFWS) and sequential backward selection (SBS), which both adhere to the same underlying principle of the selection process. SFWS begins with an empty feature subset and progressively adds features from the collection based on their ability to enhance the estimator performance. This process continues until a predetermined terminating criteria is met.

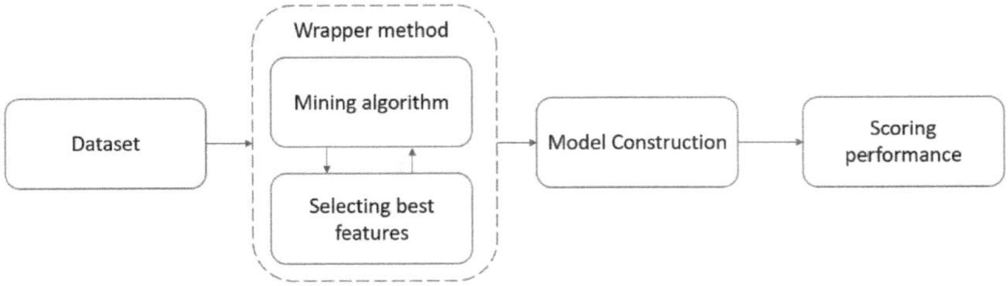

FIGURE 10.6 Wrapper process. Wrapper method process flow.

On the other hand, SBS operates in reverse by removing features from the dataset. It starts with the entire feature set and iteratively eliminates the feature that yields the best improvement in estimator performance when removed. The resultant subset, obtained after the completion of the process, represents the most optimal set of features to be utilized for training a classification model. It is worth noting that the terminating criteria for both SFWS and SBS methods remain consistent throughout the selection process.

Gacav et al. (2016) utilized SFS as a technique to extract a highly significant subset of distance vectors from facial expressions, with the aim of improving classification accuracy. Their approach yielded remarkable results, achieving an impressive mean class recognition accuracy of 89.9%. Similarly, Zheng et al. (2019) utilized a hybrid approach combining the backward variation of SFS with the inclusion of Information Gain. Their study aimed to determine an optimal subset of risk factors for diabetic patients using data from the Korean National Health survey. From an initial set of 33 features, they identified 10 features as optimal for classification, resulting in a remarkable accuracy of 95.6%.

10.4.3.2 Recursive feature elimination

The RFE method involves iteratively eliminating features of the least importance from the dataset until the desired number of inputs is achieved. In contrast to the sequential nature of SFS, RFE conducts repeated performance tests on the classifier at each iteration when a feature is removed from the current dataset. These tests assign weighted coefficients to the features, enabling the identification of their ranking in terms of importance. To determine the optimal number of features, cross-validation is introduced into the RFE process. Different feature subsets are evaluated using scoring metrics, and the collection with the highest score is selected as the best subset of features. This approach enhances the selection process in RFE by incorporating cross-validation and scoring techniques to identify the most relevant features.

The most paired estimator with RFE is the SVM as seen in the work of Qiang et al. (2014), who made use of the method to construct a low-dimensional face image feature for face recognition which obtained 93.5% recognition accuracy with feature extraction from 720 to only 60. Similarly, Adorada et al. (2018) and Wang et al. (2010) have also implemented the use of an SVM-RFE feature selector to reduce the dimensionality of MicroRNA expression features of breast cancer and features involved in the prediction of subhealth state in patients respectively.

For future studies, there is also another approach to feature selection, which is the hybrid of the filter and wrapper methods. This involves performing the feature selection sequentially to obtain the benefits of each method. Previous works such as those belonging to Yuan et al. (1999) and Elmaizi et al. (2017) have demonstrated the use of hybrid techniques to outperform conventional standalone feature selection methods. Additional research and experimentation will need to be done to determine the significance of the multistep procedure on low-dimensional datasets.

10.4.4 Feature extraction methods

Another category of dimensionality reduction technique is feature extraction. The main difference between feature selection and feature extraction revolves around the state of the feature subsets from the original dataset. Feature selection selects only a subset of the best features from the original dataset while feature extraction transforms the data while still preserving as much information from the original dataset as possible (Khalid et al., 2014). Despite sharing a common goal of feature reduction for classification, these methods utilize distinct approaches, leading to different yet equally effective strategies for addressing the challenge of dimensionality reduction. In certain scenarios, the advantages of these methods may surpass those of feature selection (Pechenizkiy et al., 2004).

10.4.4.1 Principal component analysis

PCA is a form of feature extraction method most commonly used, with great results, in high-dimensional datasets. Using PCA, the features are reduced to only a few "components," making ML models able to train faster and increase accuracy (see Fig. 10.7). PCA is a statistical process that transforms data into orthogonally related variables known as principal components that are uncorrelated to one another (Ibrahim & Al-Jumaily, 2017). The number of new components is usually less than or equal to the number of original components but will never be higher, making PCA a great tool for dimensional reduction.

In their study, Song et al. (2010) demonstrated the achievement of reducing eigenvector features for facial recognition. Their aim was to enhance the accuracy of classification outcomes while ensuring computational efficiency and reducing the dataset's dimensionality. For the scope of medical dataset classification, Roopa and Asha (2019) made use of PCA as a feature extraction technique for the Pima Indians Diabetes dataset and linear regression to classify diabetes patients with 82.1% accuracy, while Kanchan and Kishor (2016) compared classification algorithms of SVM, Naïve Bayes, and decision trees before and after feature extraction with PCA. SVM proved to be the most performant of the three types.

10.4.5 Classification algorithms

Classification algorithms are a vital part of ML. They are ML models for performing classification predictions based on a given training dataset in which the program "learns" how to correctly classify classes of the target

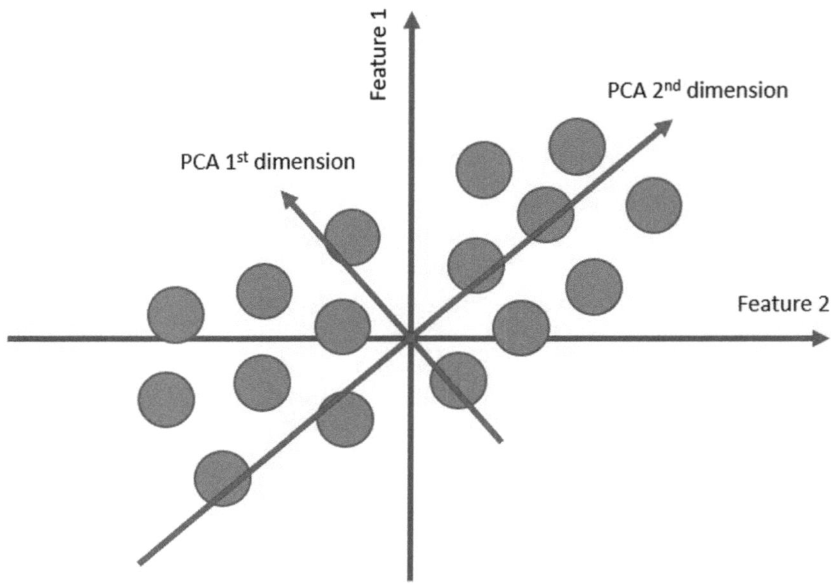

FIGURE 10.7 Principal component analysis. Representation of PCA application on two features.

output. The learning process is divided into two main kinds—supervised and unsupervised learning. In supervised learning, models were given input data that are already labeled with the desired target class in which an approximate mapping function is constructed to produce an accurate prediction for newly added values (Kavakiotis et al., 2017; Saravanan & Sujatha, 2019). This is opposed to unsupervised learning, where the input data are unlabeled and the model output depends on highlighting the distributions of the data and clustering groups of unknown target variables (Kavakiotis et al., 2017).

Three supervised learning classifying algorithms are discussed in the following section based on their merits of advantages that are suitable for the classification problems posed.

10.4.5.1 Artificial neural network

The ANN is a widely used model in ML, aiming to replicate the intricate functioning of the human brain (refer to Fig. 10.8). The ANN learns from experience and continually improves its predictive accuracy by minimizing errors during each iteration. Typically, supervised learning is employed, where labeled input data and known outputs are used for training.

During the training process, the weights of each neuron in the hidden layer are adjusted iteratively. The objective is to minimize the disparity between the network's output and the desired output, using various optimization techniques. Ultimately, the output is represented by a single Boolean neuron, indicating either true or false.

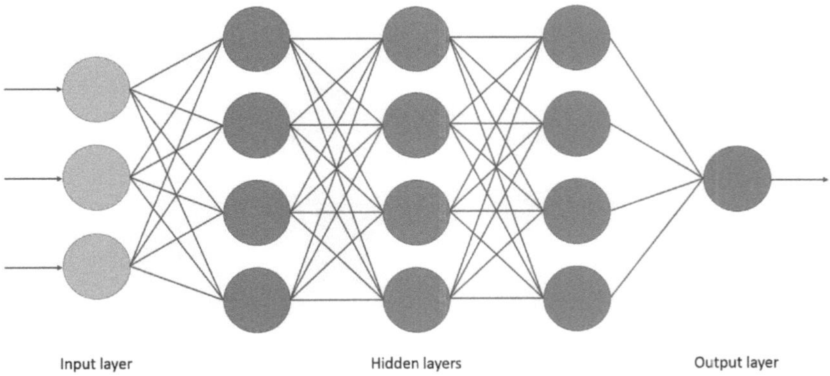

FIGURE 10.8 Artificial neural network. Representation of an artificial neural network (ANN).

In previous studies, Jayalakshmi and Santhakumaran (2010) employed an ANN as the predictive model for diabetes mellitus using the same dataset. On the other hand, Dey et al. (2008) achieved a high accuracy score when predicting diabetes cases from their local dataset. It is worth noting that in both studies, the dataset underwent preprocessing steps but did not involve the implementation of any feature selection techniques.

10.4.5.2 Support vector machine

SVM is a linear classifier that separates space with a hyperplane (Adankon & Cheriet, 2009). The plane separates each region according to its own classes of attributes. SVM takes the idea of an ideal plane and maximizes the margin between these two characteristics, that is, obtaining a decision boundary that allows for a good generalization capability. Fig. 10.9 shows the visual representation of the hyperplane, support vectors, and data classes in a two-dimensional plane.

Application of SVM includes mainly classification problems where samples are categorized into their specific classes by drawing the hyperplane for the best optimal generalization. Each classification of data lies on either side of the margin, hence, creating a distinct gap of which samples can fall onto. This does not just limit to two classes as multiclass classification is also possible. The two parameters (hyperparameters C and kernel parameters) are important to achieve a good SVM performance. The former controls the strictness of error minimization while the kernel parameters define the kernel function used to map data into higher dimensional space.

SVM comes with its own set of benefits and drawbacks. SVM shows higher accuracy of classification in smaller-sized datasets and is generally very good in dealing with high-dimensional data. On the other hand, picking the correct kernel and parameters for use can be computationally intensive compared to other models that offer the same classification purposes. For instance,

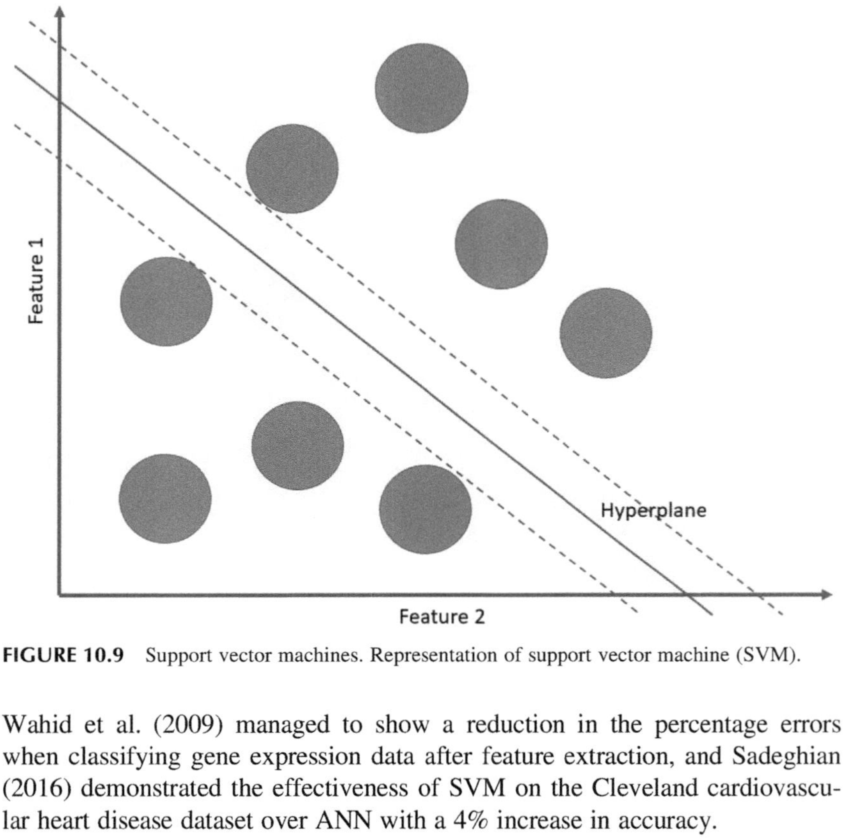

FIGURE 10.9 Support vector machines. Representation of support vector machine (SVM).

Wahid et al. (2009) managed to show a reduction in the percentage errors when classifying gene expression data after feature extraction, and Sadeghian (2016) demonstrated the effectiveness of SVM on the Cleveland cardiovascular heart disease dataset over ANN with a 4% increase in accuracy.

10.4.5.3 Random forest

As part of an ensemble classifier, RF classification algorithm uses votes of the majority to determine the classification output based on the evaluations from multiple decision trees.

A singular DT is a classification algorithm where branches are used to determine the outcome of each node that subsequently leads to a classification based on the targeted classes of the prediction. The branched paths are bound by a set of decision rules whereby each classification output results from specific combinations of input and/or conditions. RF is simply a collection of different individual DTs in which each output raises a "vote" that ultimately resolves the final output result. Fig. 10.10 shows the visual representation of the RF algorithm.

This ML model was proven to be effective because uncorrelated models can produce an ensemble prediction that is more accurate than any single individual prediction due to the omission of each tree's individual errors affecting other neighboring trees.

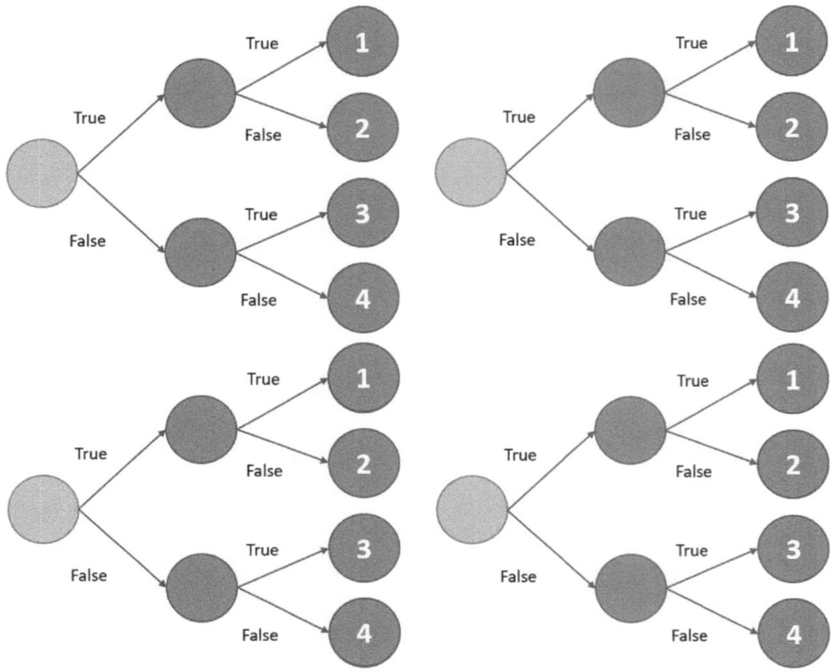

FIGURE 10.10 Random forest. Representation of random forest (RF) with voted output.

RF has been considered to be the best among these three ML models due to its reliability in classification problems. For example, Kumar and Shaikh (2017) made use of feature selection to demonstrate the effectiveness of improving the accuracies of RF classifiers on the Cleveland cardiovascular heart disease medical dataset. Their results showed a 2%–3% increase in accuracy as compared to RF classification with the unfiltered dataset. Additionally, further work has also been done in the research by Jaiswal and Samikannu (2017) with their usage of RF for selecting the best subset for kidney disease prediction and Wu et al. (2018) in using RF as the classification model for achieving the best accuracy result of 85% in comparison with other common classifiers such as SVM, LR, and Naïve Bayes.

10.4.6 Feature engineering

Another alternative to combating the problem of low-dimensional datasets is the area of manual feature engineering, a construction of relevant features to be added to the original dataset. With the addition of enough engineered features, the dataset can then be upgraded to a high-dimensional dataset depending on the ratio of features to instances. The first step includes a careful examination of the inclusion criteria for new features based on domain knowledge in the context of the dataset, followed by the processes of binning and encoding to output an engineered dataset of better quality. Fig. 10.11 shows the flow diagram of the process of feature engineering.

A notable trait of the features generated is its significant contributions toward improving the quality of the dataset as a whole which consequently improves its performances of the classification models (Nargesian et al., 2017). The disadvantage of this approach is the added upfront cost of resources in gathering reliable experts within the targeted field, most opted for the more automated approach (Garla & Brandt, 2012). Nonetheless, a limitation of many automated feature ranking methods is their univariate nature, disregarding the interdependencies that may exist among features (Saeys et al., 2007).

10.5 Examples of applications of AI biosensors

As described in the previous sections, ML has fundamentally two functions in AI-based biosensors: (1) data reduction before wireless transmission to achieve ultra-low powered AI biosensors and (2) ensuring data consistency, accuracy, and reliability (Jin et al., 2020). For the latter function, for instance, the signal output of an electrochemical biosensor is affected by various parameters such as electrodes (e.g., materials), electrical variables (e.g., voltage), electrolytes (e.g., solvent), and external factors (e.g., temperature). Combining ML with electrochemical biosensors is therefore a useful approach to predicting, describing, and optimizing the system. A glucose-oxidase biosensor (GOB) was fabricated to model its amperometric response with dependent variables under different conditions like temperature, pH, and glucose concentrations through several statistical ML methods. It was demonstrated that using ANN and SVM for regression has shown a promising very low prediction error of the biosensor outcome. Furthermore, the relationship between the available predictors (temperature, benzoquinone, pH, and glucose concentration) and the response produced a nonlinear behavior. Their interactions required optimization for maximum signal output.

Ahuja et al. (2019) integrated ML in the development of an impedance spectroscopy-based biosensor to rapidly examine the patient response to drugs in targeted cancer therapy to improve classification accuracy. The portable device allowed for the determination of the percentage of live and dead tumor cells in

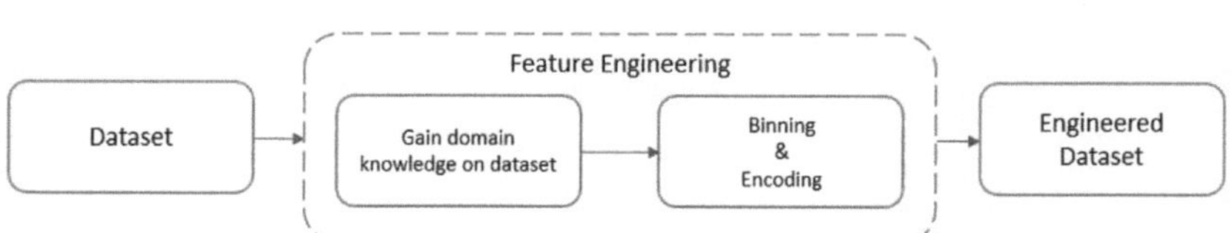

FIGURE 10.11 Feature engineering. Flow diagram of feature engineering.

the patient's sample upon treatment with antibody-conjugated drugs. Using high-dimensional data and supervised ML, the group explored the use of phase change acquired from multifrequency impedance cytometry as features of SVM classifiers. The phase change showed a trend with negative and positive at lower and higher frequencies, respectively where the quantitative change in phase for various cell types permitted improved accuracy classification. In another similar work, Zheng et al. (2018) reported a Raman spectroscopy biosensor, which is based on the indirect determination of vibrational states of samples, in combination with SVM for the detection of thyroid dysfunction. Based on serum samples taken from patients and healthy volunteers, profile and peak intensities of their serum Raman spectra demonstrated high similarity between groups while subtle differences were possible to determine through a robust data analysis algorithm. Thus to improve diagnostic efficiency and still maintain accuracy, artificial fish coupled with a uniform design (AFUD) algorithm was established to optimize the SVM parameters. With an average accuracy of 30 discriminant results was to be 82.74%, the results exhibited that the proposed serum Raman spectroscopy technique integrated with the AFUD-SVM discriminant model was an excellent noninvasive and low-cost approach for the detection of thyroid dysfunction.

Epilepsy affects the quality of one's life, especially in children where its unpredictable occurrence is distressing and often leads to social isolation. Therefore an automated seizure alarm is especially useful for generalized tonic-clonic seizures (GTCSs) with a high risk of unexpected death. Wearable devices with the capability to detect GTCS with 90%−96% sensitivity could markedly improve this situation. Combining both scalp electroencephalography (EEG) and deep learning, the prediction of seizures with a sensitivity of 79% has been reported. Tang et al. (2021) evaluated the automatic seizure detection performance of ML algorithms across nine epileptic seizure types utilizing multisignal biosensors worn on either their wrists or ankles. Body temperature, electrodermal activity, accelerometry (ACC), and photoplethysmography providing blood volume pulse (BVP) were recorded. Combining ACC and BVP modalities with CNN algorithms performed better with AUC-ROC of 0.752 than single data modality when applied to a complete dataset. Their study demonstrated this technique was able to automatically detect various epileptic seizure types. With the availability of commercial medical-grade wearable and mobile smart devices like smartwatches, this research effort opened an excellent opportunity to examine the suitability of various data modalities for determining seizure. US Food and Drug Administration has provided its first approval and medical clearance for a smartwatch that employs advanced ML to monitor seizures (Voelker, 2018). The commercially available medical-grade smartwatch (Embrace and E4 wristbands) provides detection for GTCS with a sensitivity of 92%−100%. Improvement in algorithm and accessibility of large dataset contributed to lower daily false alarm rate initial ~2 down to 0.2−1 false alarm (Regalia et al., 2019).

Cancer is a disease characterized by abnormal and uncontrolled growth of cells contributing to the development of tumors. According to the World Health Organization (WHO), cancer is the second leading contributor to global mortality with approximately 9.6 million deaths mortality across the globe. While the most common types among men are lung, prostate colorectal, stomach, and liver cancer, breast, colorectal, lung, cervical, and thyroid cancer have affected women. Early diagnosis of cancer is essential to ensure a good prognosis. A systematic analysis of various research studies and relevant papers by Amethiya et al. (2022) has led to conclude that efficient detection of breast cancer via the Fuzzy Extreme Learning Machine - Radial Basis Function (ELM-RBF) ML algorithm was determined to be the superior model for breast cancer detection allowing 98.05% accuracy (Amethiya et al., 2022). Interestingly, this model is applicable to other kinds of cancers requiring little effort.

Another investigation to develop automatic scoring of Parkinsonian tremors was accomplished by using DT, DA, kNN, RF, and SVM to predict the Unified Parkinson's Disease Rating Scale (UPDRS) (Jeon et al., 2017). The study included a wearable sensor integrated with an accelerometer and a gyroscope to determine the tremor signals of 85 patients to navigate the automatic scoring of the severity of the disease. The method was reported to perform well with an accuracy of 85.5% using a wearable device.

10.6 Conclusions

The exciting phenomenon of wearable biosensors in healthcare settings has reshaped its management systems. The introduction of biosensors into the healthcare system has eased the burden associated with finance and manpower. Despite the excellent progress made in medical biosensor research, key challenges remain. One of these is the lack of accuracy and reliability of these sensors owing to data errors/faults has greatly undermined their performance, especially in real-world situations. Hence, there is a significant demand to address this aspect. This is achievable through the application of AI, notably ML to improve healthcare services. ML has been creatively employed for data fusion of multiple biosensors, noise correction, and adaptation to variations in biosensor responses.

There are multitudes of methods and approaches in the field of research of ML that are capable of classification. Among that, feature selection (such as SFWS, SBS, and RFE) and feature extraction (PCA), which reduce the dataset to only contain important features, are found to be viably preprocessing approaches in producing a higher quality dataset, which results in better classification. Coupling these preprocessing techniques with classification methods (such as ANN, SVM, and RF) shows the positive effect of their application in data classification collected from biosensors. The use of ML in wearable biosensors has therefore provided merits in the automation of data extraction, processing, and analysis produced by biosensors. However, as information is transmitted wirelessly, a wide range of anomalies could present affecting their seamless operation. ML algorithms able to combat this issue by reducing or eliminating these anomalies thus improving biosensor data, which is significant in medical areas.

In summary, anomalies in WBNs can significantly impact the accuracy of collected data. By incorporating AI and ML, such as anomaly detection and data imputation algorithms, the quality of biosensor data can be improved. These technologies provide automated and efficient methods for identifying anomalies, ensuring data integrity, and enhancing the reliability of WBNs.

References

Adankon, M. M., & Cheriet, M. (2009). *Support vector machine* (pp. 1303–1308). Springer Science and Business Media LLC. Available from https://doi.org/10.1007/978-0-387-73003-5_299.

Adorada, A., Permatasari, R., Wirawan, P.W., Wibowo, A., & Sujiwo, A. (2018). Support vector machine – Recursive feature elimination (SVM-RFE) for selection of microRNA expression features of breast cancer. In *2018 2nd International conference on informatics and computational sciences, ICICoS 2018* (pp. 165–168). Institute of Electrical and Electronics Engineers Inc. https://doi.org/10.1109/ICICOS.2018.8621708.

Ahuja, R., Chug, A., Kohli, S., Gupta, S., & Ahuja, P. (2019). In *International conference on pervasive computing advances and applications – PerCAA 2019* (Vol. 152, pp. 341–348). https://doi.org/10.1016/j.procs.2019.050.008.

Akan, & Kılıç. (2012). The effect of feature extraction techniques on diagnosis of diabetes. In *20th Signal processing and communications applications conference* (pp. 18–20).

Amethiya, Y., Pipariya, P., Patel, S., & Shah, M. (2022). *Intelligent Medicine*, *2*(2), 69–81. Available from https://doi.org/10.1016/j.imed.2021.08.004.

Ang, Q., Wang, W., Liu, Z., & Li, K. (2010). Explored research on data preprocessing and mining technology for clinical data applications. In *ICIME 2010 – 2010 2nd IEEE international conference on information management and engineering* (Vol. 5, pp. 327–330). https://doi.org/10.1109/ICIME.2010.5477660.

Angara, S., Robinson, M., & Guillén-Rondon, P. (2019). Convolutional neural networks for breast cancer histopathological image classification. In *4th International conference on big data and information analytics: Theories, algorithms and applications in data science, BigDIA 2018*. Institute of Electrical and Electronics Engineers Inc. https://doi.org/10.1109/BigDIA.2018.8632800.RO.

Attallah, O. (2022). *Digital Health, 8*, 20552076221124430. https://doi.org/10.1177/20552076221124432.

Avižienis, A., Laprie, J. C., & Randell, B. (2004). *Dependability and its threats: A taxonomy, . IFIP advances in information and communication technology* (Vol. 156, pp. 91–120). New York LLC: Springer. Available from https://doi.org/10.1007/978-1-4020-8157-6_13.

Bai, L., Wang, Z., Shao, Y.-H., & Deng, N.-Y. (2014). A novel feature selection method for twin support vector machine. *Knowledge-Based Systems*, *59*, 1–8. Available from https://doi.org/10.1016/j.knosys.2014.01.025.

Balakrishnan, S., Narayanaswamy, R., Savarimuthu, N., & Samikannu, R. (2008). SVM ranking with backward search for feature selection in type II diabetes databases. In *Conference proceedings – IEEE international conference on systems, man and cybernetics* (pp. 2628–2633). https://doi.org/10.1109/ICSMC.2008.4811692.

Banik, P.P., Saha, R., & Kim, K.D. (2019). Fused convolutional neural network for white blood cell image classification. In *1st International conference on artificial intelligence in information and communication, ICAIIC 2019* (pp. 238–240). Institute of Electrical and Electronics Engineers Inc. https://doi.org/10.1109/ICAIIC.2019.8669049.

Beniwal, S., & Arora, J. (2012). Classification and feature selection techniques in data mining. *International Journal of Engineering Research & Technology*, *1*, 1–6.

Brownfield, M.I., Fayez, A.S., Nelson, T.M., & Davis IV, N. (2006). Cross-layer wireless sensor network radio power management. In *IEEE wireless communications and networking conference, WCNC* (Vol. 2, pp. 1160–1165). Institute of Electrical and Electronics Engineers Inc. https://doi.org/10.1109/wcnc.2006.1683634.

Chipara, O., Lu, C., Bailey, T.C., & Roman, G.C. (2010). Reliable clinical monitoring using wireless sensor networks: Experiences in a step-down hospital unit. In *SenSys 2010 – Proceedings of the 8th ACM conference on embedded networked sensor systems* (pp. 155–168). https://doi.org/10.1145/1869983.1869999

Cios, K. J., & William Moore, G. (2002). Uniqueness of medical data mining. *Artificial Intelligence in Medicine*, *26*(1–2), 1–24. Available from https://doi.org/10.1016/s0933-3657(02)00049-0.

DeepaLakshmi, S., & Velmurugan, T. (2016). Empirical study of feature selection methods for high dimensional data. *Indian Journal of Science and Technology*, *9*(39). Available from https://doi.org/10.17485/ijst/2016/v9i39/90599.

Dey, R., Bajpai, V., Gandhi, G., & Dey, B. (2008). Application of artificial neural network (ANN) technique for diagnosing diabetes mellitus. In *IEEE region 10 colloquium and 3rd international conference on industrial and information systems, ICIIS 2008*. https://doi.org/10.1109/ICIINFS.2008.4798367.

Dutta, D., Paul, D., & Ghosh, P. (2019). Analysing feature importances for diabetes prediction using machine learning. In *2018 IEEE 9th annual information technology, electronics and mobile communication conference, IEMCON 2018* (pp. 924–928). Institute of Electrical and Electronics Engineers Inc. https://doi.org/10.1109/IEMCON.2018.8614871.

Elmaizi, A., Merzouqi, M., Sarhrouni, E., Hammouch, A., & Nacir, C. (2017). Hybridization of filter and wrapper approaches for the dimensionality reduction and classification of hyperspectral images. In *Proceedings – 3rd International conference on advanced technologies for signal and image processing, ATSIP 2017*. Institute of Electrical and Electronics Engineers Inc. https://doi.org/10.1109/ATSIP.2017.8075549.

Elnahrawy, E., & Nath, B. (2003). Cleaning and querying noisy sensors. In *Proceedings of the second ACM international workshop on wireless sensor networks and applications, WSNA 2003* (pp. 78–87). Association for Computing Machinery (ACM). https://doi.org/10.1145/941350.941362.

Gacav, C., Benligiray, B, & Topal, C (2016). Sequential forward feature selection for facial expression recognition. In 24th signal processing and communication application conference, (pp. 16–19). IEEE.

Garla, V. N., & Brandt, C. (2012). Ontology-guided feature engineering for clinical text classification. *Journal of Biomedical Informatics*, *45*(5), 992–998. Available from https://doi.org/10.1016/j.jbi.2012.04.010.

Gungor, V. C., & Hancke, G. P. (2009). Industrial wireless sensor networks: Challenges, design principles, and technical approaches. *IEEE Transactions on Industrial Electronics*, *56*(10), 4258–4265. Available from https://doi.org/10.1109/TIE.2009.2015754.

Haleem, A., Javaid, M., Singh, R. P., & Suman, R. (2021). Telemedicine for healthcare: Capabilities, features, barriers, and applications. *Sensors International*, *2*, 100117. Available from https://doi.org/10.1016/j.sintl.2021.100117.

Huo, H., Xu, Y., Bilen, C.C., & Zhang, H. (2009). Coexistence issues of 2.4GHz sensor networks with other RF devices at home. In *Proceedings – 2009 3rd International conference on sensor technologies and applications, SENSORCOMM 2009* (pp. 200–205). https://doi.org/10.1109/SENSORCOMM.2009.40.

Ibrahim, M. F. I., & Al-Jumaily, A. A. (2017). *PCA indexing based feature learning and feature selection. 2016 8th Cairo international biomedical engineering conference, CIBEC 2016* (pp. 68–71). Institute of Electrical and Electronics Engineers Inc.. Available from https://doi.org/10.1109/CIBEC.2016.7836122.

Ingelrest, F., Barrenetxea, G., Schaefer, G., Vetterli, M., Couach, O., & Parlange, M. (2010). SensorScope: Application-specific sensor network for environmental monitoring. *ACM Transactions on Sensor Networks*, *6*(2). Available from https://doi.org/10.1145/1689239.1689247.

Jaiswal, J. K., & Samikannu, R. (2017). *Application of random forest algorithm on feature subset selection and classification and regression. Proceedings - 2nd World Congress on Computing and Communication Technologies, WCCCT 2017* (pp. 65–68). Institute of Electrical and Electronics Engineers Inc.. Available from https://doi.org/10.1109/WCCCT.2016.25.

Jayalakshmi, T., & Santhakumaran, A. (2010). A novel classification method for diagnosis of diabetes mellitus using artificial neural networks. In *DSDE 2010 – International conference on data storage and data engineering* (pp. 159–163). https://doi.org/10.1109/DSDE.2010.58.

Jeon, H., Lee, W., Park, H., Lee, H. J., Kim, S. K., Kim, H. B., Jeon, B., & Park, K. S. (2017). Automatic classification of tremor severity in Parkinson's disease using a wearable device. *Sensors (Switzerland)*, *17*(9). Available from https://doi.org/10.3390/s17092067.

Jian, Y., Kuang, H., Ren, C., Ma, Z., & Wang, H. (2021). A novel framework for image-based malware detection with a deep neural network. *Computers & Security*, *109*, In press.

Jin, Y., Yang, H., Wu, J. W., Chen, W., Zhang, S., Duan, W., & Virology, G. (2020). Epidemiology, pathogenesis, and control of COVID-19. *Viruses*, *12*. Available from https://doi.org/10.3390/v12040372.

Kahn, J.M., Katz, R.H., & Pister, K.S. J. (1999). Next century challenges: Mobile networking for dust. In *Proceedings of the annual international conference on mobile computing and networking, MOBICOM* (Vols. 1999-, pp. 271–278). Association for Computing Machinery. https://doi.org/10.1145/313451.313558.

Kanchan, B.D., & Kishor, Mahale (2016). Study of machine learning algorithms for special disease prediction using principal of component analysis. In *International conference on global trends in signal processing, information computing and communication (ICGTSPICC)* (pp. 22–24).

Karlof, C., & Wagner, D. (2003). *Secure routing in wireless sensor networks: Attacks and countermeasures. Proceedings of the 1st IEEE international workshop on sensor network protocols and applications, SNPA 2003* (pp. 113–127). Institute of Electrical and Electronics Engineers Inc.. Available from https://doi.org/10.1109/SNPA.2003.1203362.

Kavakiotis, I., Tsave, O., Salifoglou, A., Maglaveras, N., Vlahavas, I., & Chouvarda, I. (2017). Machine learning and data mining methods in diabetes research. *Computational and Structural Biotechnology Journal*, *15*, 104–116. Available from https://doi.org/10.1016/j.csbj.2016.12.005.

Khalid, S., Khalil, T., & Nasreen, S. (2014). *A survey of feature selection and feature extraction techniques in machine learning. Proceedings of 2014 science and information conference, SAI 2014* (pp. 372–378). Institute of Electrical and Electronics Engineers Inc.. Available from https://doi.org/10.1109/SAI.2014.6918213.

Kumar, S. S., & Shaikh, T. (2017). *Empirical evaluation of the performance of feature selection approaches on random forest. 2017 international conference on computer and applications, ICCA 2017* (pp. 227–231). Institute of Electrical and Electronics Engineers Inc.. Available from https://doi.org/10.1109/COMAPP.2017.8079769.

Langendoen, K., Baggio, A., & Visser, O. (2006). *Murphy loves potatoes: Experiences from a pilot sensor network deployment in precision agriculture. Proceedings of the 20th international conference on parallel and distributed processing* (p. 174) IEEE Computer Society, −174.

Lim, S.A., & Ahmed, M.U. (2016). In *Food biosensors* (p. 0). The Royal Society of Chemistry. https://doi.org/10.1039/9781782623908-00001.

Liu, K., Zhou, X., Feng, Y., & Liu, J. (2012). Clinical data preprocessing and case studies of POMDP for TCM treatment knowledge discovery. In *2012 IEEE 14th international conference on e-Health networking, applications and services, Healthcom 2012* (pp. 10–14). https://doi.org/10.1109/HealthCom.2012.6380056.

Lu, H.C., Loh, E.W., & Huang, S.C. (2019). The classification of mammogram using convolutional neural network with specific image preprocessing for breast cancer detection. In *2019 2nd International conference on artificial intelligence and big data, ICAIBD 2019* (pp. 9–12). Institute of Electrical and Electronics Engineers Inc. https://doi.org/10.1109/ICAIBD.2019.8837000.

Nargesian, F., Samulowitz, H., Khurana, U., Khalil, E.B., & Turaga, D. (2017). Learning feature engineering for classification. In *IJCAI international joint conference on artificial intelligence* (Vol. 0, pp. 2529–2535). https://doi.org/10.24963/ijcai.2017/352.

Ni, K., Ramanathan, N., Chehade, M. N. H., Balzano, L., Nair, S., Zahedi, S., Kohler, E., Pottie, G., Hansen, M., & Srivastava, M. (2009). Sensor network data fault types. *ACM Transactions on Sensor Networks*, *5*(3), 1–29. Available from https://doi.org/10.1145/1525856.1525863.

Pechenizkiy, M., Tsymbal, A., & Puuronen, S. (2004). PCA-based feature transformation for classification: Issues in medical diagnostics. In *Proceedings of the IEEE symposium on computer-based medical systems* (Vol. 17, pp. 535–540).

Qiang, X., Wei, J.W., & Liu. (2014). Face image feature selection based on gabor feature and recursive feature elimination. In *Sixth international conference on intelligent human-machine systems and cybernetics* (pp. 26–27).

Raghunathan, V., Schurgers, C., Park, S., & Srivastava, M. B. (2002). Energy-aware wireless microsensor networks. *IEEE Signal Processing Magazine*, *19*(2), 40–50. Available from https://doi.org/10.1109/79.985679.

Raymond, D. R., & Midkiff, S. F. (2008). Denial-of-service in wireless sensor networks: Attacks and defenses. *IEEE Pervasive Computing*, *7*(1), 74–81. Available from https://doi.org/10.1109/MPRV.2008.6.

Regalia, G., Onorati, F., Lai, M., Caborni, C., & Picard, R. W. (2019). Multimodal wrist-worn devices for seizure detection and advancing research: Focus on the empatica wristbands. *Epilepsy Research*, *153*, 79–82. Available from https://doi.org/10.1016/j.eplepsyres.2019.02.007.

Roopa, H., & Asha, T. (2019). A linear model based on principal component analysis for disease prediction. *IEEE Access*, *7*, 105314–105318. Available from https://doi.org/10.1109/ACCESS.2019.2931956.

Sadeghian, P.N. A.A. A. (2016). Application of multilayer perceptron neural networks and support vector machines in classification of healthcare data. In *Future technologies conference (FTC) 2016* (pp. 6–7).

Saeys, Y., Inza, I., & Larrañaga, P. (2007). A review of feature selection techniques in bioinformatics. *Bioinformatics (Oxford, England)*, *23*(19), 2507–2517. Available from https://doi.org/10.1093/bioinformatics/btm344.

Saravanan, R., & Sujatha, P. (2019). *A state of art techniques on machine learning algorithms: A perspective of supervised learning approaches in data classification. Proceedings of the 2nd international conference on intelligent computing and control systems, ICICCS 2018* (pp. 945–949). Institute of Electrical and Electronics Engineers Inc.. Available from https://doi.org/10.1109/ICCONS.2018.8663155.

Song, F., Guo, Z., & Mei, D. (2010). Feature selection using principal component analysis. In *Proceedings–2010 International conference on system science, engineering design and manufacturing informatization, ICSEM 2010* (Vol. 1, pp. 27–30). https://doi.org/10.1109/ICSEM.2010.14.

Srivastava,S., Prabhu,S., Ramesh,S., Pratapneni,S., Abraham,A., & Bhandary,S.V. (n.d.). Visualizing the indicators of diabetic retinopathy learnt by convolutional neural networks. In *2017 IEEE international conference on computational intelligence and computing research (ICCIC)* (pp. 1–3). https://doi.org/10.1109/ICCIC.2017.8524578.

Szewczyk, R., Mainwaring, A., Polastre, J., Anderson, J., & Culler, D. (2004). *An analysis of a large scale habitat monitoring application. SenSys'04 - Proceedings of the second international conference on embedded networked sensor systems* (pp. 214–226). Association for Computing Machinery. Available from https://doi.org/10.1145/1031495.1031521.

Tanenbaum, A., & Steen. (2002). *Distributed systems* (Vol. 2). Prentice Hall.

Tang, J., El Atrache, R., Yu, S., Asif, U., Jackson, M., Roy, S., Mirmomeni, M., Cantley, S., Sheehan, T., Schubach, S., Ufongene, C., Vieluf, S., Meisel, C., Harrer, S., & Loddenkemper, T. (2021). *Epilepsia*, *62*(8), 1807–1819. Available from https://doi.org/10.1111/epi.16967.

Tolle, G., Polastre, J., Szewczyk, R., Culler, D., Turner, N., Tu, K., Burgess, S., Dawson, T., Buonadonna, P., Gay, D., & Hong, W. (2005). *A macroscope in the redwoods. SenSys 2005 – Proceedings of the 3rd international conference on embedded networked sensor systems* (pp. 51–63). Association for Computing Machinery. Available from https://doi.org/10.1145/1098918.1098925.

Vijayan, V., & Anjali, C. (2015). (pp. 122–127). https://doi.org/10.1109/RAICS.2015.7488400.

Visalakshi, S., & Radha, V. (2015). A literature review of feature selection techniques and applications: Review of feature selection in data mining. In *2014 IEEE international conference on computational intelligence and computing research, IEEE ICCIC 2014*. Institute of Electrical and Electronics Engineers Inc. https://doi.org/10.1109/ICCIC.2014.7238499.

Voelker, R. (2018). *JAMA: The Journal of the American Medical Association*, *319*(11), 1086. Available from https://doi.org/10.1001/jama.2018.1809, –1086.

Wahid, C.M. M., Ali, A.B. M.S., & Tickle, K. (2009). Impact of feature selection on support vector machine using microarray gene expression data. In *2009 2nd International conference on machine vision, ICMV 2009* (pp. 189–193). https://doi.org/10.1109/ICMV.2009.46.

Wang, L.M., Chen, J.X., Pei, Y., Zhao, X., Cui, H.T., & Cui, H.Z. (2010). Feature selection and prediction of sub-health state using SVM-RFE. In *Proceedings – International conference on artificial intelligence and computational intelligence, AICI 2010* (Vol. 3, pp. 199–202). https://doi.org/10.1109/AICI.20100.280.

Wang, Y., Attebury, G., & Ramamurthy, B. (2006). A survey of security issues in wireless sensor networks. *IEEE Communications Surveys and Tutorials*, *8*(2), 2–22. Available from https://doi.org/10.1109/COMST.2006.315852.

Wei, S., Zhao, X., & Miao, C. (2018). *A comprehensive exploration to the machine learning techniques for diabetes identification, . IEEE world forum on internet of things, WF-IoT 2018 - Proceedings* (Vol. 2018, pp. 291–295). Institute of Electrical and Electronics Engineers Inc.. Available from https://doi.org/10.1109/WF-IoT.2018.8355130.

Werner-Allen, G., Lorincz, K., Welsh, M., Marcillo, O., Johnson, J., Ruiz, M., & Lees, J. (2006). Deploying a wireless sensor network on an active volcano. *IEEE Internet Computing*, *10*(2), 18–25. Available from https://doi.org/10.1109/MIC.2006.26.

Wong, C. K., Ho, D. T. Y., Tam, A. R., Zhou, M., Lau, Y. M., Tang, M. O. Y., Tong, R. C. F., Rajput, K. S., Chen, G., Chan, S. C., Siu, C. W., & Hung, I. F. N. (2020). Artificial intelligence mobile health platform for early detection of COVID-19 in quarantine subjects using a wearable biosensor: Protocol for a randomised controlled trial. *BMJ Open*, *10*(7). Available from https://doi.org/10.1136/bmjopen-2020-038555.

Woo, A., & Culler, D.E. (2001). A transmission control scheme for media access in sensor networks. In *Proceedings of the annual international conference on mobile computing and networking, MOBICOM* (pp. 221–235). Association for Computing Machinery (ACM). https://doi.org/10.1145/381677.381699.

Wu, Y., Wang, H., & Wu, F. (2018). *Automatic classification of pulmonary tuberculosis and sarcoidosis based on random forest, . Proceedings - 2017 10th international congress on image and signal processing, bioMedical engineering and informatics, CISP-BMEI 2017* (Vols. 2018-, pp. 1–5). Institute of Electrical and Electronics Engineers Inc.. Available from https://doi.org/10.1109/CISP-BMEI.2017.8302280.

Yuan, H., Tseng, S.S., Gangshan, W., & Fuyan, Z. (1999). Two-phase feature selection method using both filter and wrapper. In *Proceedings of the IEEE international conference on systems, man and cybernetics* (Vol. 2). IEEE.

Zheng, H., Park, H. W., Li, D., Park, K. H., & Ryu, K. H. (2019). *A Hybrid Feature Selection Approach for Applying to Patients with Diabetes Mellitus: KNHANES 2013–2015. NICS 2018 – Proceedings of 2018 5th NAFOSTED conference on information and computer science* (pp. 110–113). Institute of Electrical and Electronics Engineers Inc.. Available from https://doi.org/10.1109/NICS.2018.8606864.

Zheng, X., Lv, G., Du, G., Zhai, Z., Mo, J., & Lv, X. (2018). Rapid and low-cost detection of thyroid dysfunction using raman spectroscopy and an improved support vector machine. *IEEE Photonics Journal*, *10*(6), 1–12. Available from https://doi.org/10.1109/jphot.2018.2876686.

Zhou, G., He, T., Krishnamurthy, S., & Stankovic, J.A. (2004). Impact of radio irregularity on wireless sensor networks. In *MobiSys 2004 – Second international conference on mobile systems, applications and services* (pp. 125–138). Association for Computing Machinery. https://doi.org/10.1145/990064.990081.

Zou, Y., & Chakrabarty, K. (2007). Redundancy analysis and a distributed self-organization protocol for fault-tolerant wireless sensor networks. *International Journal of Distributed Sensor Networks*, *3*(3), 243–272. Available from https://doi.org/10.1080/15501320600781078.

Chapter 11

Commercial potential of biosensors

Rafaela S. Andre[1], Rafaela C. Sanfelice[2], Mardoqueu M. da Costa[3], Luiza A. Mercante[4], Daniel S. Correa[1] and Adriana Pavinatto[3]

[1]*Nanotechnology National Laboratory for Agriculture (LNNA), Embrapa Instrumentação, São Carlos, SP, Brazil,* [2]*Science and Technology Institute, Federal University of Alfenas, Poços de Caldas, MG, Brazil,* [3]*Scientific and Technological Institute of Brazil University, Universidade Brasil, São Paulo, SP, Brazil,* [4]*Institute of Chemistry, Federal University of Bahia (UFBA), Salvador, BA, Brazil*

11.1 Introduction to biosensors development and commercialization in medical applications

Commercial biosensors have revolutionized medical diagnosis by enabling the diagnosis of varied diseases. Unlike conventional diagnostics, which involves laborious and time-consuming procedures, biosensors offer a streamlined alternative that can significantly reduce testing time and improve patient outcomes. Since 1975, when the first glucose biosensor became available on the market by Yellow Springs Instruments (Yoo & Lee, 2010), significant efforts have been made in research and technology development for the early diagnosis of diseases. Focus has been given to diseases that cause high mortality or affect people's health conditions all over the world, such as diabetes, cancer, HIV, and COVID-19 (Corman et al., 2021).

During the development process, once the biosensor reaches the required performance (in terms of limit of detection, selectivity, reproducibility, stability, etc.) and is successful in testing using real samples, it can then be translated to a commercial stage, where the analytical abilities in real matrices, its robustness, and its economic viability are of fundamental importance. Also, before reaching the market, tests and approval, or even authorization for emergency use (seen for COVID-19 tests, for example) by the Food and Drug Administration (FDA) in USA (or similar agencies in the case of other countries) is mandatory. In this sense, the World Health Organization (WHO) has put forward a set of guidelines referred to as "ASSURED" to assist in the development of diagnostics platforms, where the devices should reach the following characteristics: be

Biosensors in Precision Medicine. DOI: https://doi.org/10.1016/B978-0-443-15380-8.00012-6

affordable, sensitive, specific, user-friendly, rapid and robust, equipment-free, and deliverable to end users (Sachdeva et al., 2021; Urdea et al., 2006). In this context, the search for handheld instruments with suitable performance for diagnosis has demanded combined efforts in biotechnology, nanotechnology, and electronic areas where the emerging technologies of lab-on-a-chip microdevices (Arshavsky-Graham & Segal, 2022; Lakhera et al., 2022), nanosensors (Mahato et al., 2018), lateral flow (Liu et al., 2021), and wearable (Kim et al., 2019; Min et al., 2021; Sempionatto et al., 2022) technologies offer opportunities for the development of new biosensors with improved performance, allowing tremendous advances in point-of-care (POC) diagnostic tests development (Liu et al., 2021; Mahato et al., 2018; Sachdeva et al., 2021). Thereby, remarkable improvements in research have preceded the development and commercialization of biosensors, as illustrated in Fig. 11.1.

Over the past two decades, there has been a notable increase in the research focused on the advancement of biosensors, leading to a larger number of devices reaching the market, with a focus on users easy access to clinical diagnosis (Cagnani et al., 2022). For instance, a milestone in this area was the "GlucoWatch" commercialization by Cygnus (Redwood City, CA, USA) in the 2000s (Tierney et al., 2000), which is considered the first commercial wearable biosensor for noninvasive glucose determination. Although many advances in wearable health devices for vital signs monitoring have been proposed since the launch of the wireless heart rate monitor by Polar Electro in the 1980s, the GlucoWatch was the first focused on molecule detection. In the 2010s, the R&D investments focused on device miniaturization, resulting in smaller, more affordable, and more sensitive test monitors, specifically blood glucose meters. Nowadays, significant endeavors are underway toward the commercialization of minimally or noninvasive biosensing alternatives; however, nanotechnology-based biosensors have yet to make a mark. Meanwhile, a sharp increase in wearable electronics has been

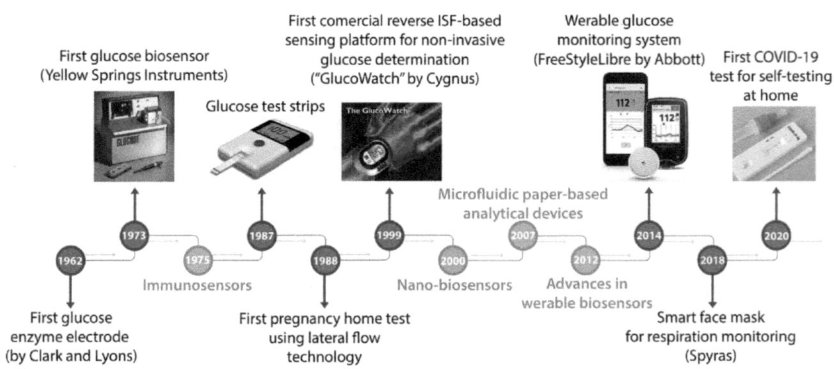

FIGURE 11.1 Timeline of significant milestones in the development of commercial biosensors for precision medicine.

witnessed in the last decade, reflected by a projected market size of billions of USD (Yin & Wang, 2022).

Several companies have been analyzing the global biosensor's market size, as shown in Table 11.1. The numbers show an estimated average value of USD 27.4 billion for 2022 and a projected growth of USD 50.4 billion by 2030 in the field, exhibiting an average compound annual growth rate (CAGR) of 8% in the forecast period (2022–2030). Such estimative was performed by the companies listed in the table, following different methodologies such as historical statistics, underlying market forces, statistical model (correlation and regression analysis), weighted moving average, industry expert feedback, times series forecast, and data reporting. Briefly, the reports show that the global market is divided into five principal regions, namely North America, with a market share of about 40%, followed by Europe, Asia Pacific, Latin America, and the Middle East and Africa, the two last which comprise together the fifth region (Vashist, 2017). The main types of commercial biosensors found are electrochemical, which has a share of around 70%, followed by thermal, optical, and piezoelectric ones. End users and/or customers are mainly interested in POC solutions that have a share of around 50%, followed by home healthcare diagnostics, research laboratories, security, and biodefense, food industry, and others (Kost et al., 2008). All this information can be found on the websites listed in Table 11.1.

Based on the growing interest in the field, in this chapter, we will discuss commercially available biosensors for clinical diagnosis or health monitoring, their performance, and the required technology.

11.2 Commercially available biosensors for medical diagnoses

Encouraged by the population's high demand for health care, the need for continuous monitoring of chronic diseases (e.g., diabetes), (Bozorgmehr et al., 2011) and the vast improvements in (bio)technology and electronics fields, which allows the development of robustness and economically viable devices, companies around the world are increasingly interested in the development and commercialization of POC biosensors. POC is defined as testing at or near the point of patient care (even by the patient themselves), wherever medical care is needed, to facilitate immediate evidence-based medical decisions that improve patient outcomes. Table 11.2 shows many types of commercial clinical biosensors, including POC devices. As can be observed, different types of monitoring mechanisms and technologies have been employed. Among them, lateral flow and wearable devices stand out for being the most used nowadays. Regarding the analyte, glucose is by far the most targeted analyte of available commercial biosensors. Based on this information, in the next sections, emphasis will be given to commercial biosensors based on lateral flow and wearable technologies, including those

TABLE 11.1 Summary of biosensor's market size analysis reported by cited companies.

Market size value in 2022 (USD billion)	Revenue forecast in 2030 (USD billion)	Growth rate (%)	Company	Research Report available in:
27.5	49.8	7.9	Markets and Markets	(https://www.marketsandmarkets.com/Market-Reports/biosensors-market-798.html)
26.9	49.8	8	Grand View Research	(https://www.grandviewresearch.com/industry-analysis/biosensors-market)
29.6	49.6	7.7	Precedence Research	(https://www.precedenceresearch.com/biosensors-market)
28.5	58	7.4	GM Insights	(https://www.gminsights.com/industry-analysis/biosensors-market)
28.5	54.2	8.5	Verified Market Research	(https://www.verifiedmarketresearch.com/product/biosensors-market/)
29.3	49.6	7.7	Globe News Wire	(https://www.globenewswire.com/en/news-release/2022/05/05/2437016/0/en/Biosensors-Market-Size-to-Surpass-US-49-6-Billion-by-2030.html)
27.1	49.3	7.8	P&S Market Research	(https://www.psmarketresearch.com/market-analysis/biosensors-market)
22.6	38.7	7.9	Market Research Future	(https://www.marketresearchfuture.com/reports/biosensor-market-1228)
26.5	54.1	9.3	Data Bridge Market Research	(https://www.databridgemarketresearch.com/reports/global-biosensors-market)
27.5	50.7	8	Emergen Research	(https://www.emergenresearch.com/industry-report/biosensors-market)

TABLE 11.2 Selected clinical biosensors commercially available and their characteristics.

Company	Product	Analyte/biofluid	Monitoring mechanism/technology	Sensitivity	Website
Yellow Springs Instruments, USA	23A	Glucose/blood	Electrochemistry	1–45 mM	Discontinued by the manufacturer
Abbott, USA	FreeStyle Libre	Glucose/ISF	Electrochemistry/wearable	20–500 mg/dL	(https://www.freestylelibre.us/)
	FreeStyle Freedom Lite			20–500 mg/dL	
	FreeStyle Life InsuLinx			20–500 mg/dL	
	Precision Xtra Overview			–	
Roche, Switzerland	Accu-Chek Aviva Plus	Glucose/blood	Electrochemistry	–	(https://www.accu-chek.com/test-strips/aviva-plus-test-strips)
	Accu-Chek Aviva Nano		Electrochemistry	–	(https://www.accu-chek.ca/en/meter-systems/aviva-nano)
	Accu-Chek Active		Photometric/reflectance	10–600 mg/dL	(https://www.accu-chek.com.br/monitores-de-glicemia/active)

(*Continued*)

TABLE 11.2 (Continued)

Company	Product	Analyte/biofluid	Monitoring mechanism/technology	Sensitivity	Website
Bayer, Canada	Countour Plus	Glucose/blood	Electrochemistry	–	(https://www.diabetes.ascensia.com.br/products/contour-plus/)
	Countour Next One		Electrochemistry	–	(https://www.diabetes.ascensia.pt/products/contour-next-one/)
	Countour Next EZ		Electrochemistry	–	(https://www.ascensiadiabetes.com/products/contour-next-ez/)
Cygnus Inc.,USA	GlucoWatch	Glucose/ISF	Electrochemistry/wearable	–	discontinued by the manufacturer
Prediktor Medical, Norway	BioMKR	Glucose/blood	Near-infrared spectroscopy, bioimpedance/wearable	–	(https://www.prediktormedical.com/)
MediWise, England	GlucoWise	Glucose/blood	Radiofrequency/wearable	–	(https://glucowise.com)
Dexcom, USA	Dexcom G6 (rt-CGM[a])	Glucose/ISF	Electrochemistry/wearable	–	(https://www.dexcom.com/global)
Senseonics, USA	Eversense	Glucose/ISF	Nonenzymatic electrochemical Fluorescence/wearable	–	(https://www.eversensediabetes.com/)

Alere,USA	TestPack PLUS hCG Combo	hCG/urine or serum	Immunoassay/lateral flow	25 mIU/mL	(https://www.globalpointofcare.abbott/en/product-details.html)
	Alere hCG Easy Pregnancy test	hCG/urine	Immunoassay/lateral flow	25 mIU/mL	
	Alere hCG Combo cassette (20/10 MIU/ML)		Immunoassay/lateral flow	10 mIU/mL in serum, 20 mIU/mL in urine	
	Alere hCG dipstick		Immunoassay/lateral flow	25 mIU/mL	
	Alere hCG combo Cassette (25 MIU/ML)	hCG/urine or serum	Immunoassay/lateral flow	25 mIU/mL	
Clearblue, USA	Clearblue Digital	hCG/urine	–	–	(https://www.clearblue.com)
	Clearblue Plus		–	–	
Coretests, China	HCG pregnancy test (midstream)	hCG/urine midstream	Immunoassay/lateral flow	25 mIU/mL	(https://www.coretests.com/product/show_107.html)
	HCG pregnancy test (cassette)	hCG/urine or serum or plasma	Immunoassay/lateral flow	25 mIU/mL	
	HCG pregnancy test (strip)	hCG/urine	Immunoassay/lateral flow	25 mIU/mL	

(*Continued*)

TABLE 11.2 (Continued)

Company	Product	Analyte/biofluid	Monitoring mechanism/technology	Sensitivity	Website
Germaine Laboratories, USA	AccuHome pregnancy	hCG/urine midstream	Immunoassay/lateral flow	25 mIU/mL	(https://www.germainelabs.com/product/accuhome-pregnancy-1-test/)
	AimStep pregnancy	hCG/urine	Immunoassay/lateral flow	20 mIU/mL	(https://www.germainelabs.com/product/aimstep-pregnancy-30-box/)
	hCG Urine Controls	Qualitative hCG/urine	Immunoassay	50 mIU/mL	(https://www.germainelabs.com/product/hcg-urine-controls-2x5mL/)
Quidel, USA	QuickVue hCG Combo Test	hCG/urineor serum	Immunoassay/lateral flow	25 mIU/ml	(https://www.quidel.com/immunoassays/rapid-pregnancy-tests/quickvue-hcg-combo-test)
	QuickVue hCG urine Test	hCG/urine	Immunoassay	25 mIU/mL	(https://www.quidel.com/immunoassays/rapid-pregnancy-tests/quickvue-hcg-urine-test)
	Sofia hCG [b]FIA		Fluorescent immunoassay/lateral flow	20 mIU/mL	(https://www.quidel.com/immunoassays/rapid-pregnancy-tests/sofia-hcg)

Abbott,USA	BinaxNOWInfluenza A & B	Influenza A and B/ nasopharyngeal swab and nasal swab	Immunoassay/lateral flow	Influenza A 70%–89%; Influenza B 50%–69%	(https://www.globalpointofcare.abbott/pt/product-details/binacnow-influenza-a-and-b.html)
	PanBioTM COVID-19/FLU A&B Rapid Panel	COVID-19/ Influenza A and B/ nasopharyngeal swab	Lateral flow	–	(https://www.globalpointofcare.abbott/en/product-details/panbio-covid-19-flu-ab-rapid-panel.html)
	Bioline Influenza Ag A/B/A(H1N1)	Influenza A, B, and A(H1N1) types/ nasal, throat, or nasopharyngeal swab	Lateral flow	Influenza A 62%; Influenza B 85% Influenza A (H1N1) 77%	(https://www.globalpointofcare.abbott/pt/product-details/bioline-influenza-ag-aba.html)
Alere,USA	Alere BinaxNOWInfluenza A & B	Influenza A and B/ nasal swab or nasopharyngeal swab	Lateral flow	Influenza A 70%–89%; Influenza B 50%–69%	(https://cliawaived.com/alere-binaxnow-ae-influenza-a-b-cards.html)
	Alere Influenza A/B/H1N1	Influenza A, B, and A(H1N1) types/ nasal, throat, or nasopharyngeal swab	Lateral flow	Influenza A 97%; Influenza B 91%	(https://www.lojamasterdiagnostica.com.br/alere-influenza-a-b-h1n1-h3n2-pr-62-346769.html)

(Continued)

TABLE 11.2 (Continued)

Company	Product	Analyte/biofluid	Monitoring mechanism/technology	Sensitivity	Website
Coris BioConcept, Belgium	Influ A + B *K*-SeT	Influenza A and B/nasopharyngeal specimens (swab, aspirate, and wash)	Lateral flow	Influenza A 88%; Influenza B 76%	(https://www.corisbio.com/products/influenza-a-and-b-viruses)
Directigen EZ Flu A + B, USA	BD Directigen EZ Flu A + B	Influenza A and B/nasopharyngeal washes/aspirates, nasopharyngeal swabs, and throat swabs	Lateral flow	–	(https://www.fishersci.com/shop/products/bd-directigen-ez-flu-a-b-test-kit/B256050)
SA Scientific Inc, USA	SAS FLUAlert A & B	Influenza A and B/nasal aspirates and nasal washes	Lateral flow	–	(https://myfisherstore.com/singapore/sas-trade-flualert-a-and-b-fissa046231.html)

Quidel,USA	QuickVue Influenza A + B Test	Influenza A and B/nasal swab or nasopharyngeal swab	Lateral flow	Influenza A 77%–94%; Influenza B 60%–82%	(http://www.quidel.com/immunoassays/rapid-influenza-tests)
	Sofia 2 Flu + SARs Antigen [b]FIA	Influenza A/B and SARS-CoV-2/nasal swab or nasopharyngeal swab	Lateral flow	Influenza A 90%–97%; Influenza B 89% 90% SARS-CoV-2: 95%	
	Sofia Influenza A + B [b]FIA	Influenza A and B/nasal swab or nasopharyngeal swab, nasopharyngeal aspirate/wash	Lateral flow	Influenza A 90%–97%; Influenza B 88%–97%	
	Solana Influenza A + B Assay	Influenza A and B/nasal swab or nasopharyngeal swab	Lateral flow	Influenza A 99%; Influenza B 100%	
	Lyra Influenza A + B Assay		Lateral flow	–	

(*Continued*)

TABLE 11.2 (Continued)

Company	Product	Analyte/biofluid	Monitoring mechanism/technology	Sensitivity	Website
Abott,USA	Real-Time HIV-1	HIV-1/EDTA plasma	Molecular (RT-PCR[c])	40 copies/mL for 1.0 mL sample volume 40 copies/mL for 0.6 mL sample volume 75 copies/mL for 0.5 mL sample volume 150 copies/mL for 0.2 mL sample volume	(https://www.molecular.abbott/us/en/products/infectious-disease/realtime-hiv-1-viral-load)
Siemens Healthcare, Germany	VERSANT HIV-1 RNA 1.5 assay (KPCR)			–	(https://www.siemens-healthineers.com/br/laboratory-diagnostics/molecular-diagnostics/molecular-diagnostics-in-vitro-diagnostics/versant-hiv-1-rna-1-assay)
	VERSANT HIV-1 v3.0		Branched-DNA technology	–	(https://www.siemens-healthineers.com/nl/molecular-diagnostics/molecular-diagnostics-in-vitro-diagnostics/versant-hiv-1-rna-3-assay)

Roche Molecular System,USA	Cobas HIV-1 Test		Molecular (RT-PCR[c])	≥95%	(https://diagnostics.roche.com/global/en/products/params/cobas-hiv-1-test.html)
		cobas AmpliPrep/COBAS TaqMan v2.0			(https://diagnostics.roche.com/global/en/products/params/cobas-ampliprep-cobas-taqman-hiv-1-test-v2-0.html)
	Cobas HIV-1/HIV-2 Qualitative Test	HIV-1/EDTA plasma, serum, and dried blood spot (DBS)		EDTA plasma HIV 1 M: 12.6 cp/mL HIV-2: 27.9 cp/mL Serum HIV-1M: 12.1 cp/mL HIV-2: 23.4 cp/mL DBS HIV-1M: 255 cp/mL HIV-2: 984 cp/mL	(https://diagnostics.roche.com/global/en/products/params/cobas-hiv-1-hiv-2-qualitative-test.html#productSpecs)

(*Continued*)

TABLE 11.2 (Continued)

Company	Product	Analyte/biofluid	Monitoring mechanism/technology	Sensitivity	Website
Abbott Diagnostic, USA	ID NOWTM COVID-19	SARS-CoV-2 and Influenza A and B/ nasal swab	Molecular (RT-PCR[c])/ lateral flow	84%	(https://www.globalpointofcare.abbott/pt/product-details/id-now-covid-19-ww.html)
	BinaxNOW TMCOVID-19 Antigen Self-Test	SARS-CoV-2/nasal swab	Immunoassay/antibody/ lateral flow	92%	(https://www.globalpointofcare.abbott/en/product-details/binaxnow-covid-19-antigen-self-test-us.html)
	Abbott RealTime SARS-COV-2	SARS-CoV-2/nasal swab, nasopharyngeal and oropharyngeal swabs	Molecular (RT-PCR[c])/ lateral flow	–	(https://www.molecular.abbott/us/en/products/infectious.disease/RealTime.SARS-CoV-2-Assay)
	PANBIOCOVID-19 Antigen Self-Test	SARS-CoV-2/nasal swab	Immunoassay/antibody/ lateral flow	96%	(https://www.globalpointofcare.abbott/pt/product-details/panbio-ovid-19-antigen-selftest.html)

Abbott Molecular, USA	TaqPathTM COVID-19 Combo Kit	SARS-CoV-2/ nasopharyngeal, oropharyngeal, anterior nasal, and mid-turbinate swabs, nasopharyngeal aspirate, and bronchoalveolar lavage specimens	Molecular (RT-PCR[c])/ lateral flow	–	(https://www.fishersci.com/shop/products/taqpath-covid-19-combo-kit/A47814)
Applied Biosystems, USA	Xpert Xpress SARS-CoV-2	SARS-CoV-2/ nasopharyngeal swab, nasal swab, and nasal wash/ aspirates	Molecular (NAAT[d])/lateral flow	–	(https://www.cepheid.com/pt_BR/tests/Critical-Infectious-Diseases/Xpert-Xpress-SARS-CoV-2)
AccessBio, USA	Cobas SARS-CoV-2 test	SARS-CoV-2/nasal, nasopharyngeal and oropharyngeal swab	Molecular (RT-PCR[c])/ lateral flow	100%	(https://diagnostics.roche.com/global/en/products/params/cobas-sars-cov-2-test.html)
ThermoFisher Scientific, USA	Accula SARS-CoV-2	SARS-CoV-2/nasal or nasal mid-turbinate swab	Molecular (NAAT[d])/lateral flow	–	(https://www.fishersci.com/shop/products/accula-sars-cov-2-test-kit/COV4100)

(Continued)

TABLE 11.2 (Continued)

Company	Product	Analyte/biofluid	Monitoring mechanism/technology	Sensitivity	Website
Cellex Inc., USA	qSARS-CoV-2 IgG/IgM	SARS-CoV-2/human serum and plasma	Immunoassay/antibody IgG/IgM (LFA[e])/lateral flow	94%	(https://www.nldiagnostica.com.br/n/coronavirus/)
Lungene, China	Self-Use COV-ID-19 Nasal	SARS-CoV-2/nasal swab	Immunoassay/antibody/lateral flow	–	(https://www.globalsources.com/Urine-parameter/Self-Use-Nasal-Rapid-Test-Cassettes-1182050070p.html)
Quidel Corporation, USA	CareStart COVID-19 IgG/IgM	SARS-CoV-2/human serum and plasma	Antibody IgG/IgM (LFA[e])/lateral flow	90%	(https://accessbio.net/products/covid-19-detection-kits/carestart-covid-19-lgm-lgg)
	Sofia 2 SARs Antigen bFIA	SARS-CoV-2/nasal swab	Immunoassay/antibody/lateral flow	98%	(https://www.quidel.com/immunoassays/rapid-sars-tests/sofia-sars-antigen-fia)

Roche Diagnostics, Switzerland	Elecsys Anti-SARS-CoV-2	SARS-CoV-2/human serum and plasma	Immunoassay/antibody/ lateral flow	99%	(https://diagnostics.roche.com/global/en/products/params/elecsys-anti-sars-cov-2.html)
Coris BioConcept, Belgium	COVID-19 Ag Respi-Strip	SARS-CoV-2/ nasopharyngeal swab	Immunoassay/lateral flow	–	(https://www.corisbio.com/products/covid19)
	COVID-19 Ag K-SeT	SARS-CoV-2/nasal swab	Immunoassay/lateral flow	–	(https://www.corisbio.com/products/covid19-ag-k-set)

[a] *rt-CGM, real-time continuous glucose monitoring.*
[b] *FIA, fluorescent immunoassay.*
[c] *RT-PCR, reverse transcription polymerase chain reaction.*
[d] *NAAT, nucleic acid amplification test.*
[e] *LFA, lateral flow assay.*

developed for glucose detection. Besides, biosensors developed for SARS-CoV-2 have gained more importance nowadays due to the recent COVID-19 outbreak and are also presented here.

11.2.1 Lateral flow–based biosensors

The most prevalent POC diagnostic platform on the market is the lateral flow assay, which consists of a paper-based device (Antiochia, 2021). In general, paper-based tests can be classified into three categories, as illustrated in Fig. 11.2A: adhesion tests (dipstick tests), microfluidic biosensors, and lateral flow assays. The first type consists of a chemically active paper strip that can detect the presence of several analytes, such as glucose, uric acid, and proteins (Park et al., 2016). In paper-based microfluidic biosensors, a small volume of fluid is transported by capillarity through microscaled multichannels on the paper substrate. Typical biofluids compatible with this type of device include total blood, serum, urine, and nasal swabs (Kim et al., 2020).

Specifically, lateral flow assays (LFAs) have been commercially available for a long time, consisting in a low-cost, simple, rapid, and portable technology (Koczula & Gallotta, 2016). As a rapid-action test that can be used as a POC with or without a reader, it is easy to use and interpret, requiring little training, and being easily deployed anywhere (Koczula & Gallotta, 2016; Liu et al., 2021; Liu et al., 2022; Mahato et al., 2018; Sharafeldin & Davis, 2021). In short, LFA is a straightforward diagnostic device used to confirm the presence or absence of a particular analyte, such as contaminants in drinking water and foods, and pathogens or biomarkers present in humans or animals. The most common and popular example of this type of sensor is the rapid pregnancy tests that can be purchased at conventional pharmacies (Wild & Kodak, 2013). Depending on the recognition elements used, LFAs can be categorized into two different types: lateral flow immunoassay (LFIA), which uses antibodies, proteins, and hormones as specific recognition elements, or nucleic acid lateral flow assay (NALFIA), which uses nucleic acids as recognition elements (Koczula & Gallotta, 2016).

Fig. 11.2B shows the main components of an LFA-type sensing platform and its detection mechanism, exemplified by a typical pregnancy test, where it is possible to confirm pregnancy when two lines appear on the rapid test display. The test is performed by placing the sample in an absorbent region, called "sample pad," and then the absorbed liquid moves through a release pad by capillary action toward the region, where the detection occurs. In addition, the platform presents control lines and an absorbent pad for retaining excess samples, avoiding backflow. Fig. 11.2C illustrates the analyte recognition mechanism. When the sample is added to the sample pad (Fig. 11.2C-i), which is impregnated with chemical components such as buffer salts and surfactant, it chemically reacts with the detection system. The sample then migrates through the conjugate release pad (Fig. 11.2C-ii),

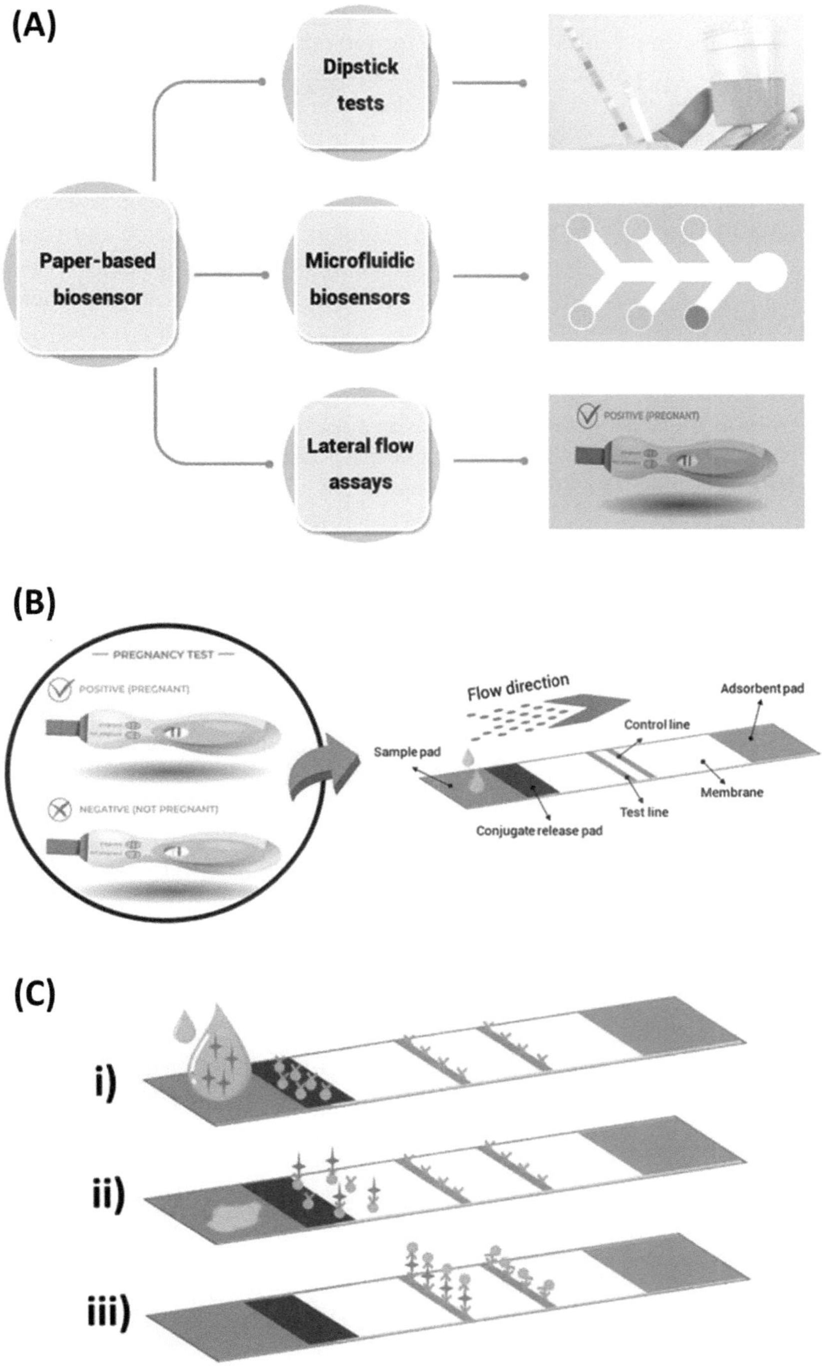
(A)
Paper-based biosensor
Dipstick tests
Microfluidic biosensors
Lateral flow assays
POSITIVE (PREGNANT)
(B)
PREGNANCY TEST
POSITIVE (PREGNANT)
NEGATIVE (NOT PREGNANT)
Flow direction
Adsorbent pad
Control line
Sample pad
Membrane
Test line
Conjugate release pad
(C)
i)
ii)
iii)

(*Continued*)

where specific antibodies for the analyte under analysis are found and are bound to other molecules responsible for providing color or fluorescence to the test line (if the result is positive). The system containing the sample, antibody, and label particles moves to the detection zone (Fig. 11.2C-iii), which is formed by a porous membrane and specific biological components for each type of analyte, whose function is to react with the complex formed between the analyte and the conjugated antibody. In the detection zone, it is possible to find the test line with the receptors immobilized on the paper, causing the label to be found in this specific area. The control line is created by immobilizing a secondary capture molecule, which can capture labeled detection reagents and produces a visible signal, indicating the proper flow of the sample through the device. While the colored trace in the control line indicates adequate liquid flow across the strip and functional bioreceptors, the trace in the test line indicates a positive result (or in some specific tests, the absence of color is indicative of a positive result). After passing through the detection zone, the liquid continues to flow through the device due to capillary forces. To maintain this movement, an absorbent pad is placed at the end of the strip to remove excess reagents and prevent the backflow of the liquid, which could impair the interpretation of the result (Koczula & Gallotta, 2016). Depending on the type of analysis, additional test lines can also be added to the detection zone, allowing the detection of multiple antibodies/antigens in a single analysis.

LFA sensors have been commercialized since the 1970s and continue to grow in popularity due to their remarkable features including high specificity and sensitivity toward complex sample matrices (Andryukov, 2020).

An important milestone in the field of immunoassay was the introduction of the first pregnancy home test in 1988. The Clearview test, developed by Unipath LTD, demonstrated remarkable accuracy of up to 99% and a response time of around 3 minutes. As this type of sensor has advanced, its response time has decreased to just 1 minute. Additionally, more modern digital versions have been developed, but the principles of hormone detection have remained largely unchanged. Moreover, the platform is evolving to become an integrated diagnostic tool that is capable to incorporate cutting-edge technologies such as optical and electrochemical transducers, as well as wireless data transfer, resulting in increasingly portable devices (Van Amerongen et al., 2018).

FIGURE 11.2 (A) Classification of different types of paper-based biosensors (Antiochia, 2021); (B) Illustration of a pregnancy test, a typical lateral flow immunoassay (LFA) test strip comprising a sample pad, a conjugate release pad, a membrane with immobilized antibodies, and an adsorbent pad, attached to an inert material holder and (C) illustration of the detection mechanism of an LFA sensor. After the analyte is deposited onto the sample pad, it migrates toward the conjugate release pad, where it binds to the conjugated antibodies and then migrates to the test line.

11.2.1.1 Commercial SARS-COV-2 biosensors

The COVID-19 pandemic has brought rapid detection tests to the forefront of worldwide attention. Pathogenic agents have posed a threat to public health since ancient times, as they can spread quickly through the air, water, or food. Even before the COVID-19 pandemic, various pathogens were responsible for around 17 million deaths each year, according to the World Health Organization (WHO, 2022). The detection time of the disease-causing agent is crucial in many cases, either to accurately treat the infected individuals or to monitor the transmission of the disease. With the advancement of research based on the detection of pathogenic agents, it is currently possible to find various types of sensors on the market that enable their rapid and efficient detection, such as the case of rapid COVID-19 tests (Baker et al., 2020; Corman et al., 2021; Fulford et al., 2021).

Serology has been widely applied for detecting SARS-CoV-2. Usually, ELISA is used to detect viral antigens or cytokines (IgG/IgM). Although ELISA is generally a reliable test, results take at least a few hours, and the test equipment is not portable. This bottleneck can be particularly problematic during pandemics, which has required the development of rapid and accurate biosensors. Chip-based POC biosensors have emerged as a viable option (Chircov et al., 2020), with some studies focusing on developing an ELISA-type test based on on-chip fluorescence, but in a more compact and fluid-free device (Lin et al., 2020; Liu et al., 2020; Rasmi et al., 2021).

In terms of sensitivity and specificity rapid tests for SARS-CoV-2 detection are still considered inferior to the ELISA method, but those present advantages as allowing faster response (Al-Alawi et al., 2021; Corman et al., 2021). The need for immediate detection has driven significant growth in the research and manufacturing of portable COVID-19 biosensors in recent years. As a result, researchers and industry are seeking to improve their performance, with particular attention to their sensitivity and specificity, which are essential parameters for the application (Michael & Mina, 2020). Various types of rapid SARS-CoV-2 tests, including POC tests, are currently available in the market, with emphasis on LFA biosensors.

One of the first rapid COVID-19 tests to receive Emergency Use Authorization (EUA) from the US FDA was the Abbott ID NOW COVID-19 test (https://www.globalpointofcare.abbott/pt/product-details/id-now-covid-19-ww.html), which can provide results within 13 minutes using a portable device. The test is based on isothermal nucleic acid amplification, which amplifies the viral RNA to detectable levels in a short time. According to the literature, (Iliescu et al., 2021; Tu et al., 2021) the Abbott ID NOW COVID-19 test has shown high accuracy, with a reported sensitivity of 93.3% and specificity of 98.4% compared to PCR-based tests. The test has been widely used in the United States and other countries and has been recommended for use in various sites, including hospitals, clinics, and long-term care facilities. The company has been a significant player in the COVID-19 response, producing a variety of diagnostic tests and other products to help control the spread of the pandemic.

Another popular test for SARS-CoV-2 is the BD Veritor System, produced by Becton, Dickinson and Company (Franklin Lakes, Nova Jersey, EUA) (https://www.bd.com/en-uk/offerings/capabilities/microbiology-solutions/clinical-microbiology/point-of-care-testing/bd-veritor-plus-system). This test uses a handheld device and provides results in 15 minutes using a nasal swab sample. The Quidel Sofia 2 SARS Antigen FIA, produced by Quidel Corporation (San Diego, California, USA), is also frequently used in POC settings (https://Www.Quidel.Com/Immunoassays/Rapid-Sars-Tests/Sofia-Sars-Antigen-Fia) as a fluorescent immunoassay to detect the virus, with results available in 15 minutes. The above-cited tests, along with other similar tests commercially available, have been fundamental in helping to slow the spread of COVID-19 by quickly identifying infected individuals and enabling prompt isolation and treatment.

11.2.2 Wearable biosensors

As discussed in Chapter 9 of this book, wearable sensors are devices that can be integrated with the human body to enable remote and continuous monitoring of health and personal performance (Heikenfeld et al., 2018). Widely applied for medical purposes, wearable sensors enable data collection of multiple physiologic parameters, such as heart rate, respiration rate, blood oxygen saturation, core, and skin temperature (Kim et al., 2019; Moonen et al., 2020; Sazonov, 2021; Sempionatto et al., 2017). Besides, wearable sensors have been widely used by the world's population as smartwatches, which are wearable devices that can track movement, sleep habits, hydration levels, heart rate, and temperature (Beh et al., 2021; Chandel et al., 2021; Kamal Basha et al., 2022; Lima et al., 2022). These devices provide real-time information based on integrated onboard sensors combined with data processing. All the provided information can be accessed and recorded/followed over time through mobile apps, improving the user experience with wearable devices (Al-Emran, 2021). The market is led by big and renowned companies like Apple with Apple-watch, Fitbit (Google company with wristbands and smartwatches), and Samsung (Galaxy watches) along with new companies such as Xiaomi. Likewise, Garmin, which is well-known for manufacturing regular watches, has also gained importance in the smartwatch market, mainly Forerunner and Venu series (Casaccia et al., 2021).

Besides, the next level of wearable sensors might be a powerful tool for the early detection of viral infections, helping to prevent imminent outbreaks (Kim et al., 2019). Although several wearable sensors are under clinical evaluation, the successful conversion of these devices into a commercial version is still challenging. Current efforts on commercial wearable devices are focused on improving minimally and noninvasive glucose sensors (which will be explored in the next section), and sweat sensors as monitoring systems for applications in fitness, military, space exploration, and industrial fields (Kim et al., 2018).

The first wearable sweat sensor (WSS) applied to humans in daily activities was developed by General Electric Global Research company and tested by US Air Force Academy members for sweat electrolyte levels monitoring (sodium and potassium) in 2018 (Alizadeh et al., 2018). Several other companies are currently working to improve and commercialize WSS. One notable example is Epicore Biosystems, which has launched an epidermal microfluidic colorimetric sweat sensor named Gatorade Gx Sweat Patch (Moonen et al., 2020). Nix Biosensors and Kenzen startups are also pioneering the market through WSS and even Google company has expressed interest in this technology by purchasing FitBit from Amazon (Min et al., 2021). This growing interest from companies in WSS is driven by the global market's expectation of expansion, estimated to reach approximately USD 2.6 billion in the next decade, with a double increase in several products available (Sensors et al., 2023). It is important to mention that WSS devices developed by startups and companies are in the process of regulation and validation to enter the market. Examples include the devices called BACtrack's (BACtrack Skyn) and Alcopro (Reliable Drug and Alcohol Testing Products), for transdermal alcohol monitoring. In addition, the latter is also designed to test marijuana, cocaine, and opioids. Alcopro is based on an absorbing pad that might be kept in place over seven days and then taken to analysis in the laboratory. Another example of alcohol WSS is the transdermal ankle bracelet called SCRAM CAM, which works with wireless communication for continuous alcohol monitoring and operates based on the concentration of alcohol vapor released from the skin (SCRAM Systems).

Another important application of WSS is in the monitoring of therapeutic drug levels. In this regard, clinical validation and improvements in sensor technologies for WSS are on the way to boost the production of commercialized WSS for disease diagnostics, monitoring, and prognostics (Alizadeh et al., 2018). Eccrine Systems Inc., for instance, has brought to market an aptamer-based electrochemical sweat sensor developed through research conducted at the University of Cincinnati (Arroyo-Currás et al., 2020). The device is designed to track therapeutic drug levels on sweat, offering a convenient and noninvasive monitoring method. This is an example where the device integration with a database system and clouds will enable highly precise disease management through drug dosage predictions, replacing conventional and tedious monitoring methods. Advancements like this are possible once drug concentrations in sweat are often well correlated with that of plasma (Tai et al., 2018; Teymourian et al., 2020). For example, the effective management of Parkinson's disease can be made by monitoring drugs such as L-Dopa, which has been explored in WSS applications for future commercial devices (Goud et al., 2019). Besides, recent studies indicate the capability of current cutting-edge WSS for monitoring metabolic profiles to help prevent severe complications caused by infectious diseases and monitor vulnerable population health (Maumdjer et al., 2020; Nasiri & Khosravani, 2020).

It is important to highlight the significance of accuracy and precision when it comes to commercial WSS devices for reliable and selective detection of analytes in complex biofluids (Sempionatto et al., 2017). Such properties can be challenging once WSS are exposed to uncontrolled conditions during on-body functions involving varied temperatures, pH, ionic interactions, humidity, and biofouling. In addition to the ordinary variations in the human body and body fluid compositions, different medications and changes in eating habits can directly influence the sensor's performance. A better understanding of these conditions and their natural variabilities is important to guide the following steps in developing WSS with improved accuracy. For the development of an accurate WSS, extensive studies involving large-scale human trials are needed for reliable sensor validation (Hauke et al., 2018). Regarding the next steps, further advances are expected in integration with critical data analysis systems, database comparison, and wireless communication for easier and faster operation of the WSS.

11.2.3 Commercial glucose biosensors

Continuous glucose monitoring (CGM) devices provide a constant track of blood glucose levels, offering timely information complementary to the finger stick test (Gifford, 2013; Xue et al., 2022). Besides, CGM devices are integrated with a database system to keep previous data about glucose trends and predictions of the current stage and rate of change in glucose levels, which can be used to pave therapeutic decisions. Commercially available reliable and user-friendly CGM systems can be divided into minimally invasive and noninvasive devices. Their operation mechanism for home use is mainly based on enzymatic electrochemical sensors (Xue et al., 2022). Traditionally, an electrochemical CGM sensor operates based on oxygen consumption and its proportional dependency on glucose concentration (Zafar et al., 2022). As an example, for GlucoWatch operation, subcutaneous fluid containing glucose is extracted from the skin when a current of 3 mA is applied on the skin and attracted to an Ag/AgCl electrode by ion movements (iontophoresis). Then, a hydrogel pad containing glucose oxidase (GOx) is placed between the skin and the iontophoresis electrode to collect the fluid. Next, the glucose is catalytically oxidized by GOx generating H_2O_2 which is electrochemically detected and translated to equivalent glucose levels every 10 minutes (Geoffrey, 2009). Skin irritation and rash formation were pointed as major disadvantages of GlucoWatch, which was withdrawn from the market and had its sales suspended (see Table 11.1). Since then, many improvements have been made to commercial CGM devices in terms of size, cost, lifetime, accuracy, and operable interface. The GlucoWatch was followed by a series of CGM commercial devices such as PENDRA (Pendragon Medical Ltd.), Enlite series (Medtronic), MiniMed iPro (Medtronic), Guardian (Medtronic) https://www.medtronic.com/ca-en/diabetes/home/products/cgm-systems/guardian-connect.html, Dexcom series (Dexcom) https://www.dexcom.com, and FreeStyle (Abbott Diabetes Care) https://www.abbott.com/ and, although much improvement had been achieved, most of these devices still need more than one calibration per day.

Minimizing and eliminating daily calibrations is a priority for improving CGM operation. To overcome such limitations, Abbott introduced the FreeStyle Libre Flash to market with factory calibration and a 14-day lifetime. This sensor is considered the next generation of CGM with no need for test strips and lancets and has been referred to as Flash Glucose Monitor (FGM) since it is a system used only on demand for intermittent measurements. The device is composed of an enzymatic amperometric system containing three electrodes with a combination of an osmium complex mediator and the GOx enzyme embedded into a polymeric matrix as the working electrode (Kim et al., 2018; Peng et al., 2022). The glucose molecules from the interstitial fluid are diffused and oxidized by the GOx enzyme, releasing electrons to the mediator and transferring them to the electrode. A general representation of such process is illustrated in Fig. 11.3 (Zafar et al., 2022). The device must be attached to the arm (usually the back of the upper arm) and can be worn throughout the day and scanned by a reader providing information on glucose levels. Similarly, Medtronic launched the Enlite sensor, offering to the market a sensor that can be integrated with automated pumps for insulin delivery, being the first one to automatically regulate basal insulin delivery using the sensor's glucose data. According to the glucose data history, new devices have been equipped with alert configurations preventing hypoglycemic events (Lee et al., 2021). The alerts can be sent to a computer or smartphone with smart technologies, increasing data-sharing possibilities and providing more safety to diabetic patients.

Self-monitoring blood glucose (SMBG) has also evolved tremendously in technological terms, and, nowadays, glucose test strips can perform more

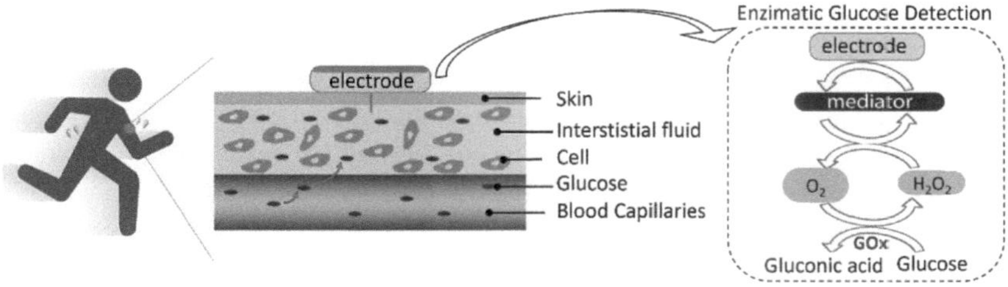

FIGURE 11.3 Schematic illustration of an enzymatic electrochemical wearable biosensor for glucose detection.

accurate assays in 5 seconds and using smaller volumes of blood (<200 nL blood) (Peng et al., 2022; Xue et al., 2022). The iBGStar is the first blood glucose meter that can be connected to an iPhone or iPod Touch and has been developed by Sanofi and AgaMatrix. The device automatically synchronizes with the appropriate app, and the users have direct access to glucose data on their smartphones. In this sense, the power of data science with virtual assistants and mobile technologies is the trend toward glucose monitoring devices and enabling ways in which patients can access and use data. Besides, smaller needles for minimally invasive monitoring will remain in line for the following improvements. At the same time, real noninvasive devices are still expected for glucose monitoring, eliminating the need to pervade the skin.

11.3 Challenges and future trends

Over the past years, many innovative biosensors have demonstrated tremendous potential for healthcare, safety/security, and environmental monitoring. These devices have become valuable for POC diagnostics, allowing the detection of different analytes including pH, metabolites (e.g., glucose, lactate, pyruvate), electrolytes (e.g., sodium, potassium, calcium), and pathogens (e.g., viruses), in biological fluids such as saliva, sweat, tears, and interstitial fluid. Indeed, incorporating nanomaterials has proven to be a powerful strategy to enhance sensor sensitivity by providing augmented signals and higher loading surface areas for the immobilization of biorecognition elements (e.g., enzymes, antibodies). Moreover, new technologies, including paper-based, inkjet printing, roll-to-roll printing, and 3D printing, have allowed the fabrication of wearable sensors. Such progress have demonstrated that biosensors have immense potential for real-world applications. However, some limitations still hamper their widespread development as commercial products.

Despite the remarkable development of lateral flow–based devices in the last few years, their sensitivity should be improved to reach the market requirements. For instance, the sensitivity of the commercial diagnostic kits based on IgG/IgM test strips needs yet to be improved. In this direction, some sensitivity enhancement strategies could be applied, such as catalytic enhancement, SERS-based recognition, enzyme-based signal improvement, and sample concentration. Moreover, multiplexed sensing, including both IgG/IgM and nucleic acid tests, could allow the detection of both early and late stages of infections, yielding more reliable and accurate results. In addition, the combination of sample treatment, preamplification, highly specific affinity molecules, and fast and easy LFA readout, at a reasonable cost, are important challenges to be overcome to enable new LFA diagnostic tests in the market.

In the case of wearable sensors, research and innovations in biocompatible, breathable, lightweight, flexible, and stretchable materials are crucial to achieve sustainable and low-cost components to enable scale-up production. Other efforts should be devoted to the improvement/development of sensing technologies, lowering analysis costs, and increasing the number/type of analytes that can be monitored. Moreover, the long-term stability of the sensing layer and the lifetime of the bioreceptors are important challenges. For example, an ideal wearable sensor should be able to monitor the analyte concentration over long periods. However, the active layer may change or even degrade during the analysis, and the use of an outer protective layer could improve the sensor's durability. Moreover, at the same time, this protection layer should not decrease the sensor sensitivity by acting as a physical barrier for the analyte. The accuracy of wearables could be improved through multiplexed sensing by using distinct transducer types and/or simultaneously measuring other analytes and/or samples on the same platform. In this direction, it is undisputable that the increased use of data collection and mining methods, machine learning, and artificial intelligence applications will be significant milestones in the commercialization of biosensors.

Another challenge is developing power units that provide continuous and prolonged energy supply. Although nanogenerators have already been demonstrated to possess promising energy harvesting performance, they still cannot supply enough sustained energy to power sensor devices. At the same time, it is essential to adopt systems with low power consumption to meet the requirements of portable applications. Additionally, advancements in mobile and wireless communication technology are expected to facilitate the development of compact and portable sensing devices, providing a wide range of opportunities for remote diagnostics.

In the near future, an extension of the diagnostic capabilities of wearable sensors is expected to pave the way for theranostic applications. Smart bandages, for example, could combine the in situ detection of bacterial infection by continuously monitoring wound pH and the administration of antibiotics or antiinflammatories locally and on demand. Another trend is related to developing devices that integrate glucose biosensors with insulin delivery systems, envisaging the possibility of effective and strict glycemic control among diabetic patients.

Acknowledgments

The authors thank the financial support from Fundação de Amparo à Pesquisa do Estado de São Paulo (FAPESP) (grant numbers: 2017/19470-8, 2018/22214-6), Fundação de Amparo à Pesquisa do Estado de Minas Gerais (FAPEMIG) (grant number: APQ-00373-18), Conselho Nacional de Desenvolvimento Científico e Tecnológico (CNPq) (grant number: 150002/2021-1) and Coordenação de Aperfeiçoamento de Pessoal de Nível Superior—Brasil (CAPES).

References

Al-Alawi, S., Al-Hinai, H., Al-Kindi, N., Al-Rashidi, M., Al-Kindi, H., Al-Shukri, I., Al-Rashdi, A., Jose, S., & Al-Jardani, A. (2021). Evaluation of four rapid antigen tests for detection of SARS-CoV-2 virus. *Oman Medical Jounal*, *36*(5). Available from https://doi.org/10.5001/OMJ.2021.106.

Al-Emran, M. (2021). Evaluating the use of smartwatches for learning purposes through the integration of the technology acceptance model and task-technology fit. *International Journal of Human-Computer Interaction*, *37*(19), 1874–1882. Available from https://doi.org/10.1080/10447318.2021.1921481.

Alizadeh, A., Burns, A., Lenigk, R., Gettings, R., Ashe, J., Porter, A., McCaul, M., Barrett, R., Diamond, D., White, P., Skeath, P., & Tomczak, M. (2018). A wearable patch for continuous monitoring of sweat electrolytes during exertion. *Lab on a Chip*, *18*(17), 2632–2641. Available from https://doi.org/10.1039/c8lc00510a.

Andryukov, B. G. (2020). Six decades of lateral flow immunoassay: From determining metabolic markers to diagnosing Covid-19. *AIMS Microbiology*, *6*(3), 280–304. Available from https://doi.org/10.3934/microbiol.2020018.

Antiochia, R. (2021). Paper-based biosensors: Frontiers in point-of-care detection of Covid-19 disease. *Biosensors*, *11*(4). Available from https://doi.org/10.3390/bios11040110.

Arroyo-Currás, N., Dauphin-Ducharme, P., Scida, K., & Chávez, J. L. (2020). From the beaker to the body: Translational challenges for electrochemical, aptamer-based sensors. *Analytical Methods*, *12*(10), 1288–1310. Available from https://doi.org/10.1039/d0ay00026d.

Arshavsky-Graham, S., & Segal, E. (2022). Lab-on-a-chip devices for point-of-care medical diagnostics. *Advances in Biochemical Engineering/Biotechnology*, *179*(2020), 247–265. Available from https://doi.org/10.1007/10_2020_127.

BACtrack Skyn. *BACtrack Skyn wearable alcohol monitor*. https://skyn.bactrack.com/.

Baker, A. N., Richards, S. J., Guy, C. S., Congdon, T. R., Hasan, M., Zwetsloot, A. J., Gallo, A., Lewandowski, J. R., Stansfeld, P. J., Straube, A., Walker, M., Chessa, S., Pergolizzi, G., Dedola, S., Field, R. A., & Gibson, M. I. (2020). The SARS-COV-2 spike protein binds sialic acids and enables rapid detection in a lateral flow point of care diagnostic device. *ACS Central Sciience*, *6*(11), 2046–2052. Available from https://doi.org/10.1021/acscentsci.0c00855.

Beh, P. K., Ganesan, Y., Iranmanesh, M., & Foroughi, B. (2021). Using smartwatches for fitness and health monitoring: The UTAUT2 combined with threat appraisal as moderators. *Behaviour & Information Technology*, *40*(3), 282–299. Available from https://doi.org/10.1080/0144929X.2019.1685597.

Bozorgmehr, K., Narayan, L., & Radhakrishna, R. (2011). Supplementary Webappendix. *Lancet*, *6736*(10), 1980–2008. Available from https://doi.org/10.1016/S0140-6736(11)61514-6. WEB-ONLY.

Cagnani, G. R., da Costa Oliveira, T., Mattioli, I. A., Sedenho, G. C., Castro, K. P. R., & Crespilho, F. N. (2022). From research to market: Correlation between publications, patent filings, and investments in development and production of technological innovations in biosensors. *Analytical and Bioanalytical Chemistry*, 0123456789. Available from https://doi.org/10.1007/s00216-022-04444-2.

Casaccia, S., Revel, G.M., Scalise, L., Cucchieri, G., & Rossi, L. (2021). Smartwatches selection: Market analysis and metrological characterization on the measurement of number of steps. In 2021 IEEE *i*nternational *s*ymposium on *m*edical *m*easurement and *a*pplication MeMeA 2021 – Conference *p*roceedings, Lausanne, Switzerland (pp. 1–5). https://doi.org/10.1109/MeMeA52024.2021.9478770.

Chandel, R. S., Sharma, S., Kaur, S., Singh, S., & Kumar, R. (2021). Smart watches: A review of evolution in bio-medical sector. *Materials Today Proceedings.*, *50*(August), 1053–1066. Available from https://doi.org/10.1016/j.matpr.2021.07.460.

Chircov, C., Grumezescu, A. M., & Andronescu, E. (2020). Biosensors-on-chip: An up-to-date review. *Molecules (Basel, Switzerland)*, 1–18.

Corman, V. M., Haage, V. C., Bleicker, T., Schmidt, M. L., Mühlemann, B., Zuchowski, M., Jo, W. K., Tscheak, P., Möncke-Buchner, E., Müller, M. A., Krumbholz, A., Drexler, J. F., & Drosten, C. (2021). Comparison of seven commercial SARS-CoV-2 rapid point-of-are antigen tests: A single-centre laboratory evaluation study. *The Lancet Microbe*, *2*(7), e311–e319. Available from https://doi.org/10.1016/S2666-5247(21)00056-2.

Fulford, T. S., Van, H., Gherardin, N. A., Zheng, S., Ciula, M., Drummer, H. E., Redmond, S., Tan, H.-X., Boo, I., Center, R. J., Li, F., Grimley, S. L., Wines, B. D., Nguyen, T. H. O., Mordant, F. L., Ellenberg, P., Rowntree, L. C., Kedzierski, L., Cheng, A. C., ... Godfrey, D. I. (2021). A point-of-care lateral flow assay for neutralising antibodies against SARS-CoV-2. *eBioMedicine*, *74*, 103729. Available from https://doi.org/10.1016/j.ebiom.2021.103729.

Geoffrey, M. G. (2009). The chemistry of commercial continuous glucose monitors. *Diabetes Technology & Therapeutics*, *11*, 307. Available from https://doi.org/10.1089/dia.2008.0133.

Gifford, R. (2013). Continuous glucose monitoring: 40 years, what we've learned and what's next. *Chemphyschem: A European Journal of Chemical Physics and Physical Chemistry*, *14* (10), 2032–2044. Available from https://doi.org/10.1002/cphc.201300172.

Goud, K. Y., Moonla, C., Mishra, R. K., Yu, C., Narayan, R., Litvan, I., & Wang, J. (2019). Wearable electrochemical microneedle sensor for continuous monitoring of levodopa: Toward Parkinson management. *ACS Sensors*, *4*(8), 2196–2204. Available from https://doi.org/10.1021/acssensors.9b01127.

Hauke, A., Simmers, P., Ojha, Y. R., Cameron, B. D., Ballweg, R., Zhang, T., Twine, N., Brothers, M., Gomez, E., & Heikenfeld, J. (2018). Complete validation of a continuous and blood-correlated sweat biosensing device with integrated sweat stimulation. *Lab on a Chip*, *18*(24), 3750–3759. Available from https://doi.org/10.1039/c8lc01082j.

Heikenfeld, J., Jajack, A., Rogers, J., Gutruf, P., Tian, L., Pan, T., Li, R., Khine, M., Kim, J., Wang, J., & Kim, J. (2018). Wearable sensors: Modalities, challenges, and prospects. *Lab on a Chip*, *18*(2), 217–248. Available from https://doi.org/10.1039/c7lc00914c.

Iliescu, F. S., Ionescu, A. M., Gogianu, L., Simion, M., Dediu, V., Chifiriuc, M. C., Pircalabioru, G. G., & Iliescu, C. (2021). Point-of-care testing—The key in the battle against SARS-CoV-2 pandemic. *Micromachines*, *12*(12), 1–33. Available from https://doi.org/10.3390/mi12121464.

Kamal Basha, N., Aw, E. C. X., & Chuah, S. H. W. (2022). Are we so over smartwatches? Or can technology, fashion, and psychographic attributes sustain smartwatch usage. *Technology in Society*, *69*(February), 101952. Available from https://doi.org/10.1016/j.techsoc.2022.101952.

Kim, J., Campbell, A. S., & Wang, J. (2018). Wearable non-invasive epidermal glucose sensors: A review. *Talanta*, *177*(August 2017), 163–170. Available from https://doi.org/10.1016/j.talanta.2017.08.077.

Kim, J., Campbell, A. S., de Ávila, B. E. F., & Wang, J. (2019). Wearable biosensors for healthcare monitoring. *Nature Biotechnology*, *37*(4), 389–406. Available from https://doi.org/10.1038/s41587-019-0045-y.

Kim, T. H., Hahn, Y. K., & Kim, M. S. (2020). Recent advances of fluid manipulation technologies in microfluidic paper-based analytical devices ($µ$PADs) toward multi-step assays. *Micromachines*, *11*(3). Available from https://doi.org/10.3390/mi11030269.

Koczula, K. M., & Gallotta, A. (2016). Lateral flow assays. *Essays in Biochemistry*, *60*(1), 111–120. Available from https://doi.org/10.1042/EBC20150012.

Kost, G.J., Tran, N.K., & Louie, R.F. (2008). *Point-of-care testing: Principles, practice, and critical-emergency-disaster medicine*. https://doi.org/10.1002/9780470027318.a0540.pub2.

Lakhera, P., Chaudhary, V., Bhardwaj, B., Kumar, P., & Kumar, S. (2022). Development and recent advancement in microfluidics for point of care biosensor applications: A review. *Biosensors & Bioelectronics X*, *11*, 100218. Available from https://doi.org/10.1016/j.biosx.2022.100218.

Lee, I., Probst, D., Klonoff, D., & Sode, K. (2021). Continuous glucose monitoring systems – Current status and future perspectives of the flagship technologies in biosensor research. *Biosensors & Bioelectronics*, *181*, 113054. Available from https://doi.org/10.1016/j.bios.2021.113054.

Lima, F. V., Kadiyala, V., Huang, A., Agusala, K., Cho, D., Freeman, A. M., & Druz, R. (2022). At the crossroads! time to start taking smartwatches seriously. *The American Journal of Cardiology*, *179*, 96–101. Available from https://doi.org/10.1016/j.amjcard.2022.06.020.

Lin, Q., Wen, D., Wu, J., Liu, L., Wu, W., Fang, X., & Kong, J. (2020). Microfluidic immunoassays for sensitive and simultaneous detection of IgG/IgM/antigen of SARS-CoV-2 within 15 Min. *Analytical Chemistry*, *2*, 0–4. Available from https://doi.org/10.1021/acs.analchem.0c01635.

Liu, Q., Zhou, X., Wu, H., & Zheng, B. (2020). Blocking-free and self-contained immunoassay platform for one-step point-of-care testing. *Biosensors & Bioelectronics*, *165*, 112394. Available from https://doi.org/10.1016/j.bios.2020.112394.

Liu, Y., Zhan, L., Qin, Z., Sackrison, J., & Bischof, J. C. (2021). Ultrasensitive and highly specific lateral flow assays for point-of-care diagnosis. *ACS Nano*, *15*(3), 3593–3611. Available from https://doi.org/10.1021/acsnano.0c10035.

Liu, Y., Luo, X., Yu, Q., Ye, L., Yang, L., & Cui, Y. (2022). Review of point-of-care platforms for diabetes: (1) sensing. *Sensors and Actuators Reports*, *4*(March). Available from https://doi.org/10.1016/j.snr.2022.100113.

Mahato, K., Maurya, P. K., & Chandra, P. (2018). Fundamentals and commercial aspects of nanobiosensors in point-of-care clinical diagnostics. *3 Biotech*, *8*(3), 1–14. Available from https://doi.org/10.1007/s13205-018-1148-8.

Maumdjer, S., Mondal, T., & Deen, M. J. (2020). Challenges in design and fabrication of flexible/stretchable carbon- and textile-based wearable sensors for health monitoring. *Sensors (Switzerland)*, *3927*(20), 1–29.

Michael, J. M., Roy, P., & Daniel, B. (2020). Larremore. Engla, Journal-2010-New England Journal. *The New England Journal of Medicine*, *108*(1), 1969–1973.

Min, J., Sempionatto, J. R., Teymourian, H., Wang, J., & Gao, W. (2021). Wearable electrochemical biosensors in North America. *Biosensors & Bioelectronics*, *172*, 112750. Available from https://doi.org/10.1016/j.bios.2020.112750.

Moonen, E. J. M., Haakma, J. R., Peri, E., Pelssers, E., Mischi, M., & den Toonder, J. M. J. (2020). Wearable sweat sensing for prolonged, semicontinuous, and nonobtrusive health monitoring. *View*, *1*(4), 20200077. Available from https://doi.org/10.1002/viw.20200077.

Nasiri, S., & Khosravani, M. R. (2020). Progress and challenges in fabrication of wearable sensors for health monitoring. *Sensors Actuators, A Physics*, *312*, 112105. Available from https://doi.org/10.1016/j.sna.2020.112105.

Park, J., Shin, J. H., & Park, J. K. (2016). Pressed paper-based dipstick for detection of foodborne pathogens with multistep reactions. *Analytical Chemistry*, *88*(7), 3781–3788. Available from https://doi.org/10.1021/acs.analchem.5b04743.

Peng, Z., Xie, X., Tan, Q., Kang, H., Cui, J., Zhang, X., Li, W., & Feng, G. (2022). Blood glucose sensors and recent advances: A review. *Journal of Innovative Optical Health Sciences*, *15*(2). Available from https://doi.org/10.1142/S1793545822300038.

Rasmi, Y., Li, X., Khan, J., Ozer, T., & Choi, J. R. (2021). Emerging point-of-care biosensors for rapid diagnosis of COVID-19: Current progress, challenges, and future prospects. *Analytical and Bioanalytical Chemistry*, *413*(16), 4137–4159. Available from https://doi.org/10.1007/s00216-021-03377-6.

Reliable Drug and Alcohol Testing Products. *Sweat collection patch for drug testing developed.* https://www.alcopro.com/.

Sachdeva, S., Davis, R. W., & Saha, A. K. (2021). Microfluidic point-of-care testing: Commercial landscape and future directions. *Frontiers Bioengineering Biotechnology*, *8*, 1–14. Available from https://doi.org/10.3389/fbioe.2020.602659.

Sazonov, E. (2021). *Front matter. Wearable sensors* (pp. i–ii). Elsevier. Available from https://doi.org/10.1016/B978-0-12-819246-7.09989-7.

SCRAM Systems. *Transdermal alcohol monitoring.* https://www.scramsystems.com/scram-international/gb/.

Sempionatto, J. R., Nakagawa, T., Pavinatto, A., Mensah, S. T., Imani, S., Mercier, P., & Wang, J. (2017). Eyeglasses based wireless electrolyte and metabolite sensor platform. *Lab on a Chip*, *17*(10), 1834–1842. Available from https://doi.org/10.1039/c7lc00192d.

Sempionatto, J. R., Lasalde-Ramírez, J. A., Mahato, K., Wang, J., & Gao, W. (2022). Wearable chemical sensors for biomarker discovery in the omics era. *Nature Review Chemistry 2022*, *6*(December), 1–17. Available from https://doi.org/10.1038/s41570-022-00439-w.

Sensors, W., Size, M., Analysis, T., By, R., Type, S., Band, F., Glasses, S., Vertical, B., Region, B. & Forecast, S. (2023). *Industry insights* (pp. 1–6). https://www.grandviewresearch.com/industry-analysis/global-wearable-sensor-market.

Sharafeldin, M., & Davis, J. J. (2021). Point of care sensors for infectious pathogens. *Analytical Chemistry*, *93*(1), 184–197. Available from https://doi.org/10.1021/acs.analchem.0c04677.

Tai, L. C., Gao, W., Chao, M., Bariya, M., Ngo, Q. P., Shahpar, Z., Nyein, H. Y. Y., Park, H., Sun, J., Jung, Y., Wu, E., Fahad, H. M., Lien, D. H., Ota, H., Cho, G., & Javey, A. (2018). Methylxanthine drug monitoring with wearable sweat sensors. *Advanced Materials*, *30*(23), 1–8. Available from https://doi.org/10.1002/adma.201707442.

Teymourian, H., Parrilla, M., Sempionatto, J. R., Montiel, N. F., Barfidokht, A., Van Echelpoel, R., De Wael, K., & Wang, J. (2020). Wearable electrochemical sensors for the monitoring and screening of drugs. *ACS Sensors*, *5*(9), 2679–2700. Available from https://doi.org/10.1021/acssensors.0c01318.

Tierney, M. J., Tamada, J. A., Potts, R. O., Eastman, R. C., Pitzer, K., Ackerman, N. R., & Fermi, S. J. (2000). The GlucoWatch® Biographer: A frequent, automatic and noninvasive glucose monitor. *Annals of Medicine*, *32*(9), 632–641. Available from https://doi.org/10.3109/07853890009002034.

Tu, Y. P., Iqbal, J., & O'leary, T. (2021). Sensitivity of Id now and Rt–Pcr for detection of Sars-Cov-2 in an ambulatory population. *Elife*, *10*, 1–23. Available from https://doi.org/10.7554/ELIFE.65726.

Urdea, A. M., Penny, L. A., Olmsted, S. S., Giovanni, M. Y., Kaspar, P., Shepherd, A., Wilson, P., Dahl, C. A., Buchsbaum, S., Moeller, G., & Burgess, D. C. H. (2006). Requirements for high impact diagnostics. *Nature*, *444*, 73–79.

Van Amerongen, A., Veen, J., Arends, H. A., & Koets, M. (2018). *Lateral Flow Immunoassays*. Elsevier Inc. Available from https://doi.org/10.1016/B978-0-12-811762-0.00007-4.

Vashist, S. K. (2017). Point-of-care diagnostics: Recent advances and trends. *Biosensors*, *7*(4), 10–13. Available from https://doi.org/10.3390/bios7040062.

WHO. *Infectious diseases kill over 17 million people a year: WHO warns of global crisis.* (2022). https://www.who.int/news/item/01-01-1996-infectious-diseases-kill-over-17-million-people-a-year-who-warns-of-global-crisis.

Wild, D., & Kodak, E. (2013). *The immunoassay handbook* (4th ed.). Elsevier. Available from https://doi.org/10.1016/C2010-0-66244-4.

Xue, Y., Thalmayer, A. S., Zeising, S., Fischer, G., & Lübke, M. (2022). Commercial and scientific solutions for blood glucose monitoring—A review. *Sensors*, *22*(2), 425. Available from https://doi.org/10.3390/s22020425.

Yin, L., & Wang, J. (2022). Wearable energy systems: What are the limits and limitations. *National Science Review*. Available from https://doi.org/10.1093/nsr/nwac060.

Yoo, E. H., & Lee, S. Y. (2010). Glucose biosensors: An overview of use in clinical practice. *Sensors*, *10*(5), 4558–4576. Available from https://doi.org/10.3390/s100504558.

Zafar, H., Channa, A., Jeoti, V., & Stojanović, G. M. (2022). Comprehensive review on wearable sweat-glucose sensors for continuous glucose monitoring. *Sensors*, *22*(2), 1–35. Available from https://doi.org/10.3390/s22020638.

Index

Note: Page numbers followed by "*f*" and "*t*" refer to figures and tables, respectively.

C

D

E

F

G

K

L

Q

R

S

T

U

V

W

X

Y

Z